NUMERICAL ASTROPHYSICS

ASTROPHYSICS AND SPACE SCIENCE LIBRARY

VOLUME 240

NUMERICAL ASTROPHYSICS

Proceedings of the International Conference on Numerical Astrophysics 1998 (NAP98), held at the National Olympic Memorial Youth Center, Tokyo, Japan, March 10–13, 1998

Edited by

SHOKEN M. MIYAMA

Theoretical Astrophysics Division,
National Astronomical Observatory,
Tokyo, Japan

KOHJI TOMISAKA

Faculty of Education and Human Sciences,
Niigata University, Japan

and

TOMOYUKI HANAWA

Department of Astrophysics, Nagoya University,
Japan

KLUWER ACADEMIC PUBLISHERS

DORDRECHT / BOSTON / LONDON

A C.I.P. Catalogue record for this book is available from the Library of Congress.

Additional material to this book can be downloaded from http://extras.springer.com.

ISBN 0-7923-5566-0

Published by Kluwer Academic Publishers,
P.O. Box 17, 3300 AA Dordrecht, The Netherlands.

Sold and distributed in North, Central and South America
by Kluwer Academic Publishers,
101 Philip Drive, Norwell, MA 02061, U.S.A.

In all other countries, sold and distributed
by Kluwer Academic Publishers,
P.O. Box 322, 3300 AH Dordrecht, The Netherlands.

Printed on acid-free paper

Contents

2. Physics of Interstellar Medium

3. Formation of Stars and Planets

4. Accretion Disks and Jets

5. Supernovae, Neutron Stars, and Black Holes

6. Stellar Atmosphere and The Solar System

II. Computer Science Oriented Sessions

7. Numerical Methods

8. Computer Science

Preface

These are the proceedings of international conference on *Numerical Astrophysics 1998 (NAP98)*, held at National Olympic Memorial Youth Center, in Tokyo, Japan in the period of March 10 – 13, 1998, and hosted by the National Astronomical Observatory, Japan (NAOJ).

In the last decade numerical simulations have grown up as a major tool for astrophysics. Numerical simulations give us invaluable information on complex systems and physical processes under extreme conditions which can be neither realized by experiments nor directly observed. Supercomputers and special purpose computers may work as very large telescopes and special purpose telescopes for theoretical astrophysics, respectively. Numerical astrophysics ranks with other tool-oriented astronomy such as radio astronomy, infrared astronomy, ultraviolet astronomy, X-ray astronomy, and γ-ray astronomy.

This conference, *NAP98*, was planned to explore recent advances in astrophysics aided by numerical simulations. The subjects of the conference included the large-scale structure formation, galaxy formation and evolution, star and planets formation, accretion disks, jets, gravitational wave emission, and plasma physics. *NAP98* had also sessions on numerical methods and computer science. The conference was attended by 184 scientists from 21 countries. We enjoyed excellent talks, posters, videos, and discussions: there are 40 oral presentations, 96 posters and 16 video presentations. We hope that these proceedings and accompanying CD-ROM replay the friendly but inspiring atmosphere of the conference.

The scientific organizing committee (SOC) consisted of Drs. A. P. Boss, J. Franco, J. F. Hawley, W. Hillebrandt, A. L. Melott, S. M. Miyama (chair), T. Nakamura, K. Nomoto, A. Nordlund, M. L. Norman, J. P. Ostriker, K. Sato, and S. L. Shapiro. We thank them and Dr. Y. Suto for very useful suggestions toward making the program of the conference scientifically more interesting. The local organizing committee (LOC) consisted of Drs. Y. Chikada, T. Hanawa, M. Hirose, S. Inutsuka, T. Kajino, N. Kikuchi, T. Kudoh, R. Matsumoto, K. Oohara, K. Shibata (chair), K. Tomisaka and T. Yokoyama. We would like to express our thanks to them for their great efforts in organizing *NAP98*. We also thank Mr. M. Ohashi, Ms. H. Suzuki, Mr. H. Yoneyama, and Ms. J. Asahi for their efforts to succeed the conference. LOC thank graduate students of the University of Tokyo, Chiba University, and Nagoya University for their assistance.

We acknowledge the sponsorship of the conference by Theoretical Astrophysics Division, Solar Physics Division, and Astronomical Data Analysis Center of NAOJ. The conference was held with financial support from the Ministry of Education, Science, Culture and Sport (Monbusho) for the promotion of international symposia. It was also supported partially by Fujitsu. The activity of LOC was financially supported in part by the Grant-in-Aid for Science Research from Monbusho (No. 09894006; PI: S.M. Miyama).

The editors of these proceedings thank Drs. Y. Suto, F. Nakamura, and LOC members for their comments to improve the proceedings. They are also indebted to Mr. K. Nobuta; without his hard work and digital techniques they could not complete these proceedings.

Nagoya, August 1998

Shoken M. Miyama (NAOJ)
Kohji Tomisaka (Niigata University)
Tomoyuki Hanawa (Nagoya University)

List of Participants

Aikawa,	Yuri	*National Astronomical Obs.*,	Japan
Ardi,	Eliani	*Kyoto University*,	Japan
Baxter,	Keith M.	*Cambridge University*,	UK
Bo-ot,	Luis Maria	*National Inst. for Fusion Science*,	Japan
Bodenheimer,	Peter	*University of California*,	USA
Botnen,	Andreas	*University of Oslo*,	Norway
Busegnies,	Yves	*Universite Libre de Bruxelles*,	Belgium
Carlsson,	Mats	*University of Oslo*,	Norway
Carraro,	Giovanni	*SISSA*,	Italy
Chen,	Pengfei	*Nanjing University*,	China
Chiba,	Takeshi	*Kyoto University*,	Japan
Chikada,	Yoshihiro	*National Astronomical Obs.*,	Japan
Chiueh,	Tzihong	*National Central University*,	Taiwan
Couchman,	Hugh	*University of Toronto*,	Canada
de Gouveia Dal Pino,	Elisabete M.	*University of Sao Paulo*,	Brasil
Den,	Mitsue	*Communications Research Lab.*,	Japan
Dorch,	Bertil F.	*The Niels Bohr Institute*,	Denmark
Dubal,	Mark	*Southampton University*,	UK
Dudorov,	Alexander E.	*Chelyabinsk State University*,	Russia
Duschl,	Wolfgang J.	*Inst. für Theoretische Astrophysik*,	Germany
Fisher,	Robert	*Univ. of California at Berkeley*,	USA
Franco,	José	*UNAM*,	Mexico
Franqueira,	Mercedes	*Univ. Complutense de Madrid*,	Spain
Fujita,	Mitsutaka	*Hokkaido University*,	Japan
Fujiwara,	Hidekazu	*Kobe University*,	Japan
Fukuda,	Hiroyuki	*Hokkaido University*,	Japan
Fukuda,	Naoya	*Nagoya University*,	Japan
Fukushige,	Toshiyuki	*University of Tokyo*,	Japan
Galloway,	David	*Australia University of Sydney*,	Australia
García-Senz,	Domingo	*Univ. Politecnica de Catalunya*,	Spain

Glazunova,	L.V.	*Odessa State University*,	Ukraine
Gouda,	Naoteru	*Osaka University*,	Japan
Habe,	Asao	*Hokkaido University*,	Japan
Hamada,	Tsuyoshi	*University of Tokyo*,	Japan
Hanawa,	Tomoyuki	*Nagoya University*,	Japan
Hansteen,	Viggo H.	*University of Oslo*,	Norway
Hawley,	John	*University of Virginia*,	USA
Hayashi,	Mitsuru	*Japan Atomic Energy Research Inst.*,	Japan
Hillebrandt,	Wolfgang	*MPI für Astrophysik*,	Germany
Hirayama,	Tomohiro	*National Astronomical Obs.*,	Japan
Hirose,	Masahito	*National Astronomical Obs.*,	Japan
Hirose,	Shigenobu	*Science University of Tokyo*,	Japan
Hoshino,	Masahiro	*Inst. Space & Astronautical Sciences*,	Japan
Hosokawa,	Tamaki	*Ibaraki University*,	Japan
Hozumi,	Shunsuke	*Shiga University*,	Japan
Ideta,	Makoto	*Kyoto University*,	Japan
Iida,	Akira	*Kobe University*,	Japan
Imaeda,	Yusuke	*National Astronomical Obs.*,	Japan
Imamura,	James N.	*University of Oregon*,	USA
Inutsuka,	Shu-ichiro	*National Astronomical Obs.*,	Japan
Ishida,	Toshihito	*Nishi-Harima Astronomical Obs.*,	Japan
Jing,	Yipeng	*University of Tokyo*,	Japan
Jones,	Thomas W.	*University of Minnesota*,	USA
Kajino,	Taka	*National Astronomical Obs.*,	Japan
Kamaya,	Hideyuki	*Kyoto University*,	Japan
Kan-ya,	Yukitoshi	*National Astronomical Obs.*,	Japan
Kato,	Seiichi	*National Astronomical Obs.*,	Japan
Kawai,	Atsushi	*University of Tokyo*,	Japan
Kiguchi,	Masayoshi	*Kinki University*,	Japan
Kikuchi,	Nobuhiro	*National Astronomical Obs.*,	Japan
Kim,	Jongsoo	*Korea Astronomy Observatory*,	Korea
Kim,	Sungsoo S.	*Korea Advanced Inst. of Sci. & Tech.*,	Korea
Klapp,	Jaime	*ININ*,	Mexico

Klein, Richard I.	*University of California*,	USA
Kobayashi, Chiaki	*University of Tokyo*,	Japan
Koda, Jin	*University of Tokyo*,	Japan
Koide, Shinji	*Toyama University*,	Japan
Korany, Badie	*NRIAG*,	Egypt
Kozyrev, Serguei	*Sci. C. Gravity Wave Studies Dulkyn*,	Russia
Kudoh, Takahiro	*National Astronomical Obs.*,	Japan
Kusano, Kanya	*Hiroshima University*,	Japan
Kusunose, Masaaki	*Kwansei Gakuin University*,	Japan
Kuwabara, Takuhito	*Chiba University*,	Japan
Kuwabara, Atsushi	*Chiba University*,	Japan
Lee, Sang Min	*Seoul National University*,	Korea
Li, Pak Shing	*University of Wyoming*,	USA
Lin, Lap Ming	*The Chinese University of Hong Kong*,	China
Lin, Lupin C. C.	*National Central University*,	Taiwan
Liu, Bulin	*Beijing Astronomical Obs.*,	China
Magara, Tetsuya	*Kyoto University*,	Japan
Makino, Junichiro	*University of Tokyo*,	Japan
Makita, Makoto	*Kobe University*,	Japan
Martos, Marco A.	*UNAM*,	Mexico
Masunaga, Hirohiko	*National Astronomical Obs.*,	Japan
Matsuda, Takuya	*Kobe University*,	Japan
Matsumoto, Tomoaki	*Nagoya University*,	Japan
Matsumoto, Ryoji	*Chiba University*,	Japan
Matsuzaki, Takaaki	*Chiba University*,	Japan
McLeod, Andrew	*Communications Research Lab.*,	Japan
Meier, David L.	*California Institute of Technology*,	USA
Miller, Kristen A.	*University of Maryland*,	USA
Miyama, Shoken M.	*National Astronomical Obs.*,	Japan
Monaghan, Joseph J.	*Monash University*,	Australia
Mori, Masao	*University of Tokyo*,	Japan
Nagai, Tomoya	*National Astronomical Obs.*,	Japan
Nagasawa, Mikio	*UNCL*,	Japan

Nagashima,	Masahiro	*Osaka University*, Japan
Nakamoto,	Taishi	*University of Tsukuba*, Japan
Nakamura,	Fumitaka	*Niigata University*, Japan
Nakamura,	Masanori	*Science University of Tokyo*, Japan
Nakamura,	Takashi	*Kyoto University*, Japan
Nakamura,	Takayoshi	*University of Tokyo*, Japan
Nakasato,	Naohito	*University of Tokyo*, Japan
Narita,	Shinji	*Doshisha University*, Japan
Nishikawa,	K.-I.	*Louisiana State University*, USA
Nisikawa,	Atuko	*Hokkaido University*, Japan
Nobuta,	Koji	*Nagoya University*, Japan
Nomoto,	Ken'ichi	*University of Tokyo*, Japan
Nordlund,	Åke	*Copenhagen University*, Denmark
Nordstrom,	Hakan C A	*National Research Establishment*, Sweden
Norman,	Michael L.	*MPI für Astrophysik*, Germany
Ogata,	Youichi	*Tokyo Institute of Technology*, Japan
Ogawa,	Tomoya	*Chiba University*, Japan
Ogochi,	Koji	*University of Tsukuba*, Japan
Ohashi,	Mitsuru	*National Astronomical Obs.*, Japan
Ohsuga,	Ken	*University of Tsukuba*, Japan
Ohsugi,	Yukimasa	*Kobe University*, Japan
Okamoto,	Takashi	*Hokkaido University*, Japan
Okazaki,	Atsuo T.	*Hokkai-Gakuen University*, Japan
Okuda,	Toru	*Hokkaido University of Education*, Japan
Oohara,	Ken-ichi	*Niigata University*, Japan
Oosato,	Takahiro	*Ibaraki University*, Japan
Patsis,	Panos	*MPI für Astronomie*, Germany
Pen,	Ue-Li	*CFA, Harvard*, USA
Pickett,	Brian K.	*NASA/Ames Research Center*, USA
Poedts,	Stefaan	*K.U.Leuven*, Belgium
Premadi,	Premana W.	*Tohoku University*, Japan
Ruffert,	Maximilian	*University of Cambridge*, UK
Saigo,	Kazuya	*Nagoya University*, Japan

Saijo, Motoyuki	*Waseda University*,	Japan
Sano, Takayoshi	*National Astronomical Obs.*,	Japan
Santillán, Alfredo J.	*UNAM*,	Mexico
Sasaki, Shin	*Tokyo Metropolitan University*,	Japan
Sekiguchi, Naosuke	*National Astronomical Obs.*,	Japan
Shapiro, Stuart L.	*Univ. Illinois at Champaign-Urbana*,	USA
Shibata, Masaru	*Osaka University*,	Japan
Shibata, Kazunari	*National Astronomical Obs.*,	Japan
Shima, Eiji	*Kawasaki Heavy Industries*,	Japan
Shimizu, Tetsuya M.	*RIKEN*,	Japan
Shinkai, Hisa-aki	*Washington Univ. St.Louis*,	USA
Staude, J.	*Astrophys. Institut Potsdam*,	Germany
Stee, Philippe	*Obs. de la Cote d'Azur - CNRS*,	France
Stone, James M.	*University of Maryland*,	USA
Sumiyoshi, Kohsuke	*RIKEN*,	Japan
Susa, Hajime	*University of Tsukuba*,	Japan
Suto, Yasushi	*University of Tokyo*,	Japan
Suzuki, Hideyuki	*KEK*,	Japan
Tagoshi, Hideyuki	*National Astronomical Obs.*,	Japan
Tajiri, Yukiko	*University of Tsukuba*,	Japan
Takabe, Hideaki	*Osaka University*,	Japan
Takahashi, Koji	*University of Tokyo*,	Japan
Takeuchi, Taku	*Tokyo Institute of Technology*,	Japan
Takizawa, Motokazu	*Kyoto University*,	Japan
Tanuma, Shunichi	*Tokyo University*,	Japan
Tatekawa, Takayuki	*Waseda University*,	Japan
Tohline, Joel E.	*Louisiana State University*,	USA
Tomisaka, Kohji	*Niigata University*,	Japan
Tonooka, Hideaki	*Chiba University*,	Japan
Toropin, Yuriy	*Moscow M.V.Lomonosov State Univ.*,	Russia
Tseng, Yao-Huan	*National Central University*,	Taiwan
Tsuchiya, Toshio	*Kyoto University*,	Japan
Tsuribe, Toru	*University of Tokyo*,	Japan

Turner,	Neal J. J.	*University of California*, USA
Uchida,	Yutaka	*Science University of Tokyo*, Japan
Umekawa,	Michihisa	*Chiba University*, Japan
Umemura,	Masayuki	*University of Tsukuba*, Japan
Valtchanov,	Ivan	*Bulgarian Academy of Sciences*, Bulgaria
Wahde,	Mattias	*NORDITA*, Denmark
Wanajo,	Shinya	*National Astronomical Obs.*, Japan
Wang,	Huaning	*National Astronomical Obs.*, Japan
Washimi,	Haruichi	*Shonan Institute of Technology*, Japan
Watari,	Shinichi	*Communications Research Lab.*, Japan
Widlund,	Lennart	*National Research Establishment*, Sweden
Yabe,	Takashi	*Tokyo Institute of Technology*, Japan
Yahagi,	Hideki	*University of Tokyo*, Japan
Yamada,	Yoshiyuki	*Kyoto University*, Japan
Yamashita,	Kazuyuki	*Chiba University*, Japan
Yan,	Yihua	*Beijing Astronomical Obs.*, China
Yano,	Taihei	*Osaka University*, Japan
Yokono,	Yasunori	*National Astronomical Obs.*, Japan
Yokosawa,	Masayoshi	*Ibaraki University*, Japan
Yokoyama,	Takaaki	*National Astronomical Obs.*, Japan
Yonehara,	Atsunori	*Kyoto University*, Japan
Yoshida,	Tatsuo	*Ibaraki University*, Japan
Yoshikawa,	Kohji	*Kyoto University*, Japan
Zhang,	Tong-jie	*Shanghai Observatory*, China
Zhang,	Chengmin	*The University of Hong Kong*, China

COSMOLOGICAL SIMULATIONS

Modelling the Formation and Evolution of Cosmic Structure

H.M.P. COUCHMAN
Department of Physics and Astronomy
University of Western Ontario
London, ON
N6A 3K7
Canada

Abstract. Studies of the formation of cosmic structure are entering a period of dramatic improvements in sophistication and verisimilitude. A wealth of observational data is becoming available, ranging from studies of galaxies, clusters and their cosmic distribution, to the high redshift universe and observations of microwave background fluctuations. These observations, together with enormous improvements in the scale and quality of the numerical computations that can be undertaken, promise very substantial progress in our understanding of the formation and evolution of cosmic structure.

Numerical simulations are the crucial theoretical probe of non-linear gravitational and hydrodynamic cosmic evolution. The talk briefly reviews the current state of numerical structure formation studies and discusses the developments in techniques which, when coupled with current supercomputers, are leading to studies of unprecedented scope and resolution. The talk concludes with comments on future directions for structure formation studies and on the challenges inherent in simulations of increasing size. Initial results from some of the first billion particle simulations are presented.

1. Introduction

Numerical simulations of cosmic structure have a number of goals. The ultimate aim of many investigations is to explore and understand the formation, distribution and morphology of galaxies. Convincing modelling of the astrophysical processes which contribute to the formation of galaxies is still

S. M. Miyama et al. (eds.), Numerical Astrophysics, 1–9.

some way in the future. Nevertheless, the simulation of very many aspects of cosmic structure has matured to a point where they are providing reliable tools for investigating processes occurring in the post-recombination universe. In particular, key aims are to distinguish world models, to constrain the initial fluctuation spectrum and to understand non-linear gravitational evolution and hydrodynamics in the cosmic context.

Traditionally, cosmological simulations have been concerned with the modelling of Large-Scale Structure, the distribution of voids, walls, clusters etc. and have been used to generate low order spatial and velocity statistics for a variety of cosmologies and initial fluctuation spectra. New instruments and observations are now providing a wealth of data spanning many different epochs and spatial scales. The talk will outline the role that simulation plays in interpreting these observations in cosmology and review the progress that has been made to date as well as providing a few pointers to the future.

2. The Post-Recombination Universe

The post recombination universe is shown schematically in Figure 1. Structure grows in the standard model from small fluctuations present at the time of recombination through the so-called "Dark Ages" to the formation of the first bound radiating objects. In standard hierarchical models structure grows through the aggregation of small objects into larger structures. Some of the key observational elements that must be described by a successful structure formation model are shown in Figure 1. Over the last decade our ability to probe to higher redshifts has increased dramatically with detailed observations of the Lyman α forest becoming available as well as very deep surveys which promise the elusive goal of detecting primæval galaxies.

In the early 1980's, the initial spectrum was constrained only within very broad limits and the distribution of galaxies and clusters was known at low redshift. Numerical structure formation studies bridged this large redshift gap with collisionless simulations to map out the general features of gravitational clustering. The detailed measurement of the fluctuation spectrum from studies of the microwave background early next decade, together with precise mapping of the galaxy distribution on large scales and the probing of structure to high redshifts will dramatically alter the nature of the gap that must be bridged by numerical simulation. With well constrained initial conditions, and a detailed understanding of the evolution of the collisionless component, the focus of many numerical studies will shift to careful investigation of the crucial hydrodynamic aspects of cosmic structure formation.

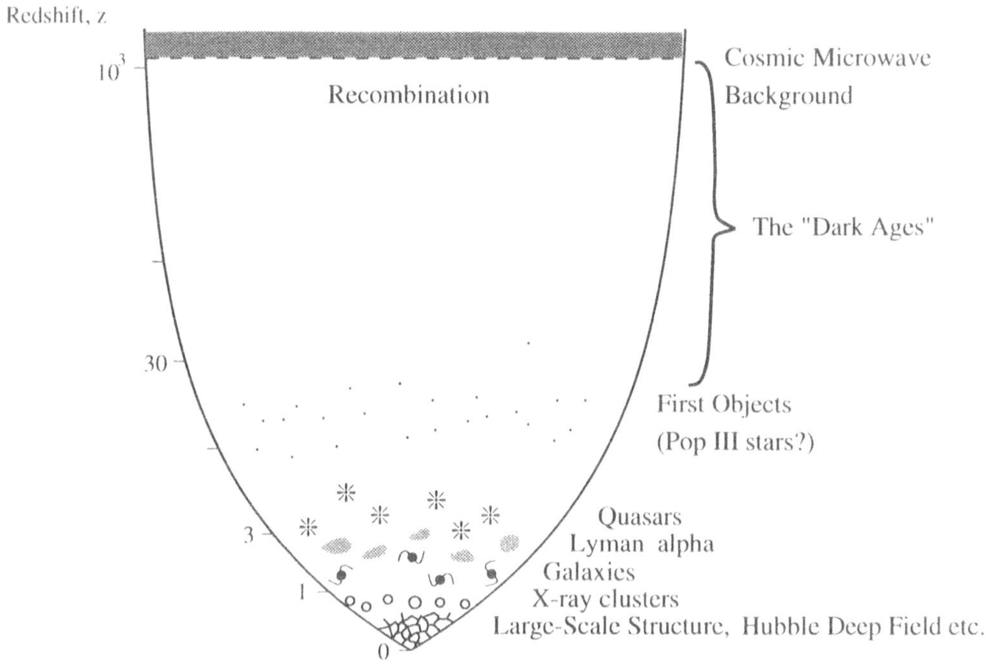

Figure 1. Schematic comoving look-back cone.

3. The Growth of Structure

Figure 2 illustrates the primary features of the growth of structure. The spectrum illustrated is typical of "CDM-like" initial conditions in which structure forms hierarchically, growing from small scales to large. For fluctuations on some scale with amplitude less than unity the growth is linear; once the amplitude of a fluctuation exceeds unity, the evolution is non-linear and numerical simulation is essential to accurately describe the structure. Numerical simulations thus have a fundamental role to play in connecting the initial linear spectrum with the non-linear regime. The further complication which hydrodynamic simulations are now beginning to address is the connection between the observed distribution of luminous matter with that of the gravitationally dominant dark matter distribution. On scales of galaxies and below, hydrodynamic simulation is an essential component of investigations of cosmic structure and of our interpretation of observational data.

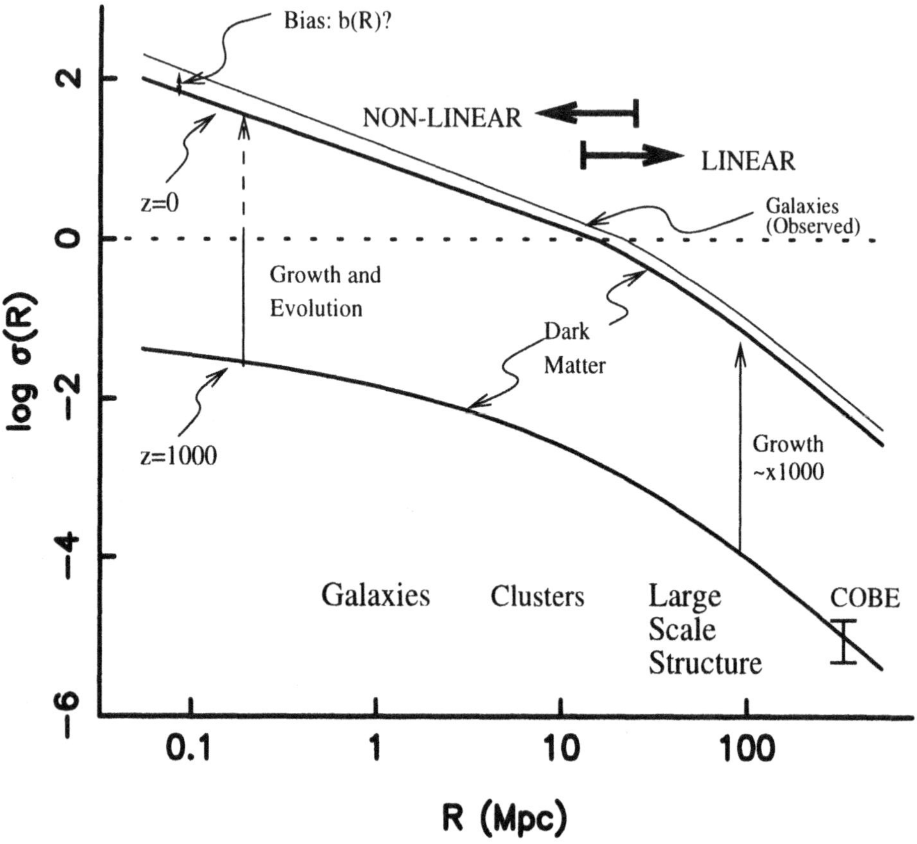

Figure 2. A schematic view of post-recombination growth of the fluctuation spectrum, here represented as the mass variance in spheres of radius R. Whilst the mass variance is less than unity, growth is linear; for larger amplitudes reliable prediction requires numerical simulation.

4. Requirements of a Numerical Simulation

Two features of gravitational collapse in the cosmic context are the presence of irregular geometries and large density contrasts. A bound object virialises with a density contrast of roughly 200 and this overdensity grows as the object maintains a constant physical size within the expanding background. A dissipated baryonic component will have a larger contrast. Lagrangian particle methods are ideally suited to accurately modelling these features. Indeed, it is difficult to see how one might easily model a collisionless component without representing it by particles. A number of techniques have been used to model the hydrodynamic component. Smoothed Particle Hydrodynamics (Monaghan 1992) is a fully Lagrangian method which fits well with gravitational particle techniques. Eulerian methods are also becoming popular in the cosmic context, but, to be successful, adaptive mesh refinement (AMR) techniques must be employed in order to faithfully resolve hydrodynamic phenomena over a wide range of densities.

The range of scales present in observed cosmic structures is very large; from sub-galactic scales to superclustering and beyond, far greater than can be modelled in a single simulation. The particle number that can be modelled sets the mass resolution of the simulation. A simple example will outline the scale of the problem. Suppose that we were to tackle the "Grand Challenge" of modelling the distribution of galaxies in a representative sample of the universe. A cubic region of the universe 200 Mpc on a side would contain of order 10^5 galaxies. Requiring a minimum of 100 mass elements (or particles) per galaxy and supposing that 10% of the total matter resides in galaxies (including their haloes) would suggest a simulation with 10^8 particles. Simulations of this size are only now becoming feasible. Useful hydrodynamic modelling of galaxies in the same context would require a resolution perhaps 10 times larger. This sort of resolution is considerably greater than that which can be achieved in hydrodynamic simulations at present. Whilst by no means all simulation programmes span this range of scales, the argument is suggestive of the mass resolution that is desirable and explains the continual push towards higher particle number by practitioners in the field.

5. Achievements of cosmological simulations

The argument of the previous section suggests that very high particle number is desirable in cosmological simulations. Figure 3 illustrates the remarkable progress that has been achieved in improving the number of particles that can be simulated. The figure shows a representative sample of leading-edge or near leading-edge simulations in the particle-number stakes at various epochs, as well as indicating the algorithmic method used to compute inter-particle forces and the general type of computer hardware used. Particle number is only one aspect of the problem of course; different algorithms have different memory requirements and compute the various types of problem with differing efficiency. Furthermore the hardware groupings are very coarse; no mention has been made of vector computers for example. Nonetheless, a dramatic improvement in attainable resolution over time is apparent as is the necessity of using parallel supercomputers to achieve the highest resolution.

Numerical simulations of structure formation are modelling simple physical laws. The key to progress is understanding the collective phenomena which are modelled. Table 1 lists several areas in cosmology which have been addressed by numerical simulation and assesses the degree of success or reliability with which the various problems have been modelled.

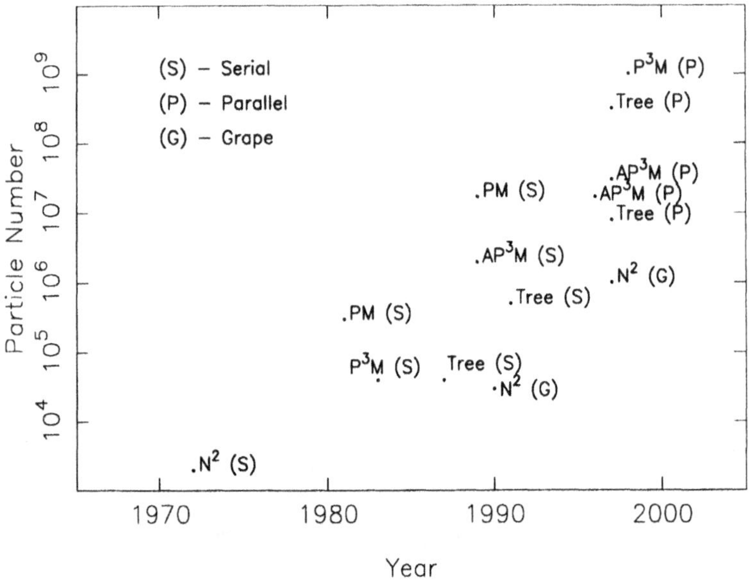

Figure 3. Sketch showing the particle number achieved in various state-of-the-art simulations for different algorithms as a function of epoch. The representative simulations illustrated were run either on serial computers, parallel computers (typically massively parallel systems) or using the special purpose GRAPE hardware.

TABLE 1. Achievement and reliability of cosmological simulations.

Achievement	Reliability
Gravitational:	
Checks of perturbation theory	Good—symbiotic relationship
Exploring validity of simple analytic models (e.g., Press–Schechter)	Reasonable—models limited
Non-linear gravitational evolution, stable clustering etc.	Reasonable—need greater dynamic range
Distinguishing different initial Ω, spectra, $\xi_{\rho\rho}$, cluster abundances etc.	Good
Comparison of predicted ξ_{gg} with observed distributions; bias etc.	Rudimentary—need reliable galaxy identification techniques—hydro
Hydrodynamic:	
Cluster gas and properties	Good
Lyman α	Fairly good—low order statistics used so far
Galaxy formation, first objects etc.	Rudimentary—need cooling and to model star formation, feedback etc.

6. The Current State-of the-Art

The continuing push for higher resolution has lead to an increasing use of massively parallel supercomputers for leading edge simulations. These computers offer both the processing speed—provided the algorithms can be efficiently parallelized—as well as providing the large total memory needed for storage of the particle data and for the computational overhead of the algorithm itself. Despite the difficulty of programming for distributed memory architectures, parallel versions of several N-body algorithms exist; in particular, versions of Tree, P^3M and AP^3M algorithms have been implemented.

Impressive performance is possible on current hardware. The Virgo (1996) collaboration uses an AP^3M algorithm which has been parallelised using Cray's proprietary data-parallel language CRAFT, to run on the T3D supercomputer. We can achieve a performance of roughly 2,500 particle updates/second/processor with excellent scaling behaviour up to 256 processors for a 17 Million particle run (for details see Pearce & Couchman 1997). The same code but without adaptive refinements (essentially P^3M) has been implemented using Cray's "shmem" library—an MPI2-like one-sided message passing paradigm—for the Cray T3E (MacFarland et al. 1998). The performance achieved in this case is 8,000 particle updates/second/processor. A 256^3 ($\sim$17 Million) particle run of approximately 2000 steps takes 10 hours on 128 processors of a Cray T3E. This is impressive performance even considering the slow down that occurs with the non adaptive P^3M implementation as clustering develops.

Storage requirements become equally impressive as the particle number increases. For collisionless simulations 6 words are required per particle or roughly 10 words for hydrodynamic simulations. Each output slice of a 256^3 particle collisionless simulation, for example, requires $\sim$ 500 Mb for storage of positions and velocities in 32 bit precision. Typically 10 to 15 output times are stored per simulation.

Two broad approaches are being pursued to achieve higher resolution. Selective volume renormalization allows very high resolution to be achieved in the investigation of galaxies, clusters or patches of the universe at high redshift for example. With proper care it is possible to model the large-scale cosmological field by using nested shells of variable mass particles and/or the application of external fields. Where uniform resolution is required over the whole computational domain there is little alternative to the straightforward calculation with very large particle number. Such a simulation has been undertaken recently by the Virgo collaboration. A cube of side 4000 Mpc containing 10^9 particles was evolved from a start redshift of 29. A slice of the final output of the simulation at redshift zero together with the physical parameters used is shown in Figure 4. The code used was the *shmem* P^3M code described above running on 512 processors of the T3E in Garching, Germany. The full simulation of 500 steps took 70 hours. Each output is ~25Gb with a total dataset of ~ 1/3 Tb. With this size dataset, analysis is a challenge, with much of the post-processing, of necessity, being done on the the T3E or large machines local to it.

7. Conclusions

There are two broad conclusions to this talk. First, the promise of secure knowledge of the initial fluctuation spectrum together with improved reliability of collisionless simulations suggests that in the next few years—barring upsets to the overall cosmological picture and current understanding of structure formation—we will have a good understanding of the evolution and structure of the dark matter distribution. This will provide the basis from which we can undertake detailed hydrodynamical simulations of the formation of galaxies and cosmic structure. Observations at high redshift are providing detailed information which will place significant constraints on the details of structure formation models as realised by numerical simulations.

The second conclusion is computational. As is evident from the data presented above, the attainable resolution has increased by 4 orders of magnitude in under two decades. With the advent, and increasing use, of massively parallel computers the simulation of very large numbers of particles is becoming feasible. This allows us to model a range of scales where studies—particularly hydrodynamic studies—of very many aspects of cosmic structure will begin to have detailed predictive power.

References

MacFarland, Tom, Couchman, H.M.P., Pearce, F.R. and Pichlmeier, Jakob, (1998) A New Parallel P^3M Code for Very Large-Scale Cosmological Simulations, *New Astronomy*, submitted (`astro-ph/9805096`)

Monaghan, J.J. (1992) *Annual Reviews of Astronomy and Astrophysics*, **30**, 543

Pearce, F.R. and Couchman, H.M.P. (1997) Hydra: An parallel adaptive grid code, *New Astronomy*, **v 2 n 5**, p 411

Virgo: Colberg, J.M., Couchman, H.M.P., Efstathiou, G.P., Frenk, C.S., Jenkins, A., Nelson A.H., Peacock, J.A., Pearce, F.R., Thomas, P. A. and White, S.D.M. (1996)

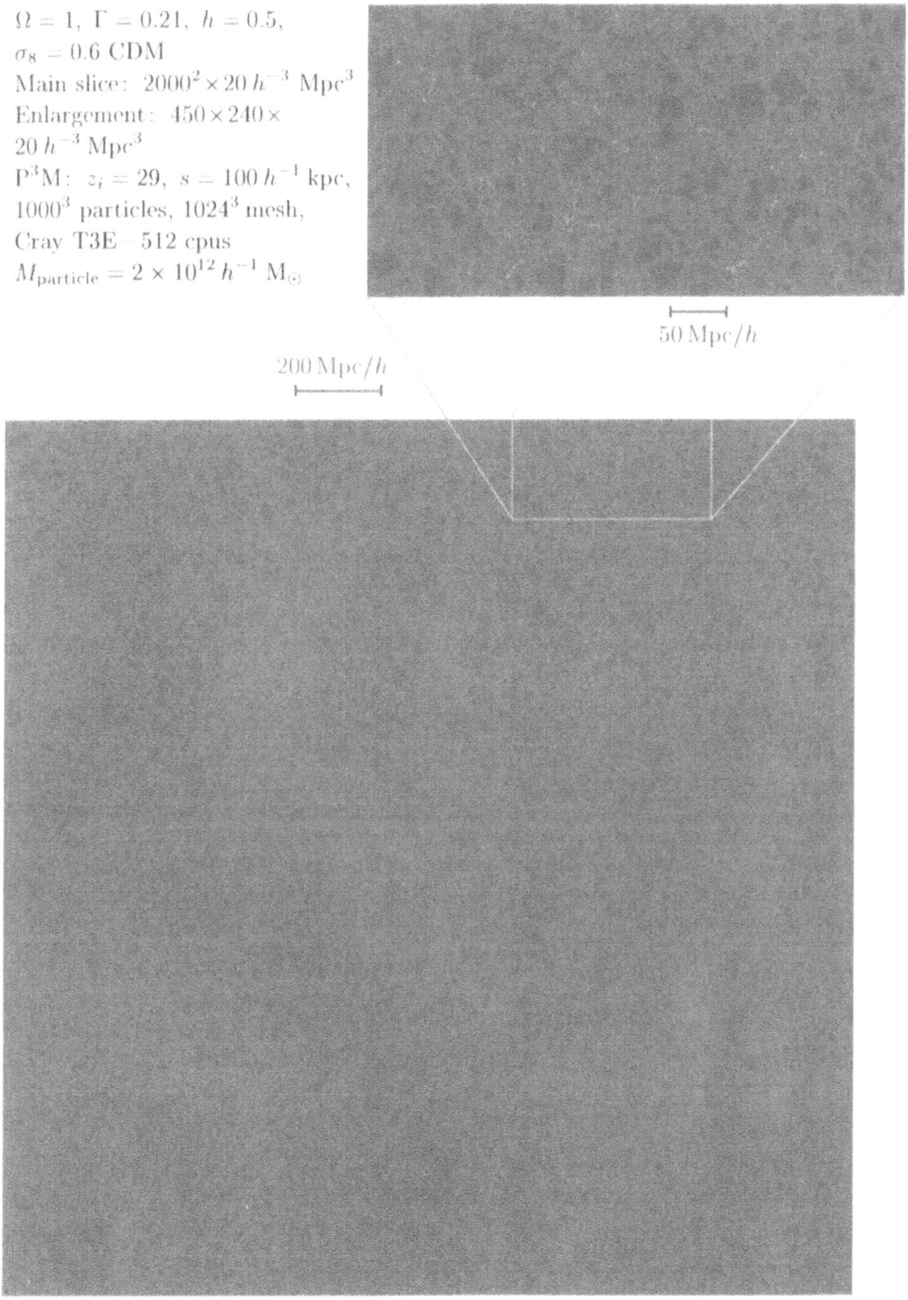

Figure 4. A slice through the 10^9 particle simulation recently completed by the Virgo consortium. The cosmology simulated is standard Cold Dark Matter but with a modified spectrum (shape parameter, $\Gamma = 0.21$).

EXPLORING REDSHIFT-SPACE CLUSTERING WITH COSMOLOGICAL N-BODY SIMULATIONS

Y. SUTO, H. MAGIRA AND Y.P. JING
Department of Physics, and Research Center for the Early Universe, The University of Tokyo, Tokyo 113-0033, Japan

1. Introduction

Cosmological N-body simulations have played an important part in describing and understanding the nonlinear gravitational clustering in the universe. In this context, a paper by Miyoshi and Kihara (1975) is truly pioneering, and should be briefly described here. They carried out a series of cosmological N-body experiments with $N = 400$ *galaxies* (particles) in an expanding universe. The simulation was performed in numerical comoving cube with a periodic boundary condition. As the title of the paper "*Development of the correlation of galaxies in an expanding universe*" clearly indicates, they were interested in understanding why observed galaxies in the universe exhibit a characteristic correlation function of $g(r) = (r_0/r)^s$. In fact, Totsuji and Kihara (1969) had already found that $r_0 = 4.7h^{-1}$Mpc and $s = 1.8$ is a reasonable fit to the clustering of galaxies in the Shane – Wirtanen catalogue. One of the main conclusions of Miyoshi and Kihara (1975) is that "*The power-type correlation function $g(r) = (r_0/r)^s$ with $s \approx 2$ is stable in shape; it is generated from motionless galaxies distributed at random and also from a system with weak initial correlation*".

Several years after Totsuji and Kihara (1969) published the paper, Peebles (1974) and Groth and Peebles (1977) reached the same conclusion, which seems to have motivated several cosmological N-body simulations all over the world (e.g., Aarseth, Gott and Turner 1979; Efstathiou 1979; Doroshkevich et al. 1980). At present, nearly a quarter century later, the cosmological N-body simulations are armed by (i) significant progress in computational methods and resources (e.g., H.P. Couchman, in these proceedings), (ii) physically motivated initial conditions on the basis of inflationary models and the idea of dark matter dominated universe (e.g., Davis et al. 1985), and (iii) reliable statistical measures to quantify the clustering

S. M. Miyama et al. (eds.), Numerical Astrophysics, 11–18.

of the universe at $z \sim 0$. Naturally they have been established as important tools in exploring the nonlinear gravitational evolution of three-dimensional collisionless system in the expanding universe. Nevertheless it is quite remarkable that all the basic ingredients of cosmological N-body simulations were already fully described in the pioneering paper by Miyoshi and Kihara (1975).

A widely accepted point of view in cosmology is that gravitational instability in a cold dark matter dominated universe is a reasonably successful model of the cosmic structure formation at $z \sim 0$. On the basis of this, it is natural to move from understanding of the *current universe* to exploring the evolution and clustering of the universe *at high redshifts*. The present talk focuses on two topics in this line; cosmological redshift distortion (Matsubara and Suto 1996; Nakamura, Matsubara, and Suto 1998; Magira, Jing, Matsubara and Suto 1998) and cosmological implications of the strong clustering of Lyman-break galaxies at $z \sim 3$ (Jing and Suto 1998).

2. Cosmological redshift distortion

The three-dimensional distribution of galaxies in the redshift surveys differ from the true one since the distance to each galaxy cannot be determined by its redshift z only; for $z \ll 1$ the peculiar velocity of galaxies, typically $\sim (100-1000)$km/sec, contaminates the true recession velocity of the Hubble flow, while the true distance for objects at $z < 1$ sensitively depends on the (unknown and thus assumed) cosmological parameters. This hampers the effort to understand the true distribution of large-scale structure of the universe. Nevertheless such redshift-space distortion effects are quite useful since through the detailed theoretical modeling, one can derive the peculiar velocity dispersions of galaxies as a function of separation, and also can infer the cosmological density parameter Ω_0, the dimensionless cosmological constant λ_0, and the spatial biasing factor b of galaxies and/or quasars, for instance. Here we point out the importance of such redshift distortion induced by the geometry of the universe.

Let us consider a pair of objects located at redshifts z_1 and z_2 whose redshift difference $\delta z \equiv z_1 - z_2$ is much less than the mean redshift $z \equiv (z_1 + z_2)/2$. Then the ***observable*** separations of the pair parallel and perpendicular to the line-of-sight direction, $s_\parallel$ and $s_\perp$, are given as $\delta z/H_0$ and $z\delta\theta/H_0$, respectively, where H_0 is the Hubble constant and $\delta\theta$ denotes the angular separation of the pair on the sky. The cosmological redshift-space distortion originates from the anisotropic mapping between the redshift-space coordinates, $(s_\parallel, s_\perp)$, and the real comoving ones $(x_\parallel, x_\perp) \equiv (c_\parallel s_\parallel, c_\perp s_\perp)$; $c_\perp$ is written as $c_\perp = H_0(1+z)D_A/z$ in terms of the angular diam-

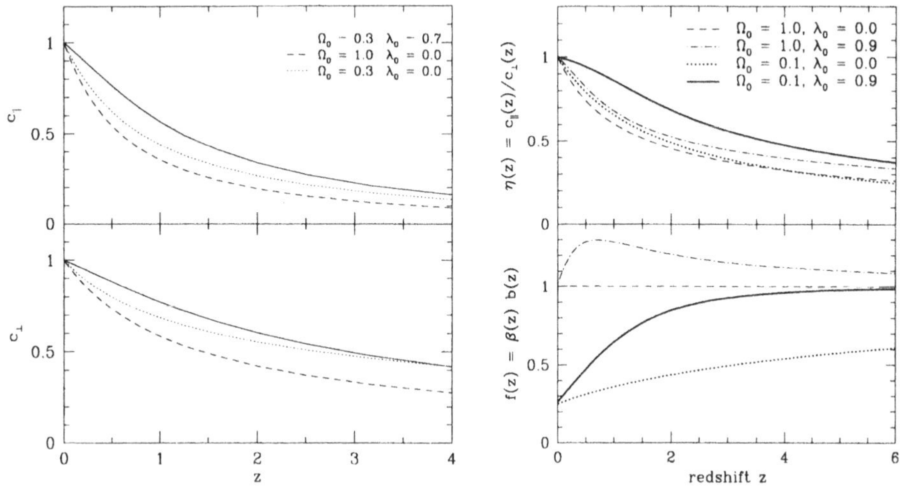

Figure 1. *Left:* Behavior of $c_{\parallel}(z)$ and $c_{\perp}(z)$ for $(\Omega_0,\ \lambda_0) = (0.3, 0.7)$ in *solid line*, $(1.0, 0.0)$ in *dashed line*, and $(0.3, 0.0)$ in *dotted line*. *Right:* $\eta(z) = c_{\parallel}(z)/c_{\perp}(z)$ (*upper panel*), and $f(z) = \beta(z)b(z)$ (*lower panel*) for $(\Omega_0,\ \lambda_0) = (1.0, 0.0)$ in *dashed line*, $(1.0, 0.9)$ in *dot-dashed line*, $(0.1, 0.9)$ in *dotted line*, and $(1.0, 0.9)$ in *thick solid line*.

eter distance D_A, and

$$c_{\parallel}(z) = \frac{H_0}{H(z)} = \frac{1}{\sqrt{\Omega_0(1+z)^3 + (1-\Omega_0-\lambda_0)(1+z)^2 + \lambda_0}}. \tag{1}$$

Thus their ratio becomes

$$\frac{x_{\parallel}(z)}{x_{\perp}(z)} = \frac{c_{\parallel}(z)}{c_{\perp}(z)}\frac{\delta z}{z\delta\theta} \equiv \eta(z)\frac{\delta z}{z\delta\theta}. \tag{2}$$

Since $x_{\parallel}/x_{\perp}$ should approach $\delta z/(z\delta\theta)$ for $z \ll 1$, $c_{\parallel}(z)/c_{\perp}(z)$ can be regarded to represent the correction factor for the *cosmological redshift distortion* (upper panel of Fig.1). This was first pointed out in an explicit form by Alcock and Paczyński (1979). Although $\eta(z)$ is a potentially sensitive probe of Ω_0 and especially λ_0, this is not directly observable unless one has an independent estimate of the ratio $x_{\parallel}/x_{\perp}$. Here, we propose to use the clustering of quasars and galaxies in z, which should be safely assumed to be statistically isotropic, in estimating $\eta(z)$. Also we take account of the distortion due to the peculiar velocity field using linear theory.

The relation between the two-point correlation functions of quasars in redshift space, $\xi^{(s)}(s_{\perp}, s_{\parallel})$, and that of *mass* in real space, $\xi^{(r)}(x)$, can be derived in linear theory (Kaiser 1987; Hamilton 1992, 1997):

$$\xi^{(s)}(s_{\perp}, s_{\parallel}) = \left(1 + \frac{2}{3}\beta(z) + \frac{1}{5}[\beta(z)]^2\right)\xi_0(x)P_0(\mu)$$
$$- \left(\frac{4}{3}\beta(z) + \frac{4}{7}[\beta(z)]^2\right)\xi_2(x)P_2(\mu) + \frac{8}{35}[\beta(z)]^2\xi_4(x)P_4(\mu), \tag{3}$$

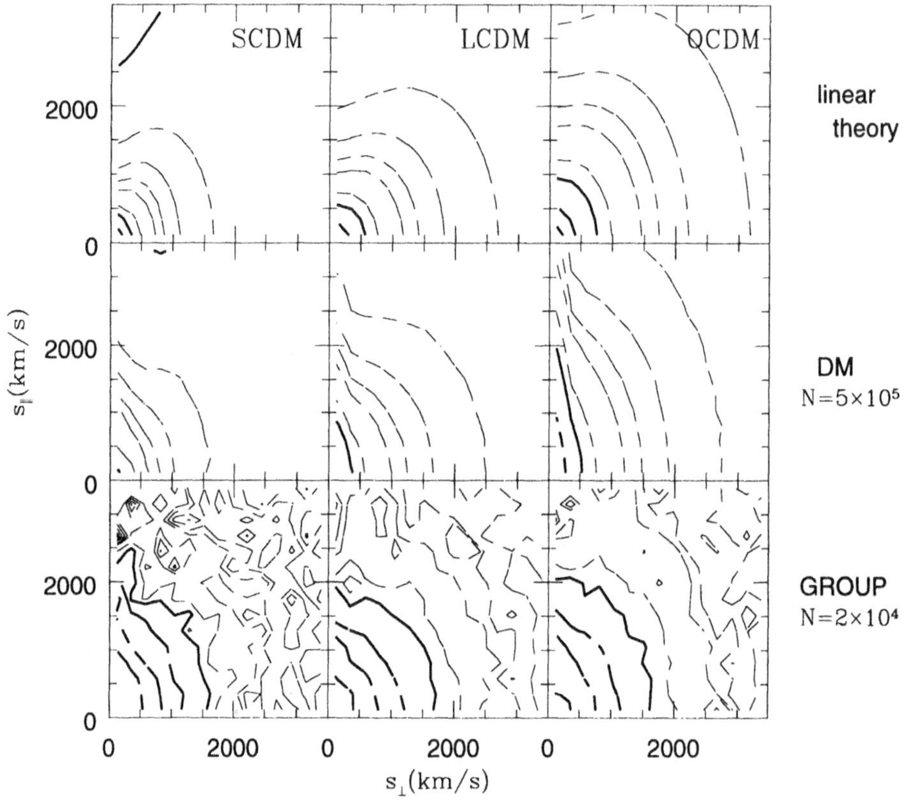

Figure 2. $\xi^{(s)}(s_\perp, s_\parallel)$ from linear theory (*Upper*) and N-body simulations with $N = 5 \times 10^5$ particles (*Middle*) and with $N = 2 \times 10^4$ massive halos of particles at $z = 2.2$ (*Lower*).

where $x \equiv \sqrt{c_\parallel{}^2 s_\parallel{}^2 + c_\perp{}^2 s_\perp{}^2}$, $\mu \equiv c_\parallel s_\parallel / x$, P_n's are the Legendre polynomials,

$$\beta(z) \equiv \frac{1}{b(z)} \frac{d \ln D(z)}{d \ln a}, \quad \xi_{2l}(x) = \frac{(-1)^l}{x^{2l+1}} \left(\int_0^x x dx \right)^l x^{2l} \left(\frac{d}{dx} \frac{1}{x} \right)^l x \xi^{(r)}(x), \quad (4)$$

and $D(z)$ is the linear growth rate (Matsubara and Suto 1996; see also Ballinger, Peacock and Heavens 1996).

In reality, however, the observable two-point correlation functions would be also contaminated by non-linear peculiar velocity as well as statistically limited by the available number of tracer objects. To incorporate the non-linear effect, we compute $\xi^{(s)}(s_\perp, s_\parallel)$ from a series of N-body simulations in CDM models with $N = 256^3$ particles in $(300h^{-1}\text{Mpc})^3$ box (Jing and Suto 1998; Magira et al. 1998; Y.P.Jing, in these proceedings).

Figure 2 shows several examples at $z = 2.2$ for representative CDM models (Table 1). The degree to the extent which one can recover the underlying correlation of small amplitudes is sensitive to the sampling rate, or the number density of objects. So Figure 2 is plotted for $N = 5 \times 10^5$ randomly selected particles and also for $N = 2 \times 10^4$ massive halos of particles from simulations.

Table 1. Simulation model parameters

Model	Ω_0	λ_0	Γ	σ_8
SCDM	1.0	0.0	0.5	0.6
OCDM	0.3	0.0	0.25	1.0
LCDM	0.3	0.7	0.21	1.0

Figure 3 plots the reduced χ^2 contour from $\xi^{(s)}(s_\perp, s_\parallel)$ of simulations. In computing the χ^2, we exclude the regions with $s_\parallel/s_\perp > 2$ which are likely to be seriously contaminated by nonlinear peculiar velocities. While Figure 3 demonstrates that the current methodology works in principle, the expected S/N is fairly low. This is largely because we adjusted the sampling rate for the high-z QSOs. The situation would be improved, though, if we apply the present methodology to a statistical sample of Lyman-break galaxies, for instance, whose number density is larger and their strong clustering is already observed (Steidel et al. 1998). We will examine the different aspect of strong clustering of the Lyman-break galaxies in the next section.

3. Clustering of Lyman-break Galaxies

Recently Steidel et al. (1998) reported a discovery of a highly significant concentration of galaxies on the basis of the distribution of 78 spectroscopic redshifts in the range $2 \le z \le 3.4$ for photometrically selected "Lyman break" objects. We examine the theoretical impact of their discovery in much greater detail using a large number of mock samples from N-body simulations. All the simulations employ 256^3 ($\approx$ 17 million) particles in a $(100h^{-1}\mathrm{Mpc})^3$ comoving box and start at redshift $z_i = 36$. We identify halos of galaxies using the Friend-Of-Friend (FOF) algorithm with a bonding length 0.2 times the mean particle separation, and assume that each halo corresponds to one Lyman break galaxy. Thus our analysis presented here properly takes account of several important and realistic effects including (i) the survey volume geometry, (ii) redshift-space distortion, (iii) selection function of the objects, (iv) fully nonlinear evolution of dark halos, and (v) finite sampling effect.

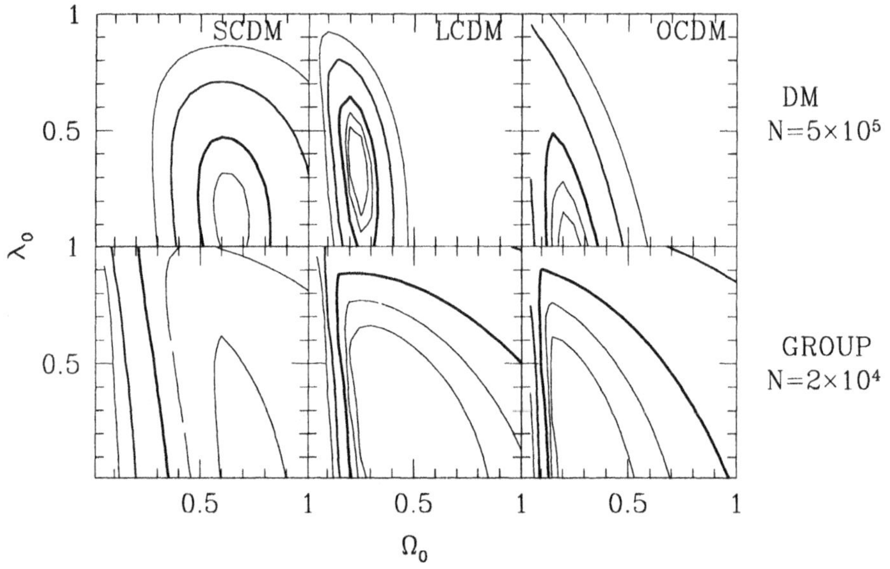

Figure 3. χ^2-contours on Ω_0 - λ_0 plane from the analysis of the data in Fig.2.

Figure 4 (left panels) plots the volume-averaged two-point correlation functions in real space $\bar{\xi}(R; M, z)$ of the halos with mass larger than M at redshift $z = 2.9$. Figure 4 (left panels) indicates that over the scales of interest the correlation functions of halos are enhanced approximately by a constant factor $b^2(M, z)$ relative to those of the dark matter. Figure 4a (right panel) plots this effective bias parameter for halos with mass larger than M, $b(> M) \equiv \sqrt{\bar{\xi}(R; M, z)/\bar{\xi}_{\text{mass}}(R; z)}$, calculated at $R = 7.5h^{-1}$Mpc and $z = 2.9$, where $\bar{\xi}_{\text{mass}}(R; z)$ is the volume-averaged correlation function of all *particles* in the simulations.

The mean number density of halos of mass larger than M (without applying the selection function) is plotted in Figure 4b (right panel). Three horizontal lines indicate the observed number density of Lyman break galaxies corresponding to our three model parameters. Note that the observed number density should be regarded as a strictly lower limit since some fraction of the galaxies might have been unobserved due to the selection criteria. Since the density of halos falls below the observed one for $M > M_{\text{max}}$, we vary the threshold mass of the halos from $10m_p$ up to M_{max} in considering the statistical significance of the clustering. We use a simple algorithm to identify a galaxy concentration in the mock sample following the procedure of Steidel et al. (1998); for each mock sample, we count galaxies within redshift bins of $\Delta z = 0.04$ centered at each galaxy and identify the redshift bin with the maximum count as the density concentration. Then we compute the probability $P_{\geqq 15}(M)$ that a mock sample

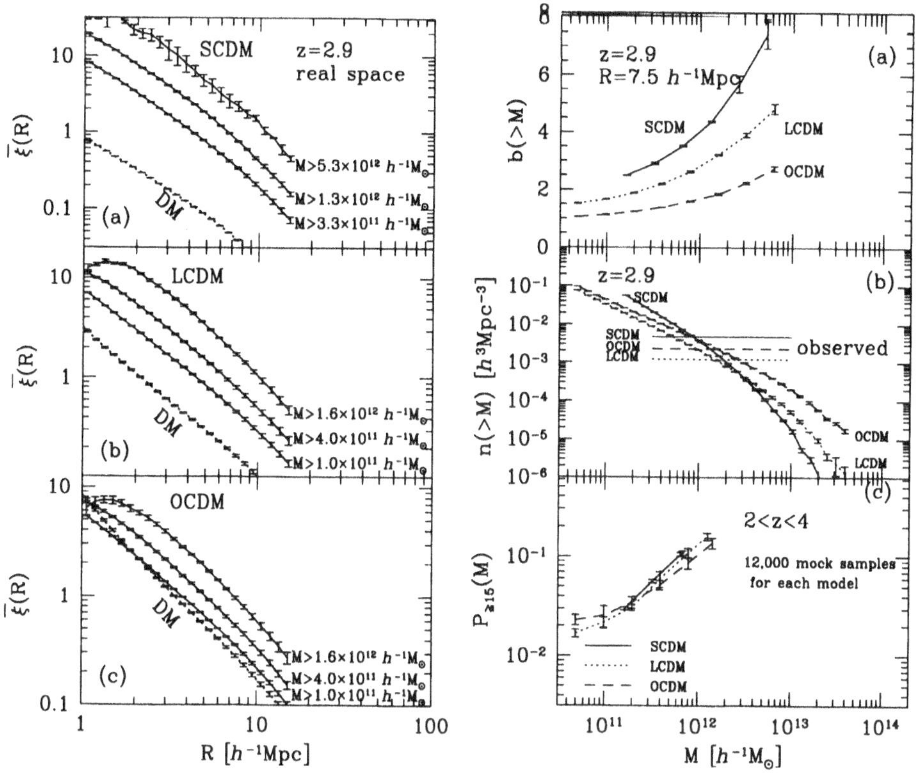

Figure 4. *Left:* Two-point correlation functions of halos at $z = 2.9$ in real space; different curves correspond to the different threshold masses M of the halos in (a) SCDM, (b) LCDM, and (c) OCDM models. Curves labeled by DM correspond to the correlation functions of all particles in the simulation. *Right:* Statistics of the halos as a function of the threshold mass; (a) bias parameter of halos at $z = 2.9$; (b) mean number density of halos in the simulations compared with the observed one of Lyman break galaxies; (c) probability of finding a concentration of at least 15 halos in a bin of $\Delta z = 0.04$.

has a concentration with at least 15 galaxies. Figure 4c (right panel) indicates that while the clustering of such objects is naturally biased with respect to dark matter, the predicted bias $1.5 \sim 3$ is not large enough to be reconciled with such a strong concentration of galaxies at $z \sim 3$ *if one similar structure is found per one field on average.* We predict one similar concentration approximately per ten fields in SCDM and per six fields in LCDM and OCDM. Therefore future spectroscopic surveys in a dozen fields (Pettini et al. 1997) are quite important in constraining the cosmological models, and may challenge all the existing cosmological models *a posteriori* fitted to the $z = 0$ universe.

4. Conclusions

Many cosmological models are known to be more or less successful in reproducing the structure at redshift $z \sim 0$ *by construction*. There are still several degrees of freedom or *cosmological parameters* appropriately *fit* the observations at $z \sim 0$ (Ω_0, σ_8, h, λ_0, $b(r,z)$). This *degeneracy* in cosmological parameters can be broken by using either accurate observational data with much better statistical significance (e.g., 2dF, Sloan Digital Sky Survey) or the data sample at higher redshifts. To extract all the potential information from the observational data requires predicting the theoretical consequences as quantitatively as possible by combining everything that we know theoretically and state-of-art numerical modeling.

We thank Takahiko Matsubara and Takahiro T. Nakamura for useful discussions. Numerical computations presented here were carried out on VPP300/16R and VX/4R at the Astronomical Data Analysis Center of the National Astronomical Observatory, Japan, as well as at RESCEU (Research Center for the Early Universe, University of Tokyo) and KEK (National Laboratory for High Energy Physics, Japan). Y.P.J. gratefully acknowledges the postdoctoral fellowship from Japan Society for the Promotion of Science. This research was supported by the Grants-in-Aid of the Ministry of Education, Science, Sports and Culture of Japan No.07CE2002 to RESCEU, and No.96183 to Y.P.J., and by the Supercomputer Project (No.97-22) of High Energy Accelerator Research Organization (KEK).

References

Aarseth, S.J., Gott, J.R. and Turner, E.L. 1979, ApJ, 228, 664.
Alcock, C., and Paczyński, B. 1979, Nature, 281, 358
Ballinger, W.E., Peacock, J.A., and Heavens, A.F. 1996, MNRAS, 282, 877
Davis, M., Efstathiou, G., Frenk, C.S., and White, S.D.M. 1985, ApJ, 292, 371.
Doroshkevich et al. 1980, MNRAS, 192, 321.
Efstathiou, G. 1979, MNRAS, 187, 117.
Groth,E.J. and Peebles,P.J.E. 1977, ApJ, 217, 385.
Hamilton, A.J.S. 1992, ApJ, 385, L5
Hamilton, A.J.S. 1997, astro-ph/9708120
Jing, Y.P. and Suto, Y. 1998, ApJ, 494, L5.
Kaiser, N. 1987, MNRAS, 227, 1
Magira, H., Jing, Y.P., Matsubara, T., and Suto, Y. 1998, in preparation.
Matsubara, T., and Suto, Y. 1996, ApJ, 470, L1
Miyoshi, K. and Kihara, T. 1975, Publ.Astron.Soc.Japan., 27, 333.
Nakamura, T.T., Matsubara, T., and Suto, Y. 1998, ApJ, 494, 13.
Peebles,P.J.E. 1974, ApJ, 189, L51.
Pettini, M. et al 1997, in the proceedings of 'ORIGINS', ed. J.M. Shull, C.E. Woodward, and H. Thronson, (San Francisco, ASP) (astro-ph/9708117).
Steidel, C.C. et al. 1998, ApJ, 492, 428.
Totsuji, H. and Kihara, T. 1969, Publ.Astron.Soc.Japan., 21, 221.

COSMOLOGICAL ADAPTIVE MESH REFINEMENT

M.L. NORMAN
Astronomy Department & NCSA,
University of Illinois, Urbana, IL 61801, USA

AND

G.L. BRYAN
Princeton University Observatory,
Princeton, NJ 08544, USA

1. Introduction

The inclusion of hydrodynamics and atomic and radiation processes into cosmological structure formation simulations is essential for many applications of interest, including galaxy formation, the structure of the intergalactic medium, formation and evolution of X-ray clusters, and cosmic reionization. A key technical challenge in such simulations is obtaining high mass and spatial resolution in collapsing structures within large cosmological volumes.

We and other members of the Grand Challenge Cosmology Consortium (GC^3) have explored Eulerian hydrodynamics methods as an alternative to SPH in cosmological simulations (Cen 1992; Anninos, Norman & Clarke 1994; Kang et al. 1994; Bryan et al. 1995; Gnedin 1995; Pen 1995). Provided grids can be constructed which achieve the necessary dynamic range, Eulerian methods have a number of distinct advantages over SPH. These are: (1) *speed*: the use of logically regular data structures avoids time-consuming nearest neighbor searches resulting in $\sim 10\times$ higher update rate; (2) *noise*: fluid is represented as a continuum, not discrete particles, eliminating Poisson noise; (3) *density sampling*: because of point (2), low density cells are computed as accurately as high density cells at the same cost; density gradients spanning many orders of magnitude can be accurately simulated with 3-4 cells per decade per dimension. (4) *integral form*: integral conservation laws are straightforward to implement for mass, momentum, energy and magnetic flux which are numerically conservative to machine roundoff. In

S. M. Miyama et al. (eds.), Numerical Astrophysics, 19–28.

addition to the above-mentioned advantages which are generic to any Eulerian method, we add the following for higher order Godunov methods such as PPM or TVD: (5) *shock capturing*: shocks are captured in 1-2 cells with correct entropy generation and non-oscillatory shocks; (6) *upwind*: wave characteristics are properly upwinded for higher fidelity and stability; (7) *low dissipation*: the use of higher order-accurate interpolation results in a very low numerical viscosity–important for angular momentum conservation in protogalactic disks. In addition, radiative transfer is most easily done on a grid. Implicit algorithms generate large sparse matrix equations for which iterative and direct linear systems solvers are available.

In the references cited above, various gridding schemes have been explored to reduce truncation error in regions of high density gradients, as invariably arise in structure formation simulations. Here, we describe a powerful method we have developed based on the adaptive mesh refinement (AMR) algorithm of Berger and Colella (1989).

2. Overview of Berger's AMR

Adaptive mesh refinement (AMR) was developed by Berger and Oliger (1984) to achieve high spatial and temporal resolution in regions of solutions of hyperbolic partial differential equations where fixed grid methods fail. Algorithmic refinements and an application to shock hydrodynamics calculations were described in Berger & Collela (1989). The hydrodynamic portion of our method is based closely on this latter paper, and we refer the reader to it for details (see also paper by Klein et al., these proceedings).

2.1. GRID HIERARCHIES

Unlike some mesh refinement schemes which move the mesh points around "rubber mesh" (e.g., Pen 1995) or subdivide individual cells, resulting in an octree data structure (e.g., Kravtsov et al. 1997), Berger's AMR utilizes an adaptive hierarchy of grid patches at various levels of resolution. Each rectangular grid patch (hereafter, simply *grid*) covers some region of space in its *parent grid* needing higher resolution, and may itself become the parent grid to an even higher resolution *child grid*. A general implementation of AMR such as ours places no restriction on the number of grids at a given level of refinement, or the number of levels of refinement. The hierarchy of grids can be thought of as a tree data structure, where each leaf is a grid.

Each grid is evolved as a separate initial boundary value (IBV) problem. Initial conditions are obtained by interpolation from the parent grid when the grid is created. Boundary conditions are obtained either by interpolation from the parent grid or by copies from abutting *sibling grids* at the same level. To simplify interpolation, grids are constrained to be aligned with their parent grid and not overlap their sibling grids. Grids at level ℓ are refined by an integer factor $R_\ell \geq 2$ relative to the grid at level $\ell - 1$.

2.2. ALGORITHM

The algorithm which creates this grid hierarchy is *local* and *recursive.* It can be written in pseudocode as follows. Here `level` is the level of refinement, and procedures are capitalized.

```
Integrate (level)
begin
  if "time for regridding" then Refine(level)
  CollectBoundaryValues(level)
  Evolve (level)
  if "level isn't finest existing" then begin
    for r = 1 to R do
      Integrate(level + 1)
    Update(level, level + 1)
  end
end
```

Consider a typical calculation which is initialized on a single, coarsely resolved grid called the *root grid*, level $\ell = 0$. The solution is integrated forward in time with procedures `CollectBoundaryValues` and `Evolve`. At each timestep, every cell is checked to determine if it requires refinement–a process known as *selection*. Selection is based either on an estimate of the local truncation error, or some simple threshold criterion, or some Boolean combination thereof. If at least one cell is flagged, then procedure `Refine` is called. It does two things. First, flagged cells are *clustered*, and minimal rectangular boundaries of these clusters are determined using the algorithm of Berger & Rigoustous (1991). Second, one or more refined $\ell = 1$ grids are allocated having these boundaries, and their initial data is computed via interpolation from the root grid. Depending on circumstance, `Refine` may also deallocate unneeded subgrids. The root grid is then advanced one coarse timestep Δt_0. The $\ell = 1$ subgrids are integrated forward in time with smaller timesteps Δt_1 until they "catch up" with the parent grid. If the hyperbolic system is linear (e.g., advection), then $\Delta t_1 = \Delta t_0/R$, and the refined grids take R timesteps for each coarse grid timestep. However, for nonlinear systems like gas dynamics, fine grid timesteps are determined by the Courant stability condition, and in general R+1 timesteps are needed to match times. This detail is not reflected in the pseudocode above. After the fine grids have been advanced to time $t + \Delta t_0$, procedure `Update` is called. It does two things. First, it injects the results of the fine grid calculation into the overlying coarse grid cells through summation or interpolation. Second, the values of coarse grid conserved quantities in cells adjacent to fine grid boundaries are modified to reflect the difference between the coarse and fine grid fluxes–a procedure known as *flux correction.*

The algorithm described above is recursive, and applies to any level in the grid hierarchy. In a multilevel grid hierarchy, the temporal integration scheme is analogous to the "W-cycle" in classic elliptic multigrid. Grids are advanced from coarse to fine with their individually determined timesteps. This order allows us to achieve second order accuracy in time on the entire hierarchy. This is accomplished by interpolating the boundary values not only in space but in time as well using time level n and n+1 values on the parent grid.

3. Cosmological AMR

Cosmological hydrodynamic simulations require a robust fluid solver capable of handling the extreme conditions of structure formation, as well as a method for evolving collisionless particles (dark matter, stars) subject to their self-consistent gravitational field, the latter requiring a solution of the Poisson equation. We briefly describe these methods here. A more detailed paper is in preparation (Bryan & Norman 1998).

3.1. COSMOLOGICAL PPM

Our fluid solver is based on the piecewise parabolic method (PPM) of Collela & Woodward (1984), suitably modified for cosmological flows. The algorithm is completely described in Bryan et al. (1995), so we merely state the essential points. PPM is a higher order-accurate version of Godunov's method, featuring third order-accurate piecewise parabolic monotonic interpolation and a nonlinear Riemann solver for shock capturing. Multidimensional schemes are built up by Strang splitting, resulting in a scheme which is formally second order-accurate in space and time. For cosmology, the conservation laws for the fluid mass, momentum, and energy density are written in comoving coordinates for a FRW spacetime with metric scale factor $a(t)$. Both the gas internal energy equation and total (internal + kinetic) energy equation are solved everywhere on the grid at all times. This *dual energy formulation* is adopted in order to have a scheme that both produces the correct entropy jump at strong shocks *and* yields accurate pressures and temperatures in the hypersonic parts of the flow. Both the conservation laws and the Riemann solver must be modified to include gravity. In order to maintain second order accuracy in time, the gravitational potential ϕ is needed at the half time level $t^{n+\frac{1}{2}}$. We use a predictor-corrector approach, wherein the particles are advanced to $t^{n+\frac{1}{2}}$

using ϕ^n, and then $\phi^{n+\frac{1}{2}}$ is computed by solving the Poisson equation, as described below. We have implemented both Lagrange + Remap (LR) and Direct Eulerian (DE) variants of our method, following the example of Colela & Woodward (1984), with comparable results. For AMR applications, we use the DE version to simplify the flux correction step.

3.2. ADDING COLLISIONLESS PARTICLES

Our method utilizes a single set of particles with comoving positions $\vec{x}_i$, proper peculiar velocities $\vec{v}_i$ and masses m_i (other characteristics may be added as needed). There is a unique, one-to-one association between particle i and a grid j at level ℓ $G_{\ell,j}$ if that particle's position lies within the grid's boundary but outside of any finer (child) grid. In other words, a particle belongs to the finest grid which contains it. We exploit this association by denoting that grid as the particle's *home* grid and store all such particles in a list along with the rest of the data connected to that grid. Note that a particle's home grid may change as it moves, requiring redistribution of particles. This disadvantage is offset by a number of factors, including (1) decreased search time for particle-grid interactions; (2) improved data encapsulation; and (3) better parallelization characteristics (see below). This association is also very natural from a physical standpoint: because the particles are indirectly connected to the solution on their home grid, they tend to share the same timestep requirement.

The particles obey Newton's equations, which in the comoving frame are:

$$\frac{d\vec{x}_i}{dt} = \frac{1}{a}\vec{v}_i, \tag{1}$$

$$\frac{d\vec{v}_i}{dt} = -\frac{\dot{a}}{a}\vec{v}_i - \frac{1}{a}(\vec{\nabla}\phi)_i \tag{2}$$

where the subscript i in the last term in Eq. 2 means the gravitational acceleration is evaluated at position $\vec{x}_i$. The gravitational potential is computed from a solution of the Poisson equation, which takes the form in comoving coordinates:

$$\nabla^2\phi = \frac{4\pi G}{a}(\rho - \bar{\rho}), \tag{3}$$

where ρ is the local comoving mass density of gas and particles, and $\bar{\rho}$ is its global average value.

These equations are finite-differenced and solved with the same timestep as the grid, to reduce bookkeeping. Equation (1) can be solved relatively simply since only quantities local to the particle are involved. We use a predictor-corrector scheme to properly time center the RHS. Equation (2) requires the gravitational acceleration at position $\vec{x}_i$. This is accomplished using the particle-mesh (PM) method. In the first step, particle masses are assigned to the grid using second order-accurate TSC (Triangular Shaped Cloud) interpolation. In the second step, the gridded particle density is summed with the gas density, and then Eq. (3) is solved on the mesh as described below. Finally, in the third step, gravitational accelerations are interpolated to the particle positions using TSC interpolation, and particle velocities are updated.

3.3. SOLVING THE POISSON EQUATION

The description above glossed over an important detail. Namely, that the gravitational field cannot be solved grid by grid, but must have knowledge of the mass distribution on the entire grid hierarchy. However, we wish to use the high spatial resolution afforded by the refined grids to compute accurate accelerations. This is accomplished using a method similar to that set out in Couchman (1991). The basic idea is to represent the acceleration on a particle as the sum of contributions from each level of the hierarchy less than or equal to the level of its home grid:

$$\vec{a_i} = \sum_{\ell'=0}^{\ell} \vec{a_i}(\ell') \tag{4}$$

where the partial accelerations $\vec{a_i}(\ell)$ are computed symbolically as follows:

$$\{\vec{x_i}, m_i\}_{G_\ell} \stackrel{TSC}{\longrightarrow} \rho_\ell(\vec{x}_{G_\ell}) \stackrel{FFT}{\longrightarrow} \hat{\rho}_\ell(\vec{k}), \tag{5}$$

$$\hat{\phi}_\ell(\vec{k}) = \hat{\rho}_\ell(\vec{k})\mathcal{G}_\ell(\vec{k}) \longrightarrow \hat{a}_\ell(\vec{k}), \tag{6}$$

$$\hat{a}_\ell(\vec{k}) \stackrel{3FFTs}{\longrightarrow} \vec{a}_\ell(\vec{x}_{G_\ell}) \stackrel{TSC}{\longrightarrow} \vec{a}_\ell(\vec{x}_i). \tag{7}$$

In the first step, the mass of *all* particles whose positions lie inside the boundaries of grid G_ℓ is assigned to the grid using TSC interpolation, and its Fourier transform is computed. In the second step the Poisson equation is solved in Fourier space using a shaped force law $\mathcal{G}_\ell(\vec{k})$ designed to reproduce a r^{-1} potential when summed over all levels. Accelerations on the grid are computed in Fourier space, and then transformed back into real space in step 3. This requires three FFTs—one for each component. Finally, the acceleration on particle i due to G_ℓ is computed using TSC interpolation. The acceleration on the fluid is computed by treating each cell as a particle with its position given by the center of mass (as determined by a tri-linear interpolation). The assignment of mass and interpolation of acceleration is done with the same TSC scheme as used for the particles.

3.4. REFINEMENT FACTOR AND CRITERIA

In our implementation, the refinement factor R_ℓ can have any integer value, and can be different on different levels. However, we have found through experimentation that $R_\ell = R = 2$ is optimal for cosmological simulations where the gravitational dynamics is dominated by dark matter. Since dark matter is represented by a fixed number of particles, the use of higher refinement factors $R \geq 3$ refines the gas grid to the point where each dark matter particle becomes an accretion center. The choice $R = 2$ maintains commensurate mass resolution in both gas and dark matter.

In order to reduce Poisson noise, we initialize a calculation with one particle per cell. We flag a cell for refinement when the baryonic mass has increased by a factor of four over its initial value. Immediately after refinement, refined cells have one half their initial mass and typically contain zero or one particle. Thus, in a collapsing structure, the ratio of baryonic to dark matter mass will vary between one half and four times its initial value.

4. Tests

We have extensively tested our code against a variety of hydrodynamic and hybrid (hydro + N-body) test problems. These include (1) linear waves and shock waves entering and exiting static subgrids in 1D and 2D at various angles; (2) a stationary shock wave in a static 3D subgrid "Shock in a Box" (Anninos, Norman & Clarke 1994); (3) Sod shock tube; (4) 1D pressureless collapse; (5) 1D Zeldovich pancake; (6) 2D pure baryonic CDM model (Bryan et al. 1995); (7) 3D adiabatic X-ray cluster "Santa Barbara cluster" (Frenk et al. 1998); and (8) 3D self-similar infall solution. In tests (3-7), AMR results were compared with unigrid results up to grid sizes of 512^3. In all cases the algorithm performs well, reproducing the reference solutions with a minimum of artifacts at coarse/fine grid interfaces. The use of upwind, second-order accurate fluxes in space and time is found to be essential in minimizing reflections.

As an example, Fig. 1 shows the code's performance on the 1D Zeldovich pancake test problem. The parameters for this problem, which includes gas, dark matter, gravity, and cosmic expansion, are given in Bryan et al. (1995). Here we compare a 3-level AMR calculation with a 256 zone uniform grid calculation. The root grid has 16 zones, and the refinement factor $R = 4$ (in 1D problems one has more latitude with R.) Thus, the AMR calculation has the same 256 zone resolution on the finest grid $\ell = 2$. We can see that the AMR algorithm places the refinements only in the central high density pancake. The solutions on the three levels of refinement match smoothly in the supersonic infalling envelope. The density maximum and the local

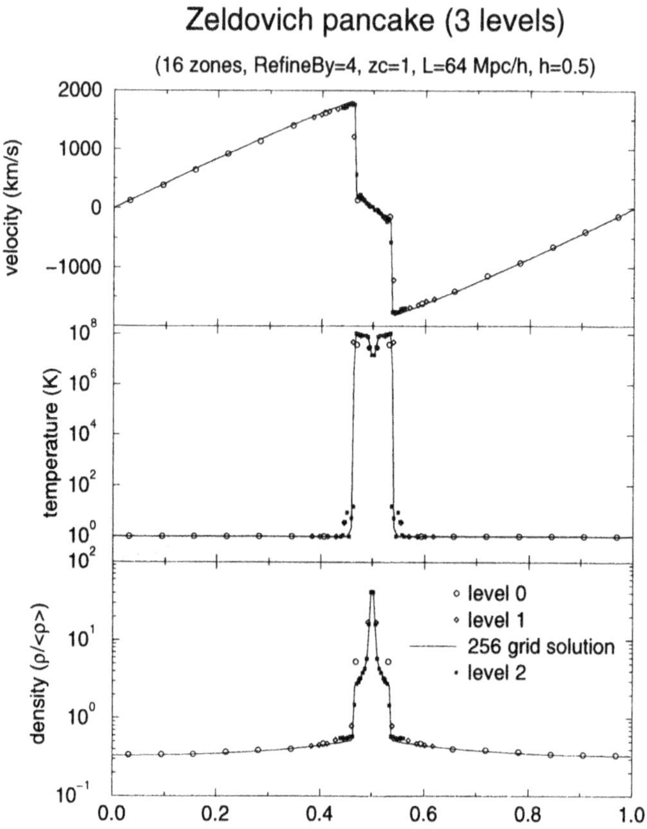

Figure 1. 1D Zeldovich pancake test problem. 3-level solution (symbols) is overlaid on 256 zone uniform grid solution (solid line).

temperature minimum at the midplane agree with the uniform grid results exactly. The shock waves at $x = 0.45$ and 0.55 are captured in two cells without post-shock oscillations. The temperature field–the most difficult to get right–exhibits a small bump upstream of the shock front. This is caused by our dual energy formulation, and marks the location where we switch from internal energy to total energy as a basis for computing pressures and temperatures. The bump has no dynamical effect as ram pressure dominates by many orders of magnitude ahead of the shock.

5. A Sample Application: First Structure Formation

We have applied our code to the formation of X-ray clusters (Bryan & Norman 1997), galaxy formation (Kepner & Bryan 1998), and the formation of the first baryonic structures in a CDM-dominated universe (Abel, Bryan & Norman 1998). In Fig. 3 of the longer version of this article on the accompanying CD-ROM (or astro-ph/9807121) we show a result from the latter calculation to illustrate the capabilities of our method.

Our ultimate goal is to simulate the formation of the first stars in the universe (Population III stars) starting from cosmological initial conditions. To resolve the protostellar cloud cores which form these stars requires a spatial dynamic range of at least five orders of magnitude in 3D. The primary coolant in zero metallicity gas for $T < 10^4$ K is H_2 line cooling. H_2 forms in the gas phase via nonequilibrium processes. Thus, in addition to gas and dark matter, we also solve a 9-species nonequilibrium chemical network for H, H^+, He, He^+, He^{++}, H^-, H_2^+, H_2 and e^- in every cell at every level of the hierarchy. This requires adding nine new field variables–one for each species–and solving the stiff reactive advection equations for the coupled network. The physical model is described in Abel et al. (1997). The numerical algorithms for cosmological reactive flow are provided in Anninos et al. (1997).

We simulate a standard CDM model in a periodic, comoving volume 128 kpc on a side. The parameters are: $\Omega_0 = 1, \Omega_b = 0.06, \sigma_8 = 0.7, H_0 = 50$ km s^{-1} Mpc^{-1}. The starting redshift is $z = 100$. The intial conditions are realized on a 3-level static nested grid hierarchy with a root grid resolution of 64^3 cells. The subgrids are centered on the most massive peak which develops. This location is determined by running the simulation once at low resolution. The $\ell = 2$ grid is made large enough so that it initially contains all the mass that eventually ends up in the condensed halo. The initial $\ell = 2$ mass resolution in the gas (dark matter) is $0.5(8) M_\odot$, respectively.

As the calculation proceeds, the AMR algorithm generates many more grids at higher levels of refinement up to a preset limit of 13 levels total. In addition to the overdensity refinement criterion described above, we also require that the numerical Jeans criterion is satisfied everywhere (Truelove et al. 1998) Nonlinear structures form in the gas by $z \approx 30$. By $z \approx 20$ sufficient H_2 has formed in the center of the virialized halo that a cooling flow ensues (Abel et al. 1998). Because of the range of mass scales which go nonlinear nearly simultaneously at these epochs, the halo is quite lumpy as structure builds up hierarchically. By $z = 18$ (Fig. 3), a highly concentrated structure has formed through the collision of two rapidly cooling blobs. A collapsing, primordial protostellar core forms as a result. The core collapses to higher densities, eventually reaching our resolution limit. The proper cell size on the $\ell = 12$ grid is $\Delta x_{12} = 0.025$ pc $\sim$ 5000 AU. The overall dynamic range achieved is $64 \times 2^{12} = 262,144$.

We stop the calculation when the Jeans criterion is violated on the $\ell = 12$ grid. This occurs in the densest cell, which has reached a baryonic overdensity of almost 10^{10} and a proper number density of more than 10^8 cm^{-3}. Above these densities, three body reactions become important, which we have not included in our model. At the end of the calculation, we have formed a parsec-sized collapsing primordial core with about $500 M_\odot$ of material, roughly equal parts gas and dark matter. The further evolution of this cloud, including the important question of fragmentation, is being studied with a separate, smaller scale AMR simulation in progress.

Acknowledgements We thank our collaborators Tom Abel and Peter Anninos for joint work cited here, as well as Sir Martin Rees for drawing our attention to the problem of first structure formation. This work was carried out under the auspices of the Grand Challenge Cosmology Consortium (GC^3), with partial funding provided by NSF grant ASC-9318185 and NASA grant NAG5-3923. Simulations were carried out using the SGI Origin2000 at the National Center for Supercomputing Applications, University of Illinois, Urbana-Champaign.

References

Abel, T., Anninos, P., Zhang, Y. & Norman, M. L. 1997, New Astronomy, 2, 181
Abel, T., Anninos, P., Norman, M. L. & Zhang, Y. 1998. Ap. J., *in press* (astro-ph/9705131)
Abel, T., Bryan, G. L. & Norman, M. L. 1998. *in preparation*
Anninos, P., Zhang. Y., Abel, T. and Norman, M. L. 1997, New Astronomy, 2, 209
Anninos, P., Norman, M. L. & Clarke, D. A. 1994. Ap. J., 436, 11
Berger, M. J. & Rigoustous, I 1991. IEEE Transactions on Systems, Man, Cybernetics, 21(5).
Berger, M. J. & Oliger, J. 1984. J. Comp. Phys. 53, 484
Berger, M. J. & Collela, P. 1989. J. Comp. Phys. 82, 64
Bryan, G. L. & Norman, M. L. 1997. in *Computational Astrophysics*, eds. D. A. Clarke & M. Fall, ASP Conference # 123
Bryan, G. L. & Norman, M. L. 1998. *in preparation*
Bryan, G. L., Norman, M. L., Stone, J. M., Cen, R. & Ostriker, J. P. 1995, Comp. Phys. Comm., 89, 149
Cen, R. 1992, Ap. J. Suppl., 78, 341
Colella, P. & Woodward, P. R. 1984. J. Comp. Phys., 54, 174
Couchman, H. 1991. Ap. J. Lett. 368, L23
Frenk, C., White, S. *et al.* 1998. Ap. J., *submitted*
Gnedin, N. Y. 1995, Ap. J. Suppl., 97, 231
Kang, H., Cen, R., Ostriker, J. P. & Ryu, D. 1994, Ap. J., 428, 1
Kepner, J. & Bryan, G. L. 1998. *in preparation*
Kravtsov, A. V., Klypin, A. A. & Khoklov, A. M. 1997. Ap. J. Suppl., 111, 73
Pen, U. 1995, Ap. J. Suppl., 100, 269
Truelove, J. K. *et al.* 1998. Ap. J., 495, 821

N-BODY SIMULATION OF DARK MATTER HALO FORMATION USING GRAPE

TOSHIYUKI FUKUSHIGE AND JUNICHIRO MAKINO
Department of General Systems Studies,
College of Arts and Sciences, University of Tokyo,
3-8-1 Komaba, Meguro-ku, Tokyo 153, Japan

Abstract. We performed N-body simulation on special-purpose computer, GRAPE-4, to investigate the structure of dark matter halos. Universal profile proposed by Navarro, Frenk, and White (1996), which has cusp with density profiles $\rho \propto r^{-1}$ in density profile, cannot be reproduced in the standard Cold Dark Matter (CDM) picture of hierarchical clustering. Previous claims to the contrary were based on simulations with relatively few particles, and substantial softening. We present simulations with particle numbers an order of magnitude higher ($N = 786,400$), and essentially no softening, and show that typical central density profiles are clearly steeper than $\rho \propto r^{-1}$.

1. Introduction

We briefly discuss the structure of dark matter halos using high-resolution N-body simulation. For more details, see Fukushige and Makino (1997).

Several simulations demonstrated that the dark matter halos formed through hierarchical clustering were not well approximated by isothermal spheres, and better fitted by a Hernquist model (Hernquist 1990) or similar models with $\rho \sim 1/r$ at the center (Dubinski and Carlberg, Navarro, Frenk and White 1996, 1997). In these simulations, halos have a power law index which changes from around -1 to -3 or -4 as the radius increases. Recently, Flores and Primack (1994) and Moore (1994) have argued that the gently rising $\rho(r)$ curves of dwarf galaxies are inconsistent with a r^{-1} central cusp, and therefore challenge the hypothesis of CDM halos.

However, whether the results of numerical simulations are physically valid or are numerical artifacts remains unclear for the following reasons.

S. M. Miyama et al. (eds.), Numerical Astrophysics, 29–32.

First, no physical mechanism has been presented so far for the formation of cusps found in simulations (White 1996). Secondly, since the mass resolution in these simulation was rather low, the central structure may have been affected by two-body relaxation effects (Quinlan 1996, Steinmetz and White 1997). Thirdly, the central structures of halos formed in these simulations are strongly affected by the potential softening used, and show rapid change in the power index of $\rho(r)$ around a radius not much larger than the potential softening length. In this paper, we have been able to separate physically real effects from numerical artifacts, through N-body simulations of hierarchical clustering with a resolution far higher than those in previous work. The number of particles we used ($N = 786,400$) is more than 10 times larger than those used in previous simulations ($N \sim 10,000 - 30,000$), and our softening is significantly smaller.

Our results show that dark halos formed through hierarchical clustering typically exhibit an inwardly decreasing temperature structure in their inner regions, with a density cusp shallower than $\rho \propto r^{-2}$, but steeper than $\rho \propto r^{-1}$.

2. Method

Initial conditions were constructed following Dubinski and Carlberg (1991). We assigned initial positions and velocities to particles in a spherical region with a radius of 2Mpc surrounding a density peak selected from a discrete realization of a standard CDM model ($H_0 = 50$ km s^{-1} Mpc^{-1} and $\Omega = 1$). The peak was chosen from a realization of the density contrast in an 8Mpc box, using COBE normalization. The peaks were found by smoothing the density with a Gaussian filter of radius 0.75 Mpc.

We followed the evolution of a density peak through direct N-body simulation. We added the local Hubble flow and integrated the orbits directly in physical space, with $N = 786,400$, an individual particle mass of $4.0 \times 10^6 M_\odot$, and a Plummer softened potential with a length of $\varepsilon =$ 0.14kpc. We started the simulation at $z \sim 46$. We did not include tidal effects from outside our 2Mpc sphere, since we are mainly interested in the core properties within 10kpc, where tidal effects from outside 2Mpc scale are negligible.

We used a 4-th order Hermite integration scheme (Makino and Aarseth 1992) with individual (hierarchical) timestep algorithm (MacMillan 1986, Makino 1991). For the force calculation, we used the GRAPE-4 (Makino et al. 1997), a special-purpose computer designed to accelerate N-body simulations using a Hermite integrator and a hierarchical timestep algorithm. Since the force calculations on GRAPE-4 are effectively of double-precision accuracy, our simulations exhibit much higher accuracy both for force cal-

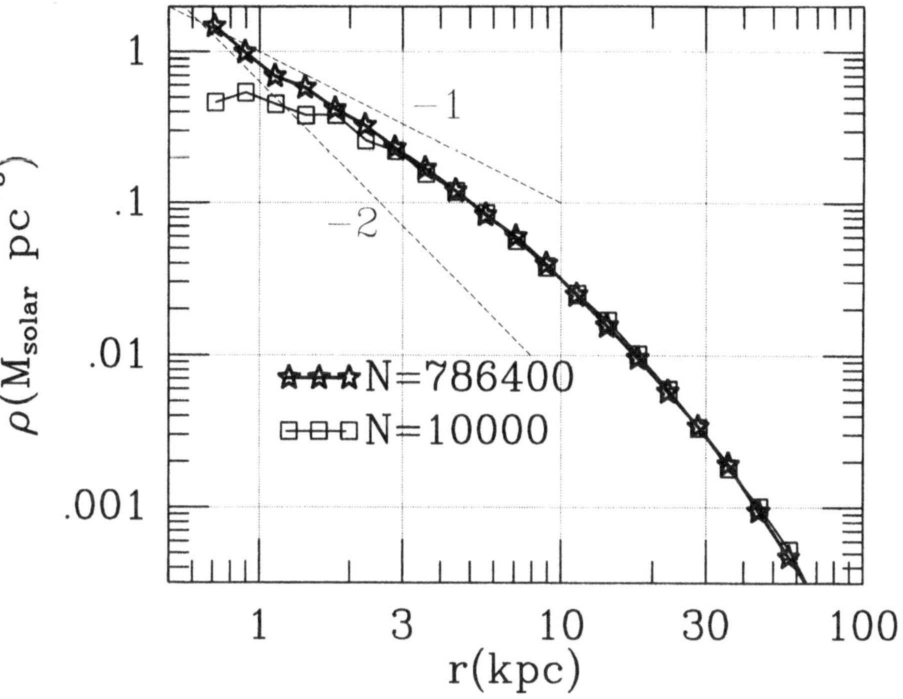

Figure 1. Density structure of the halo at $z = 1.8$. The position of the center of the halo was determined using potential minimum and averaged physical values over each spherical shell. Dashed lines indicate profiles of $\rho \propto r^{-1}$ and r^{-2}.

culations and orbit integrations than previous simulations.

3. Simulations Results

Figure 1 shows the density structure of the halo at $z = 1.8$, well after most of the mergings has already taken place (results at $z = 0$ are similar, but these earlier stages shows even less effects of two-body relaxation). We also plot a density profile obtained with lower resolution simulation ($N = 10,000$). We can see that the central density cusp obtained with lower resolution simulation is shallower.

In addition we can see a clear temperature decrease toward the center within 5kpc. This non-isothermal structure produces a density cusp shallower than $\rho \propto r^{-2}$.

4. Discussion

In this paper, we have shown that dark halos formed through hierarchical clustering have a central density cusp shallower than $\rho \propto r^{-2}$ and steeper than $\rho \propto r^{-1}$.

We offer the following interpretations for the temperature and the density structures found. The occurrence of the striking temperature inversion can be understood as follows. In bottom-up structure formation, as in CDM hierarchical clustering, a typical halo is formed through repeated merging of smaller subclumps. Each time this happens, the less concentrated clump tends to be disrupted by the tidal field of the more centrally concentrated clump. The dense core of the latter survives the merging process more or less intact, and settles down at the center of merger remnant, with locally unchanged temperature. In contrast, the temperature of the merger remnant as a whole increases since the specific binding energy of the merger remnant is almost always larger than that of its progenitors.

References

Dubinski, J. and Carlberg, R. 1991, ApJ, 378, 496
Flores, R. A. and Primack, J. R. 1994, ApJ, 427, L1
Fukushige, T. and Makino, J. 1997, ApJ, 477, L9
Hernquist, L. 1990, ApJ, 356, 359
Makino J. 1991, PASJ, 43, 859
Makino J. and Aarseth, S. J. 1992, PASJ, 44, 141
Makino, J., Taiji, M., Ebisuzaki, T., and Sugimoto, D. 1997, ApJ, 480, 432.
McMillan, S. L. W. 1986, in The Use of Supercomputers in Stellar Dynamics, eds. S. L. W. McMillan, P. Hut (Springer, Berlin), p156
Moore, B. 1994, Nature, 370, 629
Navarro, J. F., Frenk, C. S., and White, S. D. M. 1996, ApJ, 462, 563
Navarro, J. F., Frenk, C. S., and White, S. D. M. 1997, ApJ, 490, 493
Quinlan, G, D. 1996, New Astronomy, 1, 255
Steinmetz, M. H. and White, S. D. M. 1996, MNRAS, 288, 545
White, S. D. M. 1996, in Gravitational Dynamics, ed. Lahav, O., Terlevich, E., Terlevich, R. J., 121 (Cambridge University Press)

LARGE COSMOLOGICAL N-BODY SIMULATIONS AND CLUSTERING OF DARK MATTER HALOS

Y.P. JING
Research Center for the Early Universe, University of Tokyo
Bunkyo-ku, Tokyo 113, Japan

1. Introduction

Galaxy formation is one of the most important and most challenging problems in astrophysics. In the standard theories of galaxy formation, it is generally believed that the Universe is dominated by dark matters (DM), the early density perturbation generated in the inflationary era is amplified by the gravitational instability, the DM halos form via the collapse of the high density regions, and galaxies are formed from the cooled gas within the DM halos (see, e.g., H. Couchman, Y. Suto, and M. Norman in these proceedings). Obviously understanding the DM distribution is the first step to a complete understanding of the galaxy formation.

The formation of DM halos or generally the DM distribution on small scales, however, is a strongly non-linear problem. Numerical simulations seem the only way to solve this problem. Although there have been many exciting developments on analytical, semi-analytical or empirical modeling of the DM distribution, these results are generally calibrated with N-body simulations and their validity still depends on the resolution of the simulations used.

In this contribution, I will first describe the N-body simulations carried out at the University of Tokyo. The simulations have similar mass and force resolutions to those the Virgo Consortium (H. Couchman, in these proceedings) has recently obtained (except for their billion-particle simulation for a huge simulation box), but our sample is much larger. These simulations are being used to study many interesting problems of the large-scale structures in the universe. Yasushi Suto (in these proceedings) has already presented our applications of these simulations to the redshift distortion (Magira et al. 1998; in preparation) and to the strong clustering of Lyman Break galaxies at high redshift (Jing & Suto 1998). Here I will present a further interesting application, the bias parameter of the DM halos.

S. M. Miyama et al. (eds.), Numerical Astrophysics, 33–36.

2. N-body simulations

The simulations are generated with my vectorized Particle-Particle-Particle-Mesh (P^3M) code on the the supercomputer VPP300/16R at the National Astronomical Observatory of Japan. Each simulation uses 256^3 simulation particles. The code adopts the standard P^3M algorithm (Hockney & Eastwood 1981; Efstathiou et al. 1985). Twenty one simulations are generated for three representative cold dark matter (CDM) models of galaxy formation, and eighteen for six scale-free cosmologies. The CDM models are specified completely as regards the DM distribution by the density parameter Ω_0, the cosmological constant λ_0, the shape Γ and the normalization σ_8 of the linear power spectrum. The scale-free models assume an Einstein–de Sitter universe (EdS i.e. $\Omega_0 = 1$ and $\lambda_0 = 0$) and a power-law $P(k) \propto k^n$ for the linear density power spectrum. Table 1 summarizes the physical and simulation parameters used for these simulations. The CDM models with the chosen physical parameters are known to reproduce many features observed in the real Universe. The simulations of $100\,h^{-1}$Mpc box size have particle mass $m_p \approx 10^{10} M_\odot$ and force resolution $\eta = 13\,h^{-1}$kpc (the Plummer form), and therefore suit for studying galaxy formation, internal structures of DM halos and clustering of DM halos. The simulations with a larger box size ($300\,h^{-1}$Mpc) are useful for studying the cluster formation and clustering on large scales. Every model (with one box size) has 3 to 4 realizations so the cosmic variance could be properly accounted for. The scale-free models are conceptually simple and possess interesting scaling properties. Therefore they are very powerful for understanding clustering of DM, for testing analytical models, for obtaining fitting formulae, and for quantifying the numerical artifacts. Additional 8 simulations with 200^3 particles are generated for lower density models with $\Omega_0 = 0.1$ for power-law spectra $n = -1$ and $n = -2$, in order to examine the effect of cosmologies on the clustering (cf. the $\Omega_0 = 1$ models). In terms of the mass and force resolution and the number of the realizations for each model, this set of simulations stands as one of the best in the world.

With this large set of high-resolution simulations, many interesting cosmological problems can and will be investigated. As an incomplete list, we are 1) studying the DM distribution in the strongly clustering regime; 2) testing the Press-Schechter extension theories; 3) studying redshift distortion effect and investigating its potential to measure the cosmological parameters; 4) analyzing internal structures of DM halos; 5) investigating galaxy formation by hand-inputing gas physics; 6) applying the simulations to real observations. Some of the above projects are the collaborations with the members of Prof. Yasushi Suto's group at the University of Tokyo.

Table 1. List of simulations.

Model	N	Ω_0	λ_0	Γ	σ_8	m_p ($h^{-1}M_\odot$)	L (h^{-1}Mpc)	realizations
SCDM100	256^3	1.0	0.0	0.5	0.6	1.7×10^{10}	100	3
OCDM100	256^3	0.3	0.0	0.25	1.0	5.0×10^9	100	3
LCDM100	256^3	0.3	0.7	0.20	1.0	5.0×10^9	100	3
SCDM300	256^3	1.0	0.0	0.5	0.6	4.6×10^{11}	300	4
OCDM300	256^3	0.3	0.0	0.25	1.0	1.3×10^{11}	300	4
LCDM300	256^3	0.3	0.7	0.20	1.0	1.3×10^{11}	300	4
Model	N	Ω_0	λ_0	n				realizations
SF1	256^3	1.0	0.0	1.0				3
SF2	256^3	1.0	0.0	0.0				3
SF3	256^3	1.0	0.0	−0.5				3
SF4	256^3	1.0	0.0	−1.0				3
SF5	256^3	1.0	0.0	−1.5				3
SF6	256^3	1.0	0.0	−2.0				3
SF7	200^3	0.1	0.9	−1.0				2
SF8	200^3	0.1	0.0	−1.0				2
SF9	200^3	0.1	0.9	−2.0				2
SF10	200^3	0.1	0.0	−2.0				2

3. The bias parameter of the dark matter halos

The clustering of the DM halos is the basic block for modeling and understanding the clustering of galaxies and of other extragalactic objects. In my recent work (Jing 1998) I have measured the two-point correlation function of the DM halos for the scale free simulations with $n = -0.5$, -1.0, -1.5, and -2.0. The results confirmed with unprecedented accuracy that the bias of the halos b is linear in the linearly clustering regime. For a scale-free model, the bias parameter b is then expected to depend on the scaled mass $M/M_\star$ only, where $M_\star$ is a characteristic non-linear mass, unless numerical artifacts had destroyed this scaling property. In Fig. 1, I present the $b(M/M_\star)$ for different simulation outputs of each model. The excellent scaling exhibited by the simulation data assures that any numerical artifacts have negligible effect on the results of Fig. 1. The simulation results agree with the analytical formula of Mo & White (1996) only for massive halos with $M/M_\star > 1$, but are significantly higher for much less

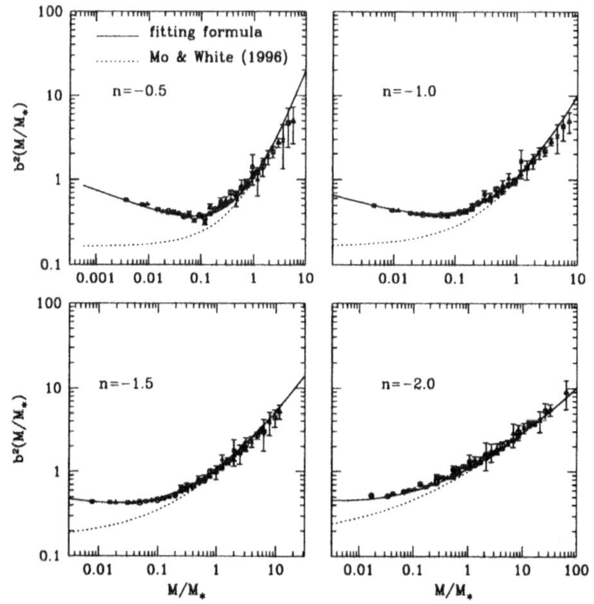

Figure 1. The $b^2(M/M_\star)$ for different evolution outputs (different symbols). The dotted lines are the analytical prediction of Mo & White (1996), and the solid lines are my fitting formula Eq. 1 which can accurately fit the simulation results.

massive halos. The simulation results could be fit very well by

$$b(M) = \left(\frac{0.5}{\nu^4} + 1\right)^{0.06-0.02n} \left(1 + \frac{\nu^2 - 1}{\delta_c}\right); \qquad \nu = (M/M_\star)^{n+3/6} \tag{1}$$

where $\delta_c = 1.68$. This fitting formula could be equally well applied to the CDM models with only a few simple modifications (Jing 1998). This finding has profound implications for the clustering studies of galaxies. Further discussions are given in Jing (1998).

Acknowledgements

I am grateful to Yasushi Suto for invaluable help and to JSPS for a fellowship. The simulations were carried out on VPP300/16R at NAOJ.

References

Efstathiou, G., Davis, M., Frenk, C.S. & White, S.D.M. 1985, ApJS, 57, 241
Hockney, R.W. & Eastwood, J.W. 1981, Computer simulations using particles. Mc Graw-Hill
Jing, Y.P., 1998, ApJ (Letters), in press
Jing, Y.P., Suto, Y., 1998, ApJ, 494, L5
Mo H.J., White S.D.M., 1996, MNRAS, 282, 347

EVOLUTION OF THE POWER SPECTRUM AND THE SELF-SIMILARITY IN THE EXPANDING ONE-DIMENSIONAL UNIVERSE

T.YANO
Department of Earth and Space Science, Graduate School of Science, Osaka University, Toyonaka, Osaka 560-0043, Japan

We have investigated the time evolution of the power spectrum in two cases of initial conditions, the scale-free case and the cut-off case. In the scale-free case, we can observe a self-similar evolution of the power spectrum. The scaled power spectra $P_*(k_*)$ coincide each time. We can roughly divide the power spectrum into three regimes. One is the linear regime ($k < k_{nl}$: regime 1). The value of the power index in this regime remains the initial power index, n. The second regime is the single-caustic regime ($k_{nl} < k < k_{snl}$: regime 2). In this regime, the power index becomes -1 and is independent of the initial conditions. This is caused by the appearance of caustics on this scale, which determine the power index of the power spectrum in this regime. The third regime is the multi-caustics regime ($k > k_{snl}$: regime 3). In this case, the distribution of the "whirlpool" in phase space determines the value of the power index. Therefore, on this scale, the power index μ has a value which depends on the initial conditions.

In the cut-off case, on the other hand, we find that there is no self-similarity on all scales. The scaled power spectrum does not coincide in all regimes. However, the spectrum coincides on scales larger than the cut-off scale. After the appearance of the first caustics, the power index in the regime of a scale smaller than the cut-off scale becomes -1. This value is caused by the appearance of caustics. As more and more caustics appear, the separation of caustics becomes smaller and smaller. Even after the appearance of many caustics, on scales smaller than the characteristic separation of caustics ($k > k_{cs}$: regime 5), the power index of the power spectrum is obtained by the density profile around the singular point. Therefore, the power index of these scales becomes -1, which can be derived according to the catastrophe theory. On the other hand, on scales larger than the characteristic separation of the caustics ($k_{cut} < k < k_{cs}$: regime 4), it is the distribution of singular points that determines the power

S. M. Miyama et al. (eds.), Numerical Astrophysics, 37–38.

index on these scales rather than the density profile around one singularity. Because in this regime, the smoothed density profile with the smoothing scale ($k_{cut} < k < k_{cs}$) determines the power index. On these scales, the distribution of the singularity occurring in the evolution of the single wave is important. Then, the power index on these scales is determined by this distribution of the singularity. Consequently, we can roughly distinguish two regimes on scales smaller than the cut-off scale k_{cut}. One is the virialized regime ($k_{cut} < k < k_{cs}$: regime 4), in which the power index has the value ν. The other one is the smallest single-caustic regime ($k > k_{cs}$: regime 5), in which the index has the value of -1 (Gouda & Nakamura 1988, 1989). Both indexes are independent of the initial conditions.

In a real situation about the evolution of the power spectrum, the initial power spectrum has a cut-off point on a certain scale. We have noticed that even if the initial power spectrum does not have a cut-off point, it decreases with the power index $n < -1$ on certain scales ($k > k_{dec}$). We can regard this scale as a cut-off scale. Therefore, the time evolution of the power spectrum is the same as the one in the cut-off case after the appearance of caustics. Subsequently, we can roughly recognize five regimes (Figure 1). Only when we consider the evolution of the power spectrum on scales larger than the cut-off scale (i.e. the first, second, and third regime), is the self-similarity satisfied.

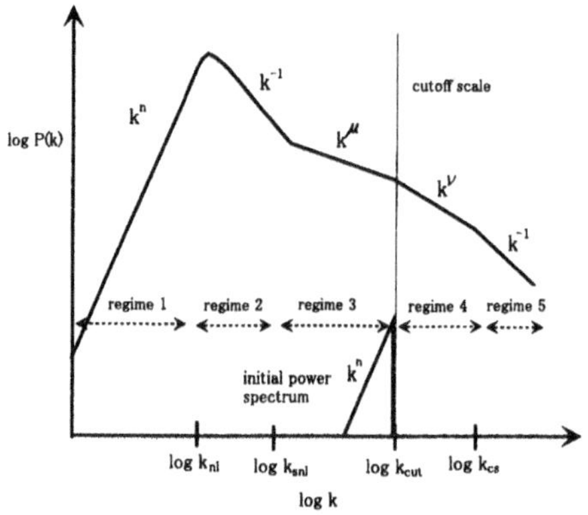

Figure 1. A schema of the power spectrum with a general initial power spectrum at a certain time after the first appearance of caustics.

References

Gouda, N. & Nakamura, T. 1988, Prog.Theor.Phys., 79, 765
Gouda, N. & Nakamura, T. 1989, Prog.Theor.Phys., 81, 633
Yano, T. & Gouda, N. 1998, ApJS, 118 inpress

LIGHT PROPAGATION IN INHOMOGENEOUS UNIVERSE

Preliminary Results from Gravitational Lensing Experiments

PREMANA W. PREMADI
Institute of Astronomy, Tohoku University
Sendai 981, Japan

HUGO MARTEL
Dept. of Astronomy, University of Texas
Austin, TX 78712, USA

AND

RICHARD A. MATZNER
Center for Relativity, University of Texas
Austin, TX 78712, USA

1. Overview

Using a multi-plane lensing method that we have developed (Premadi, Martel, & Matzner 1998), we follow the evolution of light beams as they propagate through inhomogeneous universes. We use a P^3M code to simulate the formation and evolution of large-scale structure. The resolution of the simulations is increased to sub-Megaparsec scales by using a Monte Carlo method to locate galaxies inside the computational volume according to the underlying particle distribution.

We consider cold dark matter models normalized to *COBE*, and perform a large parameter survey by varying the cosmological parameters Ω_0, λ_0, H_0, and n (the tilt of the primordial power spectrum). This parameter survey is still in progress. Table 1 gives the values of the parameters for the models we have studied so far.

2. The Ray Shooting Experiments

For each model, we perform numerous ray-tracing experiments, propagating beams of 31^2 light rays back in time up to redshifts $z = 3$ or 5. From the results of these experiments, we build statistics of the shear and magni-

S. M. Miyama et al. (eds.), Numerical Astrophysics, 39–40.

TABLE 1. The Cosmological Parameters.

Ω_0	λ_0	H_0	n	σ_8	Ω_0	λ_0	H_0	n	σ_8
1.0	0.0	65.0	0.7698	1.0	0.2	0.0	65.0	1.3188	0.5
1.0	0.0	75.0	0.7094	1.0	0.2	0.0	75.0	1.2190	0.5
1.0	0.0	65.0	0.8506	1.2	0.5	0.5	65.0	0.7808	0.8
1.0	0.0	75.0	0.7893	1.2	0.5	0.5	75.0	0.7049	0.8
0.5	0.0	65.0	0.9457	0.8	0.5	0.5	65.0	0.8807	1.0
0.5	0.0	75.0	0.8686	0.8	0.5	0.5	75.0	0.8024	1.0
0.5	0.0	65.0	1.0439	1.0	0.2	0.8	65.0	0.9326	0.6
0.5	0.0	75.0	0.9656	1.0	0.2	0.8	75.0	0.8273	0.6
0.2	0.0	65.0	1.0966	0.3	0.2	0.8	65.0	1.0702	0.8
0.2	0.0	75.0	0.9993	0.3	0.2	0.8	75.0	0.9629	0.8

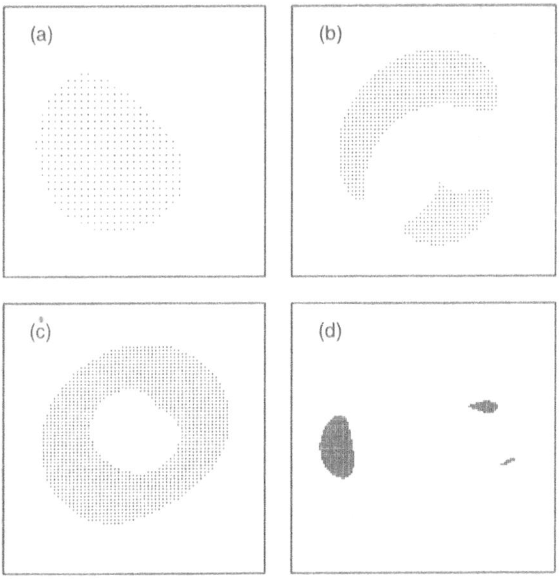

Figure 1. Images of distant circular sources.

fication of sources, as well as the properties of images. In Figure 1, we plot the images of distant circular sources, to illustrate interesting cases that occurred in some experiments: magnification and shear (Fig. 1a), double image (Fig. 1b), Einstein ring (Fig. 1c), and triple image (Fig. 1d).

References

Premadi, P.W., Martel, H., Matzner, R.A. (1998), *Astrophys. J.*, **Vol. 493**, pp. 10-27

PHOTOIONIZATION AND DYNAMICAL EVOLUTION OF MINI-PANCAKES

HAJIME SUSA
Center for Computational Physics, University of Tsukuba
1-1-1 Tennoudai, Tsukuba, Japan

1. Introduction

In the post-dark age epoch, UV radiation emitted by PopIII objects has played an important role on the thermal processes of collapsing primordial gas clouds, i.e. photoionization, UV heating, and photodissociation of H_2. These collapsing primordial gas clouds are presumably primeval galaxies. Important research carried out on this issue [Ref.1-4] assumes the quasi-static evolution of the clouds. However, in a more realistic situation, there should be shocks, and the dynamics could affect the evolution of collapsing primeval clouds. *In this paper,* we perform a 1-dimensional hydrodynamical simulation coupled with chemical reactions including H_2, radiative cooling/heating, and radiation transfer equation for ionizing photons, assuming the plane's parallel symmetry. For H_2 dissociating photons, we use the 'self-shielding function' in Ref.[5].

2. Results

In the left panel of Fig. 1, the physical states of the sheets after one dynamical time are summarized. With given initial parameters, λ and initial density, sheets evolve in different ways. The collapse of a gas cloud irradiated by UV radiation is regulated by the Jeans length with 10^4 K. This is one of the possible explanations for the biasing of galaxy formation. However, if the sheet (cloud) is sufficiently self-shielded to shut down UV heating compared to H_2 cooling, the Jeans length is reduced by a few orders of magnitude. Hence, the comparison between UV heating rate and H_2 cooling rate provides another criterion of collapse. Comparing the two rates above, we obtain critical λ_c, which is plotted in the figure. In the figure, it can be observed that "collapse" is regulated not only by the Jeans length at 10^4 K, but also by λ_c. Here, "collapsed sheet" means the one whose final Jeans length at the center is ten times smaller than its initial thickness λ. The shaded region denotes the new area where the sheet might collapse. If the sheet is cooled down to 100 K, the sheets in the shaded region become

S. M. Miyama et al. (eds.), Numerical Astrophysics, 41–42.

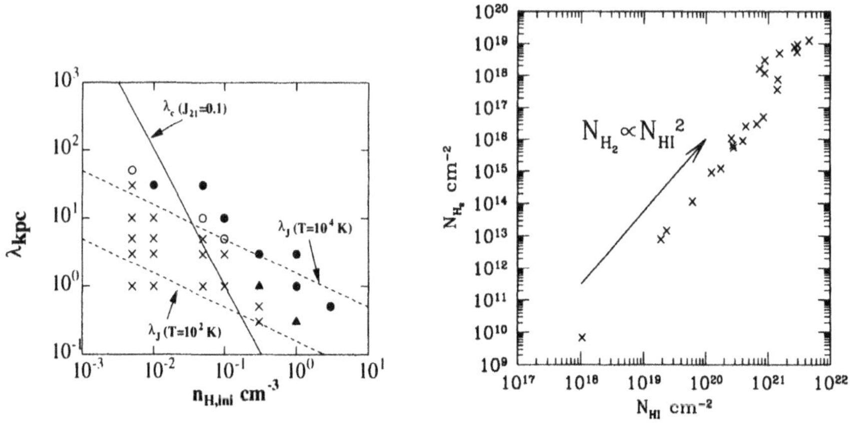

Figure 1. Left panel) Summary of physical states with the given initial parameters. The vertical axis and the horizontal axis denote the initial size and density of the sheets. The vertices denote the sheets which do not "collapse" within one dynamical time and have a high central temperature ($T \sim 10^4$ K). The triangles denote the sheets which do not collapse and have a warm central region ($T \sim 10^3$ K). The open circles denote the sheets which have collapsed and are warm. The sheets denoted by the solid circles have collapsed and are cool ($T \sim 10^2$ K). Right panel) $N_{\rm HI}$ vs. $N_{\rm H_2}$ is plotted for sheets with various initial parameters after one dynamical time.

Jeans unstable. In fact, there are some parameters according to which the sheets become Jeans unstable in our numerical calculations (denoted by the solid circles). We also plotted the physical states of the sheets after one dynamical time on $N_{\rm HI}$ vs. $N_{\rm H_2}$ plane (Right panel of Fig. 1). For low $N_{\rm HI}$ clouds, $N_{\rm H_2}$ is proportional to the square of $N_{\rm HI}$. For high $N_{\rm HI}$, $N_{\rm H_2}$ is proportional to $N_{\rm HI}$.

3. Conclusion

1) Biasing of galaxy formation is affected at high density and small size by molecular cooling. Even if the initial size is less than Jeans length at $T = 10^4$ K, sheets can collapse when they are sufficiently self-shielded.
2) We also found that the column density of H_2 of primordial sheets in UV radiation field has a remarkably simple power low relation with $N_{\rm HI}$.

Acknowledgements
This work is supported by Research Fellowships of the Japan Society for the Promotion of Science for Young Scientists, No.2370.

References

1) Haiman, Z., Rees, M.J. & Loeb, A. 1996, ApJ, **467**, 522
2) Haiman, Z., Rees, M.J. & Loeb, A. 1997, ApJ, **476**, 458
3) Corbelli, E., Galli, D. & Palla, F. 1997, ApJ, **487**, 53L
4) Kepner, J., Babul, A. & Spergel, N. 1997, ApJ, **487**, 61
5) Draine, B. T. & Bertoldi, F. 1996, ApJ, **468**, 269

3D RADIATIVE TRANSFER CALCULATIONS ON THE COSMIC REIONIZATION

M. UMEMURA, T. NAKAMOTO, & H. SUSA
Center for Computational Physics, University of Tsukuba, Tsukuba, Ibaraki 305, Japan; umemura@rccp.tsukuba.ac.jp

1. Introduction

The intergalactic medium (IGM) is thought to have been highly ionized earlier than $z \sim 4$ (Gunn & Peterson 1965; Schneider et al. 1989, 1991). However, the reionization history of the early universe has not been elucidated satisfactorily. Mainly because of the high density and possibly because of relatively weak ionizing radiation, the ionizing photons could not propagate freely at high redshifts even in linear density fluctuations, so that self-shielded neutral cores or shadowed low ionization regions could emerge. In this paper, we solve the 3D radiation transfer for UV photons in a clumpy universe, and explore the effects of cosmic reionization in detail.

2. Method

The random Gaussian density fluctuations are generated by the Zel'dovich approximation with assuming a standard CDM-dominated universe ($\Omega_{\rm CDM} = 0.95$, $\Omega_{\rm baryon} = 0.05$, $\sigma_8 = 0.7$). We consider the range of UV intensity as $10^{-2} \leq J_{21} \leq 1$ at $z > 4$, where J_{21} is the mean intensity at the hydrogen Lyman edge in units of 10^{-21} erg cm^{-2} s^{-1} Hz^{-1} sr^{-1}. We assume the frequency dependence for the uniform background radiation to be $J_\nu = J_{\nu_L}(\nu_L/\nu)$, where ν_L is the Lyman limit frequency. We calculate the ionization degree by ionization equilibrium for primordial abundance gas.

The intensity of the UV radiation field is obtained by solving the three-dimensional steady radiation transfer equation. We have adopted so called the *short characteristics* method (see the paper by Nakamoto in this volume). Here, we use basically 64^3 grids for space and 64^2 meshes for directions in this calculation. As for the frequencies, we have developed a novel

S. M. Miyama et al. (eds.), Numerical Astrophysics, 43–44.

approximation, where the ionization by the line emissions is quite properly calculated.

3. Results and Discussion

The results (including Figures) are presented in detail in the CD-ROM version. Here, we just summarize the main results.

The reionization proceeds in a patchy pattern, and ionized regions are quickly percolated at a certain epoch. Also, the ionization structure is quite sensitive to the UV background intensity. If the ionization radiation permeates the volume freely, the ionization degree is determined solely by the local density. However, the resultant distributions deviate from the prediction particularly in higher density regions. The effects emerge as the self-shielding as well as the shadowing (see also Nakamoto, Umemura, & Susa 1997).

The Lyα photons which are emitted at the reionization epoch generate near IR background radiation. The flux is at a level of $10^{-20}(\theta/\mathrm{arcmin})^2$ [W/m^2]. The resultant variance is larger by an order of magnitude at maximum than the average. The detection would be a challenging observation by 8 meter class space IR telescopes in the NGST and H2/L2 mission which are scheduled after 2007.

The results may also affect the Gunn-Peterson optical depth at $z > 4$ (Schneider, Schmidt, & Gunn 1991; Franx et al. 1997; Dey et al. 1998). The observations indicate the presence of substantial absorption at high redshifts due to neutral hydrogen in any sense. The continuum depression obviously tends to grow at higher redshifts up to 5. The trend is considered to originate in the rapid evolution of "Lyα clouds" or the rapid decrease of ionization degree at higher redshifts (the increase of Gunn-Peterson optical depth). The present 3D radiative transfer calculations show that the trend might come from the mixture. There is no distinct physical discrimination between Lyα clouds and intervening neutral hydrogen, even if they might be discriminated spectroscopically.

Numerical computations were carried out on VPP300/16R at NAOJ and CP-PACS at Center for Computational Physics, University of Tsukuba.

References

Dey, A. et al. 1998, ApJL, in press
Franx, M. et al. 1997, ApJL, 486, L75
Gunn, J. E., & Peterson, B. A. 1965, ApJ, 142, 1633
Nakamoto, T., Umemura, M., & Susa, H. 1997, *International Symposium on Supercomputing: New Horizon of Computational Science*, in press
Schneider, D. P., Schmidt, M., & Gunn, J. E. 1991, AJ, 102, 837

EFFECTS OF THE UV BACKGROUND RADIATION ON GALAXY FORMATION

M. NAGASHIMA AND N. GOUDA
Department of Earth and Space Sciencc, Graduate School of Science, Osaka University, Toyonaka, Osaka 560-0043, Japan

AND

NORIMASA SUGIURA
Department of Physics, Graduate School of Science, Kyoto University, Kyoto 606, Japan

We investigate the effects of the UV background radiation on galaxy formation with the semi-analytic model, which is an extension of Cole et al. (1994) by including of the photoionization process (Nagashima, Gouda & Sugiura 1998). In order to see this effect, we use almost the same parameters of their 'fiducial' model except for the UV radiation as chosen by Cole et al. (1994).

UV radiation permeates a gas cloud to a radius at which the ionization by the UV photons balances with the recombination of ions and electrons. For simplicity, we assume that the gas within the 'cooling radius' $r_{\rm cool}$ defined in Cole et al. (1994). The intensity of the UV radiation begins to decrease at $r_{\rm cool}$ and vanishes at $r_{\rm UV}$ by absorption of the UV photons. We estimate this radius $r_{\rm UV}$ as follows:

$$\int_{r_{\rm UV}}^{r_{\rm cool}} n_p(r) n_e(r) \alpha(T_{\rm eq}) 4\pi r^2 dr = (4\pi r_{\rm cool}^2)\pi 1.5 \times 10^5 J_{-21}, \qquad (1)$$

where n_p and n_e are the proton and electron number densities respectively, $\alpha(T_{\rm eq})$ is the recombination rate at a temperature $T_{\rm eq}$, and J_{-21} is the normalized intensity of the UV radiation, $J \equiv J_{-21} \times 10^{-21}$ ergs cm^{-2} s^{-1} sr^{-1} Hz^{-1}.

In Fig.1, we show the luminosity functions and the color distributions.

We find that the UV background radiation mainly affects low-mass objects corresponding to dwarf galaxies. If the UV radiation field exists until the present epoch, the number of dwarf galaxies becomes small, because

S. M. Miyama et al. (eds.), Numerical Astrophysics, 45–46.

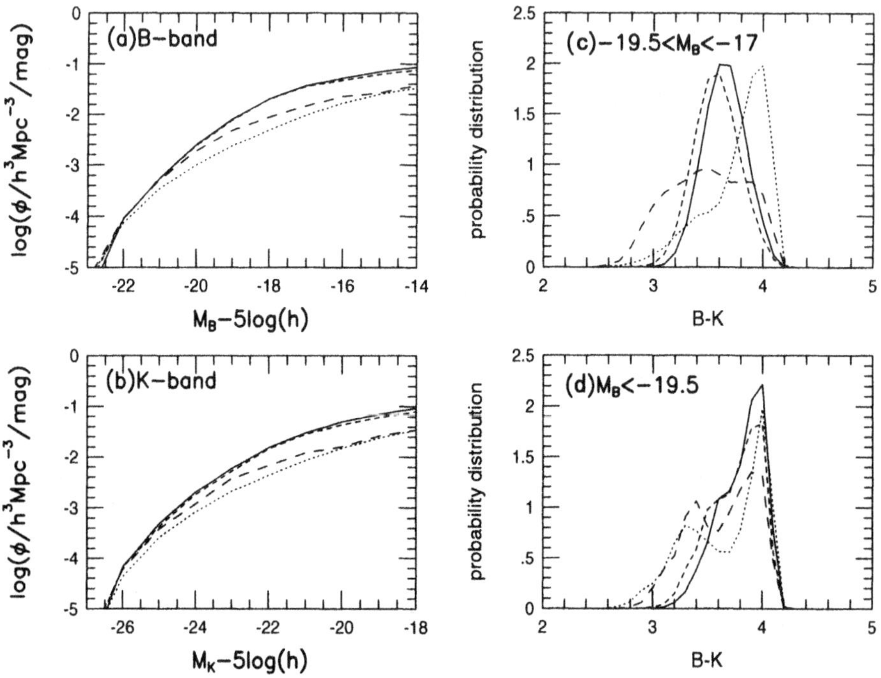

Figure 1. (a) B–band luminosity functions. (b) K–band luminosity functions. The solid lines, the dotted lines, the short-dashed lines, and the long-dashed lines show the case (1) no UV, (2) UV radiation exists in $0 \leq z \leq 5$, (3) $2 \leq z \leq 5$, and (4) $J \propto (1+z)^\gamma, \gamma = 4$ for $z \leq 2$ and $\gamma = -1$ for $z \geq 2$. Colour distributions of faint galaxies (c) and luminous galaxies (d). The types of lines are the same as (a) and (b). In all cases $J_{-21} = 10$.

the star formation is suppressed by the UV radiation. When the UV radiation vanishes at $z = 2$, the number of dwarf galaxies hardly changes but the colour becomes bluer because baryonic gas in clouds recombines and begins to form stars after the UV radiation field vanishes. These blue galaxies probably correspond to the '*faint blue galaxies*'. More realistically, in the case that the UV background evolves as $(1+z)^\gamma$, the slope of the luminosity functions in the faint-end becomes gentler and closer to the observed one, compared to that in the case of no UV background. These results suggest that the evolution of the UV background is distinguishable by investigating the redshift dependence on the luminosity function and the colour distribution of galaxies.

References

Cole, S., Aragon-Salamanca, A., Frenk, C.S., Navarro, J.F. and Zepf, S.E. (1994) A recipe for galaxy formation, *Mon. Not. R. Astro. Soc.*, **271**, pp. 781–806

Nagashima, M., Gouda, N. and Sugiura, N. (1998) Effects of the UV background radiation on galaxy formation, *preprint*, astro-ph/9804153

THE PERMEATION OF DIFFUSE UV RADIATION INTO PREGALACTIC CLOUDS

YUKIKO TAJIRI
Institute of Physics, Univ. of Tsukuba
1-1-1 Tennoudai, Tsukuba, Ibaraki, Japan

AND

MASAYUKI UMEMURA
Center for Computational Physics, Univ. of Tsukuba
1-1-1 Tennoudai, Tsukuba, Ibaraki, Japan

Abstract. Recently, a recalcitrant problem on galaxy formation has been pointed out that low-mass galaxies are overproduced as compared with observations in the context of the hierarchical bottom-up theory of galaxy formation ([4];[6]). Photoionization process has been considered as one of such mechanisms ([1];[3]). We solve the frequency-dependent radiative transfer of UV radiation to elucidate the self-shielding of pregalactic clouds from UV background radiation.

1. Radiative Transfer with Ionization Process

The equation of the ionization rate is

$$\Gamma^{\gamma}\chi_{\rm HI} + \Gamma^{ci}\chi_{\rm HI}(1-\chi_{\rm HI})n = \alpha_A(T)(1-\chi_{\rm HI})^2 n, \qquad (1)$$

where $\chi_{\rm HI}$ is the fraction of neutral hydrogen, Γ^{γ} is the photoionization rate, Γ^{ci} is the collisional ionization rate, n is the hydrogen number density, and $\alpha_A(T)$ is the total recombination coefficient. The frequency-dependent radiative transfer equation for ionizing photons is described as

$$\frac{dI_\nu}{ds} = -\chi_\nu I_\nu + \eta_\nu, \qquad (2)$$

where ν is frequency of photon, χ_ν is the extinction coefficient ($\chi_\nu = a_\nu n \chi_{\rm HI}$) and η_ν is the emissivity. If a free electron recombines directly to the ground state of hydrogen, the emitted photon has enough energy to cause further photoionization. But, when an electron is captured to an excited state of hydrogen, the emitted photon does not have enough energy

S. M. Miyama et al. (eds.), Numerical Astrophysics, 47–48.

to ionize hydrogen, because the kinetic energy of a free electron is of order 0.1 Ryd for $\sim 10^4$ K gas. Thus, the former process is regarded as scatterings which provides the emissivity, while the latter process is pure absorption. The effective scattering albedo is given by $\omega \sim 0.4$ at 10^4 K ([5]). Hence, we set $\eta_\nu = 0$ for $\nu > \nu_L$, and $\eta_{\nu_L} = h\nu\omega\alpha_A n_e n_p / 4\pi\delta\nu$ for $\nu = \nu_L$ where ν_L is the Lyman limit frequency, $\delta\nu = kT/h$ and n_p is the proton number density.

2. Numerical Results

The distributions of the HI fraction are a gradual function of radii, where we cannot recognize a clear boundary between the neutral core and the ionized envelope, and also a low but noticeable ionization is left in the self-shielded regions. We define the critical radius ($R_{\rm crit}$) at which the HI fraction at the center drops just below 0.1. We also find that the critical radius can be well fitted by a formula which is a function of the cloud mass M and the UV background intensity I_{21};

$$R_{\rm crit} = 4.1 \left(\frac{M}{10^8 M_\odot} \right)^{2/5} I_{21}^{-1/5} \text{kpc}. \tag{3}$$

3. Discussion

Here, we try to analytically estimate (A similar estimate is found in [2]) the critical radius based upon the Strömgren approximation. We have $R_{\rm crit} = 3.5\,(M/10^8 M_\odot)^{2/5}\, I_{21}^{-1/5}$kpc, by equating the number per unit time of ionizing photons which enter from the surface to the number per unit time of photons which are absorbed in the cloud, The overestimation of ionization comes from the assumption that all photons which enter into the cloud always cause ionization.

The maximum expansion radius of a top-hat density fluctuation is given by $R_{\rm max} = 10.7(M/10^8 M_\odot)^{1/3}[10/(1+z_{\rm max})]h_{50}^{-2/3}$ kpc in an Einstein-de Sitter universe, where $z_{\rm max}$ is the maximum expansion epoch, and the baryon density parameter is assumed to be 0.05. Comparing $R_{\rm max}$ with $R_{\rm crit}$ and taking into account a possibility that the UV intensity might be significantly lower at $z > 5$, we can speculate that the self-shielding could be quite effective shortly after the maximum expansion stages.

References

[1] Babul, A., & Rees, M. J. 1992, MNRAS, 255, 346
[2] Chiba, M., & Nath, B. 1994, ApJ, 436, 618
[3] Dekel, A., & Rees, M. J. 1987, Nature, 326, 455
[4] Kauffman, G., White, S. D., & Guideroni, B. 1993, MNRAS, 264, 201
[5] Osterbrock, D. E. 1989, in Astrophysics of Gaseous Nebulae and Active Galactic Nuclei (University Science Books)
[6] White, S. D. M., & Frenk, C. S. 1991, ApJ, 379, 52

A TWO-TEMPERATURE MODEL OF INTRACLUSTER MEDIUM

M. TAKIZAWA
Department of Astronomy, Faculty of Science,
Kyoto University Sakyo-ku, Kyoto, 606-8502, Japan

1. Introduction

Revealing temperature profiles of intracluster medium (ICM) is an important problem. In theoretical work concerning clusters of galaxies (CG) isothermality of ICM is often assumed. However, deviation from a flat temperature profile becomes important at lower densities in the mass estimation using X-ray data. Also in the estimation of the Hubble constant through the Sunyaev-Zel'dovich effect, non-isothermality is one important source of errors. Therefore, accurate temperature profiles of ICM are required to improve such methods.

In some CG, the observed electron temperature gradients correspond to the polytropic index, $\gamma (P \propto \rho^\gamma)$, of $\gamma = 1.5$ or even more (Markevitch 1996). On the other hand, $\gamma \simeq 1.2$ is expected for the plasma mean temperature distribution derived by the self similar solution of Bertschinger (1985) and some numerical simulations (Navarro, Frenk, & White 1995; Takizawa & Mineshige 1998; Bryan & Norman 1998, etc). There is a discrepancy between the observed values and the theoretical expectations.

To explain this discrepancy consistently, we construct a model of ICM properly incorporating the relaxation process between the ions and electrons (Takizawa 1998). For this purpose, we apply the method of Fox & Loeb (1997) to the results of numerical simulations by Takizawa & Mineshige (1988) instead of the self-similar solution.

2. Results

Figure 1a shows the radial distribution of electron temperature, $T_e(r)$ (solid line), ion temperature, $T_i(r)$ (dotted line), and mean temperature, $\bar{T}(r)$ (short dashed line), of one of our calculated CG at the present time. At

S. M. Miyama et al. (eds.), Numerical Astrophysics, 49–50.

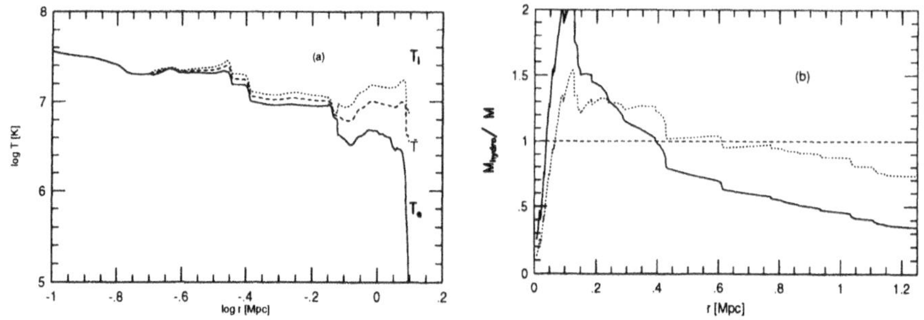

Figure 1. (a) Radial distribution of the $T_e(r)$ (solid line), $T_i(r)$ (dotted line), and $\bar{T}(r)$ (short dashed line) of one of our calculated CG. (b) The ratio of $M_{\rm hydro}$ to the actual mass. The solid line is the result based on $T_e(r)$, and the dotted line is that based on $\bar{T}(r)$.

$r < 0.6$Mpc, T_e is very close to T_i, while at $r > 0.6$Mpc, T_e is considerably lower than T_i and the discrepancy gets increased outward. At $r \simeq 1$Mpc, T_e is only half of $\bar{T}$.

To estimate the temperature gradients both for T_e and $\bar{T}$, we measure the polytropic indices, γ, in the usual way. We fit the density profile by the β-model and the temperature profile by the polytropic model. We find that the polytropic indices of $T_e(r)$ are $\gamma \sim 1.5$, which are systematically larger than those of $\bar{T}(r)$, $\gamma \sim 1.3$.

Since the observed T_e significantly deviates from $\bar{T}$ that determines the dynamics of the system, the total mass of CG is probably underestimated if hydrostatic equilibrium is calculated based on the T_e map. When ICM is assumed to be isothermal, we usually use the emissivity-weighted mean temperature, which is more like the temperature in the central high-density region. Thus, the underestimation of the mass is practically negligible. When $T_e(r)$ is used, conversely, the mass can be seriously underestimated. Figure 1b depicts the ratio of mass derived from the assumption of hydrostatic equilibrium, $M_{\rm hydro}$, to the actual mass, M. The mass derived from T_e is underestimated by almost 50% because of the lower electron temperature.

References

Bertschinger, E. 1985, ApJS, 58, 39.
Bryan, G. L., & Norman, M. L. 1997, in Computational Astrophysics, Proc. 12th Kingston Conference, eds. D Clarke & M. West (PASP), astro-ph/9710186.
Fox, D. C., & Loeb, A. 1997, ApJ, 491, 459.
Markevitch, M. 1996, ApJ, 465, L1.
Navarro, J. F., Frenk, C. S., & White, S. D. M. 1995, MNRAS, 275, 720.
Takizawa, M., & Mineshige, S. 1998, ApJ, 499, 82.
Takizawa, M. 1998, ApJ, 509, in press, astro-ph/9807003.

DYNAMICAL SIMULATION OF INTERGALACTIC HEAVY ELEMENTS

K. YAMASHITA
Chiba University
1-33 Yayoicho, Inageku, Chiba 263-8522, Japan

AND

I. MURAKAMI
National Institute for Fusion Science
Toki, 509-5292, Japan

1. Introduction

Recent observations have shown that Ly α forests are mostly associated with metal absorption lines at $z \approx 3$ [Cowie et al. (1995); Songaila & Cowie (1996)]. Ly α forests are considered to arise from density fluctuations of the intergalactic medium (IGM). Observations suggest that the IGM is already polluted by metals at such high redshift. Since metals are produced only in stars in galaxies, a specific mechanism is needed to pollute the IGM. Galactic wind caused by supernova (SN) explosions carries metals into the IGM. Therefore, if sub-galactic objects are formed everywhere in the universe and expel metals at high redshifts, the IGM could be metal-enriched (Chiba & Nath (1997); Nath & Trenthan (1997)).

We examined metal enrichment of the IGM with galactic wind using 3D hydrodynamical simulations. The model used was the CDM model of $(\Omega_0, h, \sigma_8) = (1, 0.5, 1)$. The simulation volume was $(20\text{Mpc}/h)^3$, the number of mesh and CDM particles being 128^3. Assuming the UV background given by Haardt & Madau (1996), we obtained the amount of ion of the IGM in ionization equilibrium.

2. Galaxy formation and wind

If adjacent 2^3 mesh cells satisfy all criteria that are biases for the matter density $\rho > 5\langle\rho\rangle$, the IGM density $\rho_{\rm IGM} > \langle\rho_{\rm IGM}\rangle$, the temperature $T < 2 \times 10^5$K, and the IGM flow $\nabla v < 0$, then 50% of the IGM mass is bound to be a galaxy particle at the mass center of the 2^3 cells, called galaxy

S. M. Miyama et al. (eds.), Numerical Astrophysics, 51–52.

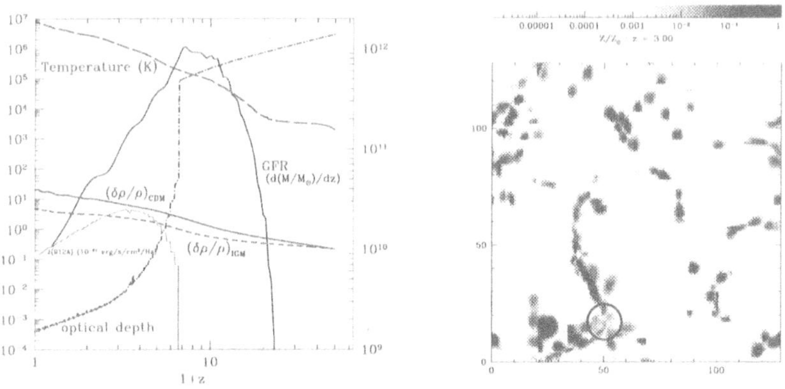

Figure 1. Left panel: Galaxy formation rate (GFR) in mass, temperature, density fluctuation for CDM and IGM, UV background, and HI optical depth as a function of redshift z. The right vertical axis is for GFR. Right panel: Snapshot of metallicity distribution at $z = 3$ in a slice of $(20\text{Mpc}/h)^2 \times 0.156\text{Mpc}/h$. A region with low metallicity but high matter density appeared in the circled area.

formation (GF). Wind occurrence is assumed to be instantaneous and one time at every GF. The wind is characterized by SN mass fraction of 11% in a galaxy, released energy of 10^{51}erg/SN, released metal of $3.2M_\odot$/SN, and mean SN mass of $20.5M_\odot$.

3. Results and conclusions

At $z = 3$, only 12% volume had metallicity $Z > 10^{-2}Z_\odot$, and the intergalactic void was metal-free. Observational information has not yet ruled out these results. Most galaxies were formed during an early epoch, $z = 15 \sim 6$, as a direct result. Infall of the surrounding primordial gas excessively attenuated the intracluster metallicity after GF at very high density regions. This effect is mostly caused by σ_8 and GF conditions but partly by a possible lack of spatial resolution, except when GF occurs quite homogeneously. A smaller σ_8 suggested from local abundance of clusters or higher GF biases would make the galaxy formation epoch later and suppress this effect; conversely, a larger σ_8 given by COBE or lower biases would make it worse.

References

Cowie, L.L., Songaila, A., Kim, T.-S., & Hu, E.M. (1995), AJ, 109, 1522.
Songaila, A., & Cowie, L.L. (1996), AJ, 112, 335.
Chiba, M., & Nath, B.B. (1997), ApJ, 483, 638.
Nath, B.B., & Trenthan, N. (1997), MNRAS, 291, 505.
Haardt, F., & Madau, P. (1996), ApJ, 461, 20.

PRIMORDIAL GAS COLLAPSING WITH H_2 COOLING

TAKAHIRO OOSATO[1] AND TATSUO YOSHIDA
Ibaraki University, Mito 310-8512, Japan
[1] *Junior Research Associate, RIKEN, Wako 351-0198, Japan*

1. Introduction

We have investigated that stability and hydrodynamics of a primordial cloud embedded in a external pressure. Whitworth (1981) and Tohline *et al.* (1987) discussed the role of cooling in the formation of clouds, assuming the spherical symmetry and polytropic equation of state. They pointed out that mildly nonlinear disturbances in cooling medium can collapse the cloud which has initially the sub-Jeans mass. We have performed hydrodynamical simulations, taking account into H_2 cooling and chemical evolution, and examined whether the cloud which has the sub-Jeans mass can collapse in the primordial gas by mild fluctuation.

2. Models and Numerical Simulations

We assume that each spherical cloud is initially in hydrostatic equilibrium and isothermal. We solve the basic hydrodynamic equations numerically by using Lagrangian Godunov method. We also take account of chemical reactions of five species H, H_2, H^+, H^-, and e^-, and determine the fraction of H_2 in order to calculate the cooling rate of H_2. The full set of reactions and rates is listed in Palla *et al.* (1983). As disturbance, we imposed an initial velocity profile onto each cloud of a form $v(r) = v_b(r/r_b)$, where r denotes the cloud radius and the suffix b the boundary, respectively. We determine the minimum of velocity required for the collapse of a cloud by numerical simulations. The results are shown in Table 1.

3. Discussion

We have confirmed that the cloud with the sub-Jeans mass can collapse in primordial gas. Furthermore, we have found that the minimum of Mach

S. M. Miyama et al. (eds.), Numerical Astrophysics, 53–54.

TABLE 1. Initial conditions and velocities required for the collapse of a cloud.

model	cloud mass ($M_\odot$)	number density at the center (cm^{-3})	Jeans mass ($M_\odot$)	the minimum of Mach number required for the collapse of a cloud
A1	10^2	10	8.1×10^3	13.1
A2	10^3	10	8.1×10^3	8.4
A3	10^4	10	8.1×10^3	6.3
B1	10^2	100	2.6×10^3	10.6
B2	10^3	100	2.6×10^3	6.4

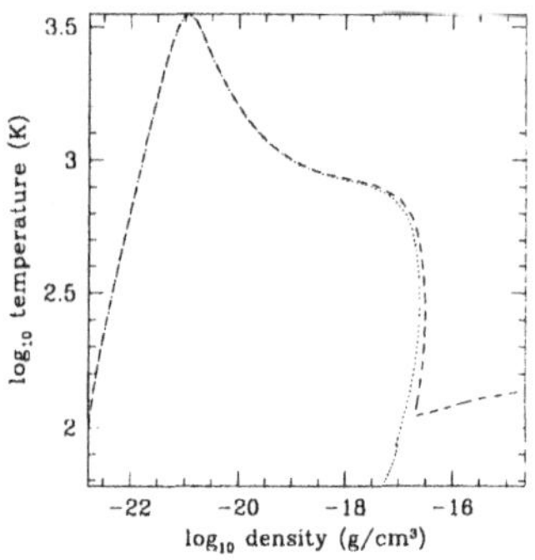

Figure 1. The thermal evolution of clouds in model A1 on the density-temperature plane. A dashed curve represents the case of collapse ($\mathcal{M}_b = 13.14$, where $\mathcal{M}_b$ is Mach number at r_b), and a dot curve the case of no collapse ($\mathcal{M}_b = 12.97$), respectively.

number required for the collapse of a cloud is larger than that of the Tohline's model, because in our simulation we consider the energy equation with H_2 cooling. In the early stage, when the dynamical time is shorter than the cooling time, the temperature increases with adiabatic compression as the cloud collapses. After the temperature reaches maximum, the cooling becomes effective, and the effective adiabatic index is less than one, as shown in Figure 1. For the Mach number higher than the minimum, the cloud can collapse. On the other hand, for the Mach number lower than the minimum, the cloud finally attains equilibrium.

References

Whitworth, A. (1981) Global gravitational stability for one-dimensional polytropes., *M.N.R.A.S.*, **195**, pp. 967–977

Tohline, J.E., Bodenheimer, P.H. and Christodoulou, D.M. (1987) The crucial role of cooling in the making of molecular clouds and stars., *ApJ*, **322**, pp. 787–794

Palla, F., Salpeter, E.E. and Stahler, S.W. (1983) Primordial star formation: the role of molecular hydrogen., *ApJ*, **271**, pp. 632–641

FRAGMENTATION OF PANCAKE COLLAPSED LAYERS FROM INITIAL FLUCTUATION OF CDM SPECTRUM

T. HOSOKAWA AND M. YOKOSAWA
Department of Physics, Ibaraki University
Bunkyo, Mito 310-8512, Japan

1. Introduction

Recent observations have revealed that the star formation rate or space density of quasars has a peak around $z = 2-3$ (Connolly *et al.* 1997, Shaver *et al.* 1996). This implies that some activities of objects for the mass scale of a galaxy are concentrated around the epoch $z \sim 3$. The thermal history of collisional gas has a critical effect on the formation of a galaxy; thus, gas dynamics with shock formation must be solved within a large dynamic range. We precisely computed the dynamics of gas and collisionless particles in a simplified situation of one-dimension, subsequently studying the typical epochs of fragmentation of first-collapsed layers and masses of fragments.

2. Model and Numerical Method

We assumed the cold dark-matter (CDM) dominated universe in the Einstein-de Sitter model. The density parameter of CDM was assumed as $\Omega_{DM} = 0.9$ and the Hubble constant as $h = H_0/100$ km s^{-1} Mpc^{-1} = 0.75. We imposed a sinusoidal density fluctuation for CDM at the epoch of decoupling ($z = 1000$). Simulations were performed with several fluctuation wavelengths between 1 kpc and 20 Mpc, their amplitudes being determined by a CDM power spectrum.

Cosmological hydrodynamic calculations were performed by the Eulerian PPM algorithm, and the Modified Euler Method was used for N-body calculations of dark-matter particles. The Poisson equation was solved with FFT in an assumption of periodic boundary conditions. Furthermore, the effect of radiative cooling from primordial gas composition was also considered.

The simulations showed that the initial fluctuations finally formed cosmological sheets at their center. We analyzed the fragmentation of the sheet

S. M. Miyama et al. (eds.), Numerical Astrophysics, 55–56.

with an energy argument (Valinia *et al.* 1997) which gave the size of fragments and the epoch of fragmentation.

3. Results

The relations between fluctuation scale and mass or diameter of fragments indicated pure power law, and the dark-matter data showed similarity. On the other hand, a characteristic scale of fragments with a mass of $10^{8-9} M_{\odot}$ and a diameter of $0.1 - 0.3$ Mpc was obtained for gas fragments. Simple relations between surface density and disk diameter,

$$\frac{\sigma}{d_{B,MGR}} \simeq \frac{\Sigma}{d_{DM,MGR}} \simeq \rho_{\rm crit},$$

for all simulated scales at fragmentation epoch were obtained, where σ and Σ were the surface densities of gas disk and DM disk, respectively, and d_{MGR} was the diameter of the disk with maximum growth rate. As Figure 1 shows, the evolution of mass of a dark-matter clump indicates the same shape as that of a spherical collapse model. On the other hand, formed gas objects have a characteristic mass of $10^{8.5-10} M_{\odot}$, there being a turnaround at $z = 3$. This shows that the formation of gas objects with subgalactic mass is active around the epoch.

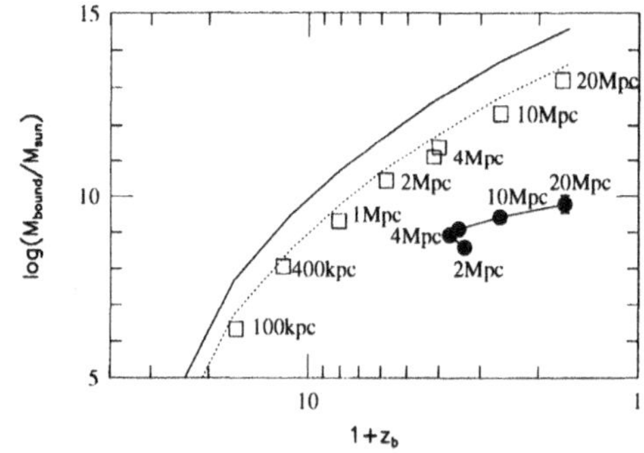

Figure 1. Bound object mass vs. bound redshift. Squares and filled circles are DM and baryonic bound mass, respectively. Solid and dotted lines are DM and baryonic bound mass assuming spherical collapse, respectively.

References

Connolly, A.J., Szalay, A.S., Dickinson, M., SubbaRao, M.U., and Brunner, R.J. (1997) The evolution of the global star formation history as measured from the Hubble Deep Field, *ApJL*, **486**, pp. 11–14

Shaver, P.A., Wall, J.V., Kellermann, K.I., Jackson, C.A., and Hawkins, M.R.S. (1996) Decrease in the space density of quasars at high redshift, *Nature*, **384**, pp. 439–441

Valinia, A., Shapiro, P.R., Martel, H., and Vishniac, E.T. (1997) Gravitational instability in collisionless cosmological pancakes, *ApJ*, **479**, pp. 46–69

FORMATION AND EVOLUTION OF GALACTIC HALOS IN CLUSTERS OF GALAXIES

T. OKAMOTO AND A. HABE
Division of Physics, Hokkaido University
N10 W8, Kitaku, Sapporo 060-0810, Japan

1. Introduction

It has been demonstrated that galaxy populations vary with the density of neighboring galaxies (Dressler, 1980) or the radius from the center of a galaxy cluster (Whitmore et al., 1993). The increase in the fraction of blue, star-forming cluster galaxies with redshift (Butcher & Oemler, 1978) has been well established as well. However, the physical mechanisms responsible for these effects are not fully understood yet.

The purpose of our study is to investigate the effects of galaxy-galaxy and galaxy-cluster interactions on cluster member galaxies and to determine when these interactions become important during cluster evolution in standard cold dark-matter cosmology. In order to study the strength of these interactions, it is necessary to examine how galactic dark halos are affected by these interactions by using high-resolution N-body simulations.

To get enough resolution, we use a multi-mass initial condition. Thus, our analyses are performed only for the high-resolution region which ends in the cluster at $z = 0$.

To follow the evolution of galactic halos, we identify the galactic halos 26 times throughout the simulation with a 0.5 Gyr time step and derive the merging history tree of galaxies directly from the N-body simulation.

2. Results and discussion

First, the merging of galactic halos is noted. A halo that has more than two ancestors in the former time stage is defined as a "merger remnant". Merging is rare in a cluster, so the merger remnant fraction of large galaxies is always less than 10% and slightly increases with redshift. On the other hand, the merger remnant fraction in pre-cluster region (high resolution

S. M. Miyama et al. (eds.), Numerical Astrophysics, 57–58.

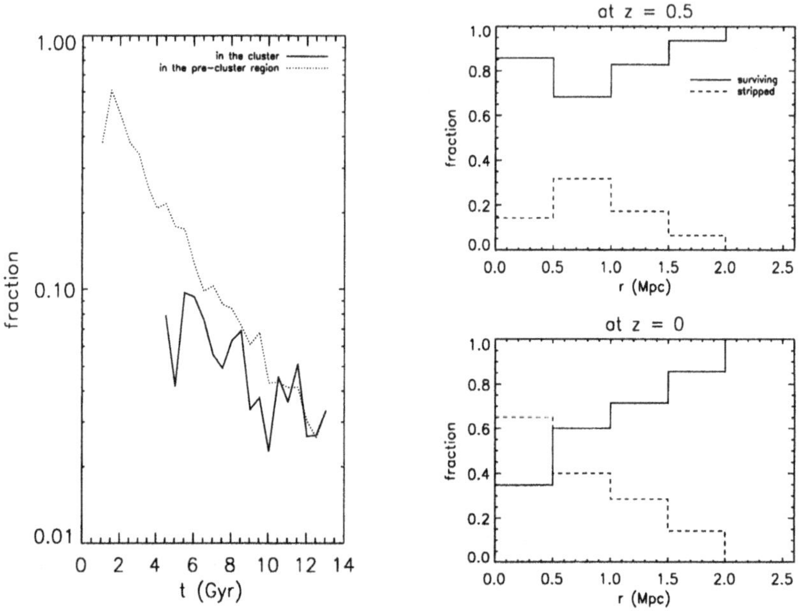

Figure 1. Left panel: Merger remnant fraction of large galaxies. Right panel: Fraction of surviving halos (solid line) and stripped galaxies (dashed line) which have typical large halos, $M_{\rm h} > 10^{11} M_{\odot}$, at $z \simeq 2$ in the 0.5 Mpc bins from the cluster center at $z \simeq 0.5$ (upper panel) and $z = 0$ (lower panel).

region of our simulation) has a large peak at $t = 1.5$ Gyr ($z \simeq 3$) (Fig. 1: left panel). Next, to show the effect of the tidal stripping on the galactic halos, we investigate whether typical large halos ($M_{\rm h} > 10^{11} M_{\odot}$) at high redshift ($z \simeq 2$) keep their halos at lower redshift. Halos containing more than ten particles (i.e., $M_{\rm h} > 10^{10} M_{\odot}$) at a given redshift are called "surviving halos", and others are called the "stripped galaxies". At $z \simeq 0.5$, a large fraction of the halos survives. In the cluster at $z = 0$, these fractions have a clear correlation with the distance from the cluster center, and, in the central region, more than 60% of galaxies completely have been stripped their halos (Fig. 1: right panel).

As we have shown, tidal stripping is the most dominant effect on galactic halos. Therefore, its effects on the luminous part of a galaxy should be examined further.

References

Butcher, H., & Oemler, A.Jr. 1978, ApJ, 301, 57.
Dressler, A. 1980, ApJ, 236, 351.
Whitmore, B.C., Gilmore, D.M., & Jones, C. 1993, ApJ, 407, 489.

COMPARISON OF SIMULATED AND OBSERVED DENSITY PROFILES IN GALAXY CLUSTER

I. VALTCHANOV, M. KALINKOV AND I. KUNEVA
Institute of Astronomy
72 Tsarigradsko Chaussee blvd, BG-1784 Sofia, Bulgaria
e-mail: ivan@astro.acad.bg, markal@astro.acad.bg

We have undertaken a detailed study of density profiles of compact groups of galaxies and rich clusters of galaxies. Some preliminary results are presented here.

We attack two problems in the study – how to define a quantity connected with the density profiles of the Hickson compact groups (HCG) and how to verify the existence of the secondary maxima in density profiles of some rich clusters of galaxies.

It is very difficult if possible to define any radial distribution of galaxies in groups, especially for the compact ones. Our effort leads to a proposal to use only one median characteristic r_{mm}.

Supposing $\mathcal{M}/\mathcal{L} = \text{const}$ for HCG galaxies we estimate weighted group centers and define the median distance r_m to the center of each group. Then a median distance r_{mm} of *all* r_m is obtained.

We used four radii – median projected galaxy separation r_s, space mean harmonic radius r_h, space virial weighted radius r_{VT} and median distance from the group center r_m and we estimated the crossing times $t_{\rm cr} = 2\, r\, H_0/\sigma_3$, where r is one of the above radii. The picture of the distribution of the crossing times is very informative about how different radii of HCGs can lead to very different crossing times. The radii r_m are very convenient for comparison between observed and simulated compact groups. The fundamental quantity r_{mm} characterizes density profiles of compact groups in the same manner as core radius characterizes density profile of galaxy clusters.

Group simulations are performed using Schechter function for masses of the galaxies. Each galaxy was evolved for 10 initial crossing times in isolation with Aarseth's NBODY2 code (Aarseth 1985). The whole group is evolved with Barnes' TREECODE (Barnes & Hut 1986) afterwards. With-

S. M. Miyama et al. (eds.), Numerical Astrophysics, 59–60.

out any specific adjustment of the initial conditions the groups evolve in well-known fashion – merging after few crossing times.

To simulate clusters of galaxies we have N particles homogeneously distributed in unit sphere with zero or $\mathbf{v} = H\mathbf{r}$ velocities. This choice of zero velocities after the turnaround is reasonable (Peebles 1969). We have performed sets of $N = 500, 1000, 2000, 4000$ and 8000 particles. NBODY2 code is used for all of cluster simulations. To test our conclusions we have performed simulations with different softening parameters $\epsilon = 0.1,\ 0.05,\ 0.01$, because this parameter is mainly responsible for "smearing" of the density distribution for the core and outer parts of the simulated clusters.

Simulated clusters were evolved for 10 initial crossing times and at different moment the spatial as well as projected density distributions were analyzed. At 1.5 initial crossing times the system passes through collapse and shortly after 4.5 crossing times it virializes and the next dynamical picture is rather uninformative.

Secondary maxima in the radial density profiles of clusters of galaxies are recognized in the last thirty years and secondary maxima appear in treatment of density profiles in many papers – Sharov (1959), Travese & Vignato (1982), Oegerle et al. (1987), there are some indications (without discussion) of secondary maxima in simulated profiles in Crone et al. (1994). Such secondary maxima appear in **all** our evolved models of clusters of galaxies using different number of points and different softening ϵ. This secondary maximum is clearly visible after the collapse phase.

The possible reason for the existence of the secondary maxima is the formation of groups of galaxies or, in general, substructures in the outer cluster regions. It cannot be attributed to spurious effect of small number of points in the outer region because the effect appears in all experiments with $N = 500 \div 8000$ at approximately the same evolutionary stage and distance from cluster center of mass.

Acknowledgments

We thank S. Aarseth, J. Barnes, P. Bode, M. Tsvetkov and D. Karadjov.

References

Aarseth, S. 1985, In: Multiple Time Scales. Eds. J.U. Brackbill & B.I. Cohen, Academic Press, New York, p. 377
Crone, M.M., Evrard, A.E., Richstone, D. 1994, ApJ, 434, 402
Peebles 1969, JRASC, 63, 4
Oegerle, W.R, Hoessel, J.G., Jewson, M.S., Oemler, A. 1974, ApJ, 194, 1
Sharov, A.S. 1959, AZh, 36, 807
Travese D., Vignato, A. 1982, A&A, 110, 238

WHERE GALAXIES COLLIDE: SIMULATION OF M81-M82-NGC3077 SYSTEM USING A PARALLELIZED TREESPH CODE

P.S. LI AND H.A. THRONSON
University of Wyoming
Department of Physics and Astronomy,
3905 University Station, Laramie, WY 82071-3905, USA

1. Introduction

The M81 galaxy has at least 15 companion galaxies. The closest companions are M82, NGC3077, and a small Magellanic dwarf galaxy, Holmberg IX. Latest radio observations of the HI content in the system indicate that M81, M82, and NGC3077 are interacting (e.g. Yun 1992, Yun et al. 1994). Holmberg IX is at the same distance as M81 (Georgiev et al. 1991) and could be a byproduct of the interaction.

Numerical simulations (e.g. Brouillet et al. 1991, Yun 1992) have modeled the interaction within the M81-M82-NGC3077 system. The huge south tidal gas arm from M81 to NGC3077 was successfully reproduced. The north gas tail is generally accepted to be a tidal gas bridge from NGC3077. If NGC3077 had plunged through the M81 gas disk, the north gas tail could be a ram-pressure stripped gas tail. However, due to the limited amount of particles used and the lack of hydrodynamics calculation for gas components in previous simulations, the true nature of the north gas tail has not been confirmed yet. In addition, the formation of the gas concentrations is believed to be related to the close encounter of the 3 galaxies, but the mechanism of the formation of these gas concentrations is still a mystery.

We develop a parallelized TreeSPH code using Message Passing Interface (MPI). We use the code to simulate the M81-M82-NGC3077 system (model mass ratio 10:1:1) with about a quarter million particles at the beginning, in which about 1/3 are gas particles. The resolution can go down to less than 100 pc for the ISM. The simulation runs on the Power Challenge Array R10000 using 8 processors at the National Center of Supercomputing Applications. About 1,400 CPU hours have been used for the whole simula-

S. M. Miyama et al. (eds.), Numerical Astrophysics, 61–62.

tion. A star formation algorithm is incorporated into the code to simulate a starburst during the encounter. New particles are generated during the simulation to mimic the formation of new stars. At the end of the simulation, there are more than half a million particles in the system.

2. Simulation Results and Conclusion

Our simulation successfully identifies the origins and reproduces the formation of gas bridges, gas concentrations, and Holmberg IX. The huge south gas bridge is reconfirmed to be a tidal arm from M81 caused by the passage of M82 with a retrograde hyperbolic orbit with respect to M81 (see figure 1 on CD-ROM). After the perigalacticon of M82, another tidal gas arm from M81, in opposite to the south gas bridge, is formed and stretches out towards M82. Material of the M81 gas disk piles up along this gas arm and the gas arm fragmented into pieces due to gravitational instability. The broken pieces are the gas concentrations observed. This is why M81 only has a single south tidal gas arm.

The north gas tail is confirmed to be a ram-pressure stripped gas tail from NGC3077 when it plunged through the M81 gas disk about 200 million years ago. NGC3077 plunging through M81 disk indeed helps the formation of Holmberg IX (see figure 2 on CD-ROM).

The galaxy model for M81 is a spiral galaxy with multiple arms. After the encounter, the stellar disk of M81 is transformed into a grand design spiral galaxy. The starburst at M82 and at the centers of M81 and NGC3077 are also reproduced.

This is the first time that a single simulation of the M81-M82-NGC3077 system can reproduce so many observed details both in optical and radio wavelengths. Crucial factors include the large number of particles used in the simulation in order to provide sufficient resolution for the formation of small features such as gas concentrations and Holmberg IX. Also, the hydrodynamical calculation included in the simulation allows an accurate interaction among gas components of the 3 galaxies.

This work is supported by the NASA ISO program and NCSA grant no. AST970013N. This work is a part of the first author's Ph.D. Thesis.

References

Brouillet, N., Baudry, A., Combes, F., Kaufman, M., and Bash, F. (1991), *A&A*, **242**, 35

Georgiev, Ts. B., Tikhonov, N. A., Karachentsv, I.D., and Bilkina, B. I. (1991), *A&AS*, **89**, 529

Yun, M.S. (1992), Ph.D. thesis, Harvard University

Yun, M.S., Ho, T.P., and Lo, K.Y. (1994), *Nature*, **372**, 530

THE FORMATION OF CUSPY DENSITY PROFILES THROUGH VIOLENT RELAXATION OF STELLAR SYSTEMS

S. HOZUMI
Faculty of Education, Shiga University
2-5-1 Hiratsu, Otsu, Shiga 520-0862, Japan

A. BURKERT
Max-Planck-Institut für Astronomie
Königstuhl 17, D-69117 Heidelberg, Germany

AND

T. FUJIWARA
Kyoto City University of Arts
Nishikyo-ku, Kyoto 610-1197, Japan

Recent *Hubble Space Telescope* observations (e.g., Gebhardt *et al.*, 1996) have revealed that elliptical galaxies do not have constant density cores but have cusps which continue toward the center until the resolution limit. Such cusps are divided into two types: one is represented by $\rho \propto r^{-n}$ with $0.5 \lesssim n \lesssim 1$ for bright elliptical galaxies and the other is approximated by $\rho \propto r^{-2}$ for faint ones. We then demonstrate that simple dissipationless collapses can produce cuspy density profiles. In particular, the effects of velocity anisotropy on the relaxed density profiles in the core are examined, because stellar systems can support velocity anisotropy unlike gaseous systems.

We carry out collapse simulations of power-law density spheres such as $\rho \propto r^{-1}$ with anisotropic velocity distributions using a phase-space solver (Fujiwara, 1983) which integrates the collisionless Boltzmann equation on the assumption of spherical symmetry. We set the initial virial ratios to be 0.1 and $10^{-1.5}$. The distribution functions are given by

$$f = \frac{\rho(r)}{\sqrt{2}\pi^{3/2}\sigma_{\rm r}\sigma_{\perp}^2} \exp\left(-\frac{u^2}{2\sigma_{\rm r}^2}\right) \exp\left(-\frac{j^2/r^2}{\sigma_{\perp}^2}\right), \tag{1}$$

where ρ is the density, $\sigma_{\rm r}$ is the radial velocity dispersion, and $\sigma_{\perp}$ is the tangential velocity dispersion with r, u, and j being the radius, radial velocity, and angular momentum, respectively. The density distribution, $\rho(r)$,

S. M. Miyama et al. (eds.), Numerical Astrophysics, 63–64.

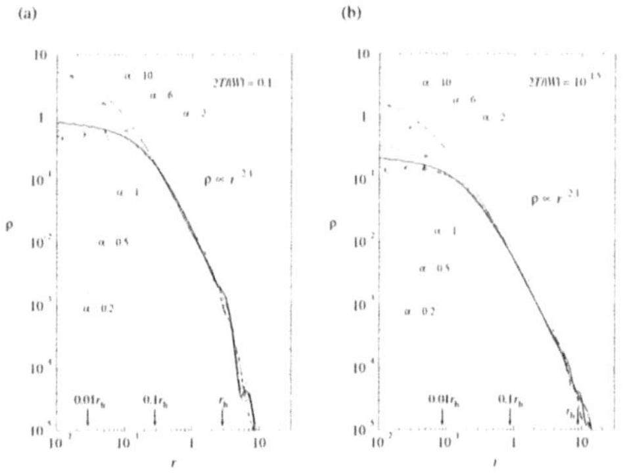

Figure 1. Relaxed density profiles for the models with (a) $2T/|W| = 0.1$ and (b) $2T/|W| = 10^{-1.5}$. The straight dotted lines show the density profiles with $\rho \propto r^{-2.1}$. The half-mass radii, $r_{\rm h}$, for $2T/|W| = 0.1$ are around 2.6 to 3.2, while they are around 7.8 to 9.7 for $2T/|W| = 10^{-1.5}$.

is given by

$$\rho(r) = (M/2\pi R_0^2)r^{-1}, \tag{2}$$

where M is the total mass, and R_0 is the radius of the sphere. We measure the velocity anisotropy by $\alpha = 2\sigma_{\rm r}^2/\sigma_\perp^2$, which is independent of radius in our models. We take $\alpha = 0.2, 0.5, 1, 2, 6$, and 10 for each value of the virial ratios.

In Figure 1, we show the final density profiles. We find from the figure that more radially anisotropic models result in more cuspy density profiles in the cores, although the density profiles at intermediate radii are well-approximated by $\rho \propto r^{-2.1}$, regardless of the virial ratios and the values of α. In particular, for the models with $\alpha = 10$, the cores can be well-described by power laws.

We can explain the density bend that occurs around the edge of the cores as follows. The maximum phase-space density (MPD) at each radius of the initial system is given by $f_{\rm m} = \rho(r)/\sqrt{2}\pi^{3/2}\sigma_{\rm r}\sigma_\perp^2$ (see equation 1), and so, $f_{\rm m}$ is proportional to $\rho(r)$, which means $f_{\rm m} \propto r^{-1}$. We postulate that this $f_{\rm m}$ is also the MPD at each radius accessible to the relaxed system. On the other hand, the relaxed density profiles show $\rho \propto r^{-2.1}$ at intermediate radii, so that $f_{\rm m} \propto r^{-2.1}$ could hold there. This indicates that there unavoidably emerges a region in which $f_{\rm m}$ for $\rho \propto r^{-2.1}$ becomes larger than that for $\rho \propto r^{-1}$. Therefore, $\rho \propto r^{-2.1}$ profiles cannot continue down to small radii as long as the initial density profile is not steeper than $\rho \propto r^{-2.1}$.

References

Fujiwara, T., 1983, PASJ, 35, 547.
Gebhardt, K., et al., 1996, AJ, 112, 105.

SIGNATURE OF DENSITY WAVE RESONANCES IN N-BODY SIMULATIONS OF SPIRAL GALAXIES

The 4:1 resonance

P.A. PATSIS
Max-Planck-Institut für Astronomie
Königstuhl 17, D-69117 Heidelberg, Germany

AND

D. KAUFMANN
Computer Sciences Co.
7700 Hubble Drive, Lanham-Seabrook, MD 20706, USA

Abstract. We study the dynamical behavior at the regions of resonances in N-body simulations with long-lived spiral structure. We find in several cases a main mode, the strong bisymmetric part of which ends at its 4:1 resonance region. This distance is found to be at 2.5→3 disk scale lengths. In real spiral galaxies this corresponds to the distance of termination of the observed prominent inner 2-armed spirals. In general other modes are found coexisting with the main one. In some cases they are found to be important in the outer part of the disks.

1. Introduction

In the present paper we employ a two-dimensional polar grid N-body code to follow the evolution of a model that generates a long-lived two-armed spiral structure in order to study the location of the end of the strong spirals relative to the location of the resonances. We report the remarkable similarity of the dynamical behavior at the 4:1 resonance in our N-body model with the dynamical behavior in the same region of the "self-consistent" stellar (Contopoulos and Grosbøl 1988, Patsis et al. 1991) and SPH gaseous (Patsis et al. 1997) response models.

2. The Simulation

Our numerical model consists of 100,000 self-gravitating particles. To calculate the forces we employ a Fourier transform method on a polar grid.

S. M. Miyama et al. (eds.), Numerical Astrophysics, 65–66.

TABLE 1. Model component parameters.

Component	Type	Mass	Scale
Disk	Exponential	6.0	3.5
Halo	Plummer	6.0	3.5
Bulge	Toomre $n = 2$	1.0	1.4

We include a gaseous component by considering 15,000 of the particles to be gas clouds having a finite collisional cross-section. The clouds are made dissipative by means of a "sticky particle" scheme as described by Schwarz (1981). Our unit length is 1 kpc, the unit of mass 10^{10} $M_{\odot}$, and the gravitational constant $G = 1$. This choice results in a time unit of approximately 4.715×10^6 years and a velocity unit of approximately 207.4 $\mathrm{km\,s^{-1}}$. Hereafter we give all quantities in these units unless otherwise specified. Table 1 shows the input parameter values for the three components. In our simulation we take $Q = 1$ for the star particles and $Q = 0$ for the gas particles. The model is evolved for approximately 3.3 billion years.

3. Results

The time evolution of our simulation can be seen in Fig. 1 (see accompanying CD-ROM). We find a long-lived bisymmetric spiral pattern with pitch angle $\approx 19°$. This pattern has a strong part ending at 2.6 exponential disk scale lengths, which corresponds to the 4:1 resonance region. Other modes are found coexisting, rotating with different pattern speeds. We calculated the radial force perturbation to be $F_r \approx 0.06$ of the axisymmetric force at the critical distance. This suggests a marginally non-linear spiral (Grosbøl 1993). Our findings confirm the results of the non-linear orbital theory (Contopoulos & Grosbøl 1988) concerning the dynamical behavior at the 4:1 resonance region for such a spiral perturbation, and underline the importance of this resonance for the dynamics of spiral galaxies.

References

Contopoulos, G., & Grosbøl, P. 1988, A&A, 197, 83
Grosbøl, P. 1993, PASP, 105, 651
Patsis, P. A., Contopoulos, G., & Grosbøl P. 1991, A&A, 243, 373
Patsis, P. A., Grosbøl, P., & Hiotelis, N. 1997, A&A, 762, 774
Schwarz, M. P. 1981, ApJ, 247, 77

THE FORMATION AND SIMULATION ANALYSIS OF A RING GALAXY

LIU BULIN
Beijing Astronomical Observatory, Chinese Academy of Sciences, Beijing, 100080, China

1. Introduction

Ring galaxies are rare celestial systems in the universe. The frequency of ring galaxies lies between 0.02% – 0.2% of the spiral galaxies, occurring mainly in the regions of low galactic density (Athanassoula & Bosma, 1985). The masses of ring galaxies are of the order of $10^{11} M_\odot$ with an estimated age of about 10^8 years (They & Spiegel, 1976). Ring galaxies have a special shape. For this reason, their origin and development have naturally been of interest.

In this paper, the formation of ring galaxies was studied using the concept of the gravitational response of interacting galaxies (Lynds & Toomre, 1976) and the gravitational collapse of the target galaxy (Miller, 1978). Numerical simulations were carried out using the tree code written by the author (Liu, 1996). The whole particle system included the target galaxy and the intruder galaxy (as a single particle), the target galaxy including two parts: disk (4000 particles) and halo (4000 particles).

2. Calculation and discussion

A tree code of N-body simulation was used in the process of formation of the ring-shape structure. We used the dimensionless differential equations of motion for the gravitational N-body problem. The equations were integrated by a finite-difference approximation time-centered leap-frog method. The Plummer model $\rho_h(r) = \frac{3}{4\pi} b^2 (r^2 + b^2)^{-\frac{5}{2}}$ (Aarseth & Fall, 1980) was used as the density-distribution function of particles in the halo. The density distribution of particles in the target disk was an exponential function $\rho_d(r) = \rho_0 e^{-0.7r}$. The initial positions of the particles were obtained by the Monte Carlo selection procedure. The velocity of the particles was gen-

S. M. Miyama et al. (eds.), Numerical Astrophysics, 67–68.

erated using a pseudo-random number of normal distribution, and their energy E was calculated. If $E \leq 0$, then the velocity was assigned to the particles; otherwise, this process was repeated again until $E \leq 0$. Thirty-four models were examined with different masses of target and intruder galaxies, intrusive angles, and relative speeds.

We analyzed the forces on a particle in the target disk. The forces were the gravitational force and the tidal force. The gravitational force included the gravitational response (corresponding to the force of wave-number $m = 0$ in azimuthal direction) and the gravitational collapse (corresponding to the force of $m > 0$). The generation of the gravitational response was caused by the triggering of the intruder galaxy in the target galaxy. Due to the gravitational collapse of the disk and the gravitational response to the intruder galaxy, a magnificent ring-shape structure arose during the process of evolution of the disk. Analysis of the ring-shape structure indicated that its lifetime was about 10^8 years, which agrees with the age of the ring galaxy estimated by Theys & Spiegel (1976). It was evidenced by numerical analysis that the gravitational response and the tidal force have equivalent functions. The gravitational force on the particle in the target disk can be divided into radial gravitational force and tangential gravitational force. The radial gravitational force, which points to the barycenter of the target disk, is an inward force which pulls the particles into the barycenter of the target disk. The radial gravitational force is called gravitational collapse. Through the evidence of numerical experiments, we think it is advisable that the gravitational response and the gravitational collapse be taken into account in the process of evolution of interacting galaxies.

This work was supported by a grant from the National Natural Science Foundation of China.

References

Aarseth, S. J. & Fall, S. M. 1980, ApJ, 236, 43
Athanassoula, E. & Bosma, A. 1985, Ann. Rev. Astron. Astrophys. 23, 147
Liu, Bulin. 1996, ACTA Astrophysica Sinica, 16, 7
Lynds, R. & Toomre, A. 1976, ApJ, 209, 382
Miller, R. H. 1978, ApJ. 223, 122
Theys, J. C. & Spiegel, E. A. 1976, ApJ, 208, 650

ORIGIN OF THE STRUCTURE DIFFERENCE BETWEEN ELLIPTICAL GALAXIES AND DWARF ELLIPTICAL GALAXIES

MASAO MORI, YUZURU YOSHII
Institute of Astronomy, University of Tokyo
Mitaka, Tokyo 118-8588, Japan

AND

KEN'ICHI NOMOTO
Department of Astronomy, University of Tokyo
Bunkyo-ku, Tokyo 113-0033, Japan

1. Introduction

According to the observations of nearby elliptical galaxies, there is a clear distinction in structural and chemical quantities between dwarf ellipticals and normal ellipticals in spite of their morphological similarity. The central concentration of dwarf ellipticals is relatively low and their luminosity profiles are best fitted by an exponential function, whereas the profiles of normal ellipticals are known to follow de Vaucouleurs' law (Faber & Lin 1983; Binggeli, Sandage & Tarenghi 1984; Ichikawa, Wakamatsu & Okamura 1986). Moreover, the color of many dwarf ellipticals becomes redder towards outer radii of the system (Vader et al. 1988; Kormendy & Djorgovski 1989), and this trend of color gradient is clearly opposite to normal galaxies. The origin of these striking features of dwarf ellipticals remains yet to be explained (for a review see Ferguson & Binggeli 1994). In particular, no attempts have ever been made to examine whether the supernova feedback mechanism is viable also in this context.

2. Models and Results

A three-dimensional N-body/SPH simulation code, combined with stellar population synthesis, is used to follow the dynamical and chemical evolution of a non-rotating protogalaxy with $10^{10} M_{\odot}$ (baryonic/dark=1/9) which originates from a 1σ CDM perturbation. This less massive galaxy receives

S. M. Miyama et al. (eds.), Numerical Astrophysics, 69–70.

significant dynamical responses from the heat input by stellar winds and supernovae.

The first starburst near the center of the system produces a supersonic spherical outflow of the gas. This outflow collides with the infalling gas and gives rise to an expanding dense shell. Then, stars begin to form in the expanding shell with its site propagating outwards with the shell. We find from the simulation that this consecutive process of star formation creates the exponential brightness profile and the inverse color gradient of the system in agreement with the observations of dwarf galaxies.

In general, the color gradient of galaxies is created by the gradient in either metallicity or age of the underlying stellar population. Simple models of chemical evolution of galaxies usually predict the negative metallicity gradient which corresponds to the color becoming redder towards the galaxy center. Since dwarf galaxies have the inverse color gradient, Vader et al. (1988) were led to interpret this observed trend in terms of the positive age gradient. We note however that stars with very low metallicities must have been formed on very short timescales and therefore no appreciable age difference results.

The above puzzling situation indicates that previous results based on simple models of chemical evolution can not be applied to small systems like dwarf galaxies. We demonstrate in this paper that dynamical modelling is the only proper way to investigate the evolution of dwarf galaxies. Successful reproduction of their basic features in our simulation suggests that the stellar energy feedback mechanism is indeed a likely mechanism against the efficient formation of low-mass galaxies in the CDM universe.

Acknowledgments: This work has been supported in part by the Research Fellowship of the JSPS for Young Scientists (6867) and COE Research (07CE2002) of the Ministry of Education, Science, Sports, and Culture in Japan. The computations were partly done on Fujitsu VPP-300 at the National Astronomical Observatory in Japan, and Fujitsu VPP-500 at the Institute of Physical and Chemical Research (RIKEN).

References

Binggeli, B., Sandage A., & Tarenghi, M., 1984, AJ, 89, 64.
Dekel, A., & Silk, J., 1986, ApJ, 303, 39.
Faber, S.M., & Lin, D.M.C., 1983, ApJL, 266, 17.
Ferguson, H.C., & Binggeli, B., 1994, A&A Rev., 6, 67.
Ichikawa, S., Wakamatsu, K., & Okamura, S., 1986, ApJS, 60, 475.
Kormendy, J., & Djorgovski, S., 1989, ARA&A, 27, 235.
Mori, M., Yoshii, T. & Nomoto, K., 1998, submitted.
Mori, M., Yoshii, Y., Tsujimoto, T., & Nomoto, K., 1997, ApJL, 487, 21.
Vader, J.P., Vigroux, L., Lachièze-Rey, M., & Souviron J., 1988, A&A, 203, 217.

FORMATION OF DWARF GALAXIES BY DELAYED STAR FORMATION

NISIKAWA ATUKO
Department of Physics, Hokkaido University,
Kita-ku, Sapporo 060, Japan

AND

NOBUO ISIBASI
Department of Information Science, Kanagawa University,
Tutiya, Hiratuka 259-12, Japan

1. Introduction

Babul & Rees (1992; MNRAS, 255, 346, hereafter BR92) proposed an idea that low-mass galaxies experiencing their initial starburst at $z \approx 1$ evolve into the present dwarf ellipticals and that the epoch of decline in the UV background determines when the initial starburst begins.

In this paper, to examine their scenario, we studied formation and evolution of dwarf galaxies by numerical simulations.

2. Models For Dwarf Galaxies

We adopted the hierarchical clustering scenario for galaxy formation in CDM-dominated universe. We made two models for the evolution of low-mass objects, that is, no-delayed formation (NDF) and delayed formation (DF) models. In the former model, star formation begins as soon as the system has been virialized and the gas has cooled. While, in the latter model, star formation begins with a latency period. The difference between two models comes from their initial gas densities. We adopted TVD-ES scheme for the gas dynamics and the shell method (Hénon) for the stellar dynamics. We assumed the spherical symmetry for all the simulations.

3. Numerical Results

We summarize our results as follows:

S. M. Miyama et al. (eds.), Numerical Astrophysics, 71–72.

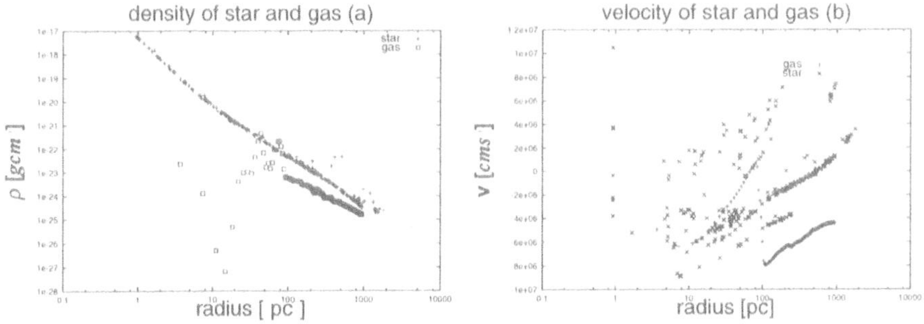

Figure 1. The density (left) and velocity (right) distributions in DF model after 3.65 $\times 10^7$ yr has passed. Stellar density distribution is approximately expressed by $\rho_* \propto r^{-3}$. This shows that gas is blown away by SN heating.

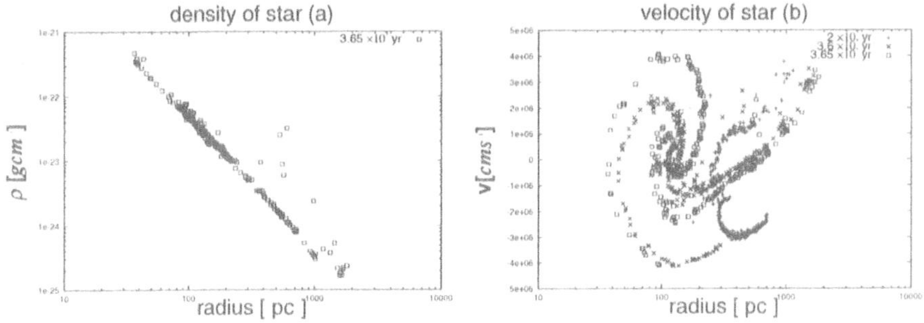

Figure 2. The same as Fig. 1 but for a model in which SN heating is artificially suppressed. Figure 2b (right) shows that the 'phase-mixing' is taking place.

1. As suggested by BR92, supernovae (hereafter SNe) heat up the system and blow gas away from the system. The formed galaxy is gas-poor. This result is in good agreement with the expectation of BR92.
2. In DF model, stellar density is more concentrated ($\rho_* \propto r^{-3}$ [Fig. 1a]) than in NDF model.
3. Mean velocity dispersion of the stellar component is approximately $\sigma_* \sim$ km s^{-1}. The distribution of gas changes rapidly after SN heating begins. As a result, the phase-space distribution of stellar component is much disturbed (Fig. 1b).
4. When SN heating is artificially suppressed, the 'phase-mixing' (Fig. 2b) drives the system. As a result, the stellar density distribution reaches $\rho_* \propto r^{-2}$ (Fig. 2a).

THE EFFECT OF SELF-GRAVITY OF GAS ON GAS FUELLING IN THE BARRED GALAXY WITH A SUPERMASSIVE BLACK HOLE

H. FUKUDA AND A. HABE
Division of Physics, Hokkaido University
Sapporo 060-0810, Japan

AND

K. WADA
National Astronomical Observatory
Mitaka, Tokyo 181-8588, Japan

We have investigated the evolution of the gas disk at the nuclear region of a barred galaxy involving a supermassive black hole (SMBH), using smoothed particle hydrodynamics (SPH) method for two dimensional non-self-gravitating hydrodynamics. We have found that the gas loses angular momentum by the effect of an additional inner Lindblad resonance (ILR) caused by the SMBH and falls into inner region and results in a nuclear gas ring. We also found, by estimating Toomre's Q value of the gas ring, that the gas ring seems to be gravitationally unstable for usual gas mass fraction in galaxies.

Next, We investigate the fate of the gas ring involving self-gravity. The gas ring is stable and behaves like the non-self-gravitating model for the surface density lower than the critical value (Fig. 1, left panel) which is expected by the above estimation of Toomre's Q value. For the surface density higher than the critical value (Fig. 1, right panel), the gas ring becomes gravitationally unstable and fragments to several gas clumps.

Then, rather diffuse gas clumps are destroyed by the tidal force of the SMBH to make a diffuse gas ring, and rather concentrated gas clumps merge to become more massive gas clumps. A part of the diffuse gas ring loses angular momentum by the torque form the massive gas clumps and falls into center. Other part of the gas ring accretes to the massive gas clumps to make the clumps grow.

The mass of the most massive gas clump is $4.2 \times 10^7 M_\odot$ in our simulation. We show that the massive gas clump collides with other smaller

S. M. Miyama et al. (eds.), Numerical Astrophysics, 73–74.

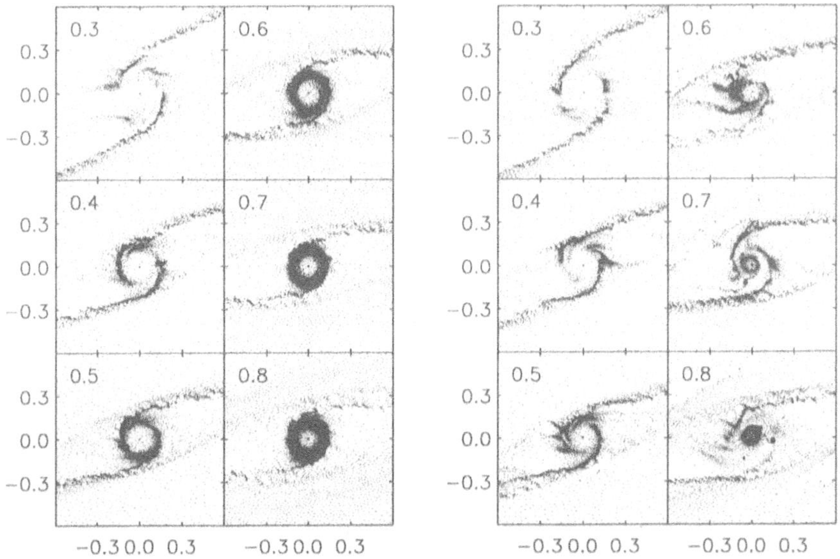

Figure 1. Time evolution of the distribution of the SPH particles in the rotating bar coordinates. Left panel is the higher surface density model, and right panel is the lower surface density model. The bar major axis is fixed horizontally on each frame. Time is represented by the number at left up of each figure. The unit of time and length is 1.75×10^8 yr and kpc, respectively.

clumps. Collision between enough massive clumps may induce active star formation. Therefore, 'super star clusters' observed in starburst galaxies may be formed from the massive gas clumps, if active star formation occurs in the massive gas clumps.

We do not consider dynamical friction of the gas clumps in our simulation, since we do treat stellar component as the static potential. Dynamical friction may affect fate of the massive gas clumps. The time scale of the massive gas clump in our simulation to spiral into the center by dynamical friction is $t_{\rm fric} = 3.4 \times 10^6 (\ln\Lambda/6)^{-1} (m_c/4.2 \times 10^7 M_\odot)^{-1}$yr where $\ln\Lambda$ is the usual Coulomb logarithm and m_c is the mass of the massive gas clump. Since the orbital period of the massive gas clump is 9.6×10^6yr, the massive gas clump is expected to fall into the center in its dynamical time scale. Therefore, the strong gas fuelling is expected. We will study this process in future paper.

FRAGMENTATION AND CORE FORMATION IN DYNAMICALLY COLLAPSING ROTATING CLOUDS

T. TSURIBE
Department of Physics, Kyoto University
Sakyo-ku, Kyoto, 606-8502, Japan

AND

S. INUTSUKA
National Astronomical Observatory
Mitaka, Tokyo, 181-8588, Japan

1. Introduction

Cold Dark Matter (CDM) models for structure formation predict the appearance of the first baryonic objects with masses $M \sim 10^5 M_\odot$ at redshifts as high as $z \sim 30$; objects with progressively higher masses assemble later[1]. Depending on the details of the cooling and angular momentum transport, these objects are expected to either fragment into smaller objects, or form central black holes exhibiting quasar activity[2]. Cosmological simulations with large amount of volume have revealed potential sites in which the first object probably forms[3]. *The subsequent evolution of these collapsing objects is, however, still in open question.* To understand the subsequent evolution, we have to clarify how and when these objects collapse and fragment. In this work, we focus on the three-dimensional hydrodynamical evolution of a single density peak, taking into account the H_2 chemistry.

2. Results

Hydrodynamical evolution of collapsing and rotating self-gravitating clouds is investigated numerically using a newly developed Lagrangian Godunov-scheme. The criterion for the collapse and fragmentation for polytropic gas clouds is obtained in case that the initial density distribution has central concentration and elongation. It is found that the fragmentation criterion is highly modified by the existence of the initial density central

S. M. Miyama et al. (eds.), Numerical Astrophysics, 75–76.

concentration[4,5]. Moreover, the initial small elongation induces the "core + spiral arm" configuration which transports angular momentum around the central core after the core formation even when the outer envelope is still dynamically accreting. As a result, the accretion onto the central core is preferred instead of the fragmentation in some cases.

We also investigated the evolution of the collapsing object including non-equilibrium chemical reactions of the pure hydrogen gas. An example of preliminary results is presented in Figure 1. The efficient cooling due to hydrogen molecules enhances the fragmentation of the collapsed disk-like object.

Numerical computations were carried out on VPP300/16R at NAOJ. Authors thank H. Susa for discussion on the H_2 chemistry.

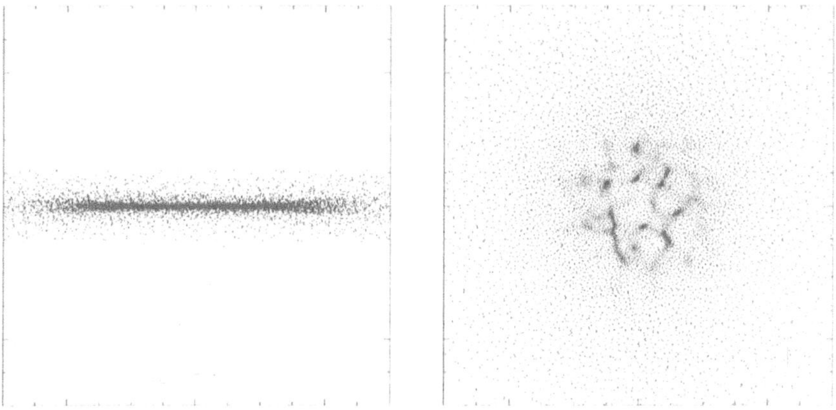

Figure 1. Edge-on and face-on view of the central region of collapsed primordial disk are shown. Box size corresponds to 240pc and baryonic mass is $M = 10^7 M_\odot$. Each fragment has a H_2 fraction of around 7×10^{-4} and temperature 160K and number density of 8×10^4. Initial number density and temperature are $n = 1.3$ and $T = 150$ K, respectively. Initial rigid rotation corresponds to $\beta = 0.3$. The effect of dark matter gravity is also included.

References

1) Tegmark, M., Silk, J., Rees, M. J., Blanchard, A., Abel, T., & Palla, F. 1997, ApJ, 474, 1
2) Eisenstein, D. J, & Loeb, A. 1995, ApJ, 443, 11
3) Abel, T., Anninos, P., Norman, M. L., & Zhang, Y. 1998, ApJ in press
4) Miyama, S. M., Hayashi, C. & Narita, S. 1984, ApJ, 279, 621
5) Boss, A. P. 1993, ApJ, 410, 157

ANISOTROPIC FOKKER-PLANCK MODELS OF STAR CLUSTERS

KOJI TAKAHASHI
Department of Earth Sciences and Astronomy,
College of Arts and Sciences, University of Tokyo,
3-8-1 Komaba, Meguro-ku, Tokyo 153, Japan

1. Introduction

Globular clusters are *collisional* stellar systems, i.e., their evolution is mainly driven by *two-body relaxation*. Since the effects of weak encounters are dominant, the dynamical evolution of globular clusters can be described by the Fokker-Planck (FP) equation.

In this paper, we investigated the cluster evolution by using numerical integration of the two-dimensional Fokker-Planck (FP) equation in energy–angular momentum space (Cohn 1979, 1980; Takahashi 1995). The two-dimensional FP models allow anisotropy of the velocity distribution of the stars. It is important to take account of the velocity anisotropy if we are interested in the *realistic* evolution of globular clusters. We report the results of FP simulations of the evolution of tidally-truncated globular clusters. In particular, we pay attention to the mass loss rates and evaporation times of clusters, the evolution of the stellar mass function, and the development of the velocity anisotropy in each mass-component.

2. Fokker-Planck Models

The distribution function $f_i(\mathbf{r}, \mathbf{v}, t)$ is defined as the number density of the stars of the i-th component in six-dimensional phase $(\mathbf{r}, \mathbf{v})$ space. If we assume spherical symmetry and dynamical equilibrium of the cluster, then the distribution function is a function of only energy (E), angular momentum (J), and time (t): $f_i = f_i(E, J, t)$. If we further assume isotropy of the velocity distribution, then $f_i = f_i(E, t)$.

It is assumed that the cluster moves on a circular orbit around the galaxy. Thus the mean density of the cluster within the tidal radius (r_t) is

S. M. Miyama et al. (eds.), Numerical Astrophysics, 77–78.

kept constant. Two different boundary conditions for the tidal truncation are adopted in anisotropic FP models (i.e., the apocenter condition and the energy condition, see Takahashi et al. 1997).

3. Results

We investigated the evolution of tidally-truncated globular clusters with multi-mass components. Initial cluster models are King models with the power-law mass functions. We refer isotropic, apocenter-condition anisotropic, and energy-condition anisotropic models, as I, Aa, and Ae models, respectively. Main results are summarized as follows.

- There is only a small difference between the I model and the Ae model in mass-loss rates. However, the mass loss in the Aa model is slower than in the other two models.
- The development of anisotropy depends on the stellar mass.
- The mass function becomes flatter as the cluster loses its mass in every models. The flattening is more significant for heavy stars.

4. Conclusions

Numerical simulations of the evolution of globular clusters by using anisotropic FP models can be performed with high numerical accuracy. Therefore more realistic FP simulations are now possible than before. In this paper, we carried out simulations of the evolution of tidally-truncated multi-mass clusters. Other important effects on the globular cluster evolution, such as tidal shocks and stellar evolution, will be incorporated into anisotropic FP models in future work. Furthermore, FP models can be applied also to the evolution of galactic nuclei containing a central massive black hole.

This paper is based on the results of collaboration work with Hyung Mok Lee (Korea). This work is supported in part by the Research for the Future Program of Japan Society for the Promotion of Science (JSPS-RFTP97P01102). A part of numerical computations were carried out at the Yukawa Institute Computer Facility.

References

Cohn, H. 1979, ApJ, 234, 1036.
Cohn, H. 1980, ApJ, 242, 765.
Takahashi, K. 1995, PASJ, 47, 561.
Takahashi, K. 1996, PASJ, 48, 691.
Takahashi, K. 1997, PASJ, 49, 547.
Takahashi, K., Lee, H.M., Inagaki, S. 1997, MNRAS, 292, 331.

NUMERICAL STUDY OF SURVIVAL CONDITION OF SUBSYSTEMS DURING COLD COLLAPSE

TOSHIO TSUCHIYA
Department of Astronomy, Kyoto University,
Kitashirakawa, Oiwake-cho, Kyoto 606-8502, Japan

Hierarchical structures are seen in various astrophysical objects in the universe, such as globular clusters in a galaxy, or galaxies in a cluster of galaxies. Such a subsystem in a virialized system forms a distinct object which is in virial equilibrium by itself. How such subsystems form and evolve in a larger scale environment is a fundamental problem of astronomy.

The disruption processes of substructures in a virialised system have been studied by many authors [1, 2, 3, 4], but only a few papers [5, 6, 7] were devoted to those processes acting in a dynamically evolving system. Since the disruption of substructures should be more effective in a collapsing system, it is important to study fundamental processes of disruption in the case.

Here, cold collapse of a cluster composed of small identical clumps, each of which is in virial equilibrium, is considered. Our main aim is to find the condition for the survival of substructure during a cold collapse.

The clumps are modeled by an equilibrium Plummer sphere, with a virial radius r_v. A cluster is composed of N_c clumps, which are distributed uniformly within a sphere. The centre of mass of each clump is located at a point which is randomly selected from the sphere and has no bulk velocity initially. Since the size of the clumps is fixed, the initial radius of the cluster $R_{\rm ini}$ is scaled by

$$\frac{R_{\rm ini}}{r_v} = 2N_c^{(n_k+5)/6}, \tag{1}$$

where n_k is a parameter. This relation corresponds to the ratio between the radii of virialised objects in the Einstein-de Sitter universe with the initial power spectrum such as $P(k) \propto k^{n_k}$ (e.g., §26 of Ref. [8]). The larger n_k corresponds to the smaller relative size of the clumps. In our simulation, the number of particles in a clump is fixed to be 512.

We examine the cases with four different values of n_k : $n_k = -3, -2, -1, 0$, and for each n_k, we assign eight different values of N_c : $N_c = 2, 4, 8, 16, 32, 64, 128, 256$, except for $n_k = 0$.

S. M. Miyama et al. (eds.), Numerical Astrophysics, 79–80.

In our numerical code, gravity is calculated by direct summation of contribution of all particles using GRAPE-3A [9]. The softening length, ϵ, and the time step of integration, δt, are set to be $\epsilon = 1/32$ and $\delta t = 1/64$, respectively. The total energy of the system is conserved within an accuracy of a few hundredth of 1% on average and at maximum 1%.

Since the clumps have no relative motion with respect to each other initially, the cluster collapses by its gravity. At the first collapse of the cluster, most of the clumps are destroyed, but some survive. From the numerical simulations, we obtained the condition for clumps to survive,

$$N_c \gg 1 \qquad \text{and} \qquad n_k \geq 1. \tag{2}$$

A simple analytic argument [10] supports the numerical result.

As an application of this result, in the CDM model of cosmological clustering [11], for example, $n_k \geq 1$. corresponds to the object larger than $2/(\Omega h^2)$Mpc. Therefore our result may suggest that in a system smaller than $2/(\Omega h^2)$Mpc, the first violent collapse is strong enough to sweep away all substructures which exist before the collapse.

Finally, for the sake of application to general objects, we give the condition in terms of other parameters. Using radii of the cluster and the clumps, for example, we derive

$$N_c \gg 1 \quad \text{and} \quad \frac{R_{\rm ini}}{r_v} \geq 2N_c^{2/3}. \tag{3}$$

With the filling factor C, which is the inverse of the fraction of the volume occupied by the clumps, the condition is

$$N_c \gg 1 \quad \text{and} \quad C \geq 8N_c. \tag{4}$$

References

1. Spitzer L., 1958, ApJ, 127, 17
2. Fall S. M., Rees M. J., 1977, MNRAS, 181, 37p
3. Aguilar L., Hut P., Ostriker J. P., 1988, ApJ, 335, 720
4. Gnedin O. Y., Ostriker J. P., 1997, ApJ, 474, 223
5. van Albada T. S., 1982, MNRAS, 201, 939
6. White S. D. M., Davis M., Efstathiou G., Frenk C. S., 1987, Nat, 330, 451
7. Frenk, C. S., White S. D. M., Davis M., Efstathiou G., 1988, ApJ, 327, 507
8. Peebles P. J. E., 1980, The Large-Scale Structure of the Universe. Princeton Univ. Press, Princeton
9. Okumura S. K. et al., 1993, PASJ, 45, 329
10. Tsuchiya, 1998, MNRAS, submitted.
11. Peebles P. J. E., 1993, Principles of Physical Cosmology, Princeton Univ. Press, Princeton

FORCE DISTRIBUTION AND RELAXATION TIME IN CLUSTERS OF GALAXIES

ELIANI ARDI, TOSHIO TSUCHIYA AND SHOGO INAGAKI
Department of Astronomy, Kyoto University
Kitashirakawa, Oiwake-cho, Kyoto 606-8502, Japan

1. Motivation

We examine the influence of nearest neighbor galactic encounters in gravitationally clustered systems by using cosmological N-body simulation and determine the two-body relaxation time in the expanding universe, which is based on the effect of two-body encounters in the clusters of galaxies.

2. Research

2.1. FORCE DISTRIBUTION

The evolution of the stochastic force distribution which comes from all the galaxies in the system and the force from the nearest galaxy before and during gravitational clustering of galaxies in the expanding universe are investigated by using cosmological N-body simulation (Aarseth, 1985).

We assume that initially galaxies are distributed under a Poisson distribution. In this case, we make comparisons from both analytical Holtsmark distributions (Chandrasekhar, 1942) and numerical results. During clustering, all of 1000 particles, each of which represents a galaxy, are examined for $\Omega = 1.0$ universe for time scales of $a/a_o = 4.0, 8.0, 16.0, 22.63, 32.0$, where a is the scale factor.

2.2. RELAXATION TIME

Some numerical calculations are conducted to describe the orbit of two-body interactions. Based on these interactions, we determine the two-body relaxation time in the Einstein-de Sitter universe (Peebles, 1980). The two-body relaxation time is presented as a function of velocity dispersion of cluster of galaxies.

S. M. Miyama et al. (eds.), Numerical Astrophysics, 81–82.

3. Results and Conclusions

3.1. FORCE DISTRIBUTION

Stochastic force distribution and nearest neighbor distribution at initial condition show a good agreement between simulation results and analytical curves. As evolution of clustering is in progress, the stochastic force distribution and the nearest neighbor distribution agree over most of the range of force. After several time scale of expansions, when clustering is strong, the contribution of force is almost generated from gravitational attraction of the nearest neighbors. We conclude that nearest neighbor galactic encounters play the main role in the dynamics of galaxy clustering in the expanding universe (Ardi&Inagaki, 1997).

3.2. RELAXATION TIME

The orbits of two interacting galaxies can be classified into two kinds: one makes a binary and the other makes a hyperbolic orbit as well as in usual scattering. The orbits which forms a binary occur when relative velocities between two galaxies are very small. In these cases, the free-fall time is used as the time scale of the orbits. In scattering orbit, relaxation time is produced by scattering process. Combining the results of the scattering time and the free-fall time as a function of the velocity dispersion of cluster of galaxies, we finally find the relaxation time $t_{\rm rel}$ (in unit of dynamical time $t_{\rm dyn}$).

Relaxation time of cluster of galaxies show three different regions. First region, cluster of galaxies whose velocity dispersion is very low, less than critical velocity dispersion in our simulation unit. These systems relax at about the dynamical time of the systems. Second region, clusters of galaxies with low velocity dispersion, $\sigma_{\rm cr} < \sigma < 10\,\sigma_{\rm cr}$ in simulation units. The effect of expansion is noticed here, where the relaxation time is proportion to square of velocity dispersion. In this case, $t_{\rm rel} < 50\,t_{\rm dyn}$. Third region, where velocity dispersion are high. It seems that expansion gives no influence. As same as in the non-expanding case, in this region the relaxation time $t_{\rm rel} \propto \sigma^3$. Relaxation time is found to be $t_{\rm rel} \geq 100\,t_{\rm dyn}$.

References

Aarseth S.J., 1985, Multiple Time Scales Academic Press, New York
Ardi E., Inagaki S., 1997, PASJ, 49, 187
Chandrasekhar S., 1942, Principles of Stellar Dynamics, The University of Chicago Press, Chicago
Peebles P.J.E., 1980, The Large Scale Structure of The Universe, Princeton University Press, Princeton

AN X-RAY MICROLENSING TEST OF AU-SCALE QUASAR DISK STRUCTURE

ATSUNORI YONEHARA, SHIN MINESHIGE
AND TADAHIRO MANMOTO
Department of Astronomy, Kyoto University,
Sakyo-ku, Kyoto 606-8502, Japan
JUN FUKUE
Astronomical Institute, Osaka Kyoiku University,
Asahigaoka, Kashiwara, Osaka 582-0026, Japan
MASAYUKI UMEMURA
Center for Computational Physics, Tsukuba University,
Tsukuba, Ibaraki 305-0006, Japan
AND
EDWIN L. TURNER
Princeton University Observatory,
Payton Hall, Princeton, NJ 08544, U.S.A.

1. Introduction

Belief in high power energy output from quasars is usually created by the combination of a supermassive black hole with a surrounding accretion disk. This belief is supported by a number of observations. However, the real vicinity of a putative black hole has not been resolved yet.

Now, we present a new method by using microlensing to determine the lower mass limit of a supermassive black hole in Q2237+0305 (e.g., Huchra et al., 1985), in which quasar microlensing events were detected (e.g., Ostensen et al., 1996).

2. X-Ray Microlens

As a source, we put an ADAF (e.g., Manmoto et al., 1997) and calculated the light curve of a microlensing event caused by caustic (Schneider et al., 1992). The resultant light curve expected by an X-ray observation is shown in Figure 1. In this figure, the section with a sharp increase directly

S. M. Miyama et al. (eds.), Numerical Astrophysics, 83–84.

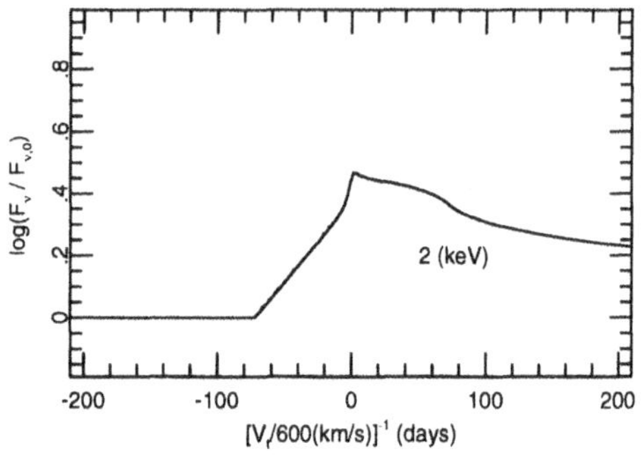

Figure 1. The light curve of microlensing at 2 (keV).

reflects the time during a caustic sweep in front of an X-ray emitting region. Although there is the observational limit, it was possible to estimate the size of the emitting region (r) by using the duration of this rise part ($r_{\rm cross}$, $\sim r$).

Since a plasma emitting X-ray is within the potential, we have $kT_{\rm e}/m_{\rm p} < GM/r$ (with k being the Boltzmann constant, $T_{\rm e}$ the electron temperature, and $m_{\rm p}$ the proton mass), namely,

$$M > 2.0 \times 10^5 M_\odot \left(\frac{r_{\rm cross}}{2.5 \times 10^{14}{\rm cm}}\right)\left(\frac{T_{\rm e}}{10^9{\rm K}}\right). \quad (1)$$

If we observe the X-rays every 6 days and assume that the transverse velocity of the caustic is $\sim 600{\rm km\ s^{-1}}$, we can achieve a spatial resolution of 2.5×10^{14}cm on the disk plane and thus set a lower mass limit of $\sim 10^5 M_\odot$ for typical $T_{\rm e} \sim 10^9$K. This method can determine the lower mass limit in very small sizes (AU-scale).

Details are shown in Yonehara et al., (1997).

References

Huchra, J., Gorenstein, M., Horine, E., Kent, S., Perley, R., Shapiro, I. I., & Smith, G. (1985) 2237+0305 - A New and Unusual Gravitational Lens, *AJ*, **90**, p691–696

Manmoto, T., Mineshige, S., & Kusunose, M. (1997) Spectrum of Optically Thin Advection-dominated Accretion Flow around a Black Hole: Application to Sagittarius A*, *ApJ*, **489**, p791–803

Ostensen, R., et al. (1996) Monitoring of the Einstein Cross with the Nordic Optical Telescope, *A&A*, **309**, p59–64

Schneider, P., Ehlers, J., & Falco, E.E. (1992) *Gravitational Lenses, 2nd ed.*, Springer-Verlag, New York.

Yonehara, A., Mineshige, T., Manmoto, T., Fukue, J., Umemura, M., & Turner, E.L. (1997) Microlens Diagnostics of Accretion Disks in Active Galactic Nuclei, *ApJ*, submitted

PHOTOIONIZATION AND MAGNETIZED WINDS: HII REGIONS AND PLANETARY NEBULAE

JOSÉ FRANCO AND GUILLERMO GARCíA-SEGURA
Instituto de Astronomía-UNAM, Mexico

1. Introduction

Photoionized nebulae are generated by hot stars located at the two ends of the stellar mass spectra (and near the extremes of their evolutionary tracks): H II regions are excited by recently formed massive stars, and Planetary Nebulae (PNe) are generated by low mass stars evolving toward the white-dwarf phase (see Osterbrock 1989). They display a variety of fascinating shapes, with irregular forms delineated by dusty filaments and bright rims, and with a complex network of velocity fields, ionization fronts, shock waves, and instabilities. At large scales, integrated over time and space, these objects can have a strong impact in the evolution of gaseous galaxies. For instance, massive stars inject large amounts of radiative energy which photoionizes and disrupts the parental clouds, setting the efficiency of star formation at galactic scales (*e.g.* Cox 1983; Franco *et al.* 1994; Shore & Ferrini 1994; Diaz-Miller *et al.* 1998). In contrast, evolved low mass stars do not have such a disrupting effect but they provide a generous gas mass return rate. Thus, PNe ejecta can maintain the gaseous component at the late stages of galaxy evolution, and are responsible for the enrichment of several heavy elements.

H II regions form a class of relatively well studied objects but there are no clear definitions of the different types. They are sometimes classified as ultracompact, compact, and extended HII regions (*e.g.* Habing & Israel 1979). Ultracompact H II regions (UCHIIs) have small sizes, of about 0.1 pc, and are located in the inner, high-pressure, parts of the parental molecular cloud (*e.g.* Kurtz *et al.* 1994; Xie *et al.* 1996). Compact H II regions (HIIs) have larger sizes, 0.1 – 0.3 pc, and are carving their way out of the parental clouds. Extended H II regions, on the other hand, have sizes of up to several parsecs and they represent the mature state of these objects.

S. M. Miyama et al. (eds.), Numerical Astrophysics, 85–94.

Thus, one can sketch a simple evolutionary link with the size growing in time and, after Strömgren (1939) and Kahn (1954) set the physical basis to their modeling, the details of the expansion has been studied with a variety of different analytical and numerical tools, and under different cloud conditions, over the last four decades (see Yorke 1986; Franco *et al.* 1989,1990; Garcia-Segura & Franco 1996). In the case of PNe, there are a variety of different morphological classes. They have been cataloged as bipolar, elliptical, point-symmetric, irregular, spherical, and quadrupolar (see Manchado *et al.* 1996). They are formed by the mass ejected at the final stages of low-mass star evolution, and the shapes are created by at least two interacting winds. Dust shells around AGB stars, the progenitors of PNe, do not show signs of asphericity and, during the transition from the AGB to the post-AGB phase, one or more physical processes responsible for the shape of these objects must be initiated. The origin of aspherical nebulae remains as one of the fundamental problems of PNe formation and evolution, and stellar rotation and magnetized winds can produce the observed morphologies.

2. H II Regions

Star clusters are formed inside molecular cloud complexes, which are composed of a collection of high-density condensations interconnected by a more diffuse intercloud medium. The internal density distributions in clouds are proportional to r^{-w}, with an average value of $w \sim 2$ (*e.g.* Arquilla & Goldsmith 1985; Gregorio Hetem *et al.* 1988), and there are also disk-like fragments (*e.g.* Torrelles *et al.* 1983; McCaughrean & O'Dell 1996). The high-density condensations, or cloud cores, have peak densities of about 10^7 cm^{-3} and peak temperatures above 10^2 K (*e.g.* Bergin *et al.* 1996; Hofner *et al.* 1996). These observed values already indicate that the gas thermal pressures are large inside the cloud cores. In addition, the existence of large non-thermal "turbulent" velocities, of several km s^{-1}, and strong magnetic fields, ranging from tens of μG to tens of mG (see Myers & Goodman 1988 and references therein), imply that the *total* pressures are obviously higher and the cores of massive molecular clouds are highly pressurized regions. The resulting *maximum* core pressures (generated by their own self-gravity) could reach values in excess of 5×10^{-6} dyn cm^{-2} (see Garcia-Segura & Franco 1996). Thus, H II region expansion in high-density cores will be affected by their large total pressures. Numerical and analytical models of H II region evolution under these conditions (*i.e.*, $n(r) \propto r^{-w}$ and disk-like density distributions, and large total pressures) provide many details of the flows (Franco *et al.* 1989, hereafter FTB89; Franco *et al.* 1990, hereafter FTB90; García-Segura & Franco 1996, hereafter GF).

For constant stellar UV fluxes and uniform ambient densities, the expansion of H II regions has well defined evolutionary phases (see Spitzer 1978 and Osterbrock 1989). During the formation phase, the radiation field creates an ionization front that, in a recombination time, reaches the initial Strömgren radius (Strömgren 1939) $R_S = \left[3\ F_\star/4\ \pi\ n_0^2\ \alpha_B\right]^{1/3}$, where n_0 is the ambient density, $F_\star$ is the total number of ionizing UV photons per unit time, $\alpha_B = 2.6 \times 10^{-13}$ cm^3 s^{-1} is the hydrogen recombination coefficient to all levels above the ground level. For a dusty cloud, the initial radius is smaller and is given by $R_{S,\rm dust} \simeq R_S\ e^{-\tau/3}$ (FTB90), where τ is the optical depth due to dust absorption. This simple formula agrees with the results of radiative transfer calculations, and indicates that the *initial* Strömgren radius is probably smaller than is usually considered (Díaz-Miller *et al.* 1998; hereafter DFS). In addition, the radiation pressure on dust grains is transferred to the ionized gas and, at late times, a central hole can be created in dusty H II regions (Mathews 1967). After the formation time, due to the pressure difference with the surrounding medium, the recently formed H II regions expand and drive a strong shock wave ahead of the ionization front. The expansion continues until the photoionized gas reaches pressure equilibrium with the ambient interstellar medium (GF). If the ionization front encounters a strong negative density gradient and overruns it, then the expansion will enter into its "champagne" phase (Tenorio-Tagle 1982; FTB90).

2.1 High Ambient Pressures and the UCH II *Class*

The photoionized region reaches its equilibrium temperature, $T_i \sim 10^4$, K in a relatively short time-scale, and expands (in a nearly isothermal fashion) into the neutral medium whose pressure is P_0. Assuming that the ion density n_i is equal to the electron density n_e (*i.e.*, for a pure hydrogen nebula), the thermal pressure of the ionized region is $P_i = 2n_i kT_i$, where k is Boltzmann's constant. The advance of the ionization front is controlled by recombinations in the photoionized gas and, for a region of radius R_{HII}, the average ion density is $< n_i >= [3\ F_\star/(4\ \pi\ \alpha_B)]^{1/2}\ R_{HII}^{-3/2}$. The corresponding internal pressure is $P_i = \left(3\ k^2 T_i^2 F_\star/\pi\ \alpha_B\right)^{1/2}\ R_{HII}^{-3/2}$. For a constant density medium and $P_i \gg P_0$, the HII radius grows as $R_{HII} \propto t^{4/7}$, and the internal pressure decreases as $P_i \propto t^{-6/7}$. The expansion continues until $P_i \rightarrow P_0$, and the internal pressure tends to a constant value. These limits show that, for a constant density medium, β evolves as a function of the pressure difference from 4/7 to zero. When pressure equilibrium is reached, the ion density is simply given by $n_{i,\rm eq} = (P_0/2kT_i) \simeq 3.6 \times 10^4 P_7 T_{\rm HII,4}^{-1}$ cm^{-3}, where $P_7 = P_0/10^{-7}$ dyn cm^{-2}, and $T_{\rm HII,4} = T_i/10^4$ K. The stagnation radius of the H II region, then, corresponds to a Strömgren radius at this equilibrium density $R_{\rm S,eq} \approx 2.9 \times 10^{-2}\ F_{48}^{1/3}\ T_{\rm HII,4}^{2/3}\ P_7^{-2/3}$ pc , where $F_{48} =$

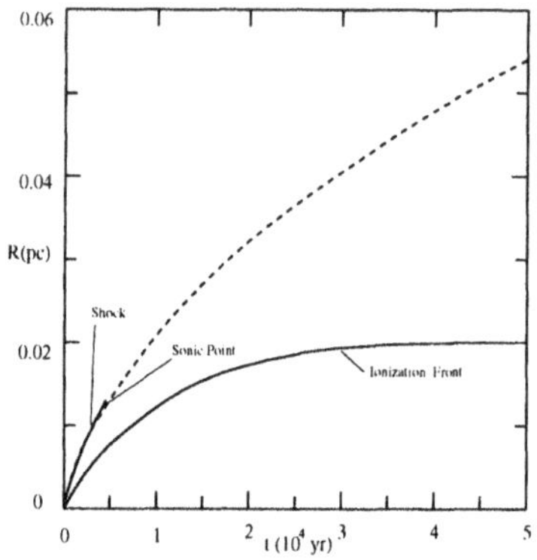

Figure 1. UCH II region evolution, for a star with $F_{48} = 1$, in a high-pressure medium with $P_7 = 1$.

$F_\star/10^{48}$ s^{-1}.

The interstellar pressure at the solar circle is about $P_{ISM} \sim 10^{-12}$ dyn cm^{-2}, and the equilibrium values for single massive stars (with $F_{48} \sim 1$) are $n_{i,\mathrm{eq}} \sim 0.4$ cm^{-3} and $R_{S,\mathrm{eq}} \sim 30$ pc (the corresponding sizes for "naked" OB associations can reach a few times 10^2 pc). The time scales at which these equilibrium values are reached, however, are very long, $t_{\mathrm{eq}} > 10^7$ yr, and most massive stars move out of the main sequence earlier. For high-pressure cores with $P_7 \sim 1$ and $r_c \sim 0.1$ pc (see GF; Xie *et al.* 1996), the resulting UCH II could be stable and long lived. The equilibrium stage is reached in time scales of the order of $t_{\mathrm{eq}} \sim 10^4$ yr, and the corresponding values are very similar to the observationally derived sizes and electron densities in ultracompact H II regions, UCH II (see Kurtz *et al.* 1994).

2.2 Density Gradients

If the star is located near the core boundary, or moves towards the edge, the ionization and shock fronts can move out of the core and the gas accelerates out of the high-density core. The relative position of the exciting star, then, defines whether the UCH II reaches pressure equilibrium or not. For static or slowly moving stars, which are born near the center of the core, the ultracompact stage can be long lasting. Otherwise, when the stars are born at a distance smaller than $R_{S,\mathrm{eq}}$ from the boundary, the ionized flow is accelerated outwards and the UCH II stage is a transient one (a possible example of this type of transient UCH II region is G29.96, which is probably younger than 10^5 yr and has a champagne flow: see Lumsden & Hoare 1996 and Watson & Hanson 1997).

The evolution in decreasing density stratifications displays a rich variety of hydrodynamical phenomena with internal shocks, receding ionization fronts, rapid evaporation of cloud cores, and the creation of new clumps in the expanding shocked shell (FTB89; FTB90; Rodriguez-Gaspar *et al.* 1995;

GF; Franco *et al.* 1998). For spherical clouds with density distributions consisting of a central core (with radius r_c and constant density n_0) and an envelope with a power-law density stratification, $n_{H_2}(r) = n_0(r/r_c)^{-w}$, the H II region growth is approximated by (FTB90)

$$R(t) \simeq R_w \left[1 + \frac{7-2w}{4}\left(\frac{12}{9-4w}\right)^{1/2}\frac{c_i t}{R_w}\right]^{4/(7-2w)} \quad (1)$$

The expansion phase has a critical exponent, $w_{crit} = 3/2$, which is independent of the initial conditions and is not affected by dust absorption. For $w < 3/2$, the interface between the ionization front and the leading shock accumulates neutral gas. For $w = w_{crit}$, the ionization front and the shock front move together without a neutral interface. For $w > 3/2$, the ionization front overtakes the shock front and proceeds to ionize the whole cloud. Once the whole ionized cloud is set into motion, the expanded core feels the strongest outwards acceleration, and there is a wave driven by the fast growing core. Thus, fast flows can be generated in steep density gradients. For instance, flow velocities between 20 and 30 km s^{-1} are generated in $w \sim 2$ distributions, and the speeds can increase to more than 100 km s^{-1} for $w > 3$. This implies that UCH II can be formed in high-pressure cores with $R_s < r_c$, but a rapid evaporation of the core occurs when $R_s > r_c$ (Rodriguez-Gaspar *et al.* 1995).

For a gaseous disk, with a density distribution along the z-axis equal to $n(z) = n_0\ G(z/H)$ [where n_0 is the density at midplane, H is the scale height, and $G(z/H)$ is the disk stratification; say, $e^{-z/H}$, e^{-z^2/H^2}, or $\mathrm{sech}^2(z/H)$], the density fall-off is steeper than $r^{-3/2}$. Thus, ionization front can become unbounded (during the expansion, however, the ionization front can recede; see FTB89). The initial shape of the ionized region is defined by photoionization equilibrium, but initial size of the H II region as a function of the azimuthal angle θ.

2.3 Fragmentation of the Shocked Shell

The shock generated by the expanding H II region collects and accelerates the ambient gas and, after radiative losses become important, a cold and dense shell forms. The ionization front is located at the inner boundary of the shell, and their evolution is subject to instabilities. In particular, the ionization front exacerbates the growth of the thin shell instability and the models performed by GF, which assume ionization equilibrium to define the location of the ionization front, show that these instabilities generate a rapid fragmentation of the shell. The shapes of the resulting fragments are similar to those observed in cometary globules and elephant trunks, and the flow instabilities appear under a wide variety of conditions. Figure 2 shows the instability for an H II region evolving in a disk-like cloud. The model

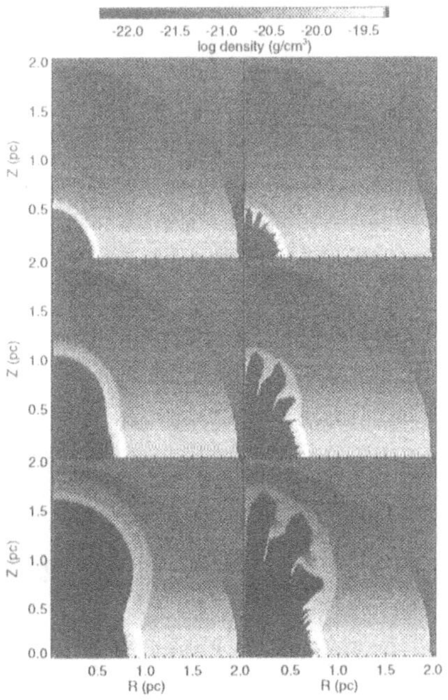

Figure 2. Expanding H II region evolving in a disk-like molecular cloud. Panels a-c; stable shell of an adiabatic shock. Panels d-f; shell fragmentation in a radiative shock.

illustrates the continuous formation of clumps by the shocked shell. The presence of a strong stellar wind also generates clumpy structures and the composite structures formed around massive stars with strong winds and expanding H II regions display a complex behavior (see Franco *et al.* 1998). The wind-driven shell forms large density clumps and the UV photon field creates long photoionized fingers in the directions that are free of clumps. Also, the radiation field with energies just below the Lyman limit creates a photodissociated region (PDR) between the H II region and the rest of the molecular cloud (*e.g.* Bertoldi & Draine 1996). This shell of HI gas, with its photodissociating front, adds new complications to the dynamical evolution (Roger & Dewdney 1992; DFS). In addition, hot stars produce large composite H II regions with PDRs, but intermediate mass stars produce only PDRs (DFS found that the the minimum stellar temperature required to create a sizable PDR is about 1.3×10^4 K).

3. Planetary Nebulae: Toroidal Fields and Collimation

It is commonly accepted that the formation of PNe involves the action of at least two interacting stellar winds (Kwok, Purton & Fitzgerald 1978; Balick

1987; Kwok 1994; Aller 1993). It has been shown by Langer (1997,1998) and García-Segura et al. (1998) that luminous blue variables (LBVs) and asymptotic giant branch (AGBs) stars can reach critical rotation. García-Segura et al. (1998) have shown that the result is a slow, massive, equatorially confined outflow. The striking similarity of the shapes of some PNe and LBV nebulae (*i.e.*, compare the "Hourglass Nebula", MyCn18, with the "Homunculus Nebula" surrounding η Carinae) indicates that the same physical processes may be responsible for their morphologies. If this is the case, the main cause of bipolarity could be due to stellar rotation. Asymmetries of the precursor AGB wind are not the only factor responsible for the shaping of PNe. Using the thin shell approximation, Begelman & Li (1992) obtained self-similar solutions of aspherical nebulae produced by a fast rotating neutron star with a *magnetized* fast wind. Their procedure did not include wind asymmetries, and the aspherical shapes were due solely to the tension of the toroidal magnetic field. This was later applied by Chevalier & Luo (1994) to the case of PNe winds and found similar steady-state aspherical results. More recently, with the aid of 2-dimensional MHD simulations in cylindrical coordinates, Różyczka & Franco (1996) found that the time-dependent evolution has a very complex behavior and jet-like features and collimated outflows can be created (see Figure 3). The 3-D computations performed by García-Segura (1997) corroborate 2-D results, with the main difference that the collimated gas can be probably subject to kink instabilities. Thus, the formation of jets and ansae are confirmed in 3-D, axis-free, calculations. Now García-Segura et al. (1998) is performing 2-D, spherical calculations, including the effects rotation and magnetic fields that seem to be responsible for the wide variety of observed nebular morphologies. The magnetic field in an outflowing wind from a rotating star has a toroidal component that decreases with distance as r^{-1}. The precise form of the off-equator distribution of this toroidal component is unimportant, provided that the field is sufficiently strong to cause a deformation in the shocked wind region (Różyczka & Franco 1996). The radial field component decreases as r^{-2}, and can be neglected, so the used field configuration obeys $\nabla \cdot B = 0$.

The main effects of the magnetic field is the elongation of the nebula in the polar direction, and the creation of jets. The generation of a jet-like outflow has been described by Różyczka & Franco (1996). First, the outer part of the magnetized shocked wind region becomes magnetically rather than thermally dominated (the magnetic energy density gets larger than the thermal energy density). Second, the tension of the toroidal field slows down the expansion in the directions perpendicular to the symmetry axis, while the expansion in the direction parallel to the axis proceeds unimpeded. Third, a flow from the equatorial parts of the shocked wind

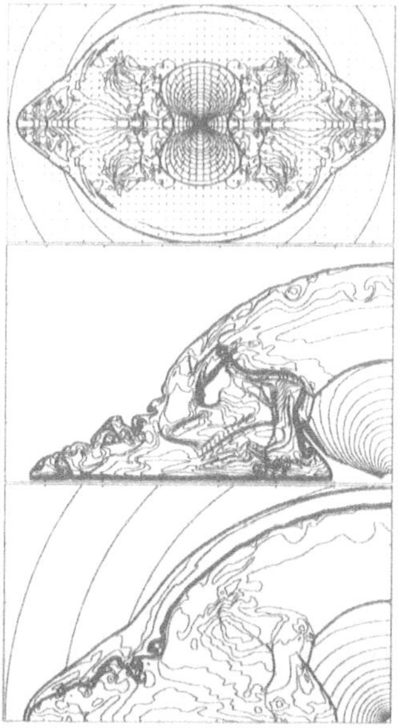

Figure 3. Interacting winds: magnetic fast wind interacting with a slow AGB wind. Both winds are spherically symmetric and the fast post-AGB wind has a toroidal magnetic field.

region toward the symmetry axis is initiated, leading to the formation of stagnation regions at the axis and to the formation of jets. The gas arriving at the polar regions of the nebula forms relatively dense blobs which can be identified with the ansae observed in PNe (such as the ones in NGC 7009; see Lopez *et al.* 1993). Perhaps the most important issue that has to be stressed here, is that magnetic collimation becomes very efficient (even for a spherically symmetric outflow) after the flow has been processed by a shock. Also, the collimation is operative up to very large distances, even when the resulting jets develop kink instabilities. Note that external anisotropic density distributions are not required but, if they occur, they can produce even more elongated structures. When stellar rotation is included in the AGB wind, a high-density region with a compressed B-field is generated near the equatorial plane (see Ignace, Cassinelli & Bjorkman 1998). In this case, all the effects described above occur faster and have a stronger impact in the evolution of the resulting nebula: the magnetic effect is enhanced, and a higher degree of collimation is achieved. Thus, a combination of rotation and a magnetic field can naturally account for some of the most interesting features in PNe. This however does not preclude that other effects (such as external density anisotropies) can take place at the same time, resulting in even richer morphological structures.

4. Summary

Molecular clouds are composed of a collection high-density clumps, with densities reaching 10^7 cm^{-3}. These dense clumps, or cloud cores, are supported by non-thermal motions and seem to be the actual sites of star formation. The initial shape and early evolution of the H II regions, then, are controled by the core density distributions. Depending on these distributions, and the corresponding stellar UV field, the photoionized regions can either reach pressure equilibrium inside the high-pressure cores (with sizes and densities similar to those observed in UCH II regions), or create blister-type H II regions with champagne flows. Disk-like cores can generate photoionized regions with a bipolar shape, and the flows are accelerated along the steeper density gradients. In all cases, the density inhomogeneities engulfed within the ionization fronts are smoothed out in short timescales, but new clumps are continuously created by the fragmentation of the dense shells generated by the shock fronts. Instabilities in the ionization/shock front drive the fragmentation of the shell, and the resulting finger-like structures can explain the existence of elephant trunks and cometary-like globules in most H II regions. Finally, when one includes the external dissociation front, a double shock structure with a complex behavior is formed. The dissociation front destroys additional molecular gas and, effectively, diminishes the productivity of star formation.

In the case of PNe, the ambient density structure is strongly modified by winds from the progenitor star. Stellar wind asymmetries and magnetic fields from rotating stars, along with the ionizing flux from the star, can create the wide range of observed PNe morphologies and magnetically collimated outflows (jets). This magnetic collimation is also operative in young stellar objects and can be responsible for the generation of protostellar jets. Perhaps the most important issue here is that magnetic collimation, and jet formation, becomes very efficient after the flow has been passed through a shock. All these features warrant further research, and future studies should shed more light on magnetic collimation in shocked regions.

We are grateful to our good friends Rossy Diaz-Miller, Tim Freyer, Beto Lopez, Norbert Langer, Mordecai-Mark Mac Low, Michal Rozyczka, and Steve Shore who have collaborated with us in different aspects of the work reported here. We also thank Frank Bertoldi for suggestions on PDRs. This work was partially supported by DGAPA-UNAM, CONACyT grants 400354-5-4843E and 400354-5-0639PE, and by a R&D CRAY Research grant. Part of the computations were performed at the Supercomputer Center of UNAM.

References

Aller, L. H. 1993, in Planetary Nebulae, ed. R. Weinberger & A. Acker (Dordrecht: Kluwer), 1

Arquilla, R. & Goldsmith, P. F. 1985, ApJ, 297, 436

Balick, B. 1987, AJ, 94, 671

Begelman, M. & Li, Z. Y. 1992, ApJ, 397, 197

Bergin, E., Snell, R. & Goldsmith, P. 1996, ApJ, 460, 343

Bertoldi, F. & Draine, B. T. 1996, ApJ, 458, 222

Bodenheimer, P., Tenorio-Tagle, G. & Yorke, H. W. 1979, ApJ, 233, 85

Chevalier, R. A., & Luo, D. 1994, ApJ, 421, 225

Cox, D. P. 1983, ApJ, 265, L61

Díaz-Miller, R. I., Franco, J. & Shore, S. N. 1998, ApJ, in press (DFS)

Franco, J., Díaz-Miller, R. I., Freyer, T. & García-Segura, G. 1998, in Astrophysics from Antarctica, ASP Conf Series, in press

Franco, J., Shore, S. N. & Tenorio-Tagle, G. 1994, ApJ, 436, 795

Franco, J., Tenorio-Tagle, G. & Bodenheimer, P. 1989, RMexAA, 18, 65 (FTB89)

Franco, J., Tenorio-Tagle, G. & Bodenheimer, P. 1990, ApJ, 349, 126 (FTB90)

García-Segura, G. & Franco, J. 1996, ApJ, 469, 171 (GF)

García-Segura, G. 1997, ApJ, 489, L189

García-Segura, G., Langer, N., Różyczka, M., Mac Low, M.-M., & Franco, J. 1998, RevMexAA, ConfSer, 7, 50

Gregorio Hetem, J., Sanzovo, G. & Lepine, J. 1988, AAS, 76, 347

Habing, H. J. & Israel, F. P. 1979, ARAA, 17, 345

Hofner, P. Kurtz, S., Churchewell, E., Wamsley, M. & Cesaroni, R. 1996, ApJ, 460, 359

Ignace, R. Cassinelli, J. & Bjorkman, J. 1998, ApJ, in press

Kahn, F. D. 1954, BAN, 12, 187

Kurtz, S., Churchwell, E. & Wood, D. O. S. 1994, ApJs, 91, 659

Kwok, S. 1994, PASP, 106, 344

Kwok, S., Purton, C. R., & Fitzgerald, P. M. 1978, ApJ, 219, L125

Langer, N. 1997, in LBVs: Massive Stars in Transition, ASP ConfSer, 120, 83

Langer, N. 1998, A&A 329, 551

Lopez, J. A., Roth, M. & Tapia, M. 1993, A&A, 267, 194

Lumsden, S. L. & Hoare, M. G. 1996, ApJ, 464, 272

Manchado, A., Guerrero, M., Stanghellini, L., & Serra-R icart, M. 1996, The IAC Morphological Catalog of Northern Galactic PNe, IAC, Spain

Mathews, W. G. 1967, ApJ, 147, 965

McCaughrean, M. J. & O'Dell, C. R. 1996, AJ, 111, 1977

Myers, P. C. & Goodman, A. A. 1988, ApJ, 326, L27

Osterbrock, D. E. 1989, Astrophysics of Gaseous Nebulae and Active Galactic Nuclei, Mill Valley: University Science Books

Rodríguez-Gaspar, J. A., Tenorio-Tagle, G. & Franco, J. 1995, ApJ, 451, 210

Roger, R. S. & Dewdney, P. E. 1992, ApJ, 385, 536

Różyczka, M. & Franco, J. 1996, ApJL, 469, L127

Shore, S. N. & Ferrini, F. 1994, Fund.CosmicPhys, 16, 1

Spitzer, L. 1978, Physical Processes in the Interstellar Medium, New York: Wiley

Strömgren, B. 1939, ApJ, 89, 526

Tenorio-Tagle, G. 1982, in Regions of Recent Star Formation, eds. R. S. Roger & P. E. Dewdney, Dordrecht: Reidel, 1

Torrelles, J. M. *et al.* 1983, ApJ, 274, 214

Watson, A. M. & Hanson, M. M. 1997, ApJ, 490, L165

Xie, T., Mundy, L. G., Vogel, S. N. & Hofner, P. 1996, ApJ, 473, L131

Yorke, H. W. 1986, ARAA, 24, 46

3-D SIMULATIONS OF THE MHD KELVIN-HELMHOLTZ INSTABILITY

T.W. JONES
University of Minnesota, USA
e-mail: twj@astro.spa.umn.edu

DONGSU RYU
Chungnam National University, Korea
e-mail: ryu@canopus.chungnam.ac.kr

AND

ADAM FRANK
University of Rochester, USA
e-mail: afrank@alethea.pas.rochester.edu

1. Introduction

Strongly sheared flows should be ubiquitous in astrophysics and will often be hydrodynamically Kelvin-Helmholtz (K-H) unstable. This can lead to turbulence and mixing. Most astrophysical environments are electrically conducting, so it is important to understand the role of magnetic fields in the K-H instability. While strong magnetic fields, through their tension, are well known to stabilize the K-H instability, the potential for much weaker fields to modify the nonlinear instability has only recently been recognized. Malagoli *et al.* (1996), Frank *et al.* (1996) and Jones *et al.* (1997) have recently carried out high resolution numerical MHD simulations of the K-H instability for periodic sections of 2.5 dimensional flows, *demonstrating clearly that a weak magnetic field can add order during the nonlinear evolution of this instability. This paper describes 3-D simulations confirming the earlier trends and showing the ability of a very weak, large-scale field to provide a significant self-organizational influence that smoothes flows.* "Weak" excludes fields creating linear stability or stability through tension during linear instability growth. "Order" means that the flows retain a more laminar character.

S. M. Miyama et al. (eds.), Numerical Astrophysics, 95–98.

The most important parameter in this outcome is the Alfvénic Mach number, $M_A = v/v_A$, with v the velocity difference across the shear layer and $v_A = B/\sqrt{\rho}$ the Alfvén speed in units giving magnetic pressure $p_b = B^2/2$. The above studies found the weak field regime for $M_A \gtrsim 5$. The initial instability is largely hydrodynamical, so "Cat's Eye" vortices form with axes in the shear layer, but perpendicular to the initial flow. Magnetic fields are initially parallel to the planes containing the shear layer, although its direction with respect to the velocity is otherwise arbitrary. In 2-D, field lines are wrapped into the Cat's Eye and stretched until tearing-mode instabilities and magnetic reconnection relax them. Jones *et al.* categorized the subsequent influence for 2.5-D flows in terms of "weak" and "very weak" fields. In the "weak field" regime, magnetic stretching around a single Cat's Eye turnover enhances the tension enough that local Alfvénic Mach numbers drop below unity; magnetic tension then competes with inertial stresses. During the reconnection that follows, there is a dynamical alignment between the magnetic and velocity fields, and cross helicity ($|v \cdot B|$) is maximized. Simultaneously, the Cat's Eyes are disrupted by Maxwell stresses, and the flow once again becomes laminar, but stable to perturbations smaller than the size of the computational box. Those authors emphasized that this disruption was magnetically driven, despite the fact that the initial $\beta = p_g/p_b \gg 1$, where p_g is the thermal gas pressure.

By contrast, "very weak fields", which in the Jones *et al.* 2.5-D simulations corresponded to initial $M_A \gtrsim 50$, cannot disrupt the Cat's Eyes. But, magnetic field within the vortices is largely annihilated by unsteady, driven reconnection. This enhances entropy and dissipation of kinetic energy, but otherwise, does not profoundly modify flow evolution.

2. 3-D Simulations

We have now extended the Jones *et al.* study to fully 3-D flows in order to understand the roles of weak fields more completely. As before we simulated flows that were initially uniform except for the velocity shear layer and a small, random velocity perturbation. The sonic Mach number was unity, with adiabatic index $\gamma = \frac{5}{3}$. The simulations were carried out in a cubic computational box, periodic in directions contained within the shear layer. Reflecting boundaries were placed above and below the shear layer. Weak initial magnetic fields were oriented 30° from the flow direction with strengths corresponding to $M_A =$ 4.3, 12.4, 43.3, 124, 433, 1.24×10^3. For comparison, $\beta = \frac{6}{5} M_A^2$ here. The ideal MHD equations were evolved using the conservative MHD TVD code described in Ryu *et al.* 1995. Grids included 64^3, 128^3 and 256^3 zones.

In 3-D the Cat's Eyes are hydrodynamically unstable; without magnetic

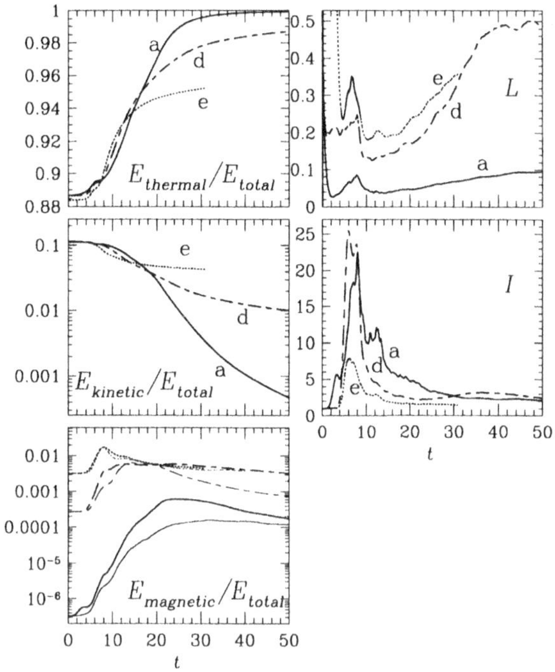

Figure 1. Evolution of some global properties. Left panels: thermal, kinetic and magnetic energy. Top right: "magnetic curvature radius", $L = [\langle B^4\rangle/\langle(B\cdot\nabla)B)\rangle^2]^{1/2}$. Bottom right: "magnetic intermittency", $I = \langle B^4\rangle/\langle B^2\rangle^2$. The sequence **a**, **d**, and **e** corresponds to increasing magnetic field, with details in the text. Magnetic energy results are shown for 128^3 (thin lines) and 256^3 (heavy lines) grids. All others are 256^3.

effects they violently break up after approximately one "eddy turnover", and in the center-of-momentum frame the flow becomes highly disordered with little evidence of the initial shear field; i.e., decaying turbulence develops. For our weakest-field MHD case ($M_A = 1.24 \times 10^3$) this behavior is also seen. All of our other cases show important differences from this, however, and evolve some degree of self-organized, shear. For the strongest field cases, which are still "weak field", but with initial $M_A \gg 1$ and $\beta \gg 1$, the flows become nearly laminar. Among other things that self-organization substantially slows the rate of kinetic energy dissipation, since it reduces the energy transfer to small, dissipation scales. This is evident in the left three panels of Figure 1, which show the evolution of thermal, kinetic and magnetic energies for three cases: $M_A = 1.24 \times 10^3$ (**a**), $M_A = 43.3$ (**d**) and $M_A = 12.4$ (**e**). In sound-crossing time units the Cat's Eye forms by about $t = 6$ and breaks up by about $t = 12$. For case **e** the velocity and magnetic fields organize then into "sheets" roughly aligned with the original shear layer; i.e., turbulence does not develop. By contrast, there is for Case **d**, a briefly chaotic flow following the break-up. In that respect Case **d** superficially resembles the hydrodynamic case or Case **a**. However, in Case **d** the flow quickly self-organizes by way of magnetic flux tubes formed across the original Cat's Eye. These tubes are parallel to the original velocity field. They also hold vorticity and magnetic helicity. Magnetic lines within them are amplified, not only by the lengthening of the tubes, but also by their twisting. Since these tubes become dynamically important before memory of the original shear pattern disappears, that shear remains to the end of the simulation, by contrast to Case **a**.

Formation and the subsequent evolution of flux tubes is signaled in the "magnetic intermittency", I, and "magnetic curvature radius", L, shown on the right side of Figure 1. Large I indicates high contrast in the magnetic field strength; i.e., the presence of magnetic "voids" and relatively intense flux tubes. That always happens during the Cat's Eye formation, but is less extreme for the strongest field Case **e**. Subsequent decay reduces the intermittency more moderate values in all cases, however. We note that for the boundary conditions used the mean vector magnetic field, $\langle \mathbf{B} \rangle$ is exactly constant. So, any net increase in magnetic energy must involve $I > 1$. Similarly, reduction in the curvature radius, L, signals formation of highly bent or twisted field regions. Again, this always happens during the Cat's Eye formation, because the field is wrapped into the Cat's Eye. It is partially relaxed during reconnection just prior to the Cat's Eye break-up. During break-up the field becomes twisted and tangled, so that L is again reduced. In Case **a**, L remains small, since there is little dynamical self-organization. The field remains mostly passive. In the other two cases, however, self-organization relaxes the fields. That difference also shows in the evolution of magnetic energy, E_{magnetic}. For Cases **d** and **e** there is relatively modest peak E_{magnetic} enhancement, which mostly disappears as the flow relaxes. However, for Case **a** E_{magnetic} increases by about 3 orders of magnitude, and remains near that level to the end. Further details and animations of Case **d** are contained in the contribution submitted for the CD ROM accompanying this volume.

We acknowledge support for this work from the NSF [TWJ] (AST96-19438, AST96-16964, INT95-11654), [AF] (AST-9702484), NASA [TWJ] (NAG5-5055), KOSEF [DR] (975-0200-006-2, 981-0203-011-2) and the Minnesota Supercomputing Institute.

References

Frank, A., Jones, T. W., Ryu, D. & Gaalaas, J. B. (1996) *The Astrophysical Journal*, **460**, pp. 777-793.

Jones, T. W., Gaalaas, J. B., Ryu, D. & Frank, A. (1997) *The Astrophysical Journal*, **482**, pp. 230-241.

Malagoli, A., Bodo, G. & Rosner, R. (1996) *The Astrophysical Journal*, **456**, pp. 708-716.

Ryu, D., Jones, T. W. & Frank, A. (1995) *The Astrophysical Journal*, **452**, pp. 785-796.

SIMULATIONS OF MHD TURBULENCE IN GIANT MOLECULAR CLOUDS

J.M. STONE
Department of Astronomy, University of Maryland
College Park, MD 20742-2421, USA

Abstract. Direct numerical simulations are used to investigate the properties of supersonic magnetohydrodynamic turbulence. Such turbulence is thought to dominate the internal dynamics of star forming regions. The decay rate of saturated turbulence is found to be large. Some of the properties of compressible MHD turbulence in a strongly magnetized medium are found to be remarkably similar to those of incompressible MHD turbulence.

1. Introduction

At the present time, stars form in our galaxy in Giant Molecular Clouds (GMCs). These clouds are characterized by several properties: they are cold, their density is strongly inhomogeneous, they are strongly magnetized, and they contain highly supersonic random internal motions (Blitz 1993). The latter property has been interpreted as meaning that the interiors of molecular clouds are dominated by compressible turbulence.

It is thought that this turbulence may help to support GMCs against collapse driven by their own self-gravity, and that this turbulence controls the highly clumpy nature of the density. However, supersonic *hydrodynamic* turbulence is known to decay rapidly (e.g. Porter, Pouquet, & Woodward 1992). It has long been suggested that the strong magnetic fields in GMCs must somehow reduce the decay rate of compressible magnetohydrodynamic (MHD) turbulence (e.g. Arons & Max 1975).

Our group (C. Gammie, E. Ostriker, and myself) has begun a project to study systematically the internal dynamics of magnetized, self-gravitating molecular clouds in two- and three-dimensions. Our method of investigation is to use direct numerical simulations of compressible turbulence in three

S. M. Miyama et al. (eds.), Numerical Astrophysics, 99–102.

circumstances: (1) randomly driven turbulence, (2) decaying turbulence from saturated initial conditions, and (3) decaying turbulence with self-gravity. In this contribution, I review some results from the first class of models.

2. Method

The numerical simulations are all performed with the MHD code ZEUS (Stone & Norman 1992). The initial conditions consist of a cubic box of size L which contains a plasma with uniform density ρ_0 threaded by a uniform magnetic field in the x-direction, $\mathbf{B} = (B_x, 0, 0)$. The sound speed in the fluid C_s is constant, so that the relevant timescales for the dynamics are the sound crossing time $t_s = L/C_s$, and the Alfven crossing time $t_A = L\sqrt{4\pi\rho_0}/B_x$. The dynamics is computed using an isothermal equation of state. We use periodic boundary conditions in each direction, and grid resolutions which vary between 32^3 and 512^3. The simulations presented here assume the magnetic field is perfectly coupled to the fluid (the ideal MHD approximation), Additionally, in all models we evolve a passive contaminant which initially fills a cylindrical volume in the center of the grid orientated with the symmetry axis parallel to $\mathbf{B}$ and with diameter and axial length equal to $L/2$. To drive turbulence in our simulations, we add random velocity perturbations $\delta\mathbf{v}$ at discrete time intervals Δt with $\Delta t/t_s = 0.001$ and with a spectrum corresponding to $|\delta v_k^2| \propto k^6 \exp(-8k/k_{pk})$. This form gives a steeply rising spectrum at small wavenumbers with an exponential cut-off near $k = k_{pk}$. We set $k_{pk} = 8$. The perturbations are normalized so that the kinetic energy input rate is a constant value $\dot{E}$. We study models with $\dot{E}/\rho_0 L^3 C_s^2 = 1000$, which corresponds to about $10L_\odot$ for typical cloud parameters (Gammie & Ostriker 1996). The magnetic field strength is specified by choosing the ratio of the square of the sound to Alfven speeds $\beta = C_s^2/V_A^2$. We study values of $\beta = 0.01, 1.0$ and ∞ (corresponding to pure hydrodynamics).

3. Results

The wave energy E_W, defined as the sum of the kinetic and magnetic energy associated with velocity and magnetic field fluctuations respectively, rises quickly at the start of our driven simulations, and then saturates at a constant value. By running simulations at resolutions of 32^3 up to 512^3, we find that the amplitude of the wave energy converges in the sense that there is little difference in its value above resolutions of 128^3. There is a very weak dependence on the amplitude of E_W with the strength of the field; strongly magnetized models are about a factor of two larger than very weakly magnetized (or purely hydrodynamic) models. At saturation, the energy input rate is equal to the decay rate. The decay rate is a factor of 3 – 5 times larger than that measured in one-dimensional simulations (Gammie & Ostriker 1996). In fact, the decay time is about one eddy turnover time at the scale at which the turbulence is driven (corresponding to the wavenumber k_{pk}), which is not significantly different than the decay time of unmagnetized turbulence.

Figure 1 plots grayscale images of the density and passive contaminant in a two-dimensional slice through the middle of a 512^3 simulation of driven MHD turbulence. The magnetic field strength used in this model is $\beta = 1.0$. The density field shows how filaments and shocks form quickly in the box. Because of the isothermal equation of state, all density discontinuities are associated with shocks (there are no contact discontinuities in an isothermal flow). It is therefore easy to see that the entire volume of the simulation is filled with shocks at late time. The RMS velocity dispersion is about 5 at late times, indicating such shocks are weak (only Mach 5, corresponding to about 1 km s^{-1} for gas at a temperature of 10 K). It is interesting to contemplate whether or not such shocks would be detectable in spectral line observations of molecular clouds.

Plotting the power spectra of the turbulence at late times also reveals several interesting features. There is a distinct difference in the slope of the power spectrum between strongly and weakly magnetized models. The former has a slope similar to that expected for a Kolmogorov spectrum, whereas the latter has a slope similar to a spectrum of discontinuities (shocks). Moreover, the spectrum of strongly magnetized turbulence is clearly anisotropic, with more power at wavenumbers perpendicular to the magnetic field than along it. The results of a Kolmogorov and anisotropic spectrum are remarkably similar to the properties of *incompressible* strong MHD turbulence (Goldreich & Sridhar 1995).

Acknowledgments. Support for this work is provided by NASA under contract number NAG53840. Computations were performed on the Cray/SGI Origin 2000 system at the NCSA. I thank my colleagues Charles Gammie and Eve Ostriker who contributed significantly to the work presented here.

References

Arons, J., & Max, C.E. 1975, ApJ196, L77

Blitz, L. 1993, in Protostars and Planets III, ed. E. Levy & J. Lunine (Tucson: Univ. Ariz. Press), 175

Gammie, C.F., & Ostriker, E.C. 1996, ApJ, 466, 814

Goldreich, P., & Sridhar, S. 1995, ApJ, 438, 763

Porter, D., Pouquet, A., & Woodward, P. 1992, Theor. Comp. Fluid Dyn., 4, 13

Stone, J.M., & Norman, M.L., 1992, ApJS, 80, 791

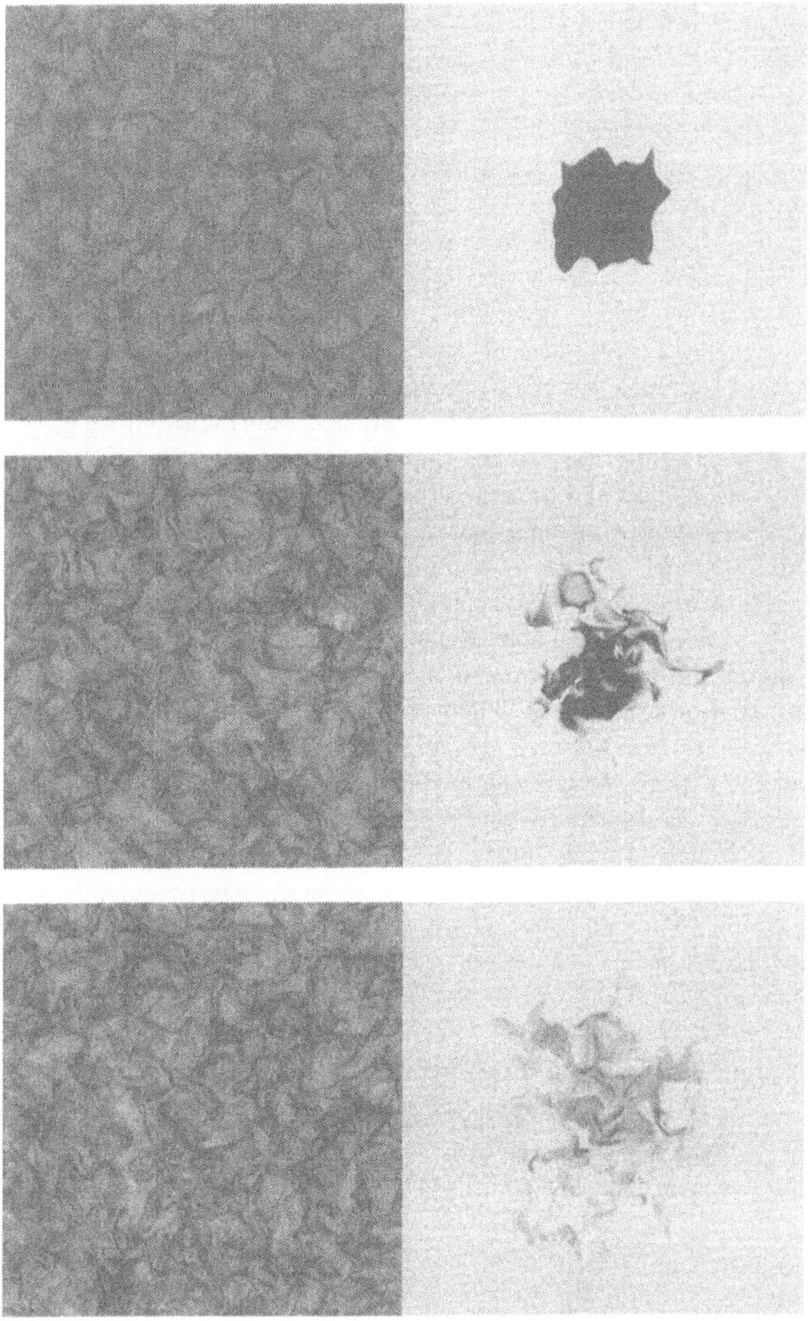

Figure 1. Images of the density (left) and passive contaminant (right) in a two-dimensional slice in the middle of a 512^3 simulation of driven MHD turbulence. The images are shown at times of 0.01 (top), 0.04 (middle) and 0.09 (bottom) sound crossing times. The magnetic field is horizontal.

STUDYING THE PHYSICS OF STAR FORMATION IN SINGLE AND INTERACTING GALAXIES

KEITH BAXTER AND PAUL ALEXANDER
MRAO, Department of Physics, Cavendish Laboratory, Cambridge, England, CB3 0HE, U.K.

We have constructed a versatile new simulation system for studying the small-scale physics of star formation (SF) and feedback within single and interacting galaxies, with an emphasis upon attempting to model the molecular clouds and their interaction with the ISM in a realistic fashion. Our model is fully self-gravitating with optional SPH. In order to attempt to ascertain the primary method of star formation within a galaxy, we have implemented models based upon:

Stochastic (Propagating) star formation; we take the propagating mechanism to be the triggering of cloud collapse by a supershell shock arising from an OB association. We assume that the probability of a shocked cloud undergoing SF will be expressible as a power series in the cloud mass: $P_{st} = \sum_i (m/m_{st})^i$ (Sleath & Alexander, 1995). Winds from OB stars are assumed to disperse a parent cloud when SF occurs.

Threshold-spontaneous SF is based upon Jeans collapse of clouds with minimum mass m_{fr}. We assume that once a cloud exceeds the minimum mass, it is subject to Jeans Collapse (star formation / cloud fragmentation) with an exponential lifetime given by

$$t(m) = \frac{t_{gc}}{1 + \log(m/m_{fr})}.$$

Spontaneous SF based upon Jeans collapse of a molecular cloud, with no minimum mass. (Kwan & Valdes, 1986).

We have embarked upon an analytic study of the outcome of cloud-cloud collisions, with possible results including coalescence, dissipative collisions, induced SF, fragmentation and heating. As a first approximation, we have implemented schemes based upon coalescence, in which two clouds merge if they approach to within the radius of the larger cloud, and dissipative collisions, in which the cloud velocities are reduced within the zero momentum frame (ZMF). Cloud growth mechanisms include growth through

S. M. Miyama et al. (eds.), Numerical Astrophysics, 103–104.

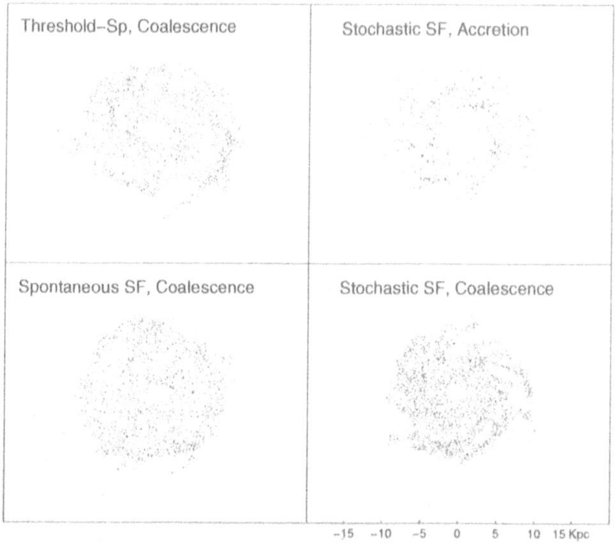

Figure 1. Positions of OB associations and SN within a steady state galaxy for the four different models.

clouds merging during low-energy, low angular momentum collisions and through accretion from the interstellar medium.

In order to perform a systematic comparison of the models, we constrained the models to have an equal Cluster Formation Rate (CFR), with the chosen CFR being 3×10^{-7} Clusters $\mathrm{Kpc}^{-2}\ \mathrm{yr}^{-1}$, a value appropriate to the Milky Way. We have derived an expression (Baxter & Alexander) constraining the CFR in terms of the cloud buildup rate (accretion / coalescence rate), and thus we may obtain the desired CFR by scaling the massgrowth parameters. The differing microphysics within each model determines the sites of star formation, and thereby the morphology of the young stellar population, and the relationship between ρ_{clouds} and ρ_*. We have obtained a Schmidt law of the form $\rho_* \propto \rho^{\eta}_{\mathrm{clouds}}$, with $\eta \approx 1.7$ for the stochastic-accretion model and $\eta \approx 1.0$ for the spontaneous - coalescence model. Spiral arms arise naturally as a result of instabilities within the disk component. Models including stochastic physics show a much stronger correlation between star formation and the spiral arms than the spontaneous models, leading to the young stellar population within stochastic models being strongly concentrated within the spiral arms.

References

Baxter, Keith, Alexander, Paul, in preparation.
Kwan, J., Valdes, F. 1986, ApJ. 315, 92
Sleath, J., Alexander, P. 1995 MNRAS, 275, 507

NUMERICAL SIMULATIONS OF COMPRESSIBLE MULTI-PHASE FLOW BY CIP-METHOD

HIDEYUKI KAMAYA
Department of Astronomy
Kyoto University, Sakyo-ku
Kyoto 606-8502, Japan
kamaya@tap.scphys.kyoto-u.ac.jp

1. Introduction

Because the interstellar medium (ISM) is not uniform but multi-phased, as is well known, ordinary hydrodynamics cannot accurately describe the dynamics of ISM. For that reason, it seemed interesting to apply the method of the multi-phase flow to describe the dynamics of ISM [Kamaya (1996, 1997a) and Kamaya & Shchekinov (1997, 1998)], a method originally developed in the field of engineering to describe any mixture. A construction law of a basic set of equations for compressible fluid was proposed (Kamaya 1997a) since there were no relevant basic sets for multi-phase flow dynamics. Adopting this construction law, the wave dynamics of ISM can be discussed with accuracy.

2. CIP-method

In this paper (see also Kamaya 1997b), two numerical results are presented. These results solve the basic set by means of the CIP-method, which was developed by Prof. Yabe and his collaborators (Yabe & Takei 1988). An example of its application to the astrophysical problem is found in Kudoh & Shibata (1997). The method is a unified scheme to describe the dynamics of both compressible and incompressible fluids, this being the major advantage of the CIP-method. For ISM, compressibility is essential. Moreover, since we needed many basic equations, a simple algorithm of the CIP-method was suitable for multi-phase flow research. target for CIP-solvers.

S. M. Miyama et al. (eds.), Numerical Astrophysics, 105–106.

3. Results

Two examples were calculated by the CIP-method: (1) The shock-tube problem in a two-phase mixture; and (2) Jackson's instability. (1) Interestingly, the shock was significantly damped because the volume-dominant diffuse phase was dragged by the dense clumps. The origin of the friction between the two phases is the momentum exchange due to the thermal process at their boundary (Kamaya & Shchekinov 1997). In the post-shock region, moreover, we observed the inhomogeneous distribution of the volume-filling factor of clumps, since the clumps were delayed to the shock wave of the diffuse medium due to their own inertia. spatial inhomogeneity of the friction force between phases, resulting in the aggregation of dense clumps in the diffuse medium. This instability occurs when the spatial distribution of the friction force is not uniform (see also Kamaya 1997b). In the context of ISM, thus, we expected interstellar cloud formation through Jackson's instability.

4. Implication

We found that an inhomogeneous distribution of the volume fraction of each interstellar phase developed through the instability of the fluidized bed (Jackson 1963). The system becomes unstable when the drag force increases at the point where the volume fraction of the dense blob increases. The number density of blobs is believed to become large through thermal instabilities in the arm of spiral galaxies. Then, interstellar clouds may easily form through Jackson's instability at the galactic arms even when the so-called galactic shock is damped. Due to this instability, the blobs may be expected to stick together through collisions where the blobs aggregate, because the m.f.p. among the dense blobs is expected to be very short. Thus, we propose a new model of interstellar cloud formation: *Fluidized-bed instability model.* This may be confirmed observationally when the HI column density oscillation is found to be as described in Kamaya (1997c).

References

Jackson, R., 1963, *Trans. Inst. Chem. Eng.*, **Vol.41** 13
Kamaya, H. 1996, *Astrophysical Journal*, **Vol.465** 769
Kamaya, H. 1997a, *Astrophysical Journal*, **Vol.474** 308
Kamaya, H. 1997b, *Astrophysical Journal*, **Vol.480** 694
Kamaya, H. 1997c, *Astrophysics and Space Science*, **Vol.253** 199
Kamaya, H. & Shchekinov, Yu. A. 1997 *Astrophysical Journal*, **Vol.486** 840
Kamaya, H. & Shchekinov, Yu. A. 1998 *Astrophysical Journal*, **Vol.501** (in press)
Kudoh, T., & Shibata, K., 1997 *Astrophysical Journal*, **Vol.476** 632
Yabe, T., & Takei, E. 1988 *Journal of Physical Society of Japan*, **Vol.57** 2598

MODELING OF A GALACTIC SPIRAL ARM AS A BORE-SHOCK COMBINATION

M.A. MARTOS[1] AND D.P. COX[2]
1. Instituto de Astronomía, UNAM
Apdo. Postal 70-264, 04510 México, D.F., México
2. University of Wisconsin-Madison, Department of Physics
1150 University Ave., Madison, WI 53706, USA

Abstract. We consider the interarm-to-arm transition for gas flow in the galactic disk, modeled as a thick, magnetized, cloudless layer of gas in hydrostatic equilibrium with external gravity from stars, and having parameters appropriate to the solar neighborhood. We neglect the self-gravity of the gas, shear, and radial variations in gravity. We show that such a transition, if supersonic, must present characteristics of both a hydraulic jump (or bore) and a shock. Our numerical simulations confirm this prediction. Modeling the spiral perturbation as a local one, we find that gas passing through it experiences sudden vertical acceleration upstream the arm, giving rise to a shock above the large vertical structure created, midplane density enhancements, and downflow regions behind the jump, as well as, sometimes, secondary jumps. Gravity waves generated in the thick disk appear to promote the formation of the marked density enhancements in the midplane

1. Results

If there is a spiral Galactic shock and in the post-shock gas the pressure is greater than hydrostatic, the gaseous disk must expand and acquire characteristics of a hydraulic jump as well. So long as the interstellar medium as whole is stiffer than isothermal, the jump will lead to arms that are thicker than the inflowing interarm region. And, as in the more familiar hydraulic bores on Earth, there will be associated vertical and circulating motions.

In the Galactic case, the strong components of interstellar pressure in magnetic fields and cosmic rays will act to make the effective γ significantly greater than 1. In fact, for a magnetic field transverse to the motion, we find that the $\gamma = 2$ is the case that better describes the situation. Using a

S. M. Miyama et al. (eds.), Numerical Astrophysics, 107–108.

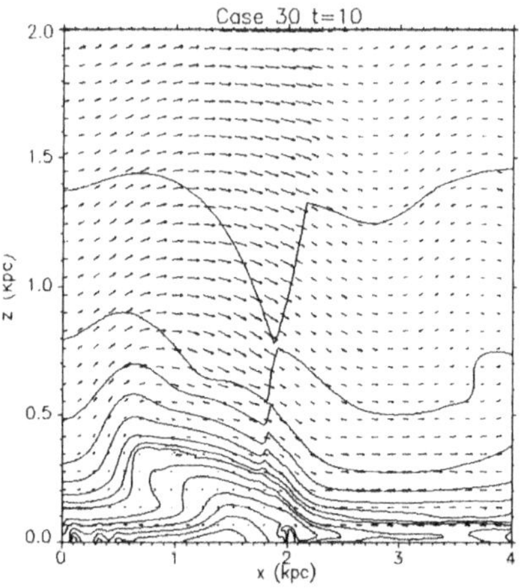

Figure 1. Density contours and velocity field for gas flowing from the left into a spiral arm. The scale of the maximum velocity arrow in the grid is 70 km s^{-1}. Elapsed time is 333 Myr.

thick disk model for the Galaxy (Martos and Cox 1994) we performed 2- and 2.5- D simulations with the ideal MHD code Zeus (Stone and Norman 1992a,b). Figure 1 shows a 400 $\times$ 200 zones simulation of gas entering the spiral arm. The x axis represents the radial coordinate along the Galactic midplane, whereas the vertical z coordinate represents distance above the Galactic midplane. The y axis is the direction of the magnetic field lines. The spiral potential is centered in the x axis with a 2 kpc extent; for local plausible values of its amplitude and the effective sound speed, the long-lived (in scales of 800 Myr) solution displayed here holds as a quasi- stable bore-shock solution only for a precise inflow speed normal to the arm in the spiral pattern frame of about 20 km s^{-1}. The essentials of the bore-shock solution are: the large increase in the effective thickness of the disk; the structure with the appearance of a breaking wave; and the diversion of the material encountering the bore to flow over it, to be accelerated at high z (of the order of 1 kpc), shock and fall back to the plane behind it.

With modest entry speeds of the order of 20−40 km s^{-1}, horizontal speeds of the order of 60−90 km s^{-1} are reached above the jump, and vertical downflows of −40 km s^{-1} for parameters appropriate to the solar neighborhood.

References

Martos, M.A. & Cox, D.P. (1994) MHD Experiments on the Thick Galactic Disk of Gas, *Numerical Simulations in Astrophysics*, ed. J. Franco, S. Lizano, L. Aguilar & E. Daltabuit, Cambridge Univ. Press., pp. 229–236.
Stone, J.M., & Norman, M.L. 1992a, ApJS, 80, 753
Stone, J.M., & Norman, M.L. 1992b, ApJS, 80, 791

FORMATION OF GMCS BY AGGREGATION IN A DISK WITH RIGID BODY ROTATION

T.J. ZHANG AND G.X. SONG
Shanghai Observatory, Chinese Academy of Sciences, China

Abstract. To study the process of formation of giant molecular clouds (GMCs) in the disk galaxy, N-body simulations were carried out based on the aggregation model. It was shown that differential rotation has a positive role in the formation of GMCs.

Key words: Galaxies: kinematics and dynamics — clouds and ISM: star formation — methods: numerical

1. Introduction

Giant molecular clouds (GMCs) in disk galaxies are main regions within which star formation takes place. For the formation of GMCs, generally speaking, there are two mechanisms, i.e., the coagulation process and the aggregation mechanism. The high resolution observation of CO shows that the emission peaks of CO are correlated, which supports the formation of GMCs by aggregation.

2. Model

A model with rigid rotation in the disk plane was established, in which $\Omega(r) = 0.05$ Myr^{-1} was taken. All clouds were randomly distributed in a ring between 3 kpc and 7 kpc with a height scale of 100 pc at the initial moment. Every cloud had a mass of $10^4\ M_\odot$ with a radius of 5.85 pc. The number of clouds was 120,000, and the effective radius of the clouds in collision was assumed to be twice the geometrical one. At the beginning, every cloud had symmetrical circular motion and dispersion velocities of 10 km s^{-1}, 7.5 km s^{-1}, and 5 km s^{-1} in the radial, circumferential, and disk thickness directions, respectively. The coefficient of restitution between the

S. M. Miyama et al. (eds.), Numerical Astrophysics, 109–110.

collision of clouds was 0.1. The self-gravity was cut off when the separation of two clouds was larger than 235 pc, and the percolation parameter was taken to be 30 pc to identify the aggregates of clouds. The disruption of GMCs induced by star formation and the spiral perturbation was excluded, and the model was run until 300 Myr.

3. Results and conclusion

At the beginning of the run, i.e., the first 150 Myr, the clumps gradually aggregated the individual clouds located near the forming aggregated clumps in both models. However, in the model with differential rotation, some clouds in the formed clumps escaped due to this differential rotation. Therefore, at the first 150 Myr, the efficiency of the formation of aggregated clumps for the rigid rotation model was higher than for the differential rotation model. With the elapse of time, the aggregated clumps exhausted the individual clouds nearby in the rigid rotation model and no further massive clumps formed. Conversely, in the model of differential rotation, this differential rotation promoted the collision of two or more aggregated clumps, and more massive clumps formed.

From the above results, it was concluded that differential rotation promotes the formation of GMCs.

Acknowledgement: This work was supported by the National Natural Science Foundation of China. In addition, TJZ is sincerely grateful for a partial fund by the HOPE Graduate Foundation of Shanghai Branch, Chinese Academy of Sciences.

References

Kwan,J., Valdes,F.(1987) The Spatial and Mass Distributions of Molecular Clouds and Spiral Structure, *Ap.J.*, **Vol. no. 315**, pp. 92–103

Song,G.X.(1991) Formation of Giant Molecular Clouds Associated with Spiral Structure, *Proceedings of Australian Astronomical Society*, **Vol. no. 9**, pp. 200–202

Rivola,A.R., Solomon,P.M. and Sanders,D.B. (1986) The Massachusetts-Stony Brook Galactic Plane CO Survey, *Ap.J.*, **Vol. no. 301**, pp. 119–123

Kahn,F.D., Song,G.X. (1994) Dynamics of Molecular Clouds on the Galactic Scale, *Ap.S.S.*, **Vol. no. 211**, pp. 127–154

Song,G.X. (1996) Role of the Self-Gravitation among Clouds in the Aggregation Formation of GMCs, *Annals of Shanghai Observatory Academia Sinica*, **Vol. no. 17**, pp. 234–239

Clifford,P., Elmegreen,B.G. (1983) A Collision Cross-section for Interactions between Magnetic Diffuse Clouds, *M.N.R.A.S.*, **Vol. no. 202**, pp. 629–646

Song,G.X. (1996) Bottom-up Model for the Formation of GMC's, *Journal of Korean Astronomical Society*, **Vol. no. 29**, pp. S165–S166

HEAD-ON COLLISIONS OF ISOTHERMAL CLOUDS

W.-T. KIM[1], S. S. HONG[1], S.-C. YOON[1], S. M. LEE[1], & J. KIM[1,2]
[1] *Department of Astronomy, Seoul National University, Korea*
[2] *Korea Astronomy Observatory, Korea*

Using our three-dimensional hydrodynamics code BTSPH (Binary Tree SPH; Kim *et al.* 1994), we have simulated the head-on collisions between two identical clouds with 8000 particles. As an initial cloud we take an isothermal polytropic sphere in hydrostatic equilibrium with an external medium. Following Nagasawa & Miyama (1987) we characterize the initial cloud by parameter A defined as $A \equiv (36\pi/125)[G^3 M^2/c_s^8]P_{\rm ext}$, where M is the cloud mass, c_s means the isothermal sound speed and $P_{\rm ext}$ represents the external pressure. The collision conditions are fully described by the parameter A and the relative collision Mach number defined as $M_{\rm rel} \equiv v_{\rm rel}/c_s$ with $v_{\rm rel}$ being the relative velocity of the two clouds. We have examined the collision outcomes for total 45 combinations of A and $M_{\rm rel}$.

The whole evolutionary process of the collision can be divided into four stages. In the beginning stage the two colliding clouds form a thin slab of density compression. Some slabs can be gravitationally unstable and collapse in this stage. In the second stage rarefaction waves develop from both boundaries of the slab, and propagate into the central part. Soon after the rarefaction waves reach the center, the slab expands along the direction of collision axis, while there is an inward motion in the plane perpendicular to the axis. In the final stage the slab either continues to expand and settles in an equilibrium with the external medium, or slows down its expansion and eventually collapses towards the collision axis.

Equi-column density contours and velocity fields at 6 selected time epochs can be seen in Figure 2 of CD-ROM for the case of $A = 0.5$ and $M_{\rm rel} = 13$. Figure 3 in the CD-ROM shows color-coded volume density in the three mutually perpendicular planes at 22 selected time epochs for the same combination of A and $M_{\rm rel}$.

The collision outcomes are classified in terms of A and $M_{\rm rel}$, and summarized in Figure 1, where the dotted-line marks Bonner-Ebert's critical configuration. In the figure the filled circles represent the cases where the

S. M. Miyama et al. (eds.), Numerical Astrophysics, 111–112.

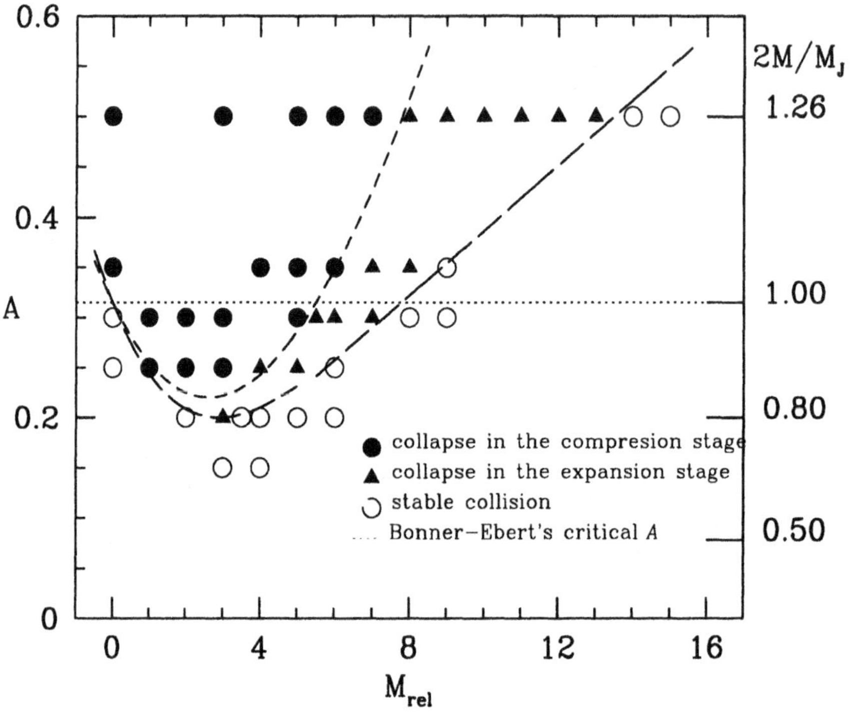

Figure 1. Final outcomes of head-on collisions.

gravitational collapse occurs in the compression stage, while the triangles are the cases of collapsing in the expansion stage.

Initially stable clouds with total mass less than the critical value become unstable after the collision, if the collision Mach number is less than ~8. In such relatively slow collisions, the duration of the compression stage becomes so long that the self-gravity has enough time to make the slab collapse. Once the slab starts to collapse, there is no rarefaction stage in the subsequent evolution. On the other hand, for high velocity collisions, the combined cloud always goes through the rarefaction and ensuing reverse expansion stages. The faster the collision velocity is, the stronger expansion the colliding clouds experience, which retards the development of gravitational instability. Even the clouds having a total mass greater than the critical mass can thus be stabilized by the collision.

References

Kim. W.-T., Hong, S. S., & Yun, H. S. 1994, JKAS, 27, 13
Nagasawa, M. & Miyama, S. M. 1987, Prog. of Theo. Phy., 78, 1250

SIMULATION OF THE AMBIPOLAR FILAMENTATION OF TURBULENT MAGNETIC FIELDS

M. FRANQUEIRA[1], M. TAGGER[2] AND A.I. GÓMEZ DE CASTRO[1]
[1]*Instituto de Astronomía y Geodesia, Fac. CC. Matemáticas, Universidad Complutense de Madrid, 28040 Madrid, Spain*
[2]*CEA, DSM/DAPNIA, Service d'Astrophysique (CNRS URA 2052) Saclay, 91191 Gif-sur-Yvette Cedex, France*

1. Description of the Mechanism

Magnetic fields play an important role in astrophysics and often dominate the behavior of ionized media. Their influence is carried out by hydromagnetic (MHD) waves which efficiently propagate perturbations, ensure a turbulent pressure or may even cause the development of instabilities (Arons & Max, 1975).

Kulsrud & Pearce (1969) showed that in the magnetized and weakly ionized interstellar medium MHD waves damp since they do not propagate in a frequency range (and thus scale) associated with the ambipolar diffusion. At low frequency (neutrals are well coupled to the ions, which are tied to the magnetic field lines), MHD waves propagate at the Alfvén speed defined by the total inertia (given by ions+neutrals). At high frequency (neutrals and ions are totally decoupled) Alfvén waves involve only the ions, which define a larger Alfvén velocity. In the intermediate range [the 'ambipolar range' (AR), between the ion-neutral and neutral-ion collision frequencies ν_{in} and ν_{ni}] the neutrals are imperfectly coupled to the ions and the waves are heavily damped by the ion-neutral friction.

When a magnetized and weakly ionized plasma presents small variations in the ionization fraction $Z = \rho_i/\rho_n$, MHD turbulence in the AR will amplify these variations (Tagger *et al.*, 1995). The turbulent velocity of the neutrals is higher in the more ionized regions, since they are better coupled to the ions. This results in a ponderomotive force (given by the average of the $\mathbf{v} \cdot \nabla \mathbf{v}$ term) driving them out of the ionized regions, so that the initial ionization inhomogeneities are strongly amplified. This is expected to result in a condensation of the flux tubes, producing a filamentary structure.

S. M. Miyama et al. (eds.), Numerical Astrophysics, 113–114.

2. Numerical Results

In order to test the efficiency of this non-linear mechanism, we wrote two numerical codes (1-D and 2-D) which simulate its behaviour. They are based on the same algorithms as the ZEUS-2D code (Stone & Norman, 1992) and solve the magnetohydrodynamics equations using the method of finite differences with a time explicit, multistep (operator split) scheme. We present here our first results.

In the 2-D code the magnetic field is initially along z, the 'vertical' direction; ρ_i and ρ_n decrease with z (supported by an ad-hoc gravity). We excite from $z = 0$ an Alfvén wave with a frequency ω of the order of ν_{ni} at mid-height: thus the wave will be strongly damped within the simulation box, but only weakly at low altitude where ν_{ni} is larger. We start with a ρ_n independent of y (the transverse direction, along which lie the perturbed velocities associated with the wave), but an ρ_i peaking at $y = 0$. In the first step reported here, we suppress the v_z in order to separate the problems. The perturbed velocity at $z = 0$ is of the order of half the Alfvén velocity. We do find a strong amplification (by a factor ~ 2 in the simulation presented here) of the ρ_i contrast along y, at the altitude where the wave is damped: the more ionized region shrinks, as expected. The magnetic flux is compressed with the ions.

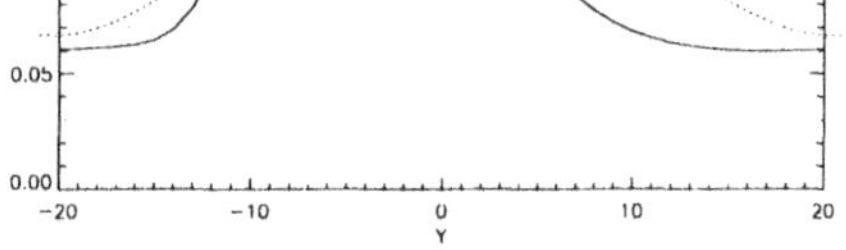

Figure 1. Initial (dotted line) and final (solid) ionization fraction profile at mid-height. The density contrast is amplified and the magnetic flux is compressed, in presence of strong Alfvenic turbulence. This should lead to a filamentation of the field lines.

References

Arons, J. & Max, C.E., 1975, ApJ. Lett. 196, L77
Kulsrud, R. & Pearce, W., 1969, ApJ. 156, 445
Stone, J.M. & Norman, M.L., 1992, ApJSS. 80, 753
Tagger, M., Falgarone, E., & Shukurov, A., 1995, A& A. 299, 940

THREE-DIMENSIONAL MHD SIMULATIONS: THE PARKER INSTABILITY UNDER A UNIFORM GRAVITY

JONGSOO KIM[1,2], S. S. HONG[2], DONGSU RYU[3], T. W. JONES[4]

[1] *Korea Astronomy Observatory, Taejon, Korea*
[2] *Department of Astronomy, SNU, Seoul, Korea*
[3] *Dept. of Astronomy & Space Science, CNU, Taejon, Korea*
[4] *Dept. of Astronomy, Univ. of Minnesota, Minneapolis, USA*

Using a newly developed isothermal MHD code (Kim et al. 1998a), we have performed three-dimensional, high-resolution (128^3 and 256^3 cells) simulations of the Parker instability under a variety of environments (e.g., under uniform or linear gravity and with or without rotation). Here, we report briefly the simulation using 256^3 cells under a uniform gravity without rotation. Details as well as other simulations will be reported elsewhere (Kim et al. 1998b). An initial magnetohydrostatic equilibrium system is composed of exponentially-decreasing isothermal gas and magnetic field (along the azimuthal direction $\hat{\mathbf{e}}_y$) under the uniform gravity (along the vertical direction $\hat{\mathbf{e}}_z$). The size of a computational box is set to be $(12H)^3$ where H is an e-folding scale height. $12H$ corresponds to the azimuthal wavelength of maximum linear growth. Random velocity perturbations are added to the initial equilibrium state.

The evolution of the instability can be divided into three phases (Fig. 1a; see also the figure in CD-ROM of three-dimensional density structures, magnetic field lines, and velocity field vectors at the three phases): linear ($0 \lesssim t \lesssim 35$), nonlinear ($35 \lesssim t \lesssim 50$), and relaxed ($50 \lesssim t$). During the linear phase, the perturbations grow exponentially with a preferred scale along the azimuthal direction but with the smallest possible scale along the radial direction ($\hat{\mathbf{e}}_x$). During the nonlinear phase, the instability is saturated and the structures become turbulent. Reconnection of magnetic field lines accompanies, which allows mass to cross the field lines. This, in turn, results in the redistribution of gas and magnetic field. The system approaches a new equilibrium in the relaxed phase. To see how gas is segregated from

S. M. Miyama et al. (eds.), Numerical Astrophysics, 115–116.

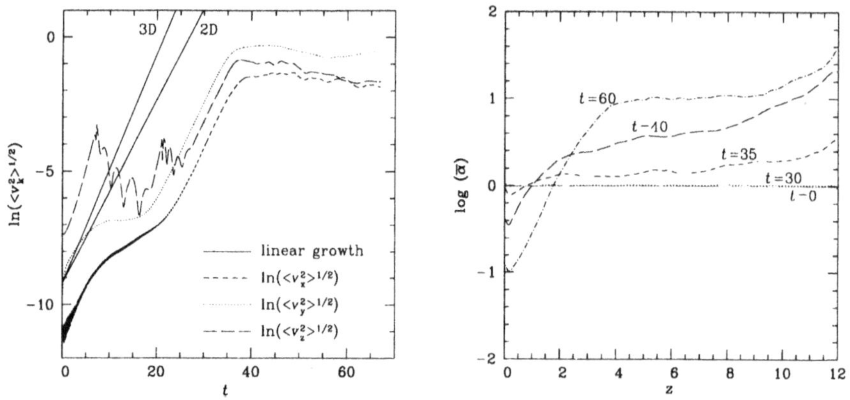

Figure 1. (a) Time evolution of the *rms* radial velocity, $<v_x^2>^{1/2}$, azimuthal velocity, $<v_y^2>^{1/2}$, and vertical velocity, $<v_z^2>^{1/2}$ (left panel). Two solid lines represent the maximum linear growth rates, 0.34 and 0.41, of the two- and three-dimensional perturbations, respectively. (b) The averaged ratio of magnetic pressure to gas pressure, $\bar{\alpha}$, at several epochs (right panel). The normalization units are a scale height H and an isothermal speed a.

magnetic field, we plot $\bar{\alpha}$, which is defined as

$$\bar{\alpha}(z;t) = \int_0^{12H}\int_0^{12H} \frac{B^2(x,y,z;t)}{8\pi a^2 \rho(x,y,z;t)} dxdy \;/\; \int_0^{12H}\int_0^{12H} dxdy, \qquad (1)$$

at several epochs in Figure 1b. Initially $\bar{\alpha}$ is equal to 1. In the linear phase, gas aggregates into magnetic valleys near the mid-plane ($z = 0$), while gas in other regions is evacuated. This decreases $\bar{\alpha}$ near the mid-plane but increases in the other regions. By the time the system has reached the relaxed stage, $\bar{\alpha}$ near the mid-plane has decreased by a factor of 10 (see the curve at $t = 60$). It is difficult to have the factor ten difference only through gas accumulation along the field lines. Reconnections have significantly contributed to the reduction of $\bar{\alpha}$. Filamentary structures form during the evolution, whose longest dimension is along vertical direction and shortest is radial, as expected from the linear stability analysis. The maximum enhancement factor in the surface density is less than 2. Since the radial dimension of the structures is too small and the density enhancement is too low, we conclude that it is difficult to regard the Parker instability *alone* as a viable formation mechanism for giant molecular clouds.

References

Kim, J., Ryu, D., Jones, T. W., & Hong, S. S. 1998a, ApJ, submitted.
Kim, J., Hong, S. S., Ryu, D., & Jones, T. W. 1998b, ApJL, submitted.

THE DISK-HALO INTERACTION VIA THE PARKER INSTABILITY

A. SANTILLAN, J. FRANCO AND M. A. MARTOS
Instituto de Astronomía & Cómputo Aplicado-DGSCA, UNAM, México

Using a collection of superbubbles (SB) and High Velocity Clouds (HVC) as perturbative agents, we model the disk-halo interaction in magnetized gaseous disks of spirals with 2-D numerical MHD simulations. The gaseous disk is assumed in magnetohydrostatic equilibrium, with the B-field oriented parallel to the disk (the x-axis). The gas properties and the gravitational acceleration within the disk are defined similar to those found at the solar circle in the Milky Way (see Martos & Cox 1994). Each perturbation is introduced individually, at time intervals of $\Delta t = 32$ Myr, and at random locations in the disk. The energy injected by each SB is 10^{53} erg, and the explosion centers are located within 100 pc from the midplane ($z = 0$). The impinging HVCs, on the other hand, have a variety of different densities and velocities (see Franco 1986, and Tenorio-Tagle et al. 1987), and collide with the disk at different angles. Thus, the energy and momentum injection per collision varies in each event. The numerical calculations were performed with the MHD code ZEUS-3D (Stone & Norman 1992a, 1992b), using the SGI ORIGIN-2000 of the Supercomputer Center at UNAM. The total extent of the computational grid, with 400×100 zones, is 16 kpc and 4 kpc in the x and z directions, respectively. The upper and lower boundaries are open, but the right and left boundaries are periodic. At midplane, the gas density and temperature are $n_0 = 1$ cm^{-3} and $T = 1.1 \times 10^4$ K, respectively, and the field intensity is $B_0 = 5$ G. Initially, gas is in hydrostatic equilibrium supported by magnetic and thermal pressures. The ratio of the magnetic-to-thermal energies, however, is not held constant with height but increases with increasing values of z (i. e., the disk is unstable to the Parker instability). The perturbations are introduced within a central box of the computational grid. Each SB has an initial energy density of 2.2×10^{13} erg/cm (equivalent to a spherical bubble with 10^{53} erg). HVCs have 120 pc in height and 200 pc long, a mass density of 8.6×10^{-2} g/cm, and different approaching velocities and incident angles. The B-field lines are distorted

S. M. Miyama et al. (eds.), Numerical Astrophysics, 117–118.

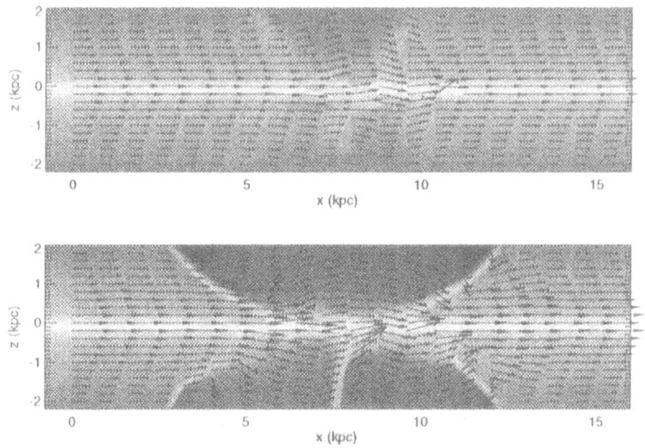

Figure 1. The plots show the density and magnetic field (indicated by arrows) structures at two different times: 193 Myr (upper panel) and 270 Myr

and compressed with each perturbation, increasing the field tension at the location of the event, and driving MHD waves that propagate to the rest of the disk and into the halo. All perturbations, then, create a complex (turbulent) network of flows that eventually trigger the Parker instability (Franco et al. 1995). The SBs expand mainly in the x-direction because the compressed B-field lines provide a strong pressure, reducing the SB growth along the z-axis. Hence, the SB energy is distributed by the MHD waves in a huge volume, but the SB mass is maintained within the disk in a well defined and small volume. In the case of HVC-disk collisions, the magnetic field prevents the cloud material from penetrating into the disk. Thus, the B-field provides an adequate coupling for the energy and momentum exchange between the disk and the halo but, for this restricted field geometry, also represents an effective shield that prevents a direct gas flow between the disk and the halo. Such a gas flow, however, can indeed occur via the Parker instability (see Figure 1), and the mass exchange between the disk and the halo is probably a complex two-step process. This work has been partially supported by grants from DGAPA-UNAM, CONACyT, and by a R&D grant from Cray Research Inc.

References

Franco, J. 1986, RMexAA, 12, 287.

Franco, J., Santillán, A. & Martos, M. A. 1995, in Formation of the Milky Way, ed. E. Alfaro & A. Delgado, Cambridge Univ. Press, 97.

Martos, M. A. & Cox, D. P. 1994, in Numerical Simulations in Astrophysics, ed. J. Franco, S. Lizano, L. Aguilar & E. Daltabuit, Cambridge Univ. Press, 229.

Stone, J. & Norman, M. 1992a,b, ApJS, 80, 753 and 791.

Tenorio-Tagle, G., Franco, J., Bodenheimer, P. & Rozyczka, M. 1987, A&A, 179, 219.

TWO-DIMENSIONAL NUMERICAL MHD SIMULATION OF MAGNETIC RECONNECTION IN GALAXY

Reconnection Triggered by Supernova and Parker Instability

S. TANUMA
Department of Astronomy, School of Science, Tokyo University
Hongo, Bunkyo 113-0033, Japan

AND

T. YOKOYAMA, T. KUDOH AND K. SHIBATA
National Astronomical Observatory
Osawa, Mitaka 181-8588, Japan

X-ray astronomy satellites detected strong X-ray emission along Galactic ridge, called Galactic Ridge X-ray Emission (GRXE), which is thermal emission of super-hot diffuse plasma. It has been revealed that GRXE plasma is composed of mainly hot (7 keV) component and cool (0.8 keV) component (Kaneda et al. 1997). Origin and nature of the hot component is not known while the cool component is made of supernova remnants. We recently showed that a supernova (or other point explosion) and the Parker instability trigger magnetic reconnection, and we proposed a possible mechanism to explain the origin of GRXE (Tanuma et al. 1998; Yokoyama & Shibata 1995). We solved 2D resistive MHD equations numerically to examine the magnetic reconnection triggered by a supernova and Parker instability. We find that the reconnection heats the interstellar gas from 0.7 keV to 8 keV if local interstellar magnetic field strength is 30 μG. The reconnection accelerates and heats the interstellar medium. Magnetic islands are generated and confine the heated plasma. Thus, we propose that the reconnection is triggered by a point explosion, Parker instability, the collision of interstellar clouds etc., and reconnection can explain X-rays from Galactic ridge and halo.

References

Kaneda, H. et al., 1997, *ApJ*, 491, 631
Tanuma, S., Yokomama, T., Kudoh, T., Matsumoto, R., Shibata, K. & Makishima, K. 1998, Proc. IAU Symp., 188, '*Hot Universe*' (in press)
Yokoyama, T. & Shibata, K., 1995, *Nature*, 375, 42

S. M. Miyama et al. (eds.), Numerical Astrophysics, 119–120.

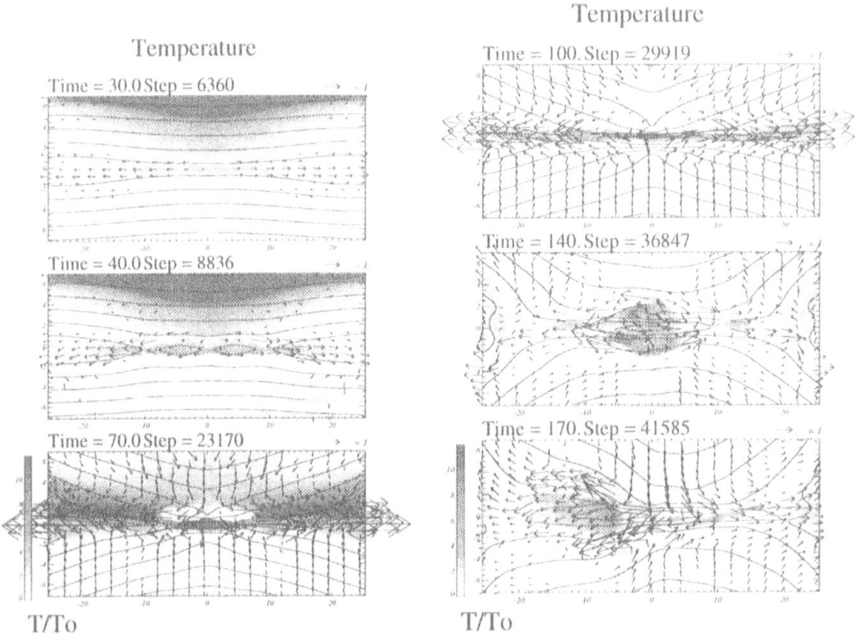

Figure 1. 2D numerical MHD simulation of the magnetic reconnection triggered by a supernova or other point-explosion.

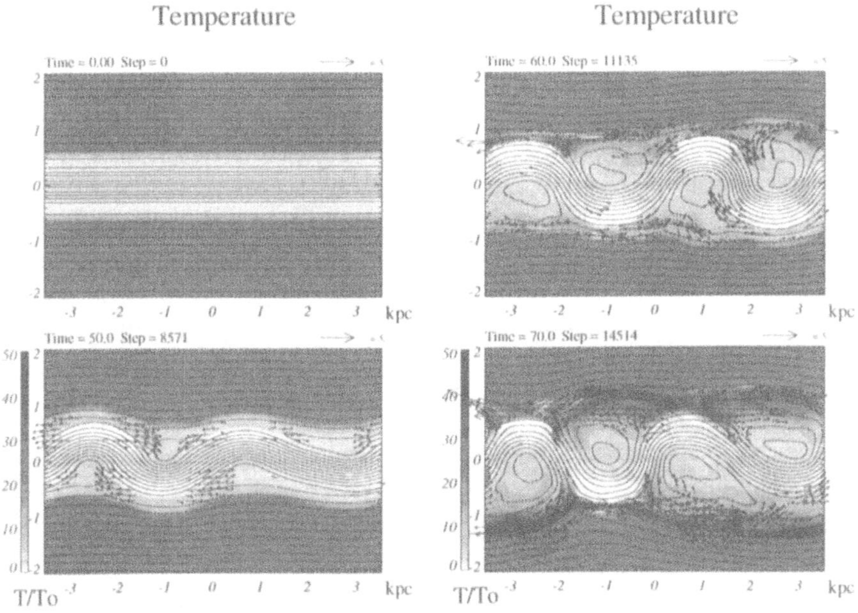

Figure 2. 2D numerical MHD simulation of the magnetic reconnection triggered by the Parker instability.

THREE-DIMENSIONAL MHD SIMULATIONS OF SELF-GRAVITATING MAGNETIZED GAS LAYER UNDER EXTERNAL PRESSURE

MICHIHISA UMEKAWA
Graduate School of Science and Technology, Chiba University, 1-33 Yayoi-cho Inage-ku Chiba 263-8522, Japan.

AND

RYOJI MATSUMOTO
Department of Physics, Faculty of Science, Chiba University

Pressure bounded gaseous slab plays important roles in sequential star formation in molecular clouds (Elmegreen and Elmegreen 1978). Through three-dimensional numerical simulations, we studied the effects of external pressure and magnetic fields on the stability of the gaseous slab. By two-dimensional nonlinear simulations of self-gravitating gaseous slab, we found: (1)When the external pressure is low, the gaseous slab fragments into collapsing high density clumps due to the Jeans instability. (2)When the external pressure is comparable to the mid-plane pressure of the slab, *incompressible mode* instability (Lubow and Pringle 1993) dominates. (3)Jeans stable clouds created by the above mechanism can merge by mutual gravitational attraction. Once the cloud mass exceeds the Jeans mass, the cloud collapses and leads to star formation.

Here, we present the results of three-dimensional simulations to study the effects of magnetic fields on the gravitational instability of the gaseous slab. We assume that plane parallel cold isothermal gaseous slab is confined by external hot, high-pressure medium. We neglect the effects of thermal conduction, viscosity, and the effects of rotation. Basic equations are self-gravitational MHD equations. We simulate two models, a non-magnetized model (model 1) with high external pressure $[P_{\rm ext}/(\rho_{00}c_s^2) = 0.91$ where $P_{\rm ext}$, ρ_{00} and c_s are the external pressure, mid plane density of the slab and sound velocity, respectively], and a magnetized model (model 2) which has high external pressure and uniform magnetic field which is parallel to the y-direction and whose plasma β is 1.0 at the mid plane of the initial slab. The left panel of Figure 1 shows a numerical result for a Jeans sta-

S. M. Miyama et al. (eds.), Numerical Astrophysics, 121–122.

ble gaseous slab. Because of the growth of the instability, the gaseous slab fragments into spherical clumps. This result is consistent with 2D simulation results. In model 2, the slab fragments into rod-like structure parallel to magnetic field lines. Figure 2 shows the time evolution of velocity at $(x, y, z) = (0.0, 3.1, 0.0)$. After $t = 10H/c_s$ where H is the scale height, velocity perturbation grows exponentially and saturates when $|v|/c_s \sim 1$. The growth rate of instability in magnetized slab (model 2) is comparable to that of the unmagnetized slab (model 1).

Numerical computations were carried out on Fujitsu VX/4R at NAOJ.

References

Elmegreen, B. G. and Elmegreen, D. M. (1978) *ApJ*, **220**, 1051–1062
Lubow, S. H. & Pringle, J. E., (1993) *MNRAS*, **263**, 701–706

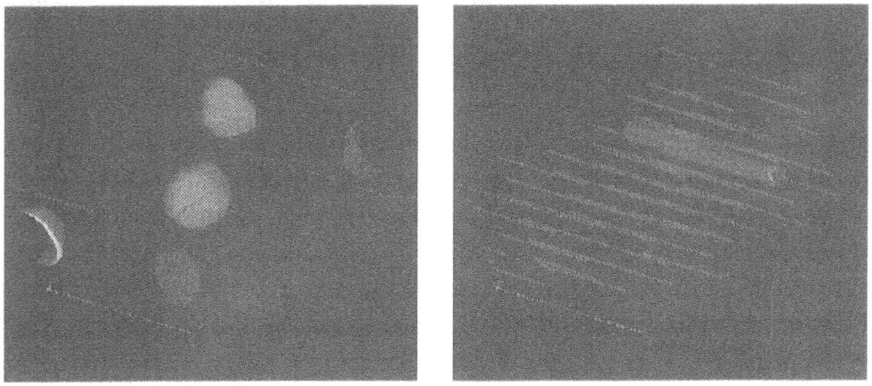

(a) (b)

Figure 1. Density structure in (a) model 1 (unmagnetized slab) at $t = 17.6$, white surfaces denote the isosurface $\rho = 1.01\rho_{00}$ and (b) model 2 (uniformly magnetized slab) at $t = 20.0$, white surface shows the isosurface $\rho = 1.20\rho_{00}$.

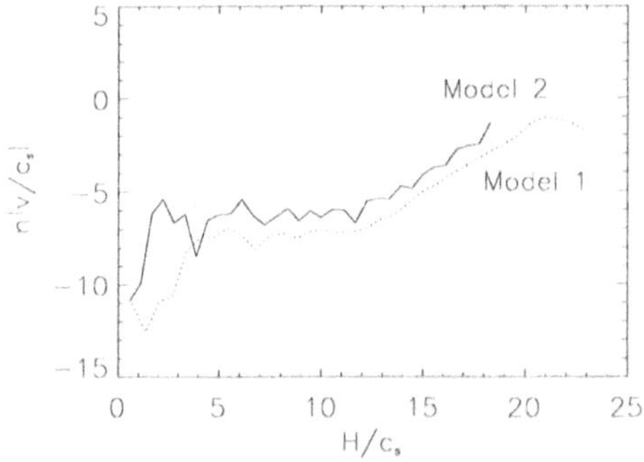

Figure 2. Growth of velocity perturbation for two models. The dotted curve and the solid curve show velocity at $(x, y, z) = (0.0, 3.1, 0.0)$ for model 1 and model 2, respectively.

FORMATION OF DISKS AND BINARIES

P. BODENHEIMER
UCO/Lick Observatory
University of California, Santa Cruz, CA 95064, USA

HAROLD W. YORKE
Jet Propulsion Laboratory
Pasadena, CA 91109, USA

AND

ANDREAS BURKERT
Max-Planck-Institute for Astronomy
D-69117 Heidelberg, Germany

1. Introduction

Two possible outcomes are considered for the collapse of a rotating core of a molecular cloud. For the first case, two-dimensional axisymmetric hydrodynamical collapse calculations with radiative transfer are presented to study the formation of a central star and a disk. For the second case, three-dimensional calculations provide insight on the process of the formation of binary and multiple systems of protostars. In general when a protostar or fragment collapses it passes through three phases. The first, the isothermal phase, corresponds to low-density gas which is optically thin to infrared radiation. The second, adiabatic, phase starts in the central regions of the protostar or fragment when it becomes dense enough to be optically thick; thus as it compresses it heats. The third, accretion, phase starts when the central regions of the collapsing fragment reach hydrostatic equilibrium and form the core of a star. Remaining infalling material accretes onto this central object, either directly or through a disk. In the case of the axisymmetric solutions presented here, the evolution passes through all three phases. The 3-D simulations, however, are limited mainly to the isothermal phase.

S. M. Miyama et al. (eds.), Numerical Astrophysics, 123–130.

2. Simulations of Disk Formation

A number of important issues should be considered in this connection, many of which are still unresolved. (1) What are the initial conditions for the formation of a star-disk system, as opposed to a binary? (2) What is the predicted appearance of the emergent spectrum during the early phases of disk formation and evolution? (3) What are the important mechanisms for angular momentum transport in a disk? At what stage is each dominant? What are the time scales? (4) What are the implications of the angular momentum transport process with regard to planet formation? (5) Will a gravitationally unstable disk formed from cloud collapse fragment, or will it transfer mass to the central object through the action of gravitational torques?

2.1. ASSUMPTIONS AND METHOD

Detailed simulations of the collapse of rotating clouds with radiation transport have previously been published by Yorke, Bodenheimer, and Laughlin (1993, 1995). The example presented here is based on a nested–grid computer code with improved physics and longer evolutionary time scale as compared with the earlier work. The code includes two-dimensional Eulerian hydrodynamics, radiation transport in the flux-limited diffusion approximation, no magnetic fields, and a radiative opacity based on the properties of dust grains, including silicates, ice-coated silicates, and carbon. When the disk becomes gravitationally unstable, as indicated by a minimum Toomre Q value of 1.5 or less, ordinary viscosity, based on the α–law prescription, is included according to the procedure described in Różyczka, Bodenheimer, & Bell (1994) and Laughlin & Bodenheimer (1994). However in this case we assume that the gravitational instability results in disk evolution and angular momentum transport such that the condition of marginal gravitational instability is maintained. Thus we adjust the value of α so that the minimum Q value remains between 1.3 and 1.5.

The innermost grid point has a size of a few AU; thus the central stellar core is not resolved. This region is modelled as a central star plus disk with luminosity equal to the sum of the contraction luminosity of the star and the accretion luminosity of disk onto star. This luminosity serves as an inner boundary condition for the radiation transfer in the grid. At selected times, frequency-dependent radiation transfer is carried out so that the emergent infrared spectrum and isophotal contours of the protostar can be determined. A ray-tracing solution of the transfer equation is obtained on a grid of lines of sight through the object, at 64 frequencies, including true absorption and scattering.

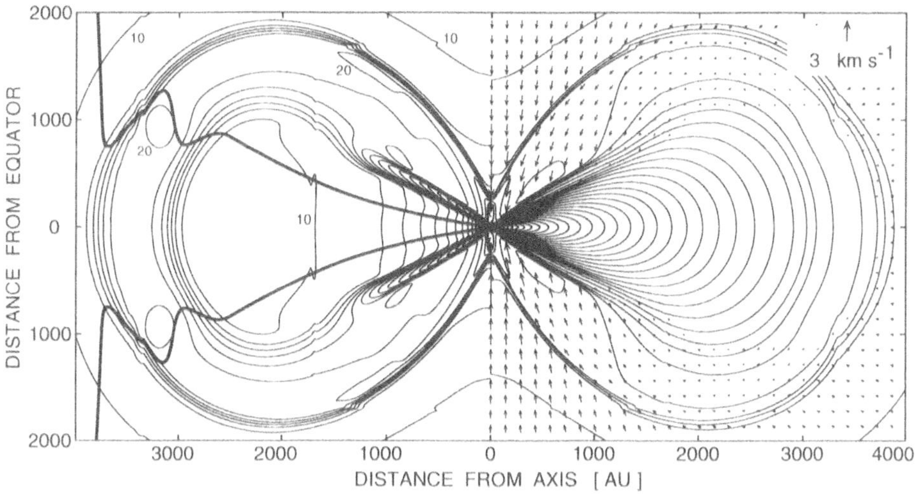

Figure 1. Disk structure in a protostar of 2 $M_\odot$ after a time of 585000 yr. (*Right*): Contours of equal density and velocity vectors in a meridional plane. (*Left*): Contours of equal gas temperature. The pair of thick lines indicate one density scale height above the equatorial plane. From Yorke & Bodenheimer (1998).

2.2. RESULTS

One example will be presented here; further details may be found in Yorke & Bodenheimer (1998) and Yorke & Kaisig (1995). The initial condition is a uniformly rotating cloud of 2 $M_\odot$, radius 2×10^{17} cm, density distribution $\rho \propto r^{-2}$. The initial ratios of thermal and rotational energies, respectively, to the absolute value of the gravitational potential energy are $\alpha = 0.39$ and $\beta = 0.008$. The specific angular momentum $R^2\Omega$ at the outer edge is 4×10^{21} cm^2 s^{-1}; all of these values are consistent with observations. The calculation is carried out on four fixed Eulerian nested grids, 124 by 124 zones each, with a factor 2 increase in resolution per grid. On the innermost grid the zone size is 2×10^{14} cm.

During most of the evolution the protostar is divided into three regions: the unresolved stellar core, the equilibrium disk, and the infalling envelope. An example of the disk structure is shown in Figure 1. Outer and inner accretion shocks are evident interior to a radius of 1000 AU. The disk becomes gravitationally unstable, and mass accretes onto the central star at a typical rate of 2×10^{-6} $M_\odot$ yr^{-1}, generating a luminosity of ≈ 10 $L_\odot$. The outer disk radius expands considerably with time as a result of angular momentum transport and accretion. After a protostellar lifetime (a few $\times 10^5$ yr) the disk mass has decreased to about 35% of the total mass and the accretion rate has also dropped considerably. Thus the calculated disk mass is higher than those observed in young stellar objects, and we conclude that some process of angular momentum transport, other than gravitational instability, must also operate during the early phases of disk evolution.

The infrared spectrum of the model shown in Figure 1 is strongly dependent on viewing angle. By this time most of the infalling material along the polar axis has collapsed onto the central star, the dust optical depth is low, and if one observes from this direction one can detect the central object and inner disk. Some optical radiation is present, and the spectrum is fairly flat in a $[\log \nu, \log(\nu L_\nu)]$ diagram from 0.8 to 20 μm. On the other hand, if one views in the equatorial plane, the central object is highly obscured and the total flux is about a factor 30 less than that observed pole-on. The spectrum has two peaks, one at about 100 μm, and another, smaller one in the optical and near infrared. This second peak is caused by scattered light from the central star and inner disk. We conclude that infrared searches for protostars are biased in the sense that pole-on objects are the most likely ones to be discovered.

3. Binary Formation by Fragmentation

The general goal of three-dimensional calculations is to explain the observed properties of binary systems, such as their frequency and their distributions of periods, secondary masses, and eccentricities. A number of formation mechanisms have been discussed (reviewed by Boss 1993a and Bodenheimer et al 1993), including fragmentation during protostar collapse, disk fragmentation, cloud–cloud collisions, capture, and fission. Here we consider only the first of these processes.

Various physical assumptions have been employed in collapse calculations, including an isothermal equation of state, an equation of state with cooling, an adiabatic or polytropic equation of state, and radiative transfer in the flux-limited diffusion approximation or the Eddington approximation. Typical initial conditions include the following assumptions:

- Cloud has an assumed geometry (spherical, or cylindrical, or oblate, or prolate

- Cloud has no infall motion and is rotating uniformly

- Cloud has a mass of about 1 solar mass and a radius of 0.05 parsec

- Cloud has a distribution of density that is somewhat centrally condensed, in agreement with observations of cores of molecular clouds

- Cloud has an initial assumed density perturbation, for example random noise or a 10% m = 2 mode

The typical 3-D hydro calculations show the formation of a binary system on an eccentric orbit with separation in the range 10–100 AU, or a multiple system on the same scale. For example, Boss (1996) and Klapp & Sigalotti (1998) have found systems with 5–10 small fragments. Miyama (1992) did adiabatic collapse calculations and found agreement with the analytical estimate for clouds of initial uniform density (with $\gamma = 7/5$), which shows that fragmentation should take place only if $\alpha < 0.09\beta^{0.2}$. Bonnell & Bate (1994), in an isothermal SPH calculation, saw the formation of an initial binary of unequal masses, evolution toward equal masses, and then formation of a second pair of fragments as a result of the spiral waves induced by the original binary. The results of many of the early fragmentation calculations need to be reexamined, because the numerical resolution was too coarse to satisfy the Jeans condition discussed by Truelove et al (1997). Rather than attempt to review a wide variety of other results, we focus on four specific questions:

- 1. A typical observed molecular cloud core is centrally condensed and has $\alpha \approx 0.4$, $\beta \approx 0.01$. Will it fragment?

- 2. Does fragmentation occur from an initial power–law density distribution (say $\rho \propto r^{-1}$) with uniform rotation?

- 3. Do finite-difference codes and SPH codes give the same results on the same fragmentation problem?

- 4. Do fragmentation results depend strongly on the form and amplitude of the assumed initial perturbation?

The answer to the first question is most clearly illustrated by a series of results by Boss (1993b). His collapse calculations, which include radiation transport, start from a Gaussian prolate density distribution with a ratio of central density to edge density of about 20. A random density perturbation is imposed. A set of models is constructed, corresponding to various points in the (α, β) plane to determine which clouds fragment and which remain as single objects. If the ratio of axes in the prolate cloud is 2:1, the cloud with the given initial condition fragments. If the ratio, on the other hand, is 1.5:1, the cloud collapses but remains single. Thus the fate of a "standard" cloud may depend sensitively on initial conditions and the numerical technique, and further studies of this case should be done.

The second and third questions were considered by Burkert, Bate, & Bodenheimer (1997). In connection with the second question it is important to know what the actual degree of central concentration of a cloud is, just before it begins to collapse. Observationally, recent work by André, Ward-Thompson, & Motte (1996) shows that pre-collapse cloud cores have density distributions close to $\rho \propto r^{-2}$ in the outer parts, but a significant fraction of the inner part of the cloud has $\rho \propto r^{-1}$. Theoretically, calculations of the evolution of clouds, as controlled mainly by ambipolar diffusion, before collapse (Tomisaka, Ikeuchi, & Nakamura 1990; Lizano & Shu 1989; Ciolek & Mouschovias 1994; Basu & Mouschovias 1994) show that the density distribution just before collapse starts is quite close to $\rho \propto r^{-2}$. With such a degree of central condensation and with uniform rotation (which is likely because of magnetic braking) fragmentation does not appear to occur. But what about the r^{-1} law? Burkert et al (1997) were able to show that, with this initial density distribution, with an initial density perturbation of the form $m = 2$ and a 10% amplitude, and with $\alpha = .25, \beta = .23$, fragmentation into four objects occurs in the inner region of the cloud, on a scale of 100 AU. The calculation was done with an Eulerian grid code, with fixed nested grids such that the resolution on the innermost grid was $10^{-3}R$, where R is the total radius of the cloud. The same problem was calculated by use of an SPH code with 200 000 particles. The resolution element in such a calculation is the smoothing length, which varies through the volume; in this case it was adjusted to be $10^{-3}R$ at a density of 10^{-12} g cm^{-3}, which is about the point where fragmentation starts. However the spatial resolution of the two codes can be quite different at other densities. Nevertheless, the fragmentation observed in the SPH calculation was very similar to that in the grid code. Only differences in small details were observed (see Burkert et al 1997). One can conclude that a typical observed density distribution in a cloud core is unstable to fragmentation, from which one can understand the fact that a large fraction of young stars are observed to be in binary systems. Note, however, that the actual simulations were carried out for values of α and β that are somewhat different from those typically observed.

For the answer to the fourth question we report on the results of some new calculations. The initial cloud had a mass of 1 $M_\odot$, a radius of 5 $\times 10^{16}$ cm, isothermal temperature of 10 K, and a centrally condensed Gaussian density distribution, corresponding to $\alpha = 0.26, \beta = 0.16$. The main grid has $64 \times 64 \times 64$ Cartesian zones. Embedded in it are three subgrids with $128 \times 128 \times 64$ zones in the x, y, and z directions, respectively. The x–values of the outer edges of the subgrids were 2.5 $\times 10^{16}$, 6.25×10^{15}, and 1.56×10^{15} cm. The resolution on the innermost grid is $R/2000$. The same configuration was run with three different initial density perturbations. The first run had no initial perturbation, other than that induced by the grid. A Fourier analysis of the growth of the modes in the azimuthal direction shows that the $m = 4$ mode grows most rapidly, followed closely by $m = 8$. The $m = 1$ and $m = 2$ perturbations stay at a very low level. Thus fragmentation is likely to be numerically induced. The result is shown in the left panel of Figure 2. A high-density condensation forms in the very center, then an

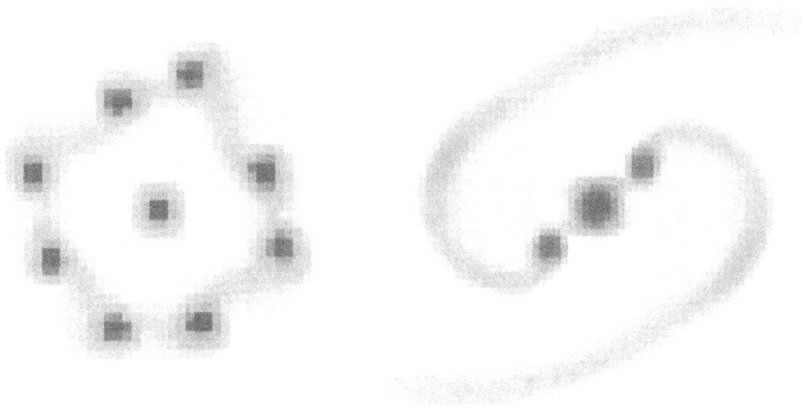

Figure 2. Gray-scale representation of the density distribution in the equatorial plane of a fragmenting cloud (*left*) with no initial density perturbation and (*right*) with an initial $m = 2$ perturbation of 10% amplitude. Black represents highest density. The box dimensions are 1.5×10^{15} by 1.5×10^{15} cm.

axisymmetric ring builds up around it, and the ring fragments into eight symmetrically located pieces. At the time corresponding to the figure the symmetry of the ring has just begun to be broken as the fragments in the unstable configuration gravitationally interact. The result is quite different if an initial $m = 2$ perturbation with 10% amplitude is applied. The Fourier analysis shows that the $m = 2$ amplitude remains almost constant in time, at 10%. The $m = 4$ mode, which is probably at least in part numerical, grows quite rapidly but remains at least an order of magnitude in amplitude below that for $m = 2$. Other modes remain at a low level until the fragmentation begins, at which time all modes increase rapidly in amplitude. The result is shown in the right panel of Figure 2. A central condensation forms, and the initial bar-like perturbation wraps up into a spiral pattern, generating two additional fragments in the process. Further details for this case can be found in Burkert & Bodenheimer (1996). The third case was run with a random initial density perturbation such that the typical mode started with an amplitude of 10^{-3} – 10^{-4}. Initially the modes grow slowly, with $m = 1$ and $m = 2$ dominating; later all modes grow rapidly with $m = 3$ slightly ahead of the rest. The result, which is not shown, is similar to that for the first run, except that the ring fragments into three pieces rather than 8. Thus the results at the time of initial fragmentation are quite different for the three runs with different initial perturbations. However no conclusions can be drawn concerning the longer-term evolution.

4. Conclusion

Some general conclusions can be summarized concerning the 3-D collapse results.

• 1. Fragmentation calculations have many possible outcomes, including binary formation, fragmentation in a surrounding disk induced by an initial binary, formation of a small cluster, formation of filaments which fragment, or the formation of a binary plus low-mass fragments.

• 2. No fragmentation calculation has been carried out long enough so that most of the material in the original cloud has collapsed to the equatorial plane. Not all of this material will eventually end up in fragments, but the final outcome in these systems, as influenced by captures, mergers, accretion, and escape of gas and stars, is still not known.

• 3. The results may depend qualitatively on numerical resolution, particularly if the Jeans condition is not satisfied. Calculations must show numerical convergence. The initial perturbation has a strong effect on the initial phase of fragmentation.

• 4. An initially highly centrally condensed cloud with uniform rotation fragments in the case $\rho \propto r^{-1}$, but the situation for steeper power laws is still unclear.

References

André, P., Ward-Thompson, D., & Motte, F. 1996, A&A, 314, 625
Basu, S., & Mouschovias, T. Ch. 1994, ApJ, 432, 720
Bodenheimer, P., Ruzmaikina, T., & Mathieu, R. 1993, in Protostars and Planets III, eds. E. Levy & J. Lunine (Tucson: Univ. of Arizona Press), p. 367
Bonnell, I., & Bate, M. R. 1994, MNRAS, 269, L45
Boss, A. P. 1993a, in The Realm of Interacting Binary Stars, eds. J. Sahade, G. McCluskey, & Y. Kondo (Dordrecht: Kluwer), p. 355
Boss, A. P. 1993b, ApJ, 410, 157
Boss, A. P. 1996, ApJ, 468, 231
Burkert, A., & Bodenheimer, P. 1996, MNRAS, 280, 1190
Burkert, A., Bate, M. R., & Bodenheimer, P. 1997, MNRAS, 289, 497
Ciolek, G. E., & Mouschovias, T. Ch. 1994, ApJ, 425, 142
Klapp, J., & Sigalotti, L. DiG. 1998, preprint
Laughlin, G., & Bodenheimer, P. 1994, ApJ, 436, 335
Lizano, S., & Shu, F. H. 1989, ApJ, 342, 834
Miyama, S. 1992, PASJ, 44, 193
Różyczka, M., Bodenheimer, P., & Bell, K. R. 1994, ApJ, 423, 736
Tomisaka, K., Ikeuchi, S., & Nakamura, T. 1990, ApJ, 362, 202
Truelove, J. K., Klein, R. I., McKee, C. F., Holliman, J. H., Howell, L. H., & Greenough, J. A. 1997, ApJ, 489, L179
Yorke, H. W., & Bodenheimer, P. 1998, ApJ, submitted
Yorke, H. W., Bodenheimer, P., & Laughlin, G. 1993, ApJ, 411, 274
Yorke, H. W., Bodenheimer, P., & Laughlin, G. 1995, ApJ, 443, 199
Yorke, H. W., & Kaisig, M. 1995, Comp. Phys. Comm., 89, 29

GRAVITATIONAL COLLAPSE AND FRAGMENTATION IN MOLECULAR CLOUDS WITH ADAPTIVE MESH REFINEMENT HYDRODYNAMICS

RICHARD I. KLEIN, R. T. FISHER, C. F. MCKEE,
J. KELLY TRUELOVE
Astronomy Department, University of California
601 Campbell Hall
Berkeley, CA 94720, USA

Abstract. We describe a powerful methodology for numerical solution of 3-D self-gravitational hydrodynamics problems with extremely high resolution. This code utilizes the technique of local adaptive mesh refinement (AMR), employing multiple grids at multiple levels of resolution. These grids are automatically and dynamically added and removed as necessary to maintain adequate resolution. This technology allows for the solution of problems in a manner that is both more efficient and more versatile than other fixed and variable resolution methods. The application of this technique to simulate the collapse and fragmentation of a molecular cloud, a key step in star formation, is discussed. Such simulations involve many orders of magnitude of variation in length scale as fragments form. In this paper we briefly describe the methodology and present an illustrative application for nonisothermal cloud collapse. We describe the numerical Jeans condition, a criterion for stability of self-gravitational hydrodynamics problems. We show the first well-resolved nonisothermal evolutionary sequence that leads to the formation of a binary system consisting of protostellar cores surrounded by distinct protostellar disks. The scale of the disks, of order 100 AU, is consistent with observations of gaseous disks surrounding single T-Tauri stars and debris disks surrounding systems such as β Pictoris.

RIK is also affiliated with Lawrence Livermore National Laboratories, L-023, Livermore, California, 94550.

S. M. Miyama et al. (eds.), Numerical Astrophysics, 131–140.

1. INTRODUCTION

Star formation ultimately involves the collapse of an interstellar molecular cloud core from an initial density of $\rho \leq 10^{-19}$ g cm^{-3} and size $\simeq 10^{17}$ cm to a final young stellar object of density $\rho \geq 10^{-1}$ g cm^{-3} and size $\simeq 10^{11}$ cm. The collapse of a portion of a cloud may lead to fragmentation, which is crucial to establishing key parameters of binary stars: the distributions of mass ratios, periods, and orbital eccentricities. It may also be essential to the formation of larger groups of stars and to the determination of the stellar initial mass function. However, this enormous dynamic range presents a formidable obstacle to obtaining an accurate numerical solution, as the flow must remain well-resolved throughout the evolution. Fixed-resolution methods cannot be used to simulate such a 3-D collapse in a practical amount of time using current computers.

In order to treat the enormous dynamic range presented by gravitational collapse calculations and address the problems associated with alternative methods such as smoothed particle hydrodynamics (SPH) and static nested-grid codes (Burkert & Bodenehimer 1993), we have developed a 3-D self-gravitational adaptive mesh refinement (AMR) code. Specifically, SPH suffers from two major drawbacks. First, it is typically very diffusive, and has great difficulty tracking shocks and contact discontinuities accurately in multi-dimensional flows. Second, although it is an adaptive method, it is only adaptive in the sense that the local smoothing length h tracks mass in a Lagrangian sense, i.e. in 3-D, $h \propto \rho^{-1/3}$. Furthermore, in the context of isothermal collapse, in which the Jeans length λ_J scales as $\lambda_J \propto \rho^{-1/2}$, the effective resolution, specified by the ratio λ_J/h varies as $\rho^{-1/6}$ – *the effective resolution of the calculation is degraded as collapse proceeds, and if the collapse is not arrested first, the calculation will eventually become underresolved.* Given that in SPH one cannot choose *a priori* which regions of the flow need further refinement, one's only recourse to increasing the resolution of a calculation is to increase the total number of particles. Lastly, static nested-grid codes suffer from the drawback of imposing a static set of grids from the beginning of the calculation, and hence are unable to resolve arbitrarily-located fragments.

Our AMR method dynamically resizes and repositions grids and inserts new, finer ones within them according to adjustable refinement criteria (unlike the static nested grid scheme of Burkert & Bodenehimer 1993). We have applied our method to great effect; prior to the addition of self-gravity, we have used AMR to study astrophysical problems including the interaction of supernova blastwaves with interstellar clouds (Klein, McKee, & Colella 1994; Klein, & McKee 1994; Klein, McKee, & Woods 1995), X-ray heated coronae and winds from accretion disks (Woods *et al.*, 1996), and the collision of interstellar clouds (Klein & Woods, 1998) with unprecedented high resolution.

We present the methodology behind our 3-D AMR self-gravity code and describe application of our work to the collapse and fragmentation of molecular clouds. Our discussion closely follows Truelove *et al.* (1998) and Klein (1998).

2. METHODOLOGY

2.1. SELF-GRAVITATIONAL HYDRODYNAMICS

The basic governing equations of our model are the Euler equations of hydrodynamics in 3-D, including effects of self-gravitation, and Poisson's equation for the gravitational potential. We term these the gravitational hydrodynamics (GHD) equations. The total non-gravitational energy per unit mass, E, is related to the internal energy per unit mass, e, by $E = e + \frac{1}{2}\mathbf{v}^2$. In turn, P is defined by an equation of state. In our isothermal work, we adopt the ideal gas law $P = (\gamma - 1)\rho e$. In recent work (Fisher, Klein, & McKee 1998) we have used a nonisothermal EOS that combines isothermal and nonisothermal components such that $P(\rho) = c_s^2\rho + K\rho^\gamma$ where $\gamma = 5/3$ and K is chosen so that the isothermal and adiabatic components balance at a critical density determined by detailed 1-D radiation-hydrodynamic calculations of Masunaga, Miyama, & Inutsuka (1998). A typical critical density for this transition is 5×10^{-14} g cm^{-3}. As written above, there are eight equations in eight unknowns. We assume periodic boundary conditions on these equations.

Our code methodology consists of three components to efficiently solve this system. The first component is a hyperbolic solver that employs an implementation of the Godunov method (see Colella & Woodward 1984 and LeVeque 1992) for solution of the Euler equations of gas dynamics. The second major component of our code methodology is an elliptic solver that utilizes an AMR multigrid method to solve Poisson's equation. These elements operate within the third component, an adaptive mesh refinement framework.

2.2. HYPERBOLIC SOLVER

2.2.1. *Overview of Godunov Implementation*

To solve the full 3-D system of hydrodynamics equations, we use an operator-split method in which we solve in each coordinate direction independently in a cyclic sequence. In 3-D, this operator splitting can be written schematically as

$$\mathbf{Q}^{n+2} = L_x L_y L_z [L_z L_y L_x (\mathbf{Q}^n)], \quad (1)$$

where the superscript n indicates the nth time level, $\mathbf{Q}^n$ indicates the state vector at the nth time level, and L_i is the update operator in the ith coordinate direction based on the higher-order Godunov method described by Colella & Woodward (1984) and Bell, Colella, & Trangenstein (1989). That is, each three-dimensional update step from t^n to t^{n+1} consists of 3 one-dimensional update steps, one per coordinate direction. At the end of the cycle from t^n to t^{n+2}, the solution is second-order accurate. The interested reader can refer to Truelove *et al.* (1998) and Klein (1998) for the details of the scheme.

2.2.2. *AMR Considerations and AMR Multigrid*

When solving a hyperbolic system on a hierarchy of computational grids, care must be taken that the solutions on finer grids are reflected on the neighboring and underlying coarser grids. Our code uses *refluxing* and *averaging down* procedures described in detail by Berger & Colella (1989) to update the solutions on these coarser grids to account for solutions on the finer grids. In refluxing, the solution in a coarse cell adjacent to a fine grid is updated after the fine grids have been advanced in time by adding the effects of a differential flux acting over Δt at its face adjacent to the refined region. This differential flux is equal to the difference between the flux at this face as computed on the coarse grid and the sum of the fluxes at this face as computed on the fine grid. The solution in a refined coarse cell is simply overwritten by averaging down the solutions on the fine cells it contains.

In contrast to the hyperbolic solution, in which only neighboring grids communicate at each time step, the solution of Poisson's equation for the gravitational potential is a much more tightly coupled process. The multigrid method is a natural choice to use as an element of the elliptic solver in an AMR calculation (Truelove *et al.* 1998).

2.3. AMR FRAMEWORK AND THE REFINEMENT CRITERION

As discussed above, the basic hydrodynamic method we use is a higher-order extension of Godunov's method of a type discussed in Colella & Woodward (1984). This algorithm is second-order accurate for smooth flow problems, and has a robust and accurate treatment of discontinuities (Klein, McKee, & Colella, 1994). We further enhance the efficiency and accuracy of our calculations by using local adaptive mesh refinement. The version of the algorithm used for the present work is rather elaborate; a detailed description is given in Berger & Colella (1989).

AMR contains five separate components. The *error estimator* is used to estimate local truncation error and will be described subsequently. The *grid generator* creates fine grid patches that cover the regions that need refinement. *Data structure routines* manage the grid hierarchy allowing access to the individual patches. *Interpolation routines* initialize a solution on a newly created fine grid and also provide the boundary conditions for integrating the fine grids. *Flux correction* routines ensure conservation at grid interfaces by modifying the coarse grid solution for coarse cells that are adjacent to a fine grid.

In AMR, the computational volume consists of a hierarchical grid structure. A base Level 0 grid fills the computational volume, discretizing it on a rectangular grid with a resolution of Δx_0 in each direction. Multiple Level 1 grids of finer resolution $\Delta x_1 = \Delta x_0/r_1$ may be embedded within it, where $r_1 = 4$ is a typical choice. Higher levels may be defined embedded in underlying levels in a similar fashion. Grids at Level L always span an integral number of cells at Level $L-1$, i.e., partial cell refinement is not permitted. Furthermore, an interior grid at Level L is always nested within a grid at Level $L-1$ such that there is a buffer region of Level $L-1$ cells surrounding it. In other words, a grid at Level L within a grid at Level $L-1$ never shares a boundary with the Level $L-1$ grid, unless it shares a boundary with the computational domain.

An integration step on this adaptive grid proceeds as follows. We are using a single timestep for all grids, determined from the most stringent timestep over the entire physical domain. The updating of the data on the locally refined grid structure is organized around the grouping of cells into rectangular grid patches, each typically containing hundred to several thousand grid cells. The AMR code passes a grid patch to an integration routine which returns the updated cell values as well as the edge-centered conservative fluxes used in the update. The overheads in both CPU and memory associated with the adaptive mesh structure have been kept quite small, relative to irregular grid schemes. Typically, 80% – 90% of the total execution time is spent advancing cells in time using the hydrodynamic and elliptic solvers, while the memory required is that needed to store two copies of the solution on all of the grids. These overheads are low because they are determined by the number of rectangles into which the AMR solution has been divided; as opposed to being determined by the number of grid cells, as is the case with irregular grid adaptive algorithms.

A key component of an AMR code is the grid generator. In our code, this procedure is divided into two steps. In the first step, a specified property is measured in each cell, and cells requiring refinement are flagged. In the second step, the distribution of flagged cells is analyzed to determine the number, sizes, and locations of grids to be inserted at the next finer level of resolution. These finer grids will always include every cell that was flagged for refinement, but they may also include additional cells that were not flagged. The degree to which the refinement is concentrated in the cells that require it is termed the *grid efficiency*. The grid efficiency is minimal when the smallest rectangular solid containing all flagged cells is refined. The grid efficiency is maximal when the only cells refined are those that were flagged.

Our general process for deciding upon a suitable level of refinement depends on a method for estimating the local truncation error of the Godunov scheme; this error estimator determines where the solution accuracy is insufficient. We estimate truncation error using Richardson extrapolation; we take data on the grid where the error is being estimated and evolve two timesteps. This solution is then combined with the data on a grid spatially coarsened by a factor of two and integrated for one timestep. The local truncation error can be shown to be proportional to the normed functional that is the difference of these two operations.

In the nonisothermal calculations presented in this paper and in the isothermal calculations of Truelove *et al.* (1997), refinement to Level 2 and above is controlled by application of the Jeans condition. This criterion can be augmented or replaced by others as necessary. The Jeans condition used to make the cell-by-cell refinement decision is a physically motivated resolution constraint discussed by Truelove *et al.* (1997). Jeans (1902 and 1928) analyzed the linearized equations of 1-D isothermal GHD for a medium of infinite extent and found that perturbations on scales larger than the Jeans length, $\lambda_J \equiv \left(\pi c_s^2/G\rho\right)^{1/2}$, are unstable. Thermal pressure cannot resist the self-gravity of a perturbation larger than λ_J, and runaway collapse results. Truelove *et al.* (1997) showed that the errors generated by numerical GHD solvers can act as unstable perturbations to the flow. In a simulation with variable resolution, cell-scale errors introduced in regions of coarser resolution can be advected to regions of finer resolution, affording these errors the opportunity to grow. The unstable collapse of numerical perturbations can lead to substantial fragments, a process termed *artificial fragmentation*, which can be avoided if λ_J is sufficiently resolved. Defining the Jeans number $J \equiv (\Delta x/\lambda_J)$, Truelove *et al.* (1997) found keeping $J \leq 0.25$ avoided artificial fragmentation in isothermal evolution of a collapse spanning 7 decades of density, the approximate range separating typical molecular cloud cores from nonisothermal protostellar fragments. The constraint that λ_J be resolved is termed the *Jeans condition*. Although it has been shown to be necessary only for isothermal evolution, it is reasonable to expect that it is necessary (although not necessarily sufficient) for nonisothermal collapse as well.

As a side effect of confining cell-sized perturbations to a length scale at which they are thermally damped, resolution of λ_J also ensures that gradients developed in isothermal flow by gravity are well resolved. Formation of structure on scales of λ_J and larger is a general feature of self-gravitational isothermal GHD flow since smaller fluctuations are damped but larger ones collapse. For example, in the self-similar solution for isothermal cylindrical collapse (Inutsuka & Miyama, 1992), the radial scale height of the cylinder is λ_J. Lack of resolution of gradients within simulated flow triggers the injection of artificial viscosity, which is generally intended to be introduced in small amounts only for numerical stability (in our Godunov scheme) or shock mediation (in many common non-Godunov schemes). Introducing excess amounts of artificial viscosity renders the problem solved different from the inviscid problem posed. Continuous resolution of λ_J, however, keeps the flow inviscid and prevents artificial slowing of gravitational collapse. It is important to note, however, that the Jeans condition is a necessary but not, in general, sufficient condition to ensure convergence.

The Jeans condition on Δx fundamentally differs from the Courant condition on Δt, although at first the two conditions might appear analogous. The Courant condition arises from a modal stability analysis of finite difference equations derived from the Euler partial differential equations (PDEs) of hydrodynamics (see, e.g., Richtmyer & Morton 1967). It is entirely a consequence of the finite differencing and has no physical counterpart in the PDEs. A modal stability analysis of finite difference equations derived from the GHD PDEs does not yield the Jeans condition, but rather a generalized Courant condition that includes effects of gravity (Truelove *et al.* 1998).

3. APPLICATIONS TO ROTATING MOLECULAR CLOUDS

In this section we briefly describe recent high-resolution results we have obtained with AMR for the gravitational collapse of nonisothermal, rotating, uniformly dense clouds (Fisher, Klein, & McKee 1998). Isothermal centrally condensed clouds (Truelove *et al.*, 1997) and isothermal uniform clouds (Truelove *et al.*, 1998) have been previously discussed and we refer the reader to these papers.

The uniform cloud is an idealized model of an astrophysical cloud but is very useful as a first approximation. The initial isothermal configuration of the uniform cloud may be completely parameterized by the dimensionless energy ratios $\alpha \equiv E_{\rm thermal}/|E_{\rm gravitational}|$ and $\beta_\Omega \equiv E_{\rm rotational}/|E_{\rm gravitational}|$.

Burkert & Bodenehimer (1993) studied a uniform cloud, with $M = 1 M_\odot$ and $R = 5 \times 10^{16}$ cm, which give $\rho_0 = 10^{-17.4}$ g cm^{-3}. Its energy ratios are $\alpha = 0.26$ and $\beta_\Omega = 0.16$. The isothermal uniform cloud thus has $c_s = 0.167$ km s^{-1} and a rotation rate $\Omega = 7.2 \times 10^{-13}$ rad s^{-1}. The cloud is perturbed by $\rho \rightarrow \rho \times [1 + 0.1 \cos(2\phi)]$, a 10% $m = 2$ mode seed perturbation. This cloud begins with $\lambda_J = 1.17R$. We have used an initial resolution of

32 cells per initial cloud radius, or R_{32}, and dynamically refined so as to ensure $J_{\text{max}} = 0.25$. In the initial configuration, the fiducial radius beyond which cloud gravity dominates perturbation pressure is $r_{\text{crit}} = 0.14R$, so that with R_{32} we still linearly resolve this scale with more than 4 cells.

3.1. NON-ISOTHERMAL UNIFORM CLOUDS

In recent work using high resolution AMR (Fisher, Klein, & McKee 1998), we followed the collapse of an initially rigidly rotating, uniform isothermal cloud. We used a two-component barotropic equation of state which makes the transition from isothermal to polytropic in a smooth fashion. The initial conditions are identical to the Burkert & Bodenehimer (1993) isothermal uniform cloud perturbed by a 10% $m = 2$ mode. In the following we discuss the evolution of this nonisothermal collapse. While a similar calculation has been recently studied by Bate & Burkert (1997), to our knowledge, none of the calculations in the literature to date have been able to follow the subsequent evolution of the fragments over dynamical (orbital) timescales, while still adhering to the Jeans criterion. However, to be fully certain of the reality of the fragmentation, it is still necessary to do a convergence study by decreasing the Jeans number (Truelove *et al.* 1997), which we have not yet done for this calculation.

Figure 1 [1] represents a grayscale 2-D slice of log density through the equatorial plane $t = 1.41 \times 10^{12}$ s after start of the collapse. The cloud has initially collapsed to an isothermal disk, and an elongated filamentary bar with the first signs of fragmentation apparent. A strong isothermal shock above the plane of the disk is soon established. After $t = 1.46 \times 10^{12}$ s (Figure 2) the isothermal bar becomes optically thick, and the accretion flow onto the bar is arrested, resulting in the growth of non-axisymmetric perturbations in the bar. Fragmentation in the bar results in the formation of binary spherical cores connected by a prominent bar. The core-bar system is embedded in a two-armed spiral, derivative of the original $m = 2$ perturbation. The binary separation decreases as the cores increase their mass by direct accretion from the low-angular momentum material of the bar, eventually leading to the dissipation of the bar. Figure 3 shows a snapshot of log density about 0.5 rotation periods later at $t = 1.51 \times 10^{12}$ s. Protostellar disks have now formed around the cores with the cores at their closest orbital separation ~ 44 AU. The disks accrete gas directly from the long spiral arms. The cores accrete by Roche Lobe overflow of the contacting protostellar disks, and each core has $0.08 M_{\odot}$ with a radius ~ 10 AU. The masses and radii of these first collapse cores are in good agreement with the detailed 1-D radiation-hydrodynamic calculations of Masunaga, Miyama, &

[1] All the figures of this paper are contained in the supplementary CD-ROM.

Inutsuka (1998). The core accretion luminosity $\sim 2 \times 10^{31}$ergs^{-1} obtained in the 3-D collapse calculations is considerably less than that found in the 1-D simulations, $\sim 4 \times 10^{32}$ergs^{-1}, due to the large angular momentum barrier in the 3-D simulations, resulting in slower accretion onto the protostellar cores.

It is important to point out that the initial rapid growth of the cores is due to accretion of matter from the bar. As the bar dissipates, the cores have ended their first phase of growth and proceed to grow more slowly by direct accretion from the surrounding protostellar disks. At $t = 1.52 \times 10^{12}$ s (Figure 4) the cores begin to separate and the surrounding disks become morphologically distinct. The protostellar disks and cores become a fully detached binary by $t = 1.56 \times 10^{12}$ s (Figure 5) with the disks attached to the long spiral. The scale of the disks, of order 100 AU, is consistent with observations of gaseous disks surrounding single T-Tauri stars (Sargent & Beckwith 1994) and debris disks surrounding systems such as β Pictoris (Artymowicz 1997). When we stopped our calculations, at $t = 1.6 \times 10^{12}$ s (Figure 6) the binary protostellar disks/core system has moved back into the surrounding spiral arms. The cores have 20% of the mass of the initial cloud and the arms have 27% of mass of the initial cloud at this time. The protostellar disks comprise about 2% of the cloud mass.

4. CONCLUSIONS

In this paper we have described a powerful new method for numerical solution of 3-D self-gravitational hydrodynamic problems. This method combines a Godunov hydrodynamics integrator with a multigrid gravity solver in an adaptive mesh refinement framework. Guided by the Jeans condition for isothermal problems, and the use of Richardson extrapolation, AMR efficiently expends computational resources only when and where the features of the flow demand it. We presented results of this methodology applied to the collapse and fragmentation of nonisothermal molecular clouds that are both initially uniform and centrally condensed, evolved over 8–9 decades of collapse and subject to initial $m = 2$ mode perturbations. We found binary fragmentation results in good agreement with published calculations after evolution over the first half of this logarithmic range. By automatically inserting ever finer grids to maintain resolution of the Jeans length, these calculations did not generate the substantial artificial viscosity found in published calculations that used fixed finest resolutions. We show the first well-resolved formation sequence of protostellar disks surrounding a pair of binary cores, and discuss the role of the bar connecting the binary pair in the accretion onto the cores. The disks formed are consistent with observations of gaseous disks surrounding T-Tauri stars.

Acknowledgements

Research on star formation for RIK and CFM is supported by a grant from NASA's Astrophysics Theory Program to the Center for Star Formation Studies. RIK acknowledges additional support under the auspices of the US Department of Energy at the Lawrence Livermore National Laboratory under contract W-7405-Eng-48. CFM's research is supported in part by NSF grant AST9530480. We wish to thank the Pittsburgh Supercomputing Center for provision of Cray C90 resources through grant AST940011P.

References

Artymowicz, P. 1997, Ann. Rev. Planet. Sci., 25, 175
Bate, M.R.& Burkert, A. 1997, M.N.R.A.S., 288, 1060
Bell, J.B., Colella, P., & Trangenstein, J. 1989, Journal of Computational Physics, 82, 362
Berger, M.J. & Colella, P. 1989, Journal of Computational Physics, 82, 64
Burkert, A. & Bodenheimer, P. 1993, M.N.R.A.S., 264, 798
Colella, P. & Woodward, P.R. 1984, Journal of Computational Physics, 54, 174
Fisher, R., Klein, R., & McKee, C. 1998, in preparation
Inutsuka, S. & Miyama, S.M. 1992, ApJ, 388, 392
Jeans, J.H. 1902, Phil. Trans. A, 199, 1
Jeans, J.H. 1928, Astronomy & Cosmogony, (London: Cambridge University Press) 1993, A&A, 273, 175
Klein, R.I. & McKee, C.F. 1994, in Numerical Simulations in Astrophysics, ed. J. Franco et al. (New York: Cambridge), 251
Klein, R.I., McKee, C.F., & Colella, P. 1994, ApJ, 420, 213
Klein, R.I., McKee, C.F., & Woods, D.T. 1995, in Physics of the Interstellar Medium and Intergalactic Medium, ed. A. Ferrara et al. (San Francisco: Astronomical Society of the Pacific), 366
Klein, R.I. & Woods, D.T. 1998, ApJ, 497, 777
Klein, R.I. 1998, Journal of Computational and Applied Mathematics, ed. H. Riffert (Elsevier Press), in press
LeVeque, R.J. 1992, Numerical Methods for Conservation Laws, (Boston: Birkhauser Verlag)
Masunaga, H., Miyama, S. & Inutsuka, S. 1998, ApJ, 495, 346
Richtmyer, R.D. & Morton, K.W. 1967, Difference Methods for Initial-Value Problems, (New York: Interscience)
Sargent, A.I, & Beckwith, S.V.W., 1994, Astroph. and Space Sci., 212, 181.
Truelove, J.K., Klein, R.I., McKee, C.F., Holliman, J.H., II, Howell, L.H., & Greenough, J.A. 1997, ApJ, 489L, 179
Truelove, J.K., Klein, R.I., McKee, C.F., Holliman, J.H., II, Howell, L.H., Greenough, J.A., & Woods, D.T. 1998, ApJ, 495, 821
Woods, D.T., Klein, R.I., Castor, J.I., McKee, C.F., & Bell, J. 1996, ApJ, 461, 767

GRAVITATIONAL CONTRACTION OF MAGNETIZED CLOUDS

KOHJI TOMISAKA
Faculty of Education and Human Sciences, Niigata University, 8050 Ikarashi-2, Niigata 950-2181, Japan

Abstract. In this paper, the dynamical contraction of magnetized clouds is studied from the run-away phase to the accretion phase throughout. Since the non-homologous gravitational contraction necessitates progressively finer spatial resolution, "nested grid technique" (Burger & Oliger 1984) is adopted. As a result, it is found that outflow driven by toroidal magnetic fields is occurred in the accretion phase, which explains molecular outflows in protostellar objects (see also Tomisaka 1998). In contrast, a rotationally supported disk is formed in a cloud core without magnetic fields.

1. Introduction

Gravitational contraction continues to be a challenging target for numerical astrophysics, since wide dynamic ranges should be treated both in densities and spatial dimensions. Let us consider an isothermal cloud supported against the self-gravity by the thermal pressure, Lorentz force, and centrifugal force. From Virial analyses and numerical studies to follow-up, it is shown that there is a maximum mass above which a cloud is not in equilibrium. Ignoring the effect of thermal pressure and centrifugal force, the maximum (critical) mass is expressed as a function of the magnetic flux threading the cloud Φ as $M_{\rm cr} \simeq 0.13\Phi/G^{1/2} \simeq 10^3 M_\odot\,(B/30\mu{\rm G})\,(R/2{\rm pc})^2$. The cloud more massive than the critical mass is called "supercritical." If a cloud is born as supercritical, it begins immediately dynamical collapse and contracts in a gravitational free-fall time-scale $\tau_{\rm ff} \simeq (4\pi G\rho)^{-1/2}$. In contrast, the cloud less massive than $M_{\rm cr}$ is called "subcritical." The subcritical clouds are considered to evolve slowly (10 times slower than $\tau_{\rm ff}$ of the cloud) loosing the magnetic flux mainly near the center of the cloud by the ambipolar diffusion. Due to the ambipolar diffusion, even an initially

S. M. Miyama et al. (eds.), Numerical Astrophysics, 141–149.

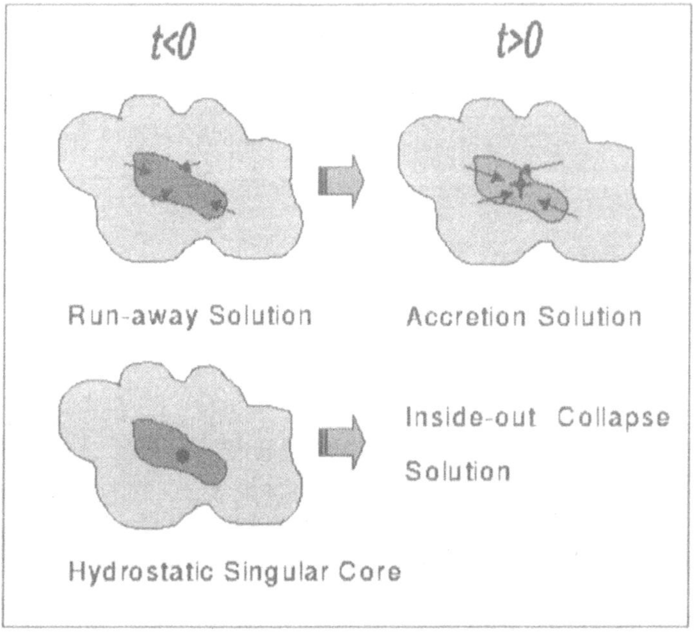

Figure 1. Relationship between run-away collapse solutions and accretion-type solutions realized in isothermal clouds. Dynamical collapse proceeds from the run-away collapse to accretion phase. In particular, when accretion occurs directly in a hydrostatic singular molecular core, the inside-out collapse solution is realized (Shu 1977).

subcritical cloud finally becomes supercritical and experiences dynamical contraction. Therefore, it is concluded that the cloud (core) has to experience the dynamical contraction stage to form stars.

As for a non-magnetic isothermal cloud, semi-analytic (Larson 1969; Penston 1969; Hunter 1977; Whitworth & Summers 1985) and numerical studies (Foster & Chevalier 1993) have revealed that the dynamical collapse of the cloud is divided into two phases: the earlier "run-away collapse" phase and the later "accretion" phase. In the former phase, the cloud shows non-homologous contraction and the central density increases greatly in a finite time scale ($\sim \tau_{\rm ff}$), while in the latter accretion phase the infalling gas accretes onto a central high-density body formed in the run-away collapse phase (Fig. 1). The latter phase corresponds to the "inside-out collapse" (Shu 1977) which is realized when accretion begins at the center of a hydrostatic cloud core. For rotating (non-magnetic) clouds (Norman, Wilson & Barton 1980; Narita, Hayashi & Miyama 1984) and magnetized (non-rotating) clouds (Scott & Black 1980; for recent progress, see Tomisaka 1995), these two phases exist. In this paper, the contraction of magnetized clouds is studied from the run-away phase to the accretion phase through-

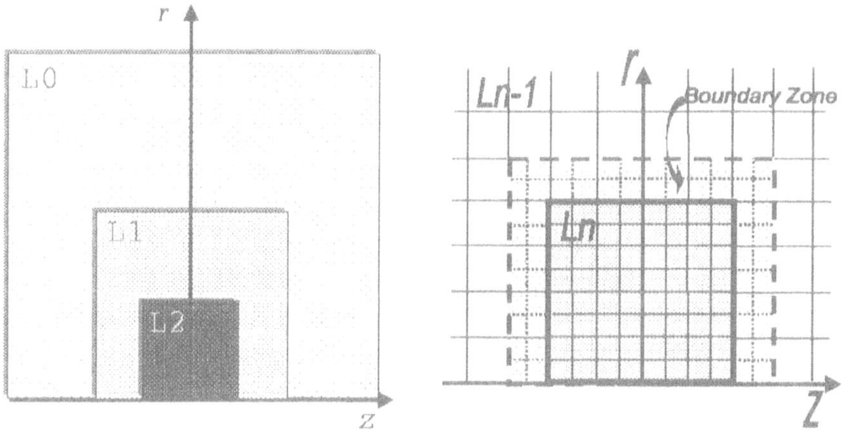

Figure 2. (a) Left: Schematic view of "nested grid method." Each level of grid has a co-center which is identical with the position of the cloud center. (b) Right: Communication between coarser grid L$n-1$ and finer grid Ln. The boundary conditions for Ln (both MHD eqs. and the Poisson eqs.) are given by interpolation of those of L$n-1$. In contrast, in the region covered by Ln, physical values of L$n-1$ are overwritten by averaging over corresponding values of Ln after the calculation of Ln has been completed. Since the Courant condition restricts the time-step, we have to calculate Ln twice and then L$n-1$ once.

out. As a result, it is found that outflow driven by toroidal magnetic fields is occurred in the accretion phase, which explains molecular outflows in protostellar objects (see also Uchida & Shibata 1985; Pudritz & Norman 1986; Ouyed & Pudritz 1997; Shu et al. 1994).

2. Model and Numerical Method

To overcome difficulties related to the wideness of dynamic range, we employ "nested grid method" (Burger & Oliger 1984). A number of grid systems with different spacings are prepared and time evolution is calculated using these grids. Center of the cloud is covered with a finer grid and the global structure is traced by a coarser grid. In this paper, we use 15 levels of grids from L0 (the coarsest) to L14 (the finest). Grid spacing of Ln is chosen as a half of that of L$n-1$ and each level of grids uses 64×64 grid points. Thus, the finest grid (L14's) spacing is $\simeq 10^{-6}$ times smaller than the box size of L0. Since the non-homologous gravitational contraction necessitates progressively finer spatial resolution, the depth of maximum level of grids increases with time. Details of the numerical method are described in separate papers (Tomisaka 1996a,b).

We begin our simulation from an infinitely long, cylindrical, rotating, isothermal cloud in hydrostatic balance, that is, the gravitational force is counterbalanced with the pressure force, the centrifugal force, and the Lorentz force. Here we assume the rotating axis coincides with the cylinder (z-)axis. As the initial state, only poloidal magnetic fields are taken into account and their strength is assumed proportional to the square root of local gas density. We add small density perturbations on it to initiate contraction. Distributions of density, magnetic flux density, and rotation speed in the radial (r-)direction are taken identical with Matsumoto, Nakamura & Hanawa (1994), although we assume no toroidal magnetic field component B_ϕ initially. The size of the numerical box is chosen to agree with the wavelength of the most unstable mode of the self-gravitational instability $\lambda_{\rm MGR}$ (Matsumoto et al. 1994) and the periodic boundary condition is applied to the upper and lower boundaries in z-direction. Quantities are scaled with the isothermal sound speed c_s, the surface pressure (or surface density) $p_s = c_s^2\rho_s$ and the corresponding scale-height $H = c_s/(4\pi G\rho_s)^{1/2}$. Typical values would be $c_s = 200{\rm m\ s^{-1}}$, $\rho_s = 100{\rm cm^{-3}}$, $H = 0.36{\rm pc}$, and thus the time-scale is normalized by the free-fall time as $\tau_{\rm ff} = (4\pi G\rho_s)^{-1/2} \simeq 1.75{\rm Myr}(\rho_s/100{\rm cm^{-3}})^{-1/2}$. Using these scalings, initial radial distributions of density, magnetic fields and angular rotation speed are written as

$$\rho(r) = (\rho_c/\rho_s)\{1 + [\rho_c/\rho_s - 2(\Omega_0\tau_{\rm ff})^2]/[8(1+\alpha)] \times (r^2/H^2)\}^{-2}\rho_s, \quad (1)$$

$$(B_z(r), B_r(r)) = ([2\alpha\rho(r)/\rho_s]^{1/2}, 0)\rho_s^{1/2}c_s, \quad (2)$$

$$\Omega(r) = \Omega_0[\rho(r)/\rho_c]^{1/4}. \quad (3)$$

As model parameters, α (the ratio of magnetic pressure to the thermal one for poloidal field component) and Ω_0 (the angular rotation speed at the center of the cloud) are chosen as $\alpha = 0.5$, $\Omega_0 = 5/\tau_{\rm ff}$ and $\rho_c = 100\rho_s$ in the present paper.

3. Results

3.1. RUN-AWAY COLLAPSE PHASE

The evolution is similar to that of magnetized cloud with no rotation (Tomisaka 1995, 1996a; Nakamura, Hanawa, & Nakano 1995). First, the cylindrical cloud breaks into prolate spheroidal fragments which is elongated along the cylinder axis (z-axis). As long as the amplitude of density perturbation is small $\delta\rho/\rho \sim 1$, the shape keeps essentially identical with the most unstable eigen-function of the gravitational instability. Next, gas begins to fall mainly along the magnetic field lines, and forms a disk which runs perpendicular to the magnetic field lines. At $t = 0.5908\tau_{\rm ff}$, at that time the central density ρ_c reaches as high as $10^5\rho_s$ (as the initial density we assume ρ_c is equal to $100\rho_s$), a shock wave is formed parallel to the disk

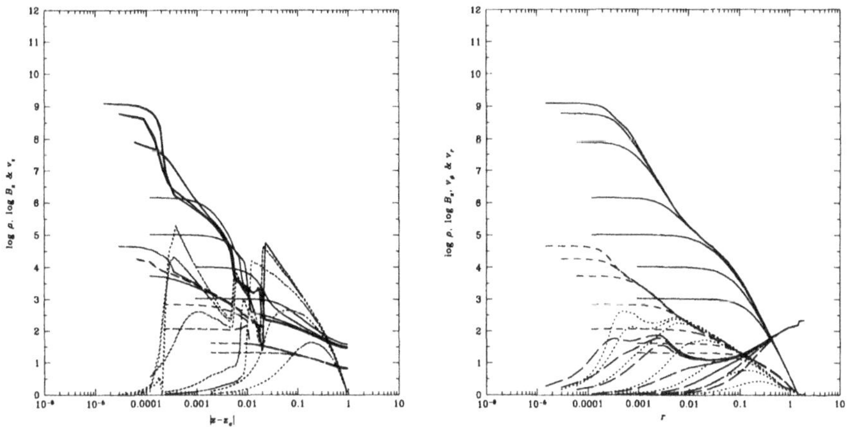

Figure 3. (a) Left: structure along the z-axis. $\log \rho(z, r=0)$ (solid lines), $|v_z(z, r=0)|$ (dotted lines), and $\log B_z(z, r=0)$ (short-dash lines) are plotted. Snapshots are made at $t = 0.5233\tau_{\rm ff}$ ($\rho_c \sim 10^3 \rho_s$), $t = 0.5725\tau_{\rm ff}$ ($\rho_c \sim 10^4 \rho_s$), $t = 0.5908\tau_{\rm ff}$ ($\rho_c \sim 10^5 \rho_s$), $t = 0.5977\tau_{\rm ff}$ ($\rho_c \sim 10^6 \rho_s$), $t = 0.5994\tau_{\rm ff}$ ($\rho_c \sim 10^8 \rho_s$), $t = 0.5996\tau_{\rm ff}$ ($\rho_c \sim 10^{8.8} \rho_s$), and $t = 0.5997\tau_{\rm ff}$ ($\rho_c \sim 10^{9.1} \rho_s$). (b) Right: structure along the equatorial plane. $\log \rho(z=0, r)$ (solid lines), $|v_r(z=0, r)|$ (dotted lines), $\log B_z(z=0, r)$ (short-dash lines), and $v_\phi(z=0, r)$ (long-dashed lines) are plotted.

(Fig. 3a). It breaks into two waves at the epoch when the central density reaches $\rho_c \sim 10^6 \rho_s$: an outer wave front propagates outwardly as a fast-mode MHD shock (seen near $z \sim 0.02H$) and an inner one does inwardly as a slow-mode MHD shock (seen near $z \sim 7 \times 10^{-3} H$) reaching the equator. The rotation angular speed Ω and toroidal magnetic field component B_ϕ also jump crossing the wave fronts. Fast rotation and thus strong B_ϕ are observed in a restricted region between these two fronts. Since the disk contracts in the radial direction, it has a larger Ω than outside the disk. In this configuration, magnetic fields transfers the angular momentum from the disk and redistributes it into the same magnetic flux tube (magnetic braking). However, from this numerical simulation it is shown that the angular momentum is confined into a region inside the fast-mode MHD shock fronts.

Although there exists a number of shock waves ($z \sim 0.02H$, $5 \times 10^{-3} H$, and $2 \times 10^{-4} H$) propagating in the z-direction, no such discontinuity is found in the r-direction (Fig. 3b). However, a kind of modulation is seen around a power-law distribution as $\rho(z=0, r) \propto r^{-2}$, which is related to the formation of multiple shock waves in the z-direction (this is found by Norman et al. (1980) for the case of rotating isothermal cloud). The run-away collapse continues till $t \simeq 0.6\tau_{\rm ff}$ and the central density would increase greatly if the isothermal equation of state continues to be valid. However, when the central density exceeds $10^{10}{\rm cm}^{-3}$, the central part of the core becomes optically thick for the thermal radiation from dusts and its temperature begins to rise.

3.2. ACCRETION PHASE

In the accretion phase, gas continues to accretes onto the newly formed opaque core. We mimic the situation by using a double polytrope composed of an isothermal one for low density as $p = c_s^2\rho$ for $\rho < \rho_{\rm crit}$ and a harder one for high density as $p = p(\rho_{\rm crit})(\rho/\rho_{\rm crit})^\Gamma$ for $\rho > \rho_{\rm crit}$. We take $\rho_{\rm crit} = 10^8\rho_s = 10^{10}{\rm cm}^{-3}(\rho_s/100{\rm cm}^{-3})$ and $\Gamma = 5/3$. By virtue of this assumption we can follow further evolution. Figure 4a shows the structure captured by L8 just after the isothermal equation of state is broken. A disk, which runs vertically, contracts radially and gas motion drags and squeezes the magnetic field lines to the center. The rotation speed v_ϕ at that time is not more than the radial infall speed v_r which is in the range of $(2-3)\times c_s$ (see Fig. 3b).

In the accretion phase, the inflow speed is accelerated at least to $(4-5)\times c_s$ and as a result magnetic field lines are strongly dragged and squeezed to the center (see Fig. 1c of Tomisaka 1996b). Beside these, the rotation speed is accelerated as the infall proceeds. Just after the state shown in Figure 1 with the highest central concentration, v_ϕ reaches $\simeq 4c_s$ at the distance of $r \sim 3\times 10^{-4}H$ inside of which a nearly spherical core is formed supported mainly by thermal pressure. The rotation velocity v_ϕ exceeds the radial velocity v_r in the accretion phase, while v_ϕ was smaller than v_r (Fig. 3b) in the run-away collapse phase. From this epoch, outflow from the disk is observed.

Figure 4 shows the structures before (a) and after (b) the outflow begins. The physical time scale between (a) and (b) is approximately equal to $\simeq 1000{\rm yr}(\rho_s/100{\rm cm}^{-3})^{-1/2}$. The directions of the magnetic fields in the disk is much affected by the radial inflow. They are pushed and squeezed to the center. It is shown that in the seeding region of the outflow the angle between a magnetic field line and the disk plane decreases from $60^\circ - 70^\circ$ (Fig. 4a) to $10^\circ - 30^\circ$ (Fig. 4b). In Figure 4b, it is indicated that the outflow, which reaches $z_s \sim \pm 0.003H \simeq \pm 200{\rm AU}(H/0.36{\rm pc})$, is confined into two expanding bubbles. In the bubble, the strength of toroidal component of magnetic fields B_ϕ is larger than that of the poloidal one $(B_z^2+B_r^2)^{1/2}$. Especially at the outer boundary of the bubble $|B_\phi|$ is $(4-5)$ times stronger than $(B_z^2+B_r^2)^{1/2}$ (see also the upper panel of Fig. 5). Further, the magnetic pressure is much larger than the thermal one. Outside of the bubble, inflow continues.

To see the seeding region of the outflow closely, we plot a close-up view captured by L10 in Figure 5. This figure shows that the outflow is ejected

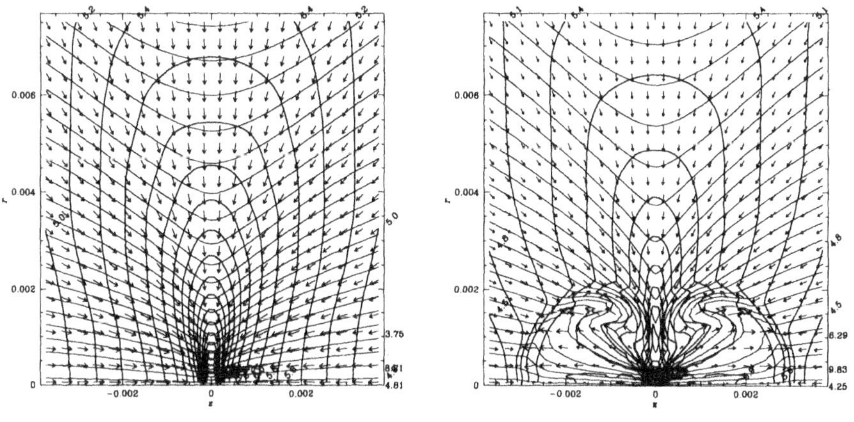

Figure 4. (a) Left: isodensity lines, magnetic field lines, and velocity vectors are plotted for the state when the central density reaches $10^{8.8}\rho_s$ ($t = 0.5996\tau_{\rm ff}$). In contrast to the usual usage, the z-axis is placed horizontally and the r-axis is vertically. (b) Right: the same as (a) but for the state when the central density reaches $10^{10}\rho_s$ ($t = 0.6002\tau_{\rm ff}$). Both are Level 8. Physical time passed between these two is equal to 1000 yr $(\rho_s/100{\rm cm}^{-3})^{-1/2}$.

from the disk near $r \simeq 5 \times 10^{-4}H \simeq 40{\rm AU}(H/0.36{\rm pc})$. This clearly shows that (1) magnetic field lines are almost parallel to the disk surface near the disk. (2) gas inflow in the disk is not disturbed by the outflow from the disk and continues to reach the central core which is supported by the thermal pressure. (3) outflow occurs along magnetic field lines whose inclination angle is in the range of 45° – 60°. This corresponding to the region where $|B_\phi|$ is dominant over the poloidal component (upper panel).

4. Discussion

Inflow and outflow rates are measured at the outer boundary of L10 numerical box as the surface integral of the mass flux density $\int \rho \mathbf{v} \cdot \mathbf{n} dS$. The inflow rate is equal to $2.7 \times 10^2 c_s^3/4\pi G \simeq 4.1 \times 10^{-5} M_\odot {\rm yr}^{-1} (c_s/200{\rm m\ s}^{-1})^3$, which is almost constant in the accretion phase. While, the outflow begins at $t \simeq 0.5998\tau_{\rm ff}$ and the outflow rate increases with time. At the stage shown in Figure 5 it reaches $83 c_s^3/4\pi G \simeq 1.2 \times 10^{-5} M_\odot {\rm yr}^{-1} (c_s/200{\rm m\ s}^{-1})^3$. It is worth notice that the outflow rate attains $\simeq 1/3$ of the inflow rate. The ratio of outflowing mass to inflowing one decreases from L10 to L8.

Linear momentum outflow rate is also measured as the surface integral of the momentum flux density $\int \rho v_z^2 dS$ over the upper and lower boundaries of L10. This increases with time from the epoch of $t \simeq 0.5998\tau_{\rm ff}$ and reaches $\simeq 330 c_s^4/4\pi G \simeq 10^{-5} M_\odot {\rm km\ s}^{-1} {\rm yr}^{-1} (c_s/200{\rm m\ s}^{-1})^4$ at the epoch shown in Figure 5. This is not inconsistent with the momentum outflow rate observed around Class 0 low-mass YSOs from ^{12}CO $J = 2 - 1$ line observations

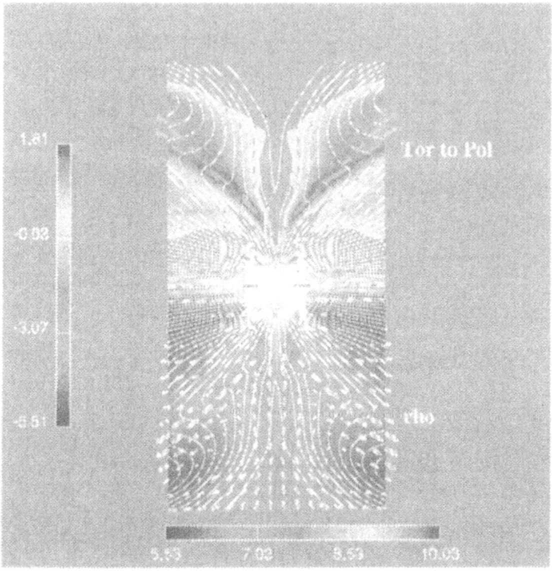

Figure 5. In the lower panel, density distribution (gray scale), magnetic field lines (white line), and velocity fields (vectors) are illustrated. In the upper panel, the ratio of toroidal magnetic pressure $B_\phi^2/8\pi$ to the poloidal one $(B_z^2 + B_r^2)/8\pi$ is plotted as well as the B-fields and velocities. This figure covers $-10^{-3}H \leq z \leq 10^{-3}H$ and $0 \leq r \leq 2 \times 10^{-3}H$, which corresponds to the central 1/16 of the area shown in Fig. 4. Original false color plate is in a supplementary CD-ROM.

(Bontemps et al. 1996) as $3 \times 10^{-6} - 5 \times 10^{-4} M_\odot \mathrm{km\ s^{-1} yr^{-1}}$.

Why does the outflow occur only in the accretion phase? Rotation velocity becomes dominant over the radial velocity only in the accretion phase. This is also indicated in a recent study using a self-similar solution for contracting isothermal rotating disk (Saigo & Hanawa 1998; see their moderately rotating model of $\omega = 0.3$). Further, toroidal magnetic fields develop only in the accretion phase.

This seems to come from the fact that the run-away collapse is fast and as a result the toroidal magnetic fields can not be developed by rotation. In other words, from numerical simulations, the angular rotation speed near the center $\Omega_c(t)$ is proportional to the central free-fall rate as $\Omega_c \simeq (0.2 - 0.4) \times (2\pi G\rho_c)^{1/2}$ (Matsumoto, Hanawa & Nakamura 1997). Therefore, the angle at which a gas element rotates in this free-fall time is as large as $\theta \sim \Omega_c/(2\pi G\rho_c)^{1/2} \sim 0.2 - 0.4$ radian. Since the disk rotates only at $\sim 0.2 - 0.4$ radian, it is concluded that the magnetic field can not be wound much in the run-away collapse phase.

In the accretion phase, we have found a rotating ring nearly in a hydrostatic balance for a model with no magnetic fields ($\alpha = 0$). Since the angular momentum in the disk is transferred to the outer region by the magnetic effect, a model with poloidal magnetic fields indicates a structure different from the non-magnetic disk. Magnetized disk consists of an innermost nearly hydrostatic core and a rotating disk.

Numerical computations were carried out on VPP300/16R at NAOJ.

References

Bachiller, R. 1996, ARAAp, 34, 111
Basu, S. 1997, ApJ, 485, 240
Bontemps, S., André, P., Terebey, S., & Cabrit, S. 1996, AAp, 311, 858
Burger, M. J., & Oliger, J. 1984, J. Comput. Phys., 53, 484
Foster, P. N., & Chevalier, R. A. 1993, ApJ, 416, 303
Hunter, C. 1977, ApJ, 218, 834
Larson, R. B. 1969, MNRAS, 145, 271
Matsumoto, T., Hanawa, T., & Nakamura, F. 1997, ApJ, 478, 569
Matsumoto, T., Nakamura, F., & Hanawa, T. 1994, PASJ, 46, 243
Nakamura, F., & Hanawa, T. 1997, ApJ, 480, 701
Nakamura, F., Hanawa, T., & Nakano, T. 1995, ApJ, 444, 770
Narita, S., Hayashi, C., & Miyama, S. M. 1984, Prog. Theor. Phys., 72, 1118
Norman, M. L., Wilson, J. R., & Barton, R. T. 1980, ApJ, 239, 968
Ouyed, R., & Pudritz, R. E. 1997, ApJ, 484, 794
Penston, M. V. 1969, MNRAS, 144, 425
Pudritz, R. E., & Norman, C. A. 1986, ApJ, 301, 571
Saigo, K., & Hanawa, T. 1998, ApJ, 493, 342
Scott, E. H., & Black, D. C. 1980, ApJ, 239, 166
Shu, F. H. 1977, ApJ, 214, 488
Shu, F. H., Najita, J., Ostriker, E., Wilkin, F., Ruden, S., & Lizano, S. 1994, ApJ, 429, 781
Tomisaka, K. 1995, ApJ, 438, 226
Tomisaka, K. 1996a, PASJ, 48, 701
Tomisaka, K. 1996b, PASJ, 48, L97
Tomisaka, K. 1998, ApJ, 502, L167
Uchida, Y., & Shibata, K. 1985, PASJ, 37, 515
Whitworth, A., & Summers, D. 1985, MNRAS, 214, 1

GRAVITATIONAL COLLAPSE OF ROTATING CLOUDS: BAR AND DISK FORMATION IN A SELF-SIMILARLY CONTRACTING CLOUD

T. MATSUMOTO
Faculty of Liberal Arts, Hosei University,
Fujimi, Chiyoda-ku, Tokyo, 183-0016, Japan

AND

T. HANAWA
Department of Astrophysics, School of Science,
Nagoya University, Chikusa-ku, Nagoya, 464, Japan

1. Introduction

Recent observations have revealed that young stars belong to binaries at high frequency as well as the main sequence stars do. Most the binary systems have been formed already in the early phase of the star formation (Mathieu 1994). According to a probable binary formation scenario, a star forming dense cloud fragments during the gravitational collapse phase. The fragments evolve into binary (or multiple stars) after merge and accretion. Many authors have studied fragmentation of collapsing clouds by numerical simulations intensively (reviewed by Bodenheimer, Ruzmaikina, & Mathieu 1993; recent work by Truelove et al. 1998). In those numerical simulations, an almost spherical cloud forms a bar at the center, and the bar fragments to form seeds of binary. The key process is the bar formation in the binary formation. We thus investigate the mechanism and the growth rate of the bar formation by three dimensional numerical simulations.

2. Model Construction and Numerical Method

We consider a gravitationally collapsing isothermal spherical cloud and approximate the structure by the Larson (1969) and Penston (1969) similarity solution. On it we imposed velocity perturbations given by,

$$v_\varphi(r, \theta, \varphi) = (\omega_0 + \delta\omega_{\varphi,0} \cos 2\varphi) \sin\theta \, r \, \hat{\Omega}(r) \,, \tag{1}$$

S. M. Miyama et al. (eds.), Numerical Astrophysics, 151–154.

and

$$v_\theta(r,\, \theta,\, \varphi) = -\delta\omega_{\theta,0}\, r\, \hat{\Omega}(r)\, \sin\theta\, \sin 2\theta\ , \tag{2}$$

in the spherical coordinates, $(r,\, \theta,\, \varphi)$. In equation (1), $\hat{\Omega}(r)$ denotes an eigen function of a spin-up mode (see Hanawa & Nakayama 1997). It is normalized so that $\hat{\Omega}(0)\, t_{\rm ff,0} = 1$, where $t_{\rm ff,0}$ denotes the initial freefall time. As the central density, ρ_c, increases owing to the collapse, the spin-up mode grows as $v_\varphi \propto {\rho_c}^{2/3}$ if $\delta\omega_{\varphi,0} = \delta\omega_{\theta,0} = 0$.

We adopt the zooming coordinates, $(x',\, y',\, z')$, in the numerical simulations. They are the same as the Cartesian coordinates except that the scale factor shortens with the time in the zooming coordinates (see Nakamura, Hanawa, & Nakano 1997 and Matsumoto, Nakamura, & Hanawa 1997 for the zooming coordinates). The zooming coordinates keep the resolution high although our simulations are restricted to the central region of the cloud. The hydrodynamical equations and Poisson equation are solved by Roe method and multigrid iteration method, respectively.

3. Results of a Typical Model

In this section, we show the results of a typical model having the model parameter, $(\omega_0,\, \delta\omega_{0,\varphi},\, \delta\omega_{0,\theta}) = (0.149,\, 1.49 \times 10^{-3},\, 0)$. In this model, the initial amplitude of velocity perturbation is appreciably small compared with the infall velocity. The collapse of the cloud is followed until the central density increases by a factor of 10^{16} and the scale length decreases by a factor of 10^8.

Figures 1 show the evolution of the dense region of the cloud by cross section in both the $x = 0$ and $z = 0$ planes. In all the stages, the central density increases in proportion to $(t_0 - t)^{-2}$. In early stages, the cloud collapses almost spherically. In the stages of $\rho_c \gtrsim 10^5$ the dense region of cloud has a disk shape. In the stages of $\rho_c \gtrsim 10^{15}$, the disk has a bar inside. The bar narrows in accordance with the increase in the central density.

The length of bar is longer than twice of Jeans length at the final stage of our simulation. The bar is thus unstable against the gravitational instability. The bar will fragment into pieces and the fragments will evolve into seeds of binary stars.

The dense part of the cloud rotates faster in accordance with the increase in the central density. We measure the spin-up by the non-dimensional angular frequency, $\Omega_c\, t_{\rm ff}$, where $t_{\rm ff}$ denotes the freefall time at that time defined by $\sqrt{3\pi/(32G\rho_c)}$. The thick dashed curve in figure 2 denotes the history of spin-up. In spherical collapse phase, the angular frequency increases in proportion to ${\rho_c}^{1/6}$ as expected by Hanawa & Nakayama (1997). When the non-dimensional angular velocity reaches $\Omega_c\, t_{\rm ff} \simeq 0.5$, the cloud

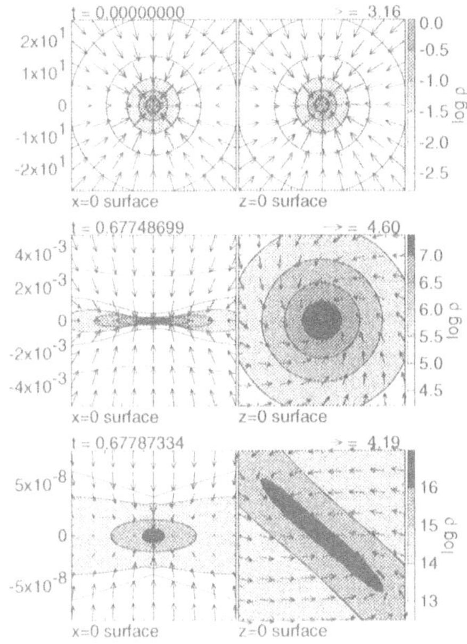

Figure 1. Density and velocity distributions in both the $x = 0$ and $z = 0$ plane for model of $(\omega_0, \delta\omega_{0,\varphi}, \delta\omega_{0,\theta}) = (0.149, 1.49 \times 10^{-3}, 0)$. The contour curves and grayness denote the density in the logarithmic scale. The arrows show the velocity distributions and are normalized separately in each figure. The numbers associated with the arrows denote the maximum velocities in each figure. Not that the lower figures are magnified more.

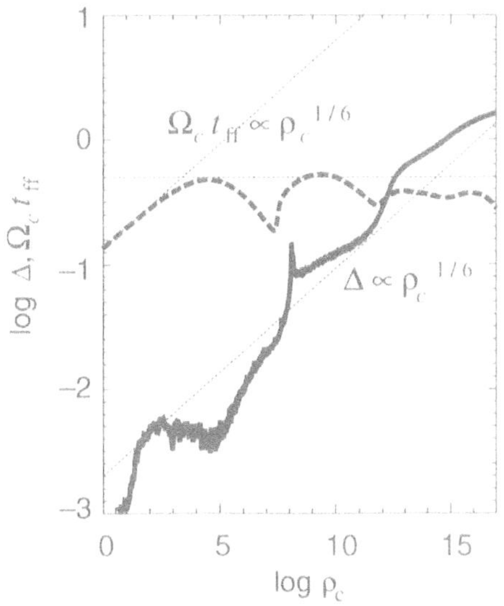

Figure 2. Time evolution of the oblateness (Δ) and spin-up $(\Omega_c\, t_{\rm ff})$ of the cloud. The thick sold and dashed curves denote the oblateness and spin-up, respectively. See text for the definitions of these values. The two thin dashed lines from the lower left to the upper right denote $\Delta \propto {\rho_c}^{1/6}$ and $\Omega_c\, t_{\rm ff} \propto {\rho_c}^{1/6}$ for comparison, respectively. The horizontal dashed line denote the $\Omega_c\, t_{\rm ff} = 0.5$, for comparison.

changes its form from sphere to disk. In disk phase, the non-dimensional angular velocity oscillates around 0.3 with considerable amplitude.

The sold thick curve in figure 2 denotes the oblateness (Δ) of the cloud defined by,

$$\Delta = \frac{2(D_{\rm major} - D_{\rm minor})}{D_{\rm major} + D_{\rm minor}}\,, \tag{3}$$

where $D_{\rm major}$ and $D_{\rm minor}$ are the lengths of the major and minor axes of the cloud in the $z = 0$ plane, respectively. The axial lengths are defined by the square root of the eigen value of $\boldsymbol{I}/M$, where M denotes the total mass in the region of $\rho \geq 0.1\rho_c$ and $\boldsymbol{I}$ does the tensor moment of inertia defined by,

$$I_{ij} = \int_{\rho(\boldsymbol{r}) \geq 0.1\rho_c} \rho(\boldsymbol{r}) r_i r_j \, dV\,, \quad i, j = 1, 2, 3\,. \tag{4}$$

The oblateness also increases roughly in proportion to ${\rho_c}^{1/6}$ while it undulates from spherical phase though disk phase to bar phase. When the oblateness is order of unity ($\Delta \simeq 1$), the cloud has a bar shape.

The growth of the perturbation is mainly due to incompressible shear flow in the dense region. The flow near the center is decomposed into the infall, rotation, and incompressible shear flow. We confirm that the incompressible shear flow grows and is responsible for the bar formation. Unless $\delta\omega_{0,\varphi} = \delta\omega_{0,\theta} = 0$, the velocity perturbation contains the incompressible shear flow. When $\delta\omega_{0,\theta} \neq 0$, the perturbation grows to form a bar along the z-axis or a disk perpendicular to the z-axis.

4. Application to Binary Star Formation

In this section we discuss binary star formation in the collapse of dense molecular cloud. The binary formation will be associated with fragmentation of a thin bar during the isothermal collapse phase.

When the initial cloud core rotates, it spins-up in proportion to $\Omega_c \, t_{\rm ff} \propto {\rho_c}^{1/6}$. When the non-dimensional angular velocity reaches ~ 0.5, a flat disk forms in the dense part during the isothermal ($n_{\rm H_2} \lesssim 10^{10}\,{\rm cm}^{-3}$) if the initial core satisfies

$$\Omega \, t_{\rm ff} \gtrsim 0.05 \left(\frac{n_{\rm H_2}}{10^4\,{\rm cm}^{-3}} \right)^{1/6} . \qquad (5)$$

The oblateness increases up to order of unity during the isothermal collapse if the initial cloud has the oblateness of,

$$\Delta \gtrsim 0.1 \left(\frac{n_{\rm H_2}}{10^4\,{\rm cm}^{-3}} \right)^{1/6} . \qquad (6)$$

When the initial cloud core satisfies with equations (5) and (6), the cloud core forms bar as its internal structure during its isothermal collapse. If the bar fragments, the fragments evolve to seeds of binary stars. The binary stars are associated with a common disk.

Numerical computations were carried out on VPP300/16R at NAOJ.

References

Bodenheimer, P., Ruzmaikina, T., & Mathieu, R. D., 1993 in *Protostars and Planets III*, ed. E. H. Levy and J. I. Lunine, pp. 125
Hanawa, T., & Nakamura, K. 1997, ApJ 484, 238
Larson, R. B. 1969, MNRAS, 145, 271
Mathieu R. D. 1994, ARAA, 32, 465
Matsumoto, T., Hanawa, T., & Nakamura, F., 1997, ApJ, 478, 569
Nakamura, F., Hanawa, T., & Nakano, T. 1995, ApJ, 444, 770
Penston, M. V. 1969, MNRAS, 144, 425
Truelove, J. K., Klein, R. I., Mckee, C. F., Holliman, J. H. II, Howell, L. H., Greenough, J. A., & Woods, D. T. 1998, ApJ, 495, 821.

THE FORMATION OF COMMON-ENVELOPE, PRE-MAIN-SEQUENCE BINARY STARS

J.E. TOHLINE, J.E. CAZES AND H.S. COHL
Louisiana State University
Department of Physics & Astronomy, 202 Nicholson Hall, Baton Rouge, LA 70803-4001 U.S.A.

1. Introduction

Recent observational investigations of the frequency of occurrence of pre-main-sequence binary stars have reinforced earlier suspicions that "binary formation is the primary branch of the star-formation process" (Mathieu 1994). As Bodenheimer *et al.* (1993) have reviewed, a number of different theories have been proposed to explain the preponderance of binary stars. Klein *et al.* (1998) show how the direct fragmentation of protostellar gas clouds may occur in early phases of collapse (at cloud densities $n \sim 10^3 - 10^{10}\text{cm}^{-3}$). But at higher densities, clouds are unable to cool efficiently upon contraction. Consequently, direct fragmentation becomes problematical. Because higher mean densities are associated with systems having shorter dynamical times, one is led to consider mechanisms other than direct cloud fragmentation for forming binary systems with orbital periods less than a few hundred years. Here we investigate whether such binaries can form by spontaneous fission of rapidly rotating protostars.

2. The Classical Fission Hypothesis

As Chandrasekhar (1969) has reviewed (see also Durisen & Tohline 1985), if protostellar objects are assumed to be self-gravitating, incompressible fluids with uniform vorticity, one can show analytically that their allowed equilibrium configurations are defined by spheroids or ellipsoids. Classically, models describing the slow contraction of rotating protostellar gas clouds have been formulated around such analytically prescribable equilibrium configurations. For example, a large, slowly rotating gas cloud with a relatively small ratio of rotational to gravitational potential energy $T/|W|$ will resem-

S. M. Miyama et al. (eds.), Numerical Astrophysics, 155–158.

ble a Maclaurin spheroid. As it contracts conserving angular momentum and mass, its evolution will proceed along the Maclaurin sequence through progressively flatter configurations of higher $T/|W|$. At a sufficiently high $T/|W|$, one finds that the axisymmetric configuration is no longer the lowest energy state available. Instead, there is an ellipsoidal configuration to which the gas cloud will prefer to evolve. Furthermore, if one follows evolution along a more and more distorted ellipsoidal sequence (such as the Jacobi sequence or any one of the Riemann sequences), one finds that eventually other configurations with even higher order surface distortions become energetically favorable. For example (see Fig. 3 of Durisen & Tohline 1985), there is a "dumbbell-binary sequence" that branches smoothly off of the Jacobi ellipsoid sequence. One might imagine, therefore, that binary stars form from the slow contraction of a rapidly rotating gas cloud along the Maclaurin, then Jacobi (or Riemann), then dumbbell-binary sequences. In reality, the picture is not this clear. Most significantly, detailed work on ellipsoidal figures of equilibrium has only been completed for incompressible fluid systems. It is not at all clear to what extent the results carry over to more realistic structures having compressible equations of state.

3. Problems Promoting the Fission Hypothesis

While examining the structure of rotating gas clouds that form the compressible analogues of Maclaurin spheroids, Ostriker & Bodenheimer (1968) showed that models with reasonable degrees of compressibility must incorporate a significant degree of differential rotation if they are to possess reasonably high values of $T/|W|$ and, therefore, be physically interesting in the context of the fission hypothesis. Employing 3D numerical hydrodynamics techniques, Durisen *et al.* (1986), Williams & Tohline (1988), and Houser *et al.* (1994) have examined the relative stability of rapidly rotating, compressible gas clouds that are initially in axisymmetric equilibrium but which reside just past a critical bifurcation point along the axisymmetric sequence according to Ostriker & Bodenheimer (1968). Invariably these simulations have shown that models with $T/|W| > 0.27$ are dynamically unstable toward the growth of a nonaxisymmetric deformation but, unlike their uniformly rotating, incompressible counterparts, the eigenmode to which these structures appear to be unstable has a spiral character.

Employing a significantly improved finite-difference simulation code and improved spatial resolution (128^3 grid zones), we recently have repeated the simulation that was first reported in Durisen *et al.* (1986). Movie1 (on the accompanying CD-ROM) shows the nonlinear development of the two-armed, spiral-mode instability. The evolution is shown in the inertial reference frame and covers 20 central initial rotation periods. Each frame of

Movie1 displays four nested isodensity contours at $\rho/\rho_{max} = 0.8, 0.4, 0.04$, and 0.004. Via the trailing spiral structure, gravitational torques are able to effectively redistribute angular momentum on a dynamical time scale; a relatively small amount of material is shed into an equatorial disk (this disk material is not visible in Movie1 because $\rho_{disk} < 0.004\rho_{max}$); and the central object (containing most of the initial object's mass) settles down into a new equilibrium configuration. Clearly, evolution to a binary star system as suggested by the classical fission hypothesis does not occur. It is primarily because simulations of this type have not produced a binary star system that the classical fission hypothesis has lost favor within the star formation community over the past decade (Bodenheimer *et al.* 1993).

4. Fission Hypothesis Revived

Interestingly, the instability illustrated by Movie1 produces a final steady-state object (hereafter referred to as the "final bar") that is dynamically stable, has a $T/|W| \sim 0.25$, and has a decidedly nonaxisymmetric structure. In many respects this final bar appears to be a compressible analog of a Riemann ellipsoid but, as Movie2 illustrates, the configuration possesses nontrivial internal motions. In the first frame of Movie2, 108 test particles have been lined up along the major axis of the final bar. Thereafter the particles are followed as they move along equatorial-plane streamlines of the flow, as viewed in a frame of reference that is rotating with the overall pattern speed of the final bar. The illustrated flow is entirely prograde and largely differential, but there is a small volume near the center of the configuration that is moving harmonically. In Movie3 we illustrate the 3D flow-field of the final bar. In the first frame of Movie3, a vertical *sheet* of test particles has been aligned with the major axis of the final bar. The subsequent motion of these particles illustrates that there is relatively little vertical fluid motion and, although it varies with R and θ, the angular velocity $\omega(\mathbf{x})$ is almost independent of z. It may be possible, therefore, to understand this and similar systems in terms of the properties of simpler, 2D nonaxisymmetric structures.

Andalib (1998) recently has developed a self-consistent-field technique that can be used to construct equilibrium models of infinitesimally thin, self-gravitating gaseous disks with (a) compressible equations of state, (b) nonaxisymmetric structures, and (c) nontrivial internal motions. By demanding that the disks have uniform vortensity $\zeta \equiv (\nabla \times \mathbf{v})/\rho$, Andalib has successfully constructed equilibrium disks with polytropic indices $0 \leq n \leq 1.3$ and minor-to-major axis ratios in the range $0.80 \geq b/a \geq 0.06$. Movie4 illustrates the internal flow of four of Andalib's compressible disks with nonaxisymmetric structures: one with fully retrograde internal motions (R); one with fully prograde internal motions (P); one with vortices sandwiched between separate regions of prograde and retrograde flow (V); and a common-envelope binary (dumbbell-shaped) configuration (D).

The similarity between the flow illustrated in Movie2 and the flow in Andalib's model P (Movie4) is striking. Apparently Andalib's model provides a good 2D analog of the 3D "final bar" that formed as a result of our fully hydrodynamic simulation of the two-armed, spiral mode instability (Movie1). Furthermore, Andalib's work demonstrates that model P is just one among a series of compressible models with nontrivial internal flows that defines a smooth elliptical-dumbbell-binary sequence. We suspect, therefore, that the final bar sits on an analogous (3D) sequence and that, if it is cooled slowly, it will evolve along the sequence to a common-envelope binary configuration such as the one illustrated by model D in Movie4. Additional support for this conjecture comes from New & Tohline (1997) who have demonstrated that stable, equal-mass common-envelope binaries can be constructed for fully 3D fluid systems with a sufficiently compressible equation of state.

In summary, it seems clear that a wide variety of rapidly rotating, nonaxisymmetric systems can be constructed with compressible equations of state. This work gives us renewed confidence that fission offers a viable route to binary star formation. Future investigations designed to model the slow cooling and contraction of initially nonaxisymmetric configurations like the final bar described above should demonstrate whether or not this scenario is correct. This research has been supported in part by the U.S. National Science Foundation (AST-9528424), and in part through grants of computing time at the San Diego Supercomputer Center and the NAVOCEANO Major Shared Resource Center.

References

Andalib, S.W. (1998), *The Structure and Stability of Selected, 2-D Self-Gravitating Systems*, Ph.D. Dissertation, Louisiana State University

Bodenheimer, P., Ruzmaikina, T. and Mathieu, R.D. (1993), in *Protostars and Planets, III*, ed. E.H. Levy and J.I. Lunine. University of Arizona Press, Tucson, AZ, U.S.A. p. 367

Chandrasekhar, S. (1969), *Ellipsoidal Figures of Equilibrium*. Yale University Press, New Haven, CT, U.S.A.

Durisen, R.H., Gingold, R.A., Tohline, J.E. and Boss, A.P. (1986), *Ap.J.*,**305**,p. 281

Durisen, R.H. and Tohline, J.E. (1985), in *Protostars and Planets, II*, ed. D.C. Black and M. Mathews. University of Arizona Press, Tucson, AZ, U.S.A. p. 534

Houser, J.L., Centrella, J.M. and Smith, S.C. (1994), *Phys. Rev. Lett.*,**72**,p. 1314

Klein, R.I, McKee, C.F. and Fisher, R. (1998), *These proceedings*

Mathieu, R.D. (1994), *Ann. Rev. Astr. Ap.*,**32**,p. 465

New, K.B.C. and Tohline, J.E. (1997), *Ap.J.*,**490**,p. 311

Ostriker, J.P. and Bodenheimer, P. (1968), *Ap.J.*,**151**,p. 1089

Williams, H.A. and Tohline, J.E. (1988), *Ap.J.*,**334**,p. 449

EVOLUTION OF ORBITAL ECCENTRICITY OF A PROTOPLANET

TAKU TAKEUCHI
Department of Earth and Planetary Sciences,
Tokyo Institute of Technology, Tokyo 152-8551, Japan

Abstract. The evolution of the orbital eccentricity of a planet due to tidal interaction with a protoplanetary disk is modeled. It is found that the eccentricity of the planet tends to increase when the mass of the planet is large enough to make a gap in the disk, i.e., is larger than about a Saturn mass. The tidal interaction with the protoplanetary disk explains the high eccentricities of the extra-solar planets.

1. Introduction

Recently, extra-solar planets have been found. These extra-solar planets can provide us key information about the formation and evolution of the planetary system. The total number of extra-solar planets is about 20, and it is natural to expect that statistical discussions on the nature of the extra-solar planets and their orbits are possible. One of the interesting arguments is about the relationship between the orbital eccentricity and the mass of the planet (Mazeh, Mayor & Latham 1997). It seems that the orbital eccentricities are distributed uniformly, from nearly circular to nearly parabolic, for planets larger than 10 times Jupiter's mass, whereas planets smaller than 10 times Jupiter's mass have small eccentricities. This feature of orbital eccentricity could be the result of the formation sequence of the planets.

In this paper, orbital evolution occurring immediately after the formation of a planet is studied in order to understand the early sequence of planet formation. At such an epoch, there remains much gas in the protoplanetary disk, and the gas interacts gravitationally with the planet to cause the orbital evolution.

S. M. Miyama et al. (eds.), Numerical Astrophysics, 159–162.

2. Calculation Procedure

Evolution of the orbital eccentricity of a planet through the gravitational interaction with the disk is strongly dependent on the properties of the disk, such as the density, temperature, and rotation velocity profiles. A planet in a disk exerts its gravity on the disk and especially modifies the density structure. When the planet is large enough, it opens a gap around its orbit in the disk. The gap formation removes the gas from the neighborhood of the planet and reduces the gravitational interaction between the planet and the disk. This results in the suppression of orbital evolution of the planet. Thus, in order to calculate the orbital evolution, we should establish how the density profile is modified.

The modification of the disk occurs in a relatively short time, i.e., the range of $10^3 - 10^4$ yr by a planet with Jupiter's mass at several AU from the central star. On the other hand, orbital evolution is a more secular process: it occurs in the range of 10^5 yr. Because of the significant difference in the time scale between these processes, these processes can be solved separately. The process of the modification of the disk is solved by numerical simulations, then the secular process of orbital evolution is solved by linear theory.

3. Gap Formation

The density structure of the disk with the planet is calculated by numerical simulations. Because the orbital evolution occurs in a much longer time range than those for the gap formation, the orbit of the planet is assumed to be circular during the simulation. I used ZEUS-2D code (Stone & Norman 1992) which can simulate the angular momentum transfer in the disk precisely. The calculations have been carried out for various masses of the planets from 0.3 to 5.6 Jupiter's mass.

After about 200 revolutions of the planet, gap formation was completed and a quasi steady state has been obtained (Figure 1). The width of the gap is related to the mass of the planet: it is larger for a more massive planet. When a quasi steady state is achieved, the surface density is averaged azimuthally, and the radial distribution of the surface density is calculated. The newly obtained radial surface density is used to evaluate the evolution of the orbital eccentricity of the planet.

4. Eccentricity Evolution

Evolution of the orbital eccentricity of the planet is calculated by linear theory (Goldreich & Tremaine 1980; Artymowicz 1993). This linear theory demonstrated that the gravitational interaction is decomposed in the

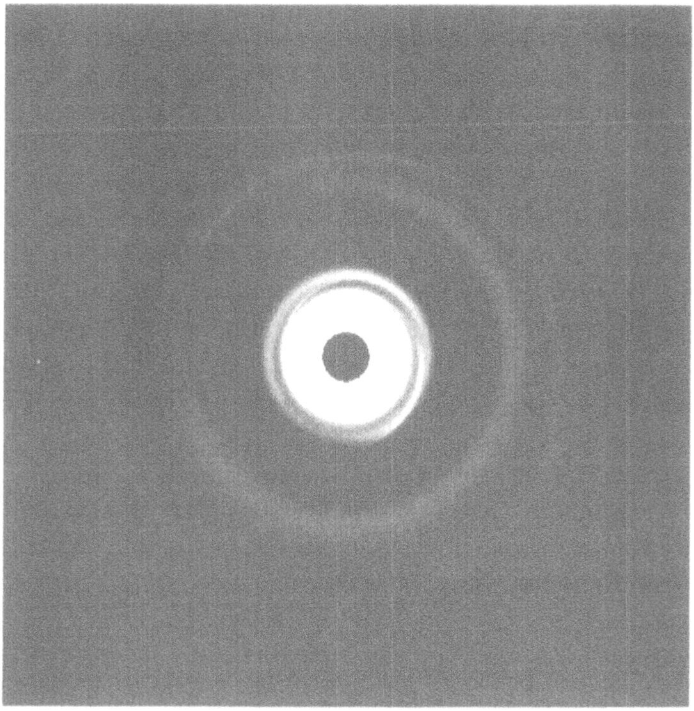

Figure 1. Surface density of a protoplanetary disk with a planet of Jupiter's mass. The planet opens a gap around its orbit.

summation of the contributions from the gas at the resonance positions:

$$\frac{de}{dt} = \sum_{\text{all resonances}} \left[\frac{de}{dt}(\sigma)\right]_{\text{each resonance}}, \tag{1}$$

where e is the orbital eccentricity of the planet. The contribution of each resonance is dependent on the surface density at the resonance position, σ. When the planet makes a gap around its orbit, then the resonances near the planet's orbit fall in the gap, and the interactions with the gas at these resonances are suppressed [$(de/dt)_{\text{each resonance}} = 0$ if $\sigma = 0$ at the resonance position]. Therefore, the rate of the eccentricity evolution is determined by the resonances survived in the disk:

$$\frac{de}{dt} = \sum_{\text{survived resonances}} \left[\frac{de}{dt}(\sigma)\right]_{\text{each resonance}}. \tag{2}$$

In order to know which resonances survive, we need the surface density profile. The surface density profiles obtained by numerical simulations were used in this study. Then, the rate of eccentricity evolution was calculated for various masses of the planets from 0.3 to 5.6 Jupiter's mass.

It is found that the eccentricity grows for any planet mass. The growth time is about 10^5 yr. This tendency of eccentricity growth is explained by competition between the following two types of resonances. The first is co-rotation resonances. The gravitational interactions through the co-rotation resonances at the edge of the gap decrease the orbital eccentricity of the planet. (The co-rotation resonances far from the gap may increase the eccentricity. However, their contributions are significantly smaller than those from the resonances at the gap edge.) On the other hand, the second type of resonance, Lindblad resonances, increase the eccentricity. The positions of the co-rotation resonances are relatively nearer to the planet than the Lindblad resonances. When a gap opens around the orbit of the planet, the resonances near the planet fall in the gap and the interactions through these resonances should be suppressed. Because the co-rotation resonances are nearer to the planet, they fall in the gap earlier and more often than the Lindblad resonances. After the gap formation, gravitational interactions through the Lindblad resonances dominate the evolution of the orbital eccentricity, and therefore, the orbital eccentricity of the planet increases.

5. Summary

It was shown that the orbital eccentricity of a planet increases through the gravitational interaction with the protoplanetary disk, if the planet opens a gap in the disk. The planet opens a gap when its mass is larger than about Saturn's mass (Takeuchi, Miyama & Lin 1996), and thus, the orbital eccentricity of a planet larger than Saturn would increase during its formation stage. This result explains the large eccentricities of the massive extra-solar planets. However, there are some extra-solar planets which are larger than Saturn and have nearly circular orbits. It seems to be difficult to explain these small eccentricities by the gravitational interaction with protoplanetary disks. In order to solve the above discrepancy and also to progress further, more observations and theoretical study are strongly needed.

Numerical computations were carried out on a VX/4R at NAOJ.

References

Artymowicz, P. 1993, ApJ 419, 166
Goldreich, P., & Tremaine, S. 1980, ApJ 241, 425
Mazeh, T., Mayor, M., & Latham, D. W. 1997, ApJ 478, 367
Stone, J. M., & Norman, M. L. 1992, ApJS, 80, 753
Takeuchi, T., Miyama, S. M., & Lin, D. N. C. 1996, ApJ 460, 832

NUMERICAL SIMULATIONS OF THE FORMATION OF SMALL PROTOSTELLAR CLUSTERS USING AN SPH CODE

JAIME KLAPP AND LEONARDO DI G. SIGALOTTI
Instituto Nacional de Investigaciones Nucleares,
Apartado Postal 18-1027, Mexico 11801 D.F.

Using a survey of 16 dense cores, Myers et al. (1991) deduced that they posses apparent elongations that would indicate prolate, rather than oblate shapes, with relative elongations that correspond to axial ratios of 2:1, but if some of them are oblate, then their axial ratios would be 3:1 to 10:1 or even larger. Myers et al. (1991) concluded that elongated cores embedded in much longer filaments are likely to be prolate if their long axes are approximately aligned with that of the filament but if their long axes are perpendicular to that of the filament, they could well be oblate.

In order to address the question of whether fragmentation of oblate cores can lead to small cluster formation, in this paper we use a smoothed particle hydrodynamics code to investigate the gravitational collapse and fragmentation of centrally condensed (Gaussian), oblate molecular clouds with varied thermal energy and axial ratios of 2:1, 4:1, and 8:1. All models start with uniform rotation and ratio of the rotational to the gravitational energy $\beta = 0.02$.

For the initial protostellar collapse models we consider values of the ratio of the thermal to the gravitational energy α of 0.48, 0.37 and 0.18 and axial ratios of 2:1, 4:1 and 8:1. This gives a total of nine initial models. In all cases, the starting point of collapse is taken to be a centrally condensed configuration, with the central density being a factor of 20 higher than the boundary density. The central condensation is defined by a Gaussian density variation in radius consistent with the average radial density profiles inferred in the inner regions of precollapse cloud cores.

All models experienced a qualitatively similar evolution regardless of the initial conditions. As a result of the initial oblate geometry, the clouds first collapse along their shorter axis to form a rotationally unsupported, intermediate central core which may be spherical, oblate spheroidal, or even disk-like depending on the value of α. The evolution proceeds with the central core undergoing rapid inward collapse and becoming progressively

S. M. Miyama et al. (eds.), Numerical Astrophysics, 163–164.

flatter. During this phase, a rotationally supported, thin pancake forms which soon after bounces and becomes highly susceptible to fragmentation.

The 2:1 oblate clouds formed low-order protostellar systems for all values of α . In particular, a binary system resulted for $\alpha = 0.48$ and a triple system formed for $\alpha = 0.37$ and 0.18. Low-order systems also formed in oblate clouds with much more extreme axial ratios for relatively high values of α. The 4:1 clouds produced a triple and a quintuple protostellar core for $\alpha = 0.48$ and 0.37, respectively, while the 8:1 clouds resulted in a quintuple system for $\alpha = 0.48$. Fragmentation into small clusters consisting of order 10 protostars was seen in the 4:1 clouds only for $\alpha = 0.18$ and in the 8:1 clouds for $\alpha = 0.37$.

The results imply that increasing the degree of cloud oblateness extends the range of the initial conditions for which fragmentation into a larger number of protostars is allowed. The number of fragments and the sizes of the calculated protostellar systems are toward the low end of the range of values inferred for real small stellar clusters (Strom et al. 1993; Chen & Tokunaga 1994), implying that they must undergo substantial orbital relaxation and possibly fragmentation into new condensations for matching the observations. The fragments form with a relatively wide spectrum of masses (0.023–4.152 $M_\odot$) and are close to become first protostellar cores. They also form with low values of α and enough rotation so that they could subfragment in the subsequent evolution.

The present models demonstrate that oblate Gaussian clouds with mild axial ratios of 2:1 can collapse and fragment into low-order systems for a wide range of values. It is believed that these systems may explain the origin of the bulk of observed binary and multiple PMS stars. The results may have implications for the formation of the small clusters of young stars observed in star-forming regions such as L1641 (Chen & Tokunaga 1994). The protostellar clusters formed in some of the models have diameters of 12000 to 33000 AU, and fragment separations ranging from 2600 to 33000 AU. These sizes are close to the lower end of the diameter range (40000 - 150000 AU) inferred for the small clusters of stars in L1641 (Strom et al. 1993). Thus matching the upper end of the observed range would require substantial orbital relaxation of the forming protostellar clusters.

This work was supported in part by the mexican Consejo Nacional de Ciencia y Tecnología (CONACyT).

References

Chen, H., & Tokunaga, A. T. 1994, ApJS, **90**, 149.
Myers, P. C., Fuller, G. A., Goodman, A. A., & Benson, P. J. 1991, ApJ, **376**, 561.
Strom, K. M., Strom, S. E., & Merrill, K. M. 1993, ApJ, **412**, 23.

FORMATION OF A DENSE FILAMENTARY CORE AND ITS FRAGMENTATION INTO PRE-PROTOSTELLAR CORES

FUMITAKA NAKAMURA
Faculty of Education, Niigata University
8050 Ikarashi-2, Niigata-shi, Niigata, 950-2181, Japan

1. Introduction

One of the most important goals in the study of star formation is to understand how stars are formed in molecular clouds. Recent observations have revealed that most molecular clouds have filamentary shapes and contain many dense cores. The dense core has typical size of the order of 0.01 – 0.1 pc, mean density of $\sim 10^{4-5}$ cm^{-3}, the temperature of $T \sim 10$ K, and mass of a few $M_{\odot}$ in low mass star forming region (Myers & Benson 1983). The dense cores often have one or more YSOs. It indicates that the dense core is the most fundamental unit that can collapse to form stars. The mass of the dense core is expected to relate to the mass of stars forming in it. It is therefore important to understand what determines the typical mass of the dense core. In this paper we thus study the collapse and fragmentation of the filamentary cloud, taking account of heating and cooling processes.

2. Numerical Results

By means of 1D hydrodynamical simulations, we follow the collapse of filamentary clouds, taking account of the following heating and cooling processes (Hasegawa 1988): (1) grain photoelectric heating, (2) cosmic ray heating, (3) CI photoionization heating, (4) H_2 formation heating, (5) gas-dust interactions, and (6) CI, CII, OI, and CO line coolings.

As a model of a filamentary cloud we consider an infinitely long cylindrical cloud in which the density is uniform along the cylinder axis. At the initial state the radial distribution of the number density is expressed as $n = n_0 \left[1 + r^2/(R_{\rm fil})^2\right]^{-2}$ and $R_{\rm fil} = \sqrt{2fkT_0/(\pi G\rho_0\mu m_{\rm H})}$, where n_0, T_0, and $f^{1/2}$ are the central number density, temperature, and the radius in unit of the equilibrium radius, $R_{\rm fil}$, respectively. When $f = 1$, the cloud is in hydrostatic equilibrium. This model is specified by the parameters, n_0, T_0, and f.

S. M. Miyama et al. (eds.), Numerical Astrophysics, 165–166.

We have pursued the evolution of various models with different initial parameters and have turned out that the evolution depends only weakly on the initial parameters. In the following, as a typical example, we show the evolution of the model with the parameters $n_0 = 500$ cm^{-3}, $T = 15$ K, and $f = 2$. Figure 1 shows the evolution of the density and temperature distributions. During the contraction the filamentary cloud keeps the nearly constant temperature of $T \sim 10 - 15$ K. When the central density (n_c) exceeds 10^4 cm^{-3}, the cloud separates into two parts, i.e., a denser spindle and a diffuse envelope. The contraction of the spindle proceeds quasistatically due to radiative cooling, while the envelope gas accretes on the spindle with a super sonic speed. Thus the shock wave is formed at the surface of the spindle, i.e., $r \sim 0.05$ pc or $n \sim 10^4$ cm^{-3}, inside which the dominant cooling transitions of CO are optically thick.

Applying a linear theory (Larson 1985), we find that the spindle is likely to fragment into dense cores by the stage that n_c reaches $\sim 10^6$ cm^{-3} (see Figure 1b). The size and mass of the dense core are estimated to be nearly equal to ~ 0.01 pc and $0.5 - 2$ M$_\odot$, respectively. These characteristics are consistent with those of observed dense cores. The dense core formed from fragmentation of the spindle may be the most fundamental unit that can collapse to form stars.

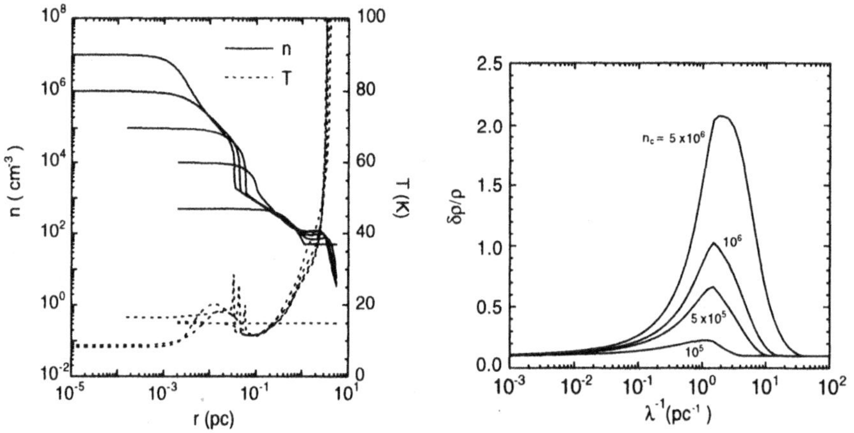

Figure 1. (a) left panel: Evolution of the number density and temperature distributions at the stages of $n_c = 500$, 10^4, 10^5, 10^6, and 10^7 cm^{-3}. (b) right panel: Evolution of the density perturbations. The solid curves denote $\delta\rho/\rho$ at the stages of $n_c \simeq 10^5$, 5×10^5, 10^6, and 5×10^6 cm^{-3}. For simplicity we take $\delta\rho/\rho|_{\rm initial\ state} = 0.1$. When $n_c \sim 10^6$ cm^{-3}, the maximum value of $\delta\rho/\rho$ exceeds unity and fragmentation is likely to occur.

References

Hasegawa, T. I. (1988) Hydrostatic Models of Bok Globules, *PASJ*, **40**, 219 – 247
Larson, R. B. (1985) Cloud Fragmentation and Stellar Masses, *MNRAS*, **214**, 379 – 398
Myers, P. C. & Benson, P. J. (1983) Dense Cores in Dark Clouds.II-NH3 Observations and Star Formation, *ApJ*, **266**, 309 – 320

DYNAMICAL CONTRACTION OF ROTATING POLYTROPIC GAS DISK

K. SAIGO, T. MATSUMOTO AND T. HANAWA
Department of Astrophysics, School of Science,
Nagoya University, Chikusa-ku, Nagoya 464-8602, Japan

We have studied gravitational collapse of rotating clouds with two-dimensional numerical simulations. In our simulations an initially filamentary cloud fragments to form a gravitationally contracting rotating gas disk. The structure of the disk depends on the initial rotation of the filamentary cloud and the stiffness of the gas. To study the latter dependence we assumed the polytropic equation of state, $P = K\rho^\gamma$, for the gas and constructed various models having different polytropic indices, γ.

We constructed an initial model by adding velocity perturbation on a rotating filamentary cloud in equilibrium. We assumed the rotation velocity in equilibrium so that the ratio of the centrifugal force to the pressure force is β:1 at any radius. Thus, our models are characterized essentially by two parameters, β and γ. The gravitationally contracting disks of our simulations are similar to those of isothermal clouds (see, e.g., Norman, Wilson, and Barton 1980). In this paper we concentrate on the dependence on γ.

The dynamically contracting disk depends critically on γ. When $\gamma < 1$, the disk thickness shortens much faster than the radius. Hence the Toomre Q-value becomes lower than unity at a late stage and ring instability occurs. When $\gamma > 1$, the collapse changes suddenly from dynamical (runaway collapse phase) to quasi-static (accretion phase). The quasi-static disk is mainly supported by the centrifugal force and accretes gas from the envelope. The central density is nearly constant in the accretion phase. The maximum central density is smaller for a larger γ and a larger β (Table 1).

Figure 1 shows the model of $\beta = 0.2$ and $\gamma = 1.2$. In this model, the equilibrium disk was formed at $t = 7.5$ (Figure 1 b). After $t = 7.5$, the equilibrium disk grows in mass and radius by accretion from the envelope (Figure 1 c, d). The grayness shows the density and the white curves denote the contour of $\sqrt{v_r^2 + v_z^2} = 0.3\, c_s$. As shown in panels (b), (c), and (d), the quasi-static disk extends mainly in the r-direction. Figure 2 shows the

S. M. Miyama et al. (eds.), Numerical Astrophysics, 167–168.

TABLE 1. The maximum central density in the accretion phase in the logarithmic scale: $\log \rho_c$

β	$\gamma = 1.05$	$\gamma = 1.10$	$\gamma = 1.20$
1.0	7.25	5.05	2.55
0.5	11.2	6.70	3.31
0.2		8.92	5.40

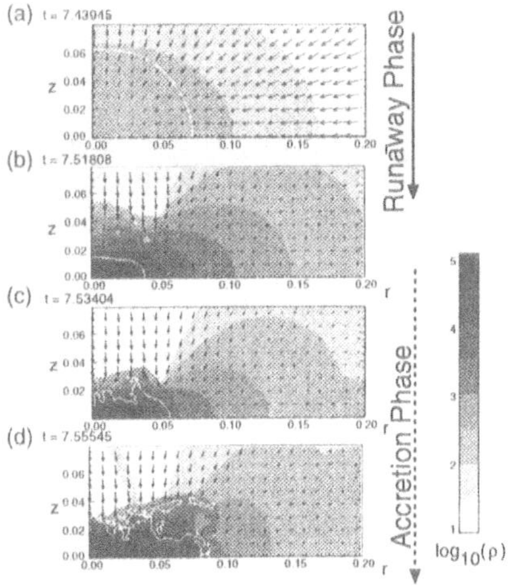

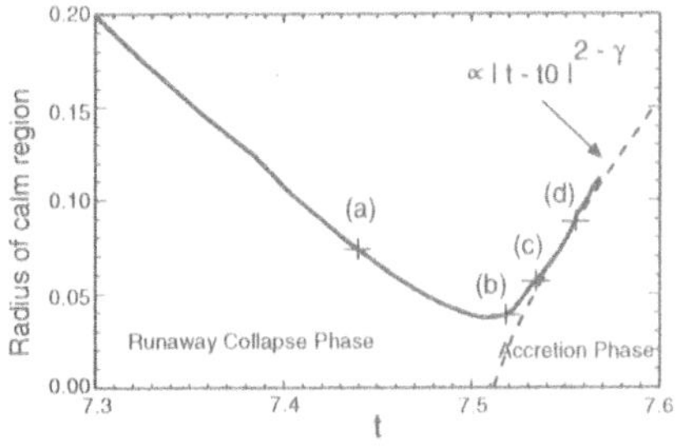

Figure 1. At the left is the evolution of the disk ($\beta = 0.2, \gamma = 1.2$).
Figure 2. At the right is the evolution of the characteristic radius, $(v_r^2 + v_z^2)^{1/2} = 0.3\, c_s$.

location of $v_r = -0.3\, c_s$ on the $z = 0$ plane as a function of the time. The radius of the equilibrium disk grows in proportional to $(t - t_0)^{2-\gamma}$. The same growth is found for $\gamma = 1$ by Saigo and Hanawa (1998). This dependence indicates that similarity law is a good approximation to the equilibrium disk (see, e.g., Yahil 1983).

References

Norman, M.L., Wilson, J.R. and Barton, R.T. (1980) A New Calculation on Rotating Protostar Collapse, *ApJ*, **239**, 968 – 981

Saigo, K. and Hanawa, T. (1998) Similarity Solution for Formation of a Circumstellar Disk through the Collapse of a Flattened Rotating Cloud, *ApJ*, **493**, 342 – 350

Yahil, A. (1983) Self-Similar Stellar Collapse, *ApJ*, **265**, 1047 – 1055

A RADIATION HYDRODYNAMICAL MODEL FOR PROTOSTELLAR COLLAPSE

HIROHIKO MASUNAGA, S. INUTSUKA AND S. M. MIYAMA
Division of Theoretical Astrophysics,
National Astronomical Observatory,
Mitaka, Tokyo 181-8588, Japan

1. Summary of the Numerical Results

In our previous work (Masunaga, Miyama, & Inutsuka, 1998), the evolutions before hydrogen molecules begin to dissociate (the so-called first collapse) were investigated. Calculations were based on an exact treatment of the radiation field in spherical symmetry by using the variable Eddington factor method with a frequency-dependent radiative transfer equation. The results are summarized as follows. [**1.**] The first core had a final mass of $\sim 0.05 M_{\odot}$, a radius of ~ 5 AU for $T_{\rm init} = 10$K, and typical opacity. When the initial temperature and opacity are given, these values for mass and radius are uniquely determined and independent both of the mass of the parent cloud core and of the initial density profile. [**2.**] The thermal evolution at the center of a collapsing cloud core was characterized by three different conditions, which are described in §2 below. [**3.**] We found the analytic expression for the radius of the first core and the accretion luminosity as functions of the central density. They were weakly dependent on the initial temperature and opacity. Those expressions were useful to understand the numerical results. [**4.**] Dense cloud cores with luminosities higher than $\sim 0.1 L_{\odot}$ possibly have young protostars inside the opaque envelopes, even if observations showed no evidence of protostellar activities.

At present, we are trying to improve our numerical code to make it suitable for the second collapse and further evolutions. The following improvements have been made since our former calculations for the first collapse. [**1.**] Partition functions for rotational and vibrational degrees of freedom for H_2 are solved to obtain the ratio of the specific heats, γ. Thus, γ is given as a function of temperature, varying from 5/3 at low T to $\leq 7/5$ at higher T. [**2.**] The dissociation of hydrogen molecules and the ionization

S. M. Miyama et al. (eds.), Numerical Astrophysics, 169–170.

of hydrogen and helium atoms have been incorporated. Dissociation of H_2 causes a further dynamical collapse of the first core, i.e., the second collapse. **[3.]** A non-ideal equation of state has been incorporated using data by Saumon et al. (1995). This new EOS enables us to take into account electron degeneracy, pressure ionization, and non-ideal Coulomb effects.

Succeeding evolutions will be investigated in subsequent work.

2. CONDITIONS FOR VIOLATION OF ISOTHERMALITY

In this section, we present a brief summary of Masunaga & Inutsuka (1999). **[1.]** The conditions for the violation of isothermality are presented through three different criteria: (1) the compressional heating rate overtakes the thermal cooling rate; (2) the optical depth for thermal radiation reaches unity; or (3) the compressional heating rate becomes comparable to the energy transport rate due to radiative diffusion. These criteria are supported by numerical results in our previous work. We found that either condition (1) or (3) above is sufficient in plausible ranges of parameters for actual molecular clouds. The common idea introduced by condition (2) has no actual importance. **[2.]** The minimum Jeans mass for fragmentation, $M_{\rm F}$, is characterized by the conditions for the violation of isothermality. We found that the typical value of $M_{\rm F}$ was $\sim 10^{-2} M_\odot$, in accordance with the commonly believed value, but $M_{\rm F}$ can increase substantially depending on $T_{\rm init}$ and κ. In particular, molecular clouds with lower metallicity yield larger $M_{\rm F}$, inversely proportionally to κ. **[3.]** We propose a simple strategy which takes into account thermal processes by radiation in hydrodynamic simulations without additional computational effort. It will be useful to pursue the hydrodynamic evolutions of collapsing clouds when adiabatic evolution takes the place of isothermal evolution. See Masunaga & Inutsuka (1999) for more detailed discussions.

References

Masunaga, H. & Inutsuka, S. 1999, ApJ., Vol. 510 (the Jan. 10 issue)
Masunaga, H., Miyama, S. M., & Inutsuka, S. 1998, ApJ., 495, 346
Saumon, D., Chabrier, G., & Van Horn, H. M. 1995 ApJS., 99, 713

RADIATION HYDRODYNAMICAL SIMULATIONS OF THE GRAVITATIONAL COLLAPSE OF FILAMENTARY GAS CLOUDS

K. OGOCHI AND T. NAKAMOTO
University of Tsukuba
Ibaraki 305-8577, Japan

1. Introduction

Some observations suggest that many molecular clouds have a filamentary shape. Also, stars seem to be formed by the fragmentation of filaments. In this fragmentation process, radiation may play an important role because it determines the temperature of the clouds. To understand this process fully, we studied the gravitational collapse of filamentary clouds using radiation hydrodynamics calculations (Stone, Mihalas, & Norman, 1992).

2. Model

A filamentary cloud is assumed to be an axisymmetric infinite cylinder. It is also assumed that the cloud does not rotate, there is no magnetic field, the opacity is gray, and the initial density distribution is given by $\rho_{\rm init}(r) = \rho_0[1 + (r/hH_0)^2]^{-2}$, where ρ_0 is the initial density at the axis, $H_0 = (2c_T^2/\pi G\rho_0)^{1/2}$ is the scale height, h is a parameter which represents a degree of deviation from the equilibrium distribution, and c_T is the isothermal sound speed. The equilibrium distribution corresponds to the $h = 1$ case. The filament contracts gravitationally if $h > 1$. In the present study, an initial assumption of $1 < h \leq 2$ and zero velocity is made.

3. Results and discussions

Our numerical simulations showed that the density distribution in the isothermal phase of the collapse could be approximated by the equilibrium distribution $\rho(r) = \rho_c[1 + (r/H_c)^2]^{-2}$, where $H_c = (2c_T^2/\pi G\rho_c)^{1/2}$ and ρ_c was the density at the symmetric axis. Notice that this resultant distri-

S. M. Miyama et al. (eds.), Numerical Astrophysics, 171–172.

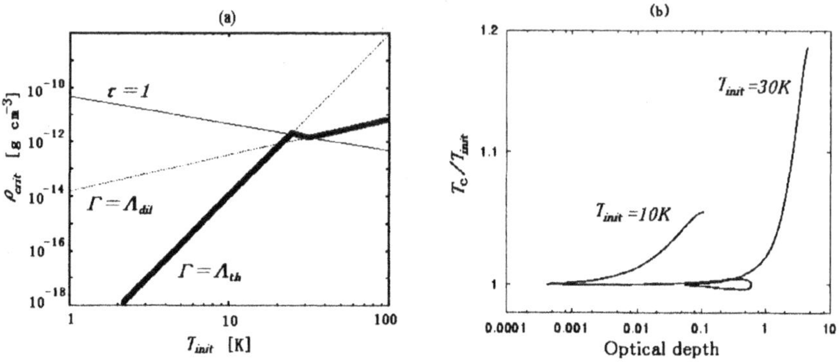

Figure 1. (a): A thick line shows the critical density as a function of initial temperature. (b): Temperature on the axis divided by the initial temperature vs. the optical depth. As the collapse proceeds, the optical depth to the axis increases. The temperature for $T_{\rm init} = 10$K case begins to rise in even an optically thin regime, $\tau \ll 1$.

bution does not depend on the parameter h. The temperature of the cloud rose when the density exceeded a certain critical value. This critical density can be estimated as follows (see also Masunaga, Miyama, & Inutsuka, 1998 for a spherical collapse case): the cloud temperature may be expected to depart from the isothermal value when the heating rate Γ exceeds the cooling rate Λ. In a case of gravitational collapse of filamentary clouds, Γ and Λ at the axis of the symmetry can be evaluated as

$$\Gamma \simeq c_T^2 \left(\frac{\pi G \rho_c}{2}\right)^{1/2}, \quad \Lambda = \begin{cases} \Lambda_{\rm th} & (\tau < 1) \\ \Lambda_{\rm dif} \simeq \Lambda_{\rm th}/\tau^2 & (\tau > 1) \end{cases},$$

where $\Lambda_{\rm th} = 4\kappa\sigma T_c^4$, T_c is the temperature at the symmetric axis, and $\tau = (\pi/4)\rho_c \kappa H_c$ is the optical depth to the axis. By equating $\Gamma = \Lambda$, we could obtain the critical density $\rho_{\rm crit}$ as a function of the initial temperature $T_{\rm init}$(Figure 1a). The estimated critical density was in good agreement with our numerical results. For instance, the deviation from the isothermal state can occur even in an optically thin cloud if the initial temperature is low enough (Figure 1b). This suggests that calculations in which a critical density is determined only by the optical depth condition may lead to significant errors.

References

Stone, J. M., Mihalas, D., & Norman, M. L. 1992, ApJS 80, 819
Masunaga, H., Miyama, S. M., & Inutsuka, S. 1998, ApJ 495, 346

GRAVITATIONAL COLLAPSE OF A FILAMENTARY MOLECULAR CLOUD TRIGGERED BY HII REGION

T. HANAWA, N. FUKUDA, T. MATSUMOTO,
AND Y. NAKAJIMA
Department of Astrophysics, Nagoya University
Chikusa-ku, Nagoya 464-8602, Japan

1. Motivation

In this paper we present a simulation of star formation triggered by HII region, i.e., dynamical influence of a newly formed star. Our model is motivated by Chandler and Carlstrom (1996) who suggested that the HII region, NGC2024, had triggered star formation in the associated molecular cloud. NGC2024 has a dark lane due to a filamentary molecular cloud running from the north to the south. The filamentary cloud is divided into northern and southern ones by ionized gas. The northern cloud contains far infrared sources, FIR 1-4, on the ridge, while the southern cloud FIR 5-7. The far infrared sources are star forming dense cores. Among the far infrared sources, FIR 4-6 are associated with outflows and are likely to be older than the rests. According to the dynamical ages of the outflows, FIR 5 is the oldest and FIR 4 is the second oldest. Since FIR4 and 5 are closest to the central star in the HII region, NGC 2024 is likely to trigger formation of the far infrared sources.

2. Simulation

We followed fragmentation and collapse of a filamentary magnetized cloud with a three-dimensional numerical simulation as a model for NGC2024. The simulation starts from a filamentary molecular cloud in magnetohydrodynamical equilibrium. The magnetic field is parallel to the cloud axis at the initial stage. We assume a point-like explosion near the filamentary cloud at the initial stage of the simulation. The explosion mimics an expanding HII region. We have made several models by changing the energy of the explosion.

S. M. Miyama et al. (eds.), Numerical Astrophysics, 173–174.

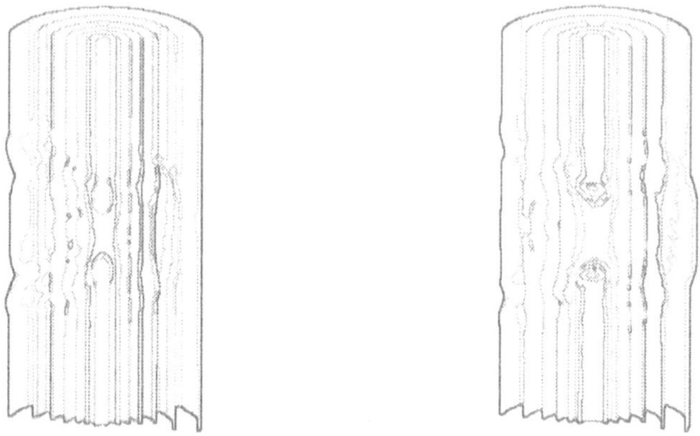

Figure 1. The left and right panels show the isodensity surfaces perspectively for the stages of $t = 8.5$ and 10.3, respectively.

Gross features of the simulation depend only weakly on the energy of the explosion. As the time proceeds, the bubble of the explosion expands and wraps up the molecular cloud. The wrapped part of the cloud becomes thinner, and the density and pressure thereof increase. Owing to the pressure increase, the gas flows along the filament axis and the filamentary cloud separates into two (see the left panel of Figure 1). Each of the two has a dense clump near the edge and the dense clump collapses into a high density disk owing to the self-gravity. Eventually the high density disk becomes isolated from the parent cloud (see the right panel of Figure 1). The filamentary cloud is expected to form gravitationally collapsed disks repeatedly. The movie of the simulation is stored in the CD-ROM.

In our simulation the self-gravity is the key for the dense core formation. When the change in the self-gravity is artificially suppressed, no dense disk forms. Our initial filamentary cloud model is unstable against the fragmentation owing to the Jeans instability. Once explosion disturbs the filamentary cloud, it fragments to form gravitationally contracting dense cores.

In this simulation we solved the ideal magnetohydrodynamical (MHD) equations and the Poisson equation. The former is solved with a TVD scheme and the latter with multi-grid iteration. The numerical simulation was carried out on VPP300/16R at NAOJ.

References

Chandler, C. J. and Carlstrom, J. E. (1996) Star Formation in the NGC2024 Molecular Ridge, *ApJ*, **vol. no. 466**, pp. 338–351.

NONAXISYMMETRIC DYNAMIC INSTABILITIES OF STAR/DISK SYSTEMS

J.N. IMAMURA
Institute of Theoretical Science, University of Oregon, Eugene, OR 97403, USA

B.K. PICKETT
NASA/Ames Research Center, MS245-3, Moffett Field, CA 94035, USA

R.H. DURISEN
Max Planck Institute for Extraterrestrial Physics, D-85740 Garching, Germany

AND

R. LINK
Department of Astronomy, Indiana University, Bloomington, IN 47405, USA

Abstract. Dynamic nonaxisymmetric instabilities of star/disk systems are studied in the linear regime. We consider perturbations with azimuthal dependence $\propto \exp(\pm im\phi)$ where $m = 1$ - 4, and ϕ is the azimuthal angle. The eigenvalues and eigenfunctions for the most dynamically unstable m-modes are found by evolving initially random infinitesimal perturbations away from equilibrium in the linear regime. For star-like objects, dynamic instability sets in for $T/|W| > 0.27$, similar to well-known case of the uniformly rotating, incompressible fluid Maclaurin spheroids. Here T is the bulk kinetic energy of the object and W is the gravitational energy of the object. For star/disk-like objects, dynamic instability sets in significantly earlier, at $T/|W| \sim 0.20$. For both star-like and and star/disk-like objects, dynamic instability sets in via an $m = 2$, *i.e.*, two-armed or bar-like mode. All models tested were stable to $m = 1$ perturbations.

In star formation, the way that mass and angular momentum are distributed in the initial pre-collapse cloud can play a large role in the outcome of the collapse (Bodenheimer 1995). In fact, a wide range of complex dynamic behaviors are possible (e.g., Bonnell & Bate 1994, Boss 1996, Burkert

S. M. Miyama et al. (eds.), Numerical Astrophysics, 175–176.

& Bodenheimer 1996). It has recently been shown, in some cases, that the global stability characteristics of the protostellar equilibrium objects which form from such collapses can also be rather sensitive to the angular momentum distribution (Imamura et al. 1995, Pickett et al. 1996). Hydrodynamic simulations in three dimensions are required to resolve the nonlinear outcome of instability. However, due to the physical and numerical complexity of such calculations, an important complementary approach is the study of these instabilities in the linear regime. Here, we report some results from a linear stability analysis of equilibrium protostellar objects that could form from pre-collapse clouds with Gaussian (G models) or singular isothermal sphere (I models) density profiles. The I models have the structure of a disk around a distinct central star, while the G models are more star-like (Pickett et al. 1997).

Our linear results for $m = 2$ in the G models are in quantitative agreement with those obtained using the Pickett et al. (1996) 3D hydrodynamics code. For a $T/|W| = 0.292$ model, the eigenvalues agree to within two digits, and greyscale plots of the density eigenfunctions are indistinguishable to the eye. G models were not studied by Toman et al. (1998) in their linear analysis; G models are intermediate between the $n' = 1$ and $3/2$ angular momentum distributions of Pickett et al. (1996) which were studied by Toman et al. (1998). Our $m = 2$ G model eigenvalues, $\omega_{osc} = 0.519$ and e-folding time $= 4.18$, are consistent with those found in Toman et al. (1998). Here, the eigenvalues are in units of CIRPs, the central initial rotation period of the equilibrium model.

The linear and nonlinear calculations for the I models agree that $m = 2$ instability sets in for lower $T/|W|$ than it does in the G objects. Beyond this, however, comparisons are problematic. The issues are: (i) In the linear calculations, the modes are global in nature, while in the nonlinear simulations they are localized. Further the eigenvalues differ widely. (ii) $m = 1$ is stable in all linear simulations, whereas at least transient growth is seen in the nonlinear code. The origin of the differences is not clear. It might be numerical in nature or due to the inherent differences between linear and nonlinear calculations. We are currently investigating this question.

References

Bodenheimer, P. (1995) ARA&A, 33, p. 199.
Bonnell, I. A. & Bate, M. R. (1994) MNRAS, 271, p. 999.
Boss, A. P. (1996) ApJ, 468, p. 231.
Burkert, A. & Bodenheimer, P. (1996) MNRAS, 280, p. 1190.
Imamura, J. N., Toman, J., Durisen, R. H., Pickett, B. K., & Yang, S. (1995) ApJ, 444, p. 360.
Pickett, B. K., Durisen, R. H., & Davis, G. A. (1996) ApJ, 458, p. 714.
Pickett, B. K., Durisen, R. H., & Link, R. (1997) Icarus, 126, p. 243.
Toman, J., Imamura, J.N., Pickett, B.K., & Durisen, R.H. (1998), ApJ, 496, 370.

THREE-DIMENSIONAL HYDRODYNAMICS OF PROTOSTELLAR DISKS: THE EFFECTS OF THERMAL ENERGETICS

BRIAN K. PICKETT AND PATRICK CASSEN
NASA/Ames Research Center
M.S. 245-3, Moffett Field, CA 94035, USA

AND

RICHARD H. DURISEN AND ROBERT LINK
Department of Astronomy, Indiana University
Swain West 319, Bloomington, IN 47405, USA

Abstract.
We use a second-order three-dimensional hydrodynamics code with self-gravity to investigate the role that thermal energetics plays in the development of nonaxisymmetric instabilities in protostellar disks. The initial axisymmetric equilibrium state is a continuous, fluid star/disk system, in which the star, the disk, and the star/disk boundary are resolved in 3D in our hydrodynamics code. An adiabatic evolution is compared to two previous simulations of the same model in which either local isentropy or local isothermality is maintained in the disk throughout the calculation. In all three cases, the model is highly unstable to multiple low-order nonaxisymmetric disturbances which induce significant mass and angular momentum transport in a few dynamical times. The star and star/disk boundary are dominated by three- and four-armed disturbances, whereas the disk is susceptible to a two-armed spiral. These disturbances saturate at moderate nonlinear amplitude in the adiabatic and isentropic evolutions; the same instabilities in the isothermal evolution lead to the disruption of the disk and concentration of material into several dense, thin arcs and arclets. Figures, tables and animations are included in the CD-ROM supplement.

Much numerical work has shown that protostellar objects are subject to nonaxisymmetric instabilities, generally due to a combination of rapid rotation, self-gravity, and thermal pressure. In this presentation, we inves-

S. M. Miyama et al. (eds.), Numerical Astrophysics, 177–178.

tigate how different assumptions about thermal energetics affect the origin and maintenance of nonaxisymmetric structure in three dimensions.

Our initial, continuous, axisymmetric equilibrium state is a cooled version of a model in the isentropic, polytropic sequence of star/disk models (the "I" models) studied in our previous work (Pickett *et al.* 1997, 1998; see also Imamura *et al.* in this volume). The equilibrium configuration represents an early stage in the axisymmetric collapse of a uniformly rotating singular isothermal sphere. The model is small in radial extent ($R_{eq} \sim 0.10$ AU), with the mass roughly evenly divided between star and rotationally supported disk. The minimum Toomre Q in the disk is 1.5. The model is shown in Figure 1; Table 1 lists physical properties.

Three simulations with our 3D hydrodynamics code (Pickett *et al.* 1996, 1998) were conducted: 1) a *locally isothermal* evolution, in which the disk temperature was held at its initial, axisymmetric value; 2) a *locally isentropic* evolution, in which the disk entropy is held fixed; and 3) an *adiabatic* evolution, in which artificial viscosity was used to treat shocks. The star was treated adiabatically in all cases. Each run begins with a random density perturbation throughout the disk; grid resolution is (128,64,16) in (r,φ,z).

Table 2 describes the simulations. We find:

1. The early growth of nonaxisymmetric modes is similar in all cases. The star and inner disk are subject to three- and four-armed Kelvin-like modes; the disk is unstable to a two-armed spiral.
2. The nonlinear development of the modes varies greatly. In the isothermal evolution, relative density amplitudes reach highly nonlinear levels ($\delta\rho/\rho \sim 80\%$ and above). The same instabilities in the isentropic evolution saturate at moderately nonlinear amplitudes, apparently from effective mode interaction (Laughlin & Różyczka 1996). In the adiabatic evolution, the amplitudes fall to the percent level or less due to heating from shocks (Figure 2).
3. Copious mass and angular momentum transport occurs (Figure 3).
4. The effects of the instabilities vary drastically. The two-armed spiral in the isothermal evolution disrupts the disk, concentrating most of the material outside the star in thin, dense arms. In the isentropic and adiabatic evolutions, the disk remains, although a strong two-armed spiral is embedded inside it. Further, heating of the low density regions in the adiabatic calculation leads to their expansion (Figures 4 and 5).

References

Laughlin, G., & Różyczka, M. (1996) ApJ, 456, 279.
Pickett, B. K., Durisen, R. H., & Davis, G. A. (1996) ApJ, 458, p. 714.
Pickett, B. K., Durisen, R. H., & Link, R. (1997) Icarus, 126, p. 243.
Pickett, B. K., Cassen, P., Durisen, R. H., & Link, R. (1998) ApJ, 504, in press.

MOLECULAR EVOLUTION IN PLANET-FORMING CIRCUMSTELLAR DISKS

Y. AIKAWA
Dept. of Physics and Astronomy, The Ohio State University
Columbus, Ohio 43210-1106, USA

T. UMEBAYASHI
Data processing Center, Yamagata University
Yamagata 990, Japan

T. NAKANO
Nobeyama Radio Obs., National Astronomical Observatory
Nobeyama, Minamisaku, Nagano 384-13, Japan

AND

S. M. MIYAMA
National Astronomical Observatory of Japan
Mitaka, Tokyo 181, Japan

1. Introduction

Theoretical study on the evolution of molecular abundance in protoplanetary disks is important from various points of view. First, it will directly show what material the bodies in planetary systems are made from. Secondly, it is indispensable for the observational study of the protoplanetary disks via molecular lines. Owing to the recent development of radio telescopes, observational study of protoplanetary disks is now in great progress. Since the molecular abundance is dependent on the physical conditions such as density, temperature and age of the disk, we can investigate the processes of planet-formation combining the theoretical study on molecular abundance with the observational data. Thirdly, molecular abundance can be a useful probe in investigating the formation processes of our own solar system. We could reducc the processes, places, and epochs of the formation of primitive bodies, such as comets, by comparing their molecular composition with the theoretical results on disk chemistry.

S. M. Miyama et al. (eds.), Numerical Astrophysics, 179–180.

2. Model

We investigate the evolution of molecular abundance in a protoplanetary disk adopting the steady accretion disk model. We solve reaction network of molecules numerically as an initial-value problem considering that the matter is gradually accreting towards the central star. We adopt the UMIST94 data base for the gas-phase reactions. We also take into account the gas-grain interaction such as grain-surface recombination and formation and desorption of ice mantle. We obtain the molecular abundance in gas phase and in ice mantle of grains as a function of time and position in the disk (Aikawa et al. 1998).

3. Results

We have found that ionization of molecules and formation of ice mantle play important roles in molecular evolution in the disk (see also Aikawa et al. 1997). In the region of surface density less than 10^2 g cm^{-2} (distance from the star $\gtrsim$ 10 AU), cosmic rays are barely attenuated even in the mid-plane of the disk and produce chemically active ions such as He^+ and H_3^+. Considerable amount of CO and N_2, which are the dominant component in the disk initially, are transformed into CO_2, CH_4, NH_3, and HCN through reactions with these ions within a few $\times 10^6$ yr. In the region where the temperature is low enough for these products to freeze onto grains, they accumulate in ice mantle. The time scale of the molecular evolution is larger if the ionization rate ζ is smaller or the grain radius a is larger in the disk than the typical values in molecular clouds ($\zeta \sim 10^{-17}$s^{-1}, $a \sim 10^{-5}$cm).

As the matter accretes towards inner warmer regions of the disk, the ice mantle evaporates. It is found that the desorbed molecules (e.g. CH_4) are transformed into less volatile molecules (e.g. hydrocarbons) by the reactions in gas phase and freeze out again. The molecular abundance both in gas phase and in ice mantle crucially depends on the temperature and thus varies significantly with the distance from the central star.

The molecular abundance derived from our model naturally reproduce the coexistence of oxidized ice and reduced ice, which is observed in comets. Our model shows that the comets formed at different regions of the disk will have different molecular composition. It is also suggested that we can estimate the temperature distribution and accretion velocity in the disk by observation of molecular lines with high spatial resolution in near future.

References

Aikawa, Y., Umebayashi, T., Nakano, T. and Miyama, S. M. 1997, ApJL, 486, L51

Aikawa, Y., Umebayashi, T., Nakano, T. and Miyama, S. M. 1998, in Faraday Discussion 109: Physics and Chemistry of Molecules and Grains in Space, in press

SEDS AND IR IMAGES OF DISKS AROUND YSOS

N. KIKUCHI, T. NAKAMOTO AND K. OGOCHI
University of Tsukuba
Tsukuba, Ibaraki 305-8577, Japan

1. Introduction

A spectral energy distribution (SED) is one of the most useful pieces of information from which physical properties of disks around young stellar objects (YSOs) can be extracted. For example, we can estimate masses, radii, and temperature of these disks by constructing phenomenological models and comparing them with observations.

However, no self-consistent model exists which can explain SEDs of so called "flat-spectrum" T Tauri stars. Disks around flat-spectrum T Tauri stars seem to emit more energy in mid to far infrared wavelengths than the standard viscous accretion disk theory would predict (Lynden-Bell & Pringle, 1974). In order to explain the flat-spectrum, Calvet *et al.* (1994) introduced an infalling envelope around a central star and a disk and showed that emission from the envelope can account for the observed large excess infrared emission. However, recent high resolution observations suggest that the flat-spectrum comes from the disk itself (Kitamura *et al.*, 1996). If the effect of heating of the disk by the envelope is taken into account, the temperature of the disk would be increased in such a way that the flat-spectrum could be explained by the disk itself (D'Alessio *et al.*, 1997).

It is likely that both emission from an envelope and heating of a disk by the envelope contribute to the flat-spectrum. Therefore, we must determine temperature distributions in the disk and in the envelope simultaneously. As a first step to constructing a fully self-consistent model of flat-spectrum T Tauri stars, we calculate temperature distributions of a disk and an envelope by solving a two-dimensional, axisymmetric radiative transfer equation. We assume a minimum mass solar nebula for a disk model (Hayashi *et al.*, 1985), which is surrounded by a uniform density envelope. It is also assumed that the heating source is only the central star, i.e., we do not take into account viscous heating of the disk. We solve time-dependent radiation moment equations forward in time until radiative equilibrium is achieved.

S. M. Miyama et al. (eds.), Numerical Astrophysics, 181–182.

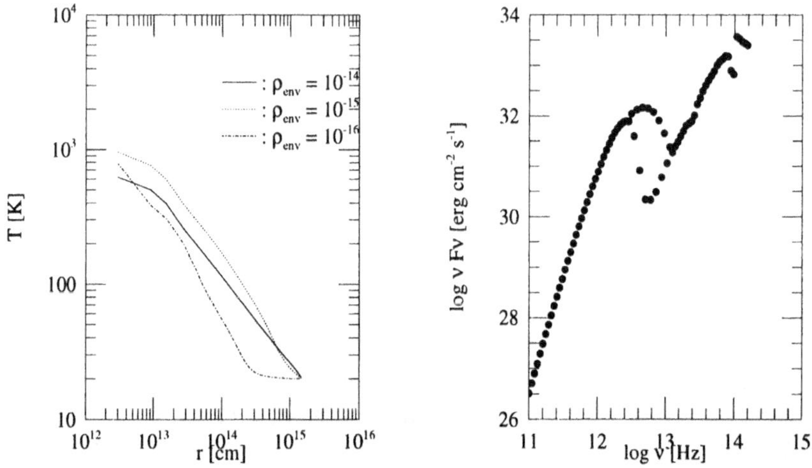

Figure 1. *Left panel*: Temperature in the midplain of the disk for the envelope density $\rho_{\rm env} = 10^{-14}$g cm^{-3} (*solid line*), $\rho_{\rm env} = 10^{-15}$g cm^{-3} (*doted line*), $\rho_{\rm env} = 10^{-16}$g cm^{-3} (*dot-dashed line*), respectively. *Right panel*: SED of the model with $\rho_{\rm env} = 10^{-15}$g cm^{-3}. Inclination angle $i = 60°$ is assumed.

The tensor Eddington factor is calculated by solving a time-independent radiative transfer equation to close the moment equations. A grey approximation is used to reduce the computational cost.

2. Results

The envelope with the optimal value of density can heat up the outer part of the disk effectively (left panel of Figure 1). The optimal value is determined by two contradicted requirements that the optical depth to the central star should be small enough for the envelope to be heated by the central star and that it should be opaque enough to have a large flux which heats the disk. An example of SEDs is also shown in right panel of Figure 1.

Numerical computations were carried out on a VX/1R at NAOJ.

References

Calvet, N., Hartmann, L.W., Kenyon, S.J., & Whitney, B.A. (1994) *Astrophys.J.* **434**, pp. 330–340

D'Alessio, P., Calvet, N., & Hartmann, L. (1997) *Astrophys.J.* **474**, pp. 397–406

Hayashi, C., Nakazawa, K., & Nakagawa, Y. (1985) In *Protostars and Planets II*, eds. D.C.Black & M.S.Matthews (Tucson: Univ.of Arizona Press), pp. 1100–1153

Kitamura, Y., Kawabe, R., & Saito, M. (1996) *Astrophys.J.Lett.* **465**, pp. 137–140

Lynden-Bell, D., & Pringle, J.E. (1974) *Mon.Not.Roy.Astron.Soc.* **168**, pp. 603–637

DRIVING PROTOSTELLAR JETS

NEAL TURNER,
UCO/Lick Observatory, University of California
Santa Cruz CA 95064, USA

MICHAL RÓŻYCZKA,
Nicolaus Copernicus Astronomical Center
Bartycka 18, 00-716 Warsaw, Poland

AND

PETER BODENHEIMER
UCO/Lick Observatory, University of California
Santa Cruz CA 95064, USA

Magnetized gas is readily lifted from accretion disks by magnetic buoyancy or Parker instability. We follow the evolution of buoyant elements after they have reached the disk surface, using axisymmetric magneto-hydrodynamic calculations. The ambient medium above the disk in our calculations is unmagnetized, differentially rotating, and in hydrostatic balance (see Figure 1).

Differentially-rotating buoyant elements initially threaded by loops of poloidal field rapidly generate toroidal field. Resulting Lorentz stresses transfer angular momentum to a small portion of the mass, while the bulk of the element loses angular momentum and falls towards the central object. For elements with initial ratios β of gas to magnetic pressure near 100, the fall is ten times slower than for $\beta = 0.3$. However throughout the range in initial β, material approaching the central object, where differential rotation is strongest, has $\beta \approx 1$. This is true despite vigorous numerical reconnection of both toroidal and poloidal field components.

Part of the magnetized material rises along the axis in a narrow stream. Driven mostly by gas pressure at the base of the flow, and increasingly by a gradient in toroidal magnetic field pressure higher up, this material at a height less than 1 AU above the central object reaches escape speed. The upflow continues for at least 1 yr at a highly variable mass rate peaking at 10^{-10} $M_{\odot}\,yr^{-1}$, or 15% of the inlet rate, smaller than seen in flows from young stellar objects. We suggest that this mechanism might produce mass rates close to those observed, from a combination of a higher assumed mass

S. M. Miyama et al. (eds.), Numerical Astrophysics, 183–184.

injection rate and a superposition of the effects of a number of buoyant elements rising from the disk at different radii.

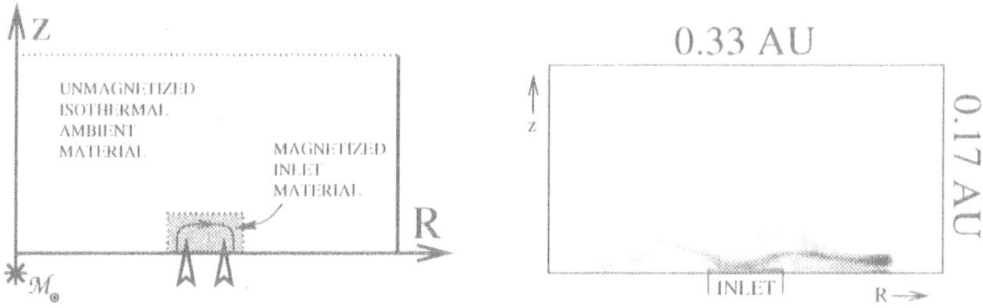

Figure 1. *(Left):* Initial conditions for the calculations. The ambient medium has temperature 2×10^6 K. On its lower boundary, density is 10^{-16} g cm^{-3}. Inlet material contains poloidal field, one loop of which is drawn. Typically β in the inlet is 10 and density 3×10^{-14} g cm^{-3}. Temperature (6 000 K) is chosen for total pressure balance with the ambient medium at inlet center. During the calculation, material is injected upwards into the inlet at 10^5 cm s^{-1}, or 1.4% of Keplerian speed at inlet center. The upper boundary is open, while others are closed. *(Right):* Distribution of total angular momentum per zone after two orbital periods at inlet center. Regions of high angular momentum are dark.

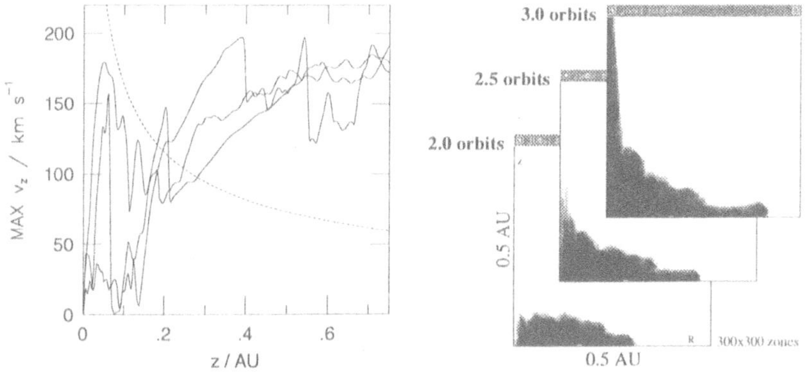

Figure 2. *(Left):* Variation of vertical velocity with height (solid lines) after 3.0, 3.1 and 3.2 orbital periods at inlet center. The velocity plotted at each height is the largest present inside radius $R = 5 \times 10^{11}$ cm. Also shown is escape speed versus height (dashed curve). The velocity at any given location is rapidly-varying, but exceeds escape speed above about 0.2 AU. Real Herbig-Haro jets have flow speeds 200–400 km s^{-1}. *(Right):* Dark areas show the progress of magnetized material up the axis from 2.0 to 3.0 orbital periods at inlet center.

N. T. was supported by NASA-GSRP grant NGT 2-52217, M. R. by KBN grant 2.P03D.004.13. We thank the Laboratory for Computational Astrophysics for providing the ZEUS-3D code, and the organizers of the 1997 Guillermo Haro Program workshop for their hospitality.

GAS ACCRETION TO THE JOVIAN PLANET IN THE PROTOPLANETARY DISK

Y. IMAEDA
*University of Tokyo, School of Sciences,
Department of Astronomy, Bunkyo-ku, Hongo,
Tokyo 113-0033, Japan*

AND

S. M. MIYAMA
*National Astronomical Observatory of Japan
Mitaka-shi, Tokyo 181-8588, Japan*

1. Introduction

After Mayor & Queloz (1995) first reported the discovery of Extrasolar Giant Planet (EGP) 51 Peg B by radial velocity observation, many EGPs have been discovered. The discovered EGPs have a different appearance from that of our solar planets. They have a Jupiter order mass ($1M_J \simeq 10^{-3}M_\odot$), but their orbital distance is quite small (2.0AU$\geq a \geq$0.05AU). On the other hand, in our solar system, Jovian-type planets can only be seen at a distant place from the Sun ($\geq$5AU). According to the formation theory of the solar system, whether a Jovian-type planet is formed or not corresponds to whether or not the mass of the protoplanet reaches a critical core mass. This critical core mass is about 10 terrestrial mass and less affected by distance from the Sun (Mizuno et al., 1980). Since, in the solar nebula, H_2O is solid farther than 3$\sim$4AU, it is reasonable to predict that there are Jovian-type planets farther than 5AU simply because of the existence of more solid material. However, EGPs have been found inside of the H_2O sublimation region.

2. Motivation for this work

How and where EGPs were formed is still unknown. However, if the protoplanetary disk has a mass about 10 times as large as the minimum mass solar nebula, it not only contains sufficient solid material to account for the formation of a critical core without H_2O ice but is also stable for self-

S. M. Miyama et al. (eds.), Numerical Astrophysics, 185–186.

gravity. Therefore, the formation of Jovian-type planets can be expected even in the inner region. However, it is not obvious that sufficient gas accretion can take place near the central star. Sekiya et al. (1988) revealed that, in the case of Jupiter, a low-density region is formed around the planet (gap region) when the mass becomes about $1M_J$. In our research, we extended this calculation to the innermost region and clarified whether or not sufficient gas accretion occurs when the EGP is formed in the inner region.

3. Calculation

We calculated 2-D hydrodynamics without self-gravity through the SPH calculation method, using 40,000 particles. For simplicity, we adopted a minimum-mass solar nebula as a disk model. We initially put $1M_J$ in the protoplanetary disk and integrated the time until quasi-equilibrium was achieved. The orbital distance was 0.4AU for the EGP and 5.2AU for Jupiter. We used the equation of state for ideal gas. The temperature was $T \propto {r_s}^{-1/2}$, where r_s was the distance from the central star.

4. Results

All figures are in the CD-ROM Edition.

5. Discussion

After several rotations, a quasi-equilibrium spiral-gap pattern was formed in both cases. The gap formed was clearer in the case of the EGP than in Jupiter's (Fig.1, 2). Such a strong gap formation was expected to prevent more gas from accreting. Indeed, when we estimated the mass accretion from Fig.3, the mass accretion per rotation was 3 times as small as in the case of Jupiter ($3.7\times10^{-3}M_J$/rot vs. $1.0\times10^{-2}M_J$/rot). Nevertheless, since dynamical time is much shorter in the inner region, the mass accretion rate in real-time scale becomes very large ($3.3 \times 10^{-1}M_J$/year vs. $8.4 \times 10^{-4}M_J$/year). Therefore, it may be concluded that at the beginning of the quasi-equilibrium stage, the mass accretion rate is large enough to account for the formation of Jovian-type planets near the central star from the point of gas accretion. However, this quasi-equilibrium state evolves in a longer viscous time-scale. Longer time integration would be needed to evaluate these effects.

6. Acknowledgement

Numerical computations were carried out on VPP300/16R at NAOJ.

References

Mayor,M. & Queloz,D. Nature 378, 355, 1995
Mizuno,H. Prog. Theo. Phys. 64, 544, 1980
Sekiya,M. ,Miyama,S.M., and Hayashi,C. Prog. Theo. Phys., 96,274, 1988

INSTABILITY AND TURBULENCE IN ACCRETION DISKS

JOHN F. HAWLEY AND STEVEN A. BALBUS
Dept. of Astronomy, University of Virginia
Charlottesville, VA 22903, USA

1. Introduction

Accretion disks are believed to be the power sources for such diverse astronomical systems as X-ray binaries, quasars, active galaxies, and cataclysmic variables. In each case, the luminosity emitted from the system comes from the gravitational energy released as gas in the disk spirals down toward the central compact object. The trajectory of an individual volume of gas as it orbits is determined by its specific angular momentum, which will be a conserved quantity of the motion in the absence of torques. Hence inspiral requires a torque. The most fundamental physical process governing the structure and evolution of accretion disks, therefore, is the extraction of angular momentum from the gas and its transfer outward through the disk.

Recently, we have made substantial progress in understanding this angular momentum transport process. The key conceptual point is the recognition that a subthermal magnetic field rapidly generates magnetohydrodynamic (MHD) turbulence via a simple linear instability. This turbulence transports angular momentum outward through the disk, allowing accretion to proceed. Although turbulence seems like a natural and straightforward transport mechanism, it turns out that the magnetic fields are essential. Not all forms of turbulence work. Purely hydrodynamic turbulence, in particular, fails to arise spontaneously in disks, is not self-sustaining (even if initiated "by hand"), and does not produce sustained outward transport of angular momentum.

In this paper we will briefly review these results. In particular, we focus on the role that numerical simulations have played in obtaining our conclusions. For a more complete review of this work see Balbus & Hawley (1998).

S. M. Miyama et al. (eds.), Numerical Astrophysics, 187–194.

2. Hydrodynamic Turbulence

Although the angular momentum in a disk would be transported outward by ordinary (molecular) viscosity, the viscosity in accretion disks is extremely small. The importance of viscosity is measured by the Reynolds number, a dimensionless quantity given by the system's characteristic length times velocity divided by the viscosity. Accretion disks are characterized by a huge Reynolds number. Far from being regarded as an obstacle to angular momentum transport, however, a disk's high Reynolds number has traditionally been seen as evidence in support of the presence of turbulence. Turbulence is, after all, a characteristic of high Reynolds number terrestrial fluids. This view goes back at least as far as the first detailed disk model developed by Crawford and Kraft (1956) who wrote that because "the Reynolds number [of the gas] is extremely high, it is quite certain that the gas is turbulent." The more cautious have long recognized that even with a high Reynolds number, it is far from obvious that turbulence develops in Keplerian disks.

Turbulence, even in a high Reynolds number flow, requires a source, that is, an instability. It is not sufficient that the viscosity be low enough to allow for small-scale, disorganized velocity fluctuations to persist. There has to be something producing those velocity fluctuations. The problem is that the differentially rotating flows in accretion disks are dynamically stable to small perturbations according to the Rayleigh criterion (angular momentum must increase outward). This means that the forces that are present in the disk (gravitational, centrifugal, and Coriolis) do not amplify small (linear) velocity fluctuations. Despite this, people clung to the possibility that nonlinear (large amplitude) instabilities might develop and produce self-sustaining turbulence. It is known from laboratory experiments, for example, that simple shear flows break down into turbulence without the presence of a linear instability. Many felt that accretion disks should behave in the same way.

This would seem like an ideal question to put to a numerical test: do accretion disk flows spontaneously break down into turbulence due to nonlinear instabilities? There are several important numerical issues to bear in mind before proceeding. First, the question is inherently three-dimensional (3D), since the known nonlinear shear flow instabilities are 3D. Second, because the Reynolds number in disks is so high, there is a huge disparity between the macroscopic lengths in the disk (e.g. the scale height) and the (physical) viscous dissipation length. If this full range of scales must be numerically resolved, one would need an enormous number of grid zones.

As it turns out, although the disk stability problem is 3D, it is not necessary to evolve the full global, large-scale accretion disk. The dynamics

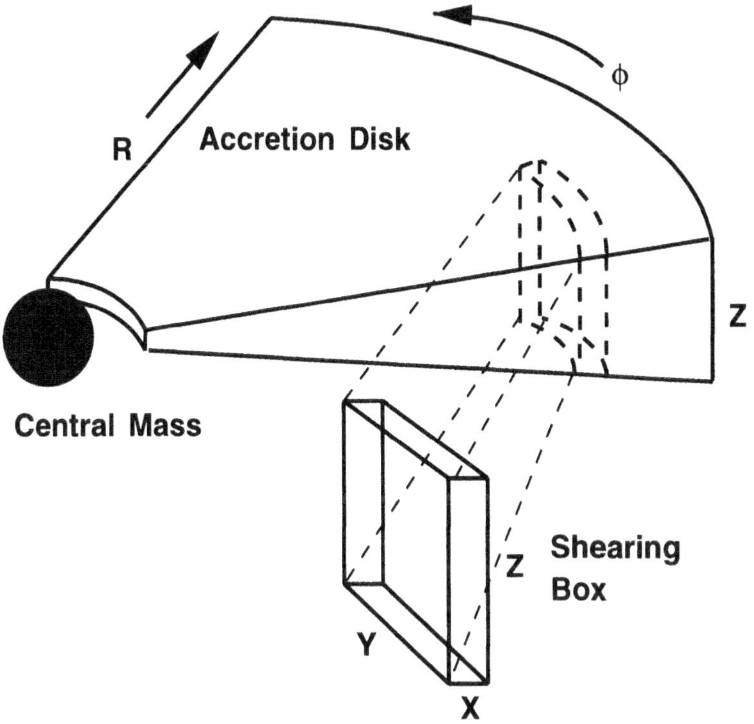

Figure 1. The local shearing box model. A small region of the full global accretion disk is mapped to a small volume that is locally Cartesian in geometry but not in dynamics: the gravitational tidal field and Coriolis forces are retained. The center of the box is considered to orbit with frequency Ω. Vertical gravity can be included to model a stratified local disk section.

of a differentially rotating fluid can more easily be studied in a local *shearing box* system (see Figure 1 and the description in Hawley, Gammie & Balbus 1995). The local model preserves the orbital dynamics while using a small, well-resolved, locally geometrically Cartesian computational domain. Long-term evolution is made possible by the use of shearing-periodic boundary conditions in the radial direction. With these boundary conditions, the radial direction is strictly periodic at $t = 0$, but a relative shear is applied as time advances. As a fluid element exits from the outer radial boundary (for example), it reappears at the inner boundary at the appropriate sheared location with its angular velocity adjusted to account for the mean shear across the box.

Even with a local model, just how much resolution is required to search for nonlinear instabilities? Since these instabilities, if they exist, are dynamically, not viscously, driven it is not necessary to resolve all lengths down to the viscous damping scale. Instead, one evolves the inviscid Euler equations with enough resolution to capture a representative range of dynamical lengths. By perturbing the computational domain with large amplitude perturbations over the full range of length scales, one should be able to observe the growth of any reasonably rapidly growing unstable modes, linear or nonlinear.

Balbus, Hawley & Stone (1996) carried out just such a series of local hydrodynamic simulations. The simulation results were understood through a moment analysis of the dynamical equations. The analysis, in turn, suggested further numerical experiments and additional flow quantities to measure. Simulation and analysis work hand in hand.

Consider an initial uniform Keplerian flow, perturbed with large amplitude velocity fluctuations and evolved forward in time. Stability is determined by whether these fluctuations grow or decay with time. Balbus et al. found that for all Keplerian simulations the initial perturbations rapidly die out and the system returns to a simple smooth differential rotation. As an important control on the results, Balbus et al. also carried out simulations of simple shear flows. These flows become unstable and develop turbulence, in stark contrast to the Keplerian results. Since the Keplerian system differs from the simple shearing system only by the addition of orbital dynamical forces, the absence of instabilities in Keplerian flows cannot be due to numerical effects associated with the use of a finite number of grid zones. Indeed, there is nothing to set a distinct physical scale to differentiate the two systems. If the shear flow develops turbulence then the Keplerian system should as well, *if* turbulence were the correct physical outcome.

This conclusion is supported by another experiment which retains the orbital dynamics but assumes that the background shear corresponds to a constant angular momentum distribution. Such a disk is marginally stable to the linear Rayleigh criterion: the epicyclic frequency is zero. By symmetry the constant angular momentum system should be equivalent to the simple shear flow; in the local analysis their moment equations are formally equivalent. This implies that the constant angular momentum system should also develop the same type nonlinear instabilities that afflict simple shear flows. Simulations confirm this expectation, and provide further evidence that the stability of Keplerian flow is due to dynamical, not numerical, effects.

Another way to check the validity of a numerical simulation (and the conclusions that follow from it) is by increasing the grid resolution. We have recently carried out a resolution study using grids ranging from a coarse 16^3 zones up to 256^3 zones. The evolution of the energy in the angular velocity perturbations for the different resolutions is displayed in Figure 2. Interestingly, the greater the resolution, the *more* rapid the decline in energy. While turbulence is present, the higher resolution simulations have

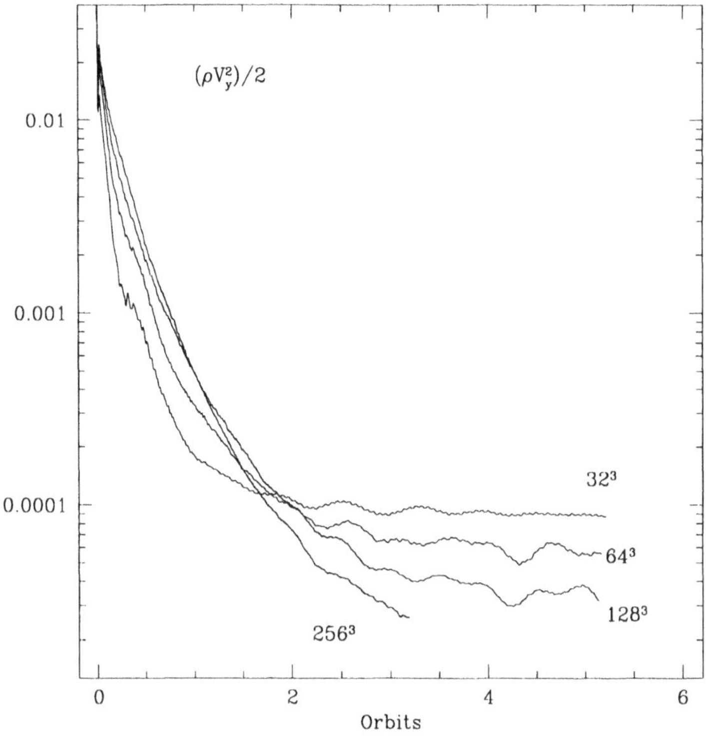

Figure 2. The time evolution of the kinetic energy in angular velocity fluctuations for hydrodynamic turbulence in Keplerian disk, in units of $(L\Omega)^2$ where L is the size of the computational domain. The curves correspond to different resolutions for the same initial conditions. They are labeled by the number of grid zones used. Instead of promoting greater turbulence, higher resolution results in more rapid decline in the δv_y fluctuation energy.

larger positive values of the Reynolds stress. The Reynolds stress transports angular momentum outward, but it also acts as a sink for energy in the angular velocity fluctuations (see Balbus et al. 1996). This appears to hasten the damping of the turbulence compared to less well-resolved simulations.

From the resolution experiments, from the direct comparison with the simple shear flow, and from the behavior of the constant angular momentum system, we conclude that Keplerian flows are hydrodynamically stable to local perturbations of any reasonable amplitude.

3. Magnetohydrodynamical Turbulence

If the source of the angular momentum transport in Keplerian disks does not arise from simple hydrodynamical turbulence, we must look elsewhere.

Because accretion disks are generally sufficiently hot to be fully ionized, they are highly conducting. Such hot, ionized plasmas are quite capable of supporting the currents that create magnetic fields. Therefore, magnetic forces are a good candidate to transport angular momentum. The idea is a venerable one: Lynden-Bell's pioneering (1969) quasar accretion disk model favored transport by magnetic stresses. What prevented a wide consensus in favor of magnetic transport was the perception that the field could not be important unless raised to an energy comparable to the disk's thermal energy. Again, it was assumed, it would be necessary to demonstrate that magnetic fields are amplified to significant strengths.

As it turns out, the presence of even a weak magnetic field fundamentally alters the stability of orbiting flow. Balbus & Hawley (1991) demonstrated through a linear analysis that a Keplerian flow is *linearly unstable* in the presence of a weak magnetic field. (The earliest version of this instability was uncovered by Velikhov [1959] in a study of magnetized Couette flows.) The disk stability criterion is quite simple: weakly magnetized disks are dynamically unstable if the *angular velocity* decreases outwards, a condition that always holds for Keplerian orbits. This *magnetorotational instability* means that disk turbulence arises naturally, just as it does in other circumstances, such as the convective turbulence that develops in the outer layers of low mass stars. In both the star and the disk, turbulence develops when a simple local linear stability criterion is violated.

Since the elucidation of the linear instability, numerical simulations have played an essential role in illuminating the instability's physical nature and nonlinear consequences. Again the local shearing box model has proven invaluable for these investigations. Hawley, Gammie & Balbus (1995) and Matsumoto & Tajima (1995) examined the evolution of uniform initial vertical and toroidal fields. Hawley, Gammie & Balbus (1996) simulated models with random initial fields. The general outcome of all such simulations is MHD turbulence and vigorous transport of angular momentum.

What limits the growth of the instability and sets the level of the turbulence (and hence angular momentum transport)? In a disk there are (at least) two principal effects which limit the field strength: local dissipation (i.e. through reconnection and resistivity), and loss of flux from the disk through buoyancy.

In the local models without vertical stratification, dissipation and the finite size of the computational domain are the limiting mechanisms. In a numerical simulation, the primary dissipation process is numerical: fields reconnect at the grid scale. To some extent this numerical effect must mimic the physical one; reconnection can occur when oppositely directed fields are brought close enough together. But finite resolution certainly exaggerates its efficacy. To gauge such numerical influences we have run many

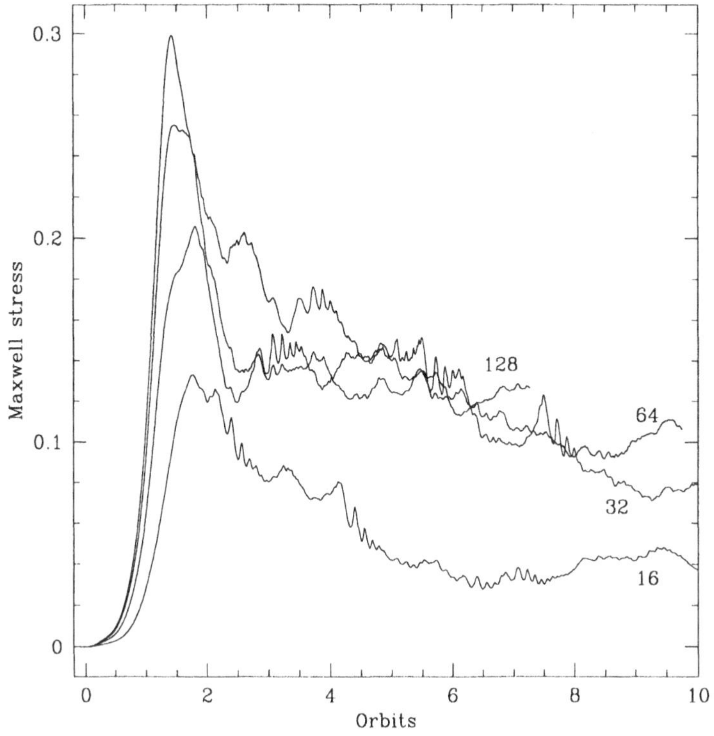

Figure 3. Maxwell Stress (as a fraction of the thermal energy) for a series of identical models at different grid resolutions. The curves are labeled by the number of grid zones in the x direction.

tests including resolution studies. Here we present a recent example. Consider a shearing box simulation initialized with a vertical field that varies sinusoidally with radius; there is no net field piercing the computational domain. This initial configuration is given a long-wavelength spectrum of velocity perturbations (identical for all resolutions) and evolved forward in time. The time evolution of the Maxwell stress in the different simulations is shown in Figure 3. Although the peak at the end of the linear growth phase shows a significant increase with resolution, the differences are much smaller once turbulence sets in. This is because the magnetic stresses are dominated by the largest scales in the box. These scales are reasonably well resolved by all but the lowest resolution model.

The issue of field loss due to buoyancy can be investigated with models that include the vertical gravity. Local models with vertical stratification have been done by Brandenburg et al. (1995) and Stone et al. (1996). An example of a vertically stratified disk simulation is included as an MPEG movie in the CD accompanying this volume. In these simulations we find that while magnetic flux is buoyantly lifted out of the disk, the rate at which this occurs is limited to about the vertical sound crossing time. The magnetorotational instability operates on shorter time scales and field is regenerated faster than it is lost due to buoyancy.

4. Conclusions

It now seems clear from the work done to date that accretion disks are fundamentally MHD systems. Further insights into the consequences of the magnetorotational instability and the nature of the resulting turbulence will require increasingly detailed numerical work. There are many potential avenues of investigation to pursue. It is important to develop full-scale global simulations in three dimensions. Although such numerical simulations are technically quite challenging, the difficulties are slowly being overcome. An example of such a global model is presented by R. Matsumoto elsewhere in these proceedings. We also need to increase the physical complexity used in disk simulations, first in the local disk domains, and eventually in the global models. Examples include radiation transport, partial ionization, resistivity and reconnection, and improved equations of state. Of course these goals (increased resolution, increased physical content) apply across the full range of astrophysical problems that are being addressed with numerical simulations. The proceedings of this workshop provide several examples of progress in all these areas, both for disks and for other astrophysical systems.

Acknowledgements

This work is supported in part by NASA grants NAG-53058, NAG5-4600, and by NSF grant AST-9423187. The computations were carried out on the Cray C90 and T3E systems of the Pittsburgh Supercomputing Center.

References

Balbus, S. A., & Hawley, J. F. 1991, ApJ, 376, 214
Balbus, S. A., & Hawley, J. F. 1998, Rev. Mod. Phys. 70, 1
Balbus, S. A., J. F. Hawley, & J. M. Stone 1996, ApJ, 467, 76
Brandenburg, A., Å. Nordlund, R. F. Stein, & U. Torkelsson 1995, ApJ, 446, 741
Crawford, J. A., & R. P. Kraft 1956, ApJ 123, 44
Hawley, J. F., C. F. Gammie, & S. A. Balbus 1995, ApJ, 440, 742
Hawley, J. F., C. F. Gammie, & S. A. Balbus 1996, ApJ, 464, 690
Lynden-Bell, D. 1969, Nature, 223, 690
Matsumoto, R., & T. Tajima 1995, ApJ, 445, 767
Stone, J. M., J. F. Hawley, C. F. Gammie, & S. A. Balbus 1996, ApJ, 463, 656
Velikhov, E. P. 1959, Soviet Phys.—JETP, 36, 995

THREE-DIMENSIONAL GLOBAL MHD SIMULATIONS OF ACCRETION DISKS AND JET FORMATION

R. MATSUMOTO
Department of Physics, Faculty of Science,
Chiba University, 1-33 Yayoi-Cho, Inage-Ku,
Chiba 263-8522, Japan

Abstract. We present the results of global magnetohydrodynamic (MHD) simulations of accretion disks threaded by large-scale magnetic fields. When the initial magnetic field is vertical, well collimated, magnetically driven jets emanate from the surface of the disk. Since the jet formation process extracts angular momentum from the disk, it enhances the accretion of disk material. Inside the disk, magnetic turbulence driven by the Balbus & Hawley instability efficiently redistributes angular momentum.

1. Introduction

When an accretion disk is threaded by large-scale poloidal magnetic fields, centrifugal force and magnetic pressure can drive outflows (Figure 1a). Theory of steady, axisymmetric MHD outflows from accretion disks has been developed by many authors (e.g., Blandford & Payne 1982; Pudritz & Norman 1986; see Kudoh & Shibata 1995, 1997 and references therein). Nonlinear, time-dependent, two-dimensional (axisymmetric) MHD simulations of magnetically driven jets were first carried out by Shibata & Uchida (1985). They showed that collimated jets are formed when magnetic twists relax along large-scale magnetic field lines. Subsequently, Uchida & Shibata (1985) and Shibata & Uchida (1986) showed in two-dimensional (2D) MHD simulations that a bipolar jet can be formed through the accumulation and relaxation of magnetic twists injected from the rotating disk (Figure 1b). The outflow is collimated along the rotation axis due to the pinch effect of the toroidal component of magnetic fields. The terminal speed of the jet generated in this mechanism was found to be the order of the Keplerian rotation speed of the disk. They called this mechanism the "sweeping mag-

S. M. Miyama et al. (eds.), Numerical Astrophysics, 195–202.

netic twist mechanism" and applied it to various astrophysical jets such as the bipolar flows in star forming regions (Uchida & Shibata 1985; Shibata & Uchida 1990) and to the Galactic center radio lobes (Uchida, Shibata & Sofue 1985; Shibata & Uchida 1987).

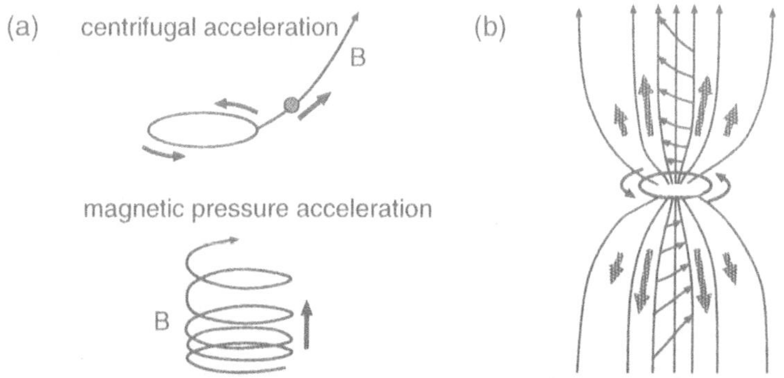

Figure 1. (a) A schematic picture of the driving mechanisms of magnetically driven jets. (b) The generation and relaxation of magnetic twists driven by the rotation of the disk. The outflows are collimated along the rotation axis due to the magnetic pinch effect.

The numerical results by Uchida & Shibata (1985) and Shibata & Uchida (1986) have been confirmed by Stone & Norman (1994) by using the ZEUS code. Matsumoto et al. (1996) applied this mechanism to jet formation from a geometrically thick disk. By using a newly developed CIP-MOCCT code, Kudoh, Matsumoto & Shibata (1998) studied the dependence of the mass accretion rate and mass outflow rate on magnetic field strength. General relativistic 2D MHD simulations of jet formation have been performed by Koide, Shibata & Kudoh (1998). Several authors have carried out global 2D MHD simulations of jet formation driven by magnetospheric interaction of the dipole magnetic field of the central gravitating object and its surrounding disk (Hirose et al. 1997; Hayashi, Shibata & Matsumoto 1996; Miller & Stone 1997; Goodson, Winglee & Böhm 1997).

The Uchida & Shibata's model of jet formation is intrinsically time-dependent because the disk gas infalls by losing angular momentum through magnetic braking. In order to obtain steady state solutions through time dependent simulations, several authors have carried out MHD simulations by fixing the boundary conditions at the surface of the disk and by neglecting the effects of disk accretion due to magnetic braking. Ustyugova et al. (1995), Romanova et al. (1997) and Ouyed & Pudritz (1997) have carried out these two-dimensional simulations for many disk rotation periods and obtained steady-like solutions. On the other hand, Ouyed, Pudritz & Stone (1997) have shown that when the initial magnetic field is uniform and parallel to the rotation axis of the disk, outflows occur episodically.

Meier et al. (1997) proposed a magnetic switch, in which the outflow speed becomes much larger than the escape speed when the Alfvén speed exceeds the escape speed. The surface conditions of the disk, however, needs to be determined self-consistently.

2. Accretion Avalanches and the Jet Formation

In this section we present typical results of two-dimensional MHD simulations of nonsteady jets (Matsumoto et al. 1996). We assume that a rotating polytropic torus with constant angular momentum distribution $L = L_0$ is imbedded in a spherical, non-rotating isothermal halo. The gravitational field is assumed to be given by a point mass M. In a cylindrical coordinate (r, φ, z), the dynamical equilibrium of the disk is described by $\Psi_g + L_0^2/(2r^2) + (n+1)P/\rho = \text{const.}$ where Ψ_g is the gravitational potential, and n is the polytropic index. We take the radius of the pressure maximum of the disk $[r = L_0^2/(GM)]$ as the reference radius r_0. The initial magnetic field is assumed to be uniform and vertical. We use the normalization $r_0 = V_{K0} = \rho_0 = 1$, where V_{K0} is the Keplerian rotation speed at $r = r_0$. The halo parameters are $1/\alpha = C_{sh}^2/(\gamma V_{K0}^2)$ and ρ_h/ρ_0 where C_{sh} and ρ_h are the sound speed and density in the halo at $(r, z) = (0, r_0)$, respectively. We use $\alpha = 1.0$ and $\rho_h/\rho_0 = 10^{-3}$. We solved the ideal MHD equations in a cylindrical coordinate by using a modified Lax-Wendroff method (Rubin & Burstein 1967) with artificial viscosity.

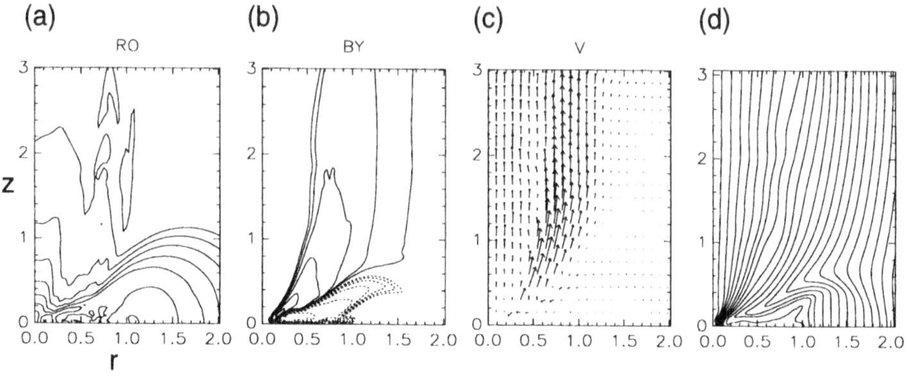

Figure 2. A result of 2.5D MHD simulation of a typical model at $t = 2\pi r_0/V_{K0}$. (a) Isocontours of density. (b) Isocontours of toroidal magnetic field component. (c) Velocity vectors. (d) Magnetic field lines.

Figure 2 shows numerical results for a typical model (model B3 in Matsumoto et al. 1996) at $t = 2\pi r_0/V_{K0}$. The model parameters are $E_{th} = (a_0/V_{K0})^2/\gamma = 0.05$, $E_{mg} = (V_{AP0}/V_{K0})^2 = 10^{-3}$, where a_0 is the sound speed and V_{AP0} is the poloidal Alfvén speed. The initial ratio of gas pressure to magnetic pressure ($\beta = P_{\rm gas}/P_{\rm mag}$) in the torus at $(r, z) = (r_0, 0)$ is $\beta_0 = 100$. The plasma β in the halo at $(r, z) = (0, r_0)$ is $\beta_h = 2.0$. After the torsional Alfvén wave generated by the rotation of the disk propagates into the corona, the surface layer of the torus loses angular momentum and infalls like an avalanche. Subsequently, the cold material in the disk surface is accelerated and ejected as a bipolar jet. The outflow is collimated along the rotation axis due to the toroidal pinch effect. The maximum speed of the jet is $V_{\rm max} = 1.7V_{K0}$.

The avalanching motion which appear in our simulation can be considered as the global version of the "two channel flow" which appeared in the nonlinear stage of the magneto-rotational (or Balbus & Hawley) instability (Balbus & Hawley 1991; Hawley & Balbus 1992). The wiggling of magnetic field lines inside the torus (see Figure 2d) is also due to the growth of the Balbus & Hawley instability. The relation between the Balbus & Hawley instability and magnetic braking has been discussed by Stone & Norman (1994) and by Matsumoto et al. (1996). Recently, Kuwabara, Shibata & Matsumoto (1998) extended this model by including the effects of resistivity and carried out 2D MHD simulations of a circumnuclear gas torus in active galactic nuclei. Numerical results indicate that magnetically driven mass accretion from the torus can fuel quasars.

3. Global 3D MHD Simulations of a Torus Threaded by Vertical Magnetic Fields

We extended the 2D cylindrical MHD code to 3D and carried out 3D simulations. The parallel performance of the code on a vector-parallel computer VPP300/16R at NAOJ is 13.7 times that of 1 processor when 15 processors are used. Numerical simulations using $(N_r, N_\varphi, N_z) = (201, 65, 240)$ grid points typically take 1 CPU hour per 10000 time step ($\sim$ one rotation period). To check the accuracy of the 3D MHD code, we simulated the growth of non-axisymmetric perturbations in differentially rotating, cylindrical plasma threaded by uniform axial magnetic fields. Numerical results agreed with those of global linear analysis (Curry & Pudritz 1996).

Next, we show results of 3D MHD simulations of jet formation from a torus. The model parameters are the same as those in model B3. We initiate the non-axisymmetric evolution by imposing perturbations for azimuthal velocity as $\delta v_\varphi = 0.01 v_\varphi \sin(m\varphi)$. Figure 3 shows numerical results when one armed ($m = 1$) perturbation is imposed. We confirmed the results of previous 2D axisymmetric simulations (Matsumoto et al. 1996) that bipolar jet is formed and that the surface layer of the disk accretes faster than the equatorial part. The avalanche flow creates the radial component of

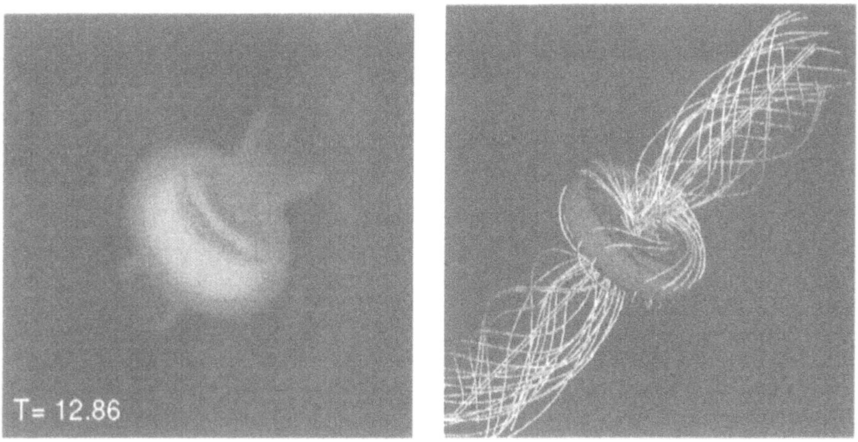

Figure 3. Results of 3D MHD simulation of a typical model with $m = 1$ perturbation. The left panel shows the volume rendered image of density distribution. The right panel shows magnetic field lines and isosurface of density.

magnetic fields which is further twisted by the differential rotation of the disk. The magnetic field lines at $t = 12.86 r_0/V_{K0}$ indicate that toroidal field components dominate inside the torus. The density isosurface shows that the dense region of the torus is deformed into a disk-shape. Due to the growth of non-axisymmetric instabilities, the magnetic field lines are bunched into helical bundles in the jet. Helical filamentary structures can also be seen in the density distribution of the jet.

Figure 4 shows the projected magnetic field lines and isocontours of β at $t = 11.4 r_0/V_{K0}$ when initially $m = 2$ perturbation is imposed. Inside the disk, accretion proceeds along spiral channels. In the innermost region of the disk where toroidal magnetic fields become dominant, spirally shaped, magnetic pressure dominated ($\beta < 1$) regions appear.

4. Global 3D MHD Simulations of a Torus Threaded by Toroidal Magnetic Fields

Figure 5 shows numerical results of a torus initially threaded by weak toroidal magnetic fields. The model parameters are the same as those in model B3 except that $\beta_0 = 100$ inside the torus. The number of grid points used in these simulations is $(N_r, N_\varphi, N_z) = (201, 65, 240)$. The solid curves show magnetic field lines and gray scale shows density distribution. Owing to the growth of the Balbus & Hawley instability, magnetic field lines get tangled. As magnetic turbulence develops, angular momentum is efficiently transported outward inside the torus, and the torus becomes flattened.

Figure 6 shows isosurfaces of $\beta = 10$ (dark gray) and $\beta = 1$ (light gray)

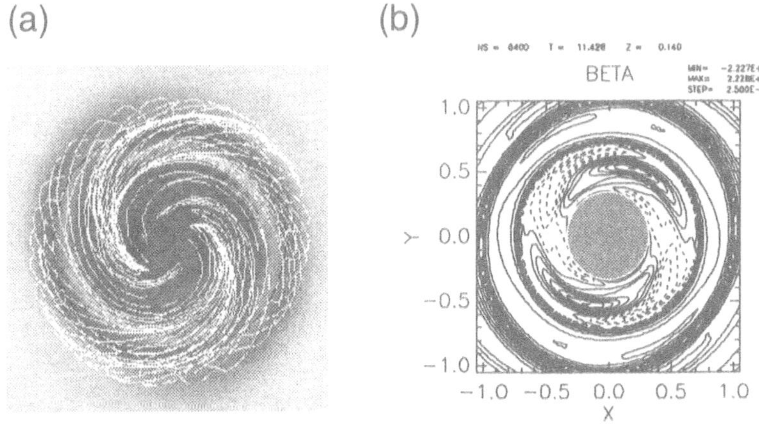

Figure 4. Results of 3D MHD simulations of a typical model with $m = 2$ perturbation. (a) Projection of magnetic field lines onto the equatorial plane at $t = 11.4r_0/V_{K0}$. Gray scale shows density distribution. (b) Isocontours of β. Dashed curves show low-β region.

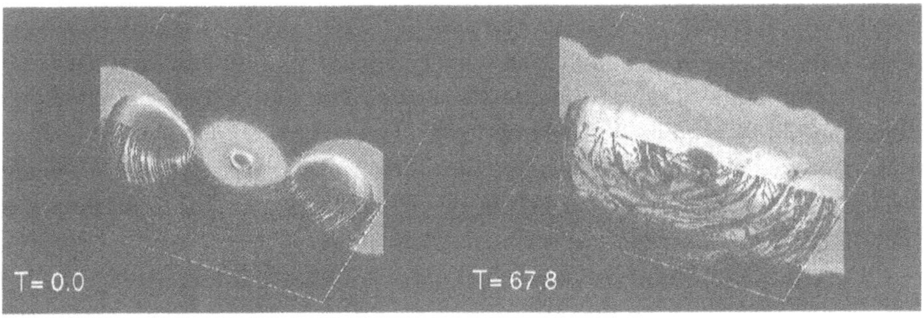

Figure 5. Numerical results of a torus initially threaded by weak ($\beta_0 = 100$) toroidal magnetic fields. The left panel shows the initial condition. Solid curves show magnetic field lines. Gray scale shows the density distribution.

for a model with $\beta_0 = 100$. As the Balbus & Hawley instability grows, magnetic fields are amplified. After 11 rotation period ($t = 67.8r_0/V_{K0}$), although the mean magnetic energy is smaller than the thermal energy, spirally shaped, low-β ($\beta < 1$) regions appear inside the torus.

The left panel of Figure 7 shows the dependence of the time development of magnetic energy $\langle B^2/(8\pi P_0)\rangle$ on initial plasma β, where $\langle\rangle$ denotes the spatial average in $0.7 < r/r_0 < 1.3$, and $0 < z/r_0 < 0.3$. When the initial magnetic field is weak, magnetic energy is amplified and approaches to a quasi-steady state with $\beta \sim 10$. These numerical results are consistent with those of the local 3D MHD simulations (Stone et al. 1996; Matsuzaki et al. 1998). The right panel of Figure 7 shows the time development of the

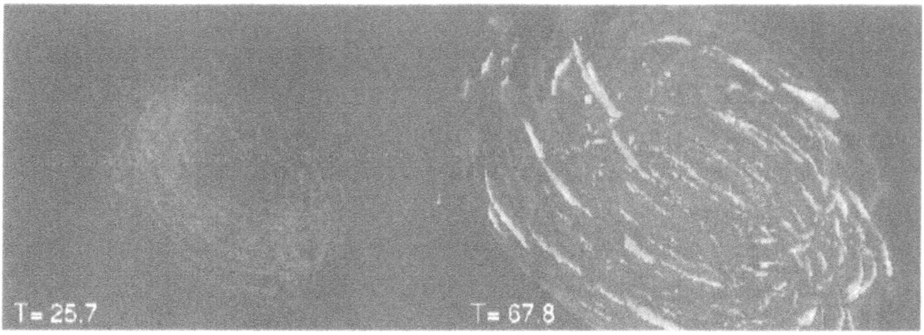

Figure 6. Isosurface of $\beta = P_{gas}/P_{mag}$. Dark gray surfaces show the surface $\beta = 10$, and light gray surfaces show the surface $\beta = 1$.

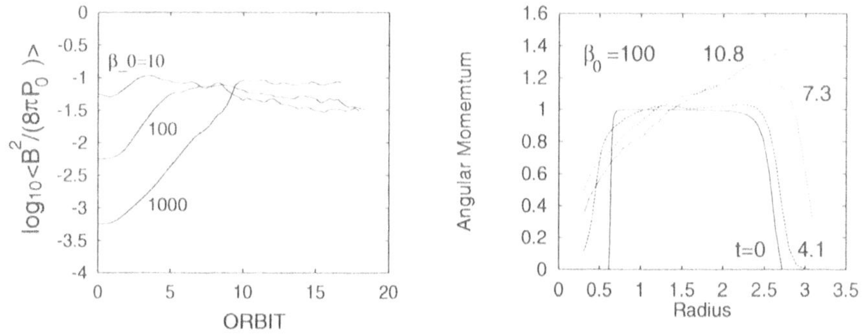

Figure 7. The left panel shows the dependence of the time history of magnetic energy on β_0. The unit of time is $2\pi r_0/V_{K0}$. The right panel shows the time evolution of angular momentum distribution

angular momentum distribution for a model with $\beta_0 = 100$. The angular momentum distribution approaches to that of Keplerian disks.

5. Summary

Recent progress in parallel computers enabled us to carry out global MHD simulations of accretion disks and jet formation with resolution capable of capturing essential physical processes. We have shown through 2D and 3D MHD simulations that when an accretion disk is threaded by large scale poloidal magnetic fields, magnetically driven jets emanate from the surface of the disk. The outflows are magnetically collimated along the rotation axis. Magnetized disks and jets can subject to global nonaxisymmetric instabilities (e.g., Curry & Pudritz 1996) and local nonaxisymmetric Balbus & Hawley instability (e.g., Hawley, Gammie & Balbus 1995). The 3D simulation results we presented here indicate that the avalanche flow breaks up into spiral channels due to the growth of non-axisymmetric modes. Spirally shaped, magnetic pressure dominated regions appear inside the disk. Since magnetic turbulence driven by the magnetic instabilities efficiently redistributes angular momentum inside the disk, a geometrically thick torus evolves toward a flattened, Keplerian accretion disk.

We thank Drs. K. Shibata, T. Kudoh, Y. Uchida and T. Tajima for discussion. Numerical computations were carried on Fujitsu VPP300/16R at NAOJ and VPP700 at Hawaii Facility, NAOJ. We thank Dr. Ogasawara of NAOJ for providing an opportunity for us to evaluate the performance of VPP700. This work is supported in part by the Grant-in-Aid of the Ministry of Education, Science, Sports and Culture, Japan (07640348).

References

Balbus, S.A., & Hawley, J.F. (1991), *ApJ*, **376**, pp.214
Blandford, R.D., & Payne, D.G. (1982), *MNRAS*, **199**, pp.883
Curry, C. & Pudritz, R.E. (1996), *MNRAS*, **281**, pp.119
Goodson, A.P., Winglee, R.M., & Böhm, K.H. (1997), *ApJ*, **489**, pp.199
Hawley, J.F., & Balbus, S.A. (1992), *ApJ*, **400**, pp.595
Hawley, J.F., Gammie, C.F., & Balbus, S.A. (1995), *ApJ*, **440**, pp.742
Hayashi, M.R., Shibata, K., & Matsumoto, R. (1996), *ApJ*, **468**, pp.L37
Hirose, S., Uchida, Y., Shibata, K., & Matsumoto, R. (1997), *PASJ*, **49**, pp.193
Koide, S., Shibata, K., & Kudoh, T. (1998), *ApJ*, **495**, pp.L63
Kudoh, T., & Shibata, K. (1995), *ApJ*, **452**, pp.L41
Kudoh, T., & Shibata, K. (1997), *ApJ*, **474**, pp.362
Kudoh, T., Matsumoto, R., & Shibata, K. (1998) *Astrophysical Journal in press*
Kuwabara, T., Shibata, K., & Matsumoto, R. (1998) in this proceedings
Matsumoto, R., Uchida, Y., Hirose, S., Shibata, K., Hayashi, M.R., Ferrari, A., Bodo, G., & Norman, C. (1996), *ApJ*, **461**, pp.115
Matsuzaki, T., Shibata, K., Tajima, T., & Matsumoto, R. (1998) in this proceedings
Meier, D.L., Edgington, S., Godon, P., Payne, D.G., & Lind, K.R. (1997), *Nature*, **388**, pp.350
Miller, K.A., & Stone, J.M. (1997), *ApJ*, **489**, pp.890
Ouyed, R., Pudritz, R.E., & Stone, J.M. (1997), *Nature*, **385**, pp.409
Ouyed, R., & Pudritz, R.E. (1997), *ApJ*, **482**, pp.712
Pudritz, R.E., & Norman, C.A. (1986), *ApJ*, **301**, pp.571
Romanova, M.M., Ustyugova, G.V., Koldoba, A.V., Chechetkin, V.M., Lovelace, R.V.E. (1997), *ApJ*, **482**, pp.708
Rubin, E.L., & Burstein, S.Z. (1967), *J. Comp. Phys.*, **2**, pp.178
Shibata, K., & Uchida, Y. (1985), *PASJ*, **37**, pp.31
Shibata, K., & Uchida, Y. (1986), *PASJ*, **38**, pp.631
Shibata, K., & Uchida, Y. (1987), *PASJ*, **39**, pp.559
Shibata, K., & Uchida, Y. (1990), *PASJ*, **42**, pp.39
Stone, J.M., & Norman, M.L. (1994), *ApJ*, **433**, pp.746
Stone, J.M., Hawley, J.F., Gammie, C.F., & Balbus, S.A. (1996), *ApJ*, **463**, pp.656
Uchida, Y., & Shibata, K. (1985), *PASJ*, **37**, pp.515
Uchida, Y., Shibata, K., & Sofue, Y. (1985), *Nature*, **317**, pp.699
Ustyugova, G.V., Koldoba, A.V., Romanova, M.M., Chechetkin, V.M., & Lovelace, R.V.E. (1995), *ApJ*, **439**, pp.L39

2.5D NONSTEADY MHD SIMULATIONS OF MAGNETICALLY DRIVEN JETS FROM ACCRETION DISKS BY USING THE CIP-MOCCT METHOD

T. KUDOH AND K. SHIBATA
National Astronomical Observatory
2-21-1 Osawa, Mitaka, Tokyo 181-8588, Japan
kudo@solar.mtk.nao.ac.jp, shibata@solar.mtk.nao.ac.jp

AND

R. MATSUMOTO
Chiba University
Inageku, Chiba 263-8522, Japan
matumoto@c.chiba-u.ac.jp

1. Introduction

Astrophysical jets have been observed in young stellar objects (YSOs), active galactic nuclei (AGNs), and some X-ray binaries (XRBs). Although the acceleration and collimation mechanisms of these jets are still not well understood, these objects are believed to have accretion disks in their central regions. One of the most promising models for jet formation is magnetic acceleration from accretion disks (Blandford & Payne, 1982; Uchida & Shibata, 1985; Shibata & Uchida, 1986; Stone & Norman, 1994; Meier et al., 1997). We present the results of 2.5-dimensional MHD simulations of jet formation by magnetic accretion disks in which both ejection and accretion of disk plasma are included self- consistently.

2. The Numerical Model

Our simulations make the following assumptions: (1) axial symmetry around the rotational axis, including the azimuthal components of a velocity and a magnetic field (i.e., 2.5 dimensional approximation); (2) ideal MHD; (3) a specific heat ratio of $\gamma = 5/3$; (4) a point-mass gravitational potential only, with disk self-gravity neglected. As an initial condition, we assume an equilibrium disk rotating in a central point mass gravitational potential. The

S. M. Miyama et al. (eds.), Numerical Astrophysics, 203–206.

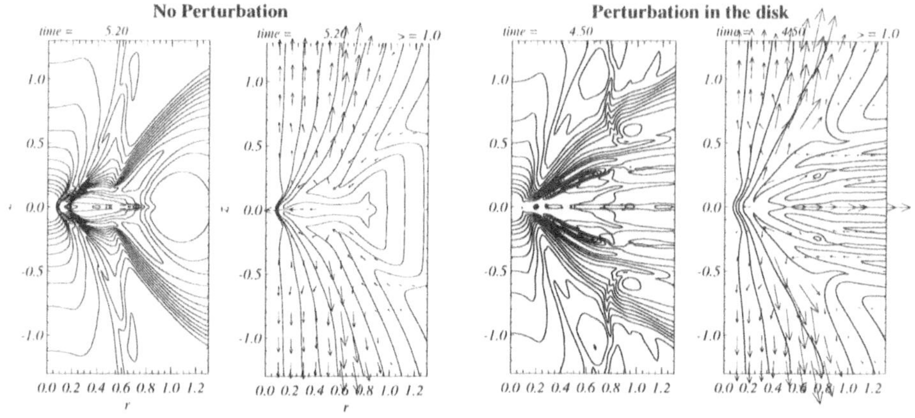

Figure 1. Density, poloidal magnetic field lines, and poloidal velocity, with (right pannel) and without perturbation in the disk (left pannel). Time $t = 2\pi \simeq 6.28$ corresponds to one Keplerian orbit at $(r, z) = (1, 0)$. Arrows show poloidal velocity vectors normalized by Keplerian velocity at $(r, z) = (1, 0)$. The parameters are $E_{\rm mg} \equiv (V_{\rm A0}/V_{\rm K0})^2 = 5.0 \times 10^{-4}$ and $E_{\rm th} \equiv (C_{\rm s0}/V_{\rm K0})^2 = 0.05$, where $V_{\rm A0}$, $C_{\rm s0}$, and $V_{\rm K0}$ are initial Alfvén velocity, initial sound velocity, and Keplerian velocity at $(r, z) = (1, 0)$, respectively.

mass distribution outside the disk is assumed to be that of the isothermal hot corona in hydrostatic equilibrium without rotation. The initial density in the corona is about 10^{-3} times smaller than the disk density.

3. Results

3.1. THE CASE OF A UNIFORM MAGNETIC FIELD IN THE DISK

For simplicity, the initial magnetic field is first assumed to be uniform and parallel to the axis of rotation. The left pannel of Fig. 1 shows the density, the poloidal magnetic field lines, and the poloidal velocity for a typical model. The initial values of plasma beta ($\beta = 8\pi p/B^2$ and the ratio of gas pressure to magnetic pressure) are $\beta \simeq 200$ at $(r, z) = (1.0, 0)$. Because the magnetic field in the disk is weak, the surface layer of the disk falls faster than the equatorial part (Matsumoto et al., 1996). The material on the disk surface is ejected as a jet. We found that nonsteady jets have very similar characteristics to steady jets: (1) The ejection point of the jets is determined by the effective potential resulting from gravitational and centrifugal forces along a field line (Blandford & Payne 1982). (2) The same scaling law of velocity as that of Michel's solution (Michel, 1969) is satisfied. (3) The dependence of the velocity (v_z) and mass outflow rate ($\dot{M}_w$) on the initial magnetic field strength (B_0) is approximately $\dot{M}_w \propto B_0$ and $v_z \propto B_0^{1/3}$. Because the velocity dependence on the field strength is not strong, the jet velocity is of the order of the Keplerian velocity of the disk for a wide range of the parameter. These results are consistent with the results of 1D steady

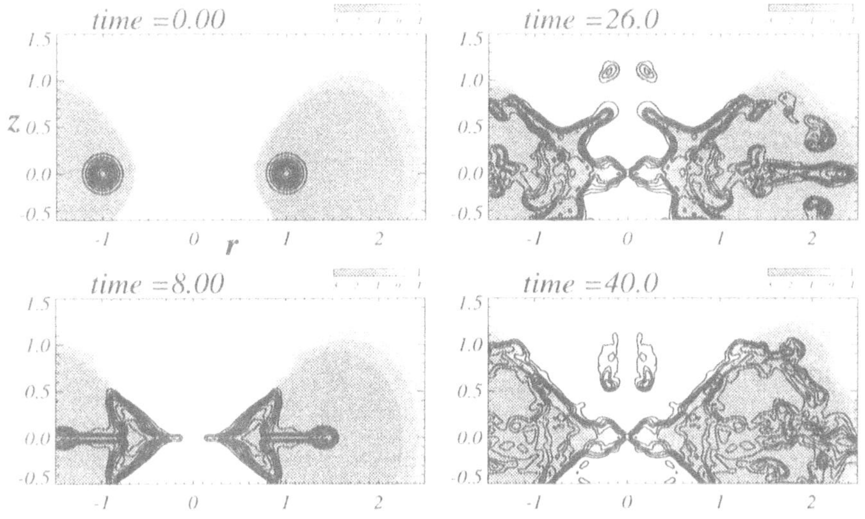

Figure 2. Time evolution of the temperature and the poloidal magnetic field lines for the case of a localized magnetic field in a disk. The initial minimum value of the plasma beta is about $\beta \simeq 145$ in the disk, and $E_{\rm th} = 0.05$.

solutions studied by Kudoh & Shibata (1997a). [The reader is referred to Kudoh, Matsumoto, & Shibata (1998) for more details.]

If inputted into the disk, the perturbation grows when the magnetic pressure is smaller than the gas pressure in the disk (Balbus & Hawley, 1991; Hawley & Balbus, 1991). The right pannel of Fig. 1 shows the density, the poloidal magnetic field lines, and the poloidal velocity for cases in which the sinusoidal large perturbation (30% of the sound velocity) is inputted into the disk. The nonlinear growth of this instability leads to a channel solution on an orbital time scale, which causes a large amount of mass accretion and mass loss rates, while the jet velocity is of the order of the Keplerian speed.

3.2. THE CASE OF A LOCALIZED MAGNETIC FIELD IN A DISK

In contrast to the case of a large-scale magnetic field, we initially assume a localized magnetic field in a disk (i.e., there is no magnetic field in the corona). Fig. 2 shows the time evolution of the temperature and the poloidal magnetic field lines for a typical model. First, Balbus & Hawley's instability causes mass accretion (time = 8.0). Second, magnetic buoyancy causes the ejection of the disk material into the corona (time = 26.0). If the coronal density is much smaller than the disk density, the ejected material behaves like a transient jet. This mechanism for jet formation is different from the Blandford-Payne mechanism.

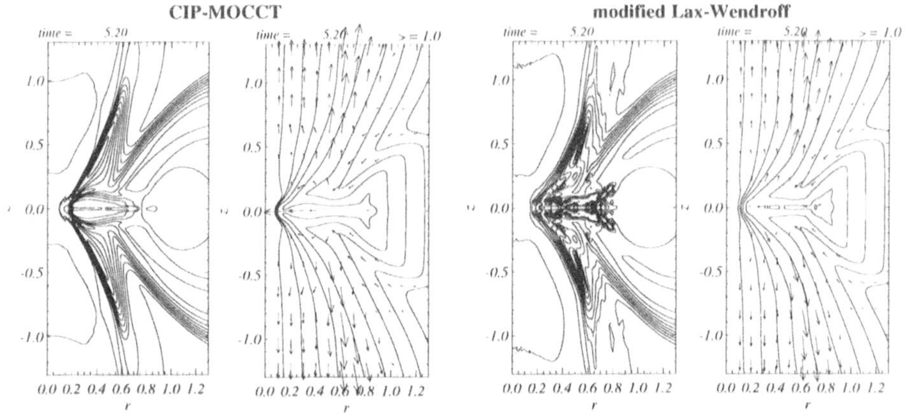

Figure 3. Temperature, poloidal magnetic field lines, and poloidal velocity, obtained by CIP-MOCCT method (left pannel) and by modified Lax-Wendroff method (right pannel).

4. Numerical Method

The numerical schemes we use are the CIP (Cubic Interpolated Propagation) method (Yabe et al., 1991) and the MOC-CT (Method of Characteristic-Constrained Transport) method (Evans & Hawley, 1988; Stone & Norman, 1992). The magnetic induction equation is solved by MOC-CT, the others being solved by CIP. [This is the 2-dimensional version of the scheme used in Kudoh & Shibata (1997b).] Fig. 3 compares that obtained by using the Lax-Wendroff method and that obtained by the CIP-MOCCT method for the same model shown in Fig. 1. The CIP method has the ability to maintain a sharp contact surface. By using the CIP method, we can easily identify the accretion disk surface within a few computational cells. Numerical computations were carried out on VPP300/16R at NAOJ.

References

Balbus, S.A., & Hawley, J.F. 1991, ApJ, 376, 214
Blandford, R. D. & Payne, D. G. 1982, MNRAS, 199, 883
Evans, C.R., & Hawley, J.F. 1988, ApJ, 332, 659
Hawley, J.F., & Balbus, S.A. 1991, ApJ, 376, 223
Kudoh, T., & Shibata, K. 1997a, ApJ, 474, 362
Kudoh, T., & Shibata, K. 1997b, ApJ, 476, 632
Kudoh, T., Matsumoto, R., & Shibata, K. 1998, ApJ in press, Nov.20
Matsumoto, R. et al. 1996, ApJ, 461, 115
Meier, D.L. et al. 1997, Nature, 388, 350
Michel, F.C. 1969, ApJ, 158, 727
Shibata, K., & Uchida, Y. 1986, PASJ, 38, 631
Stone, J. M., & Norman, M. L. 1992, ApJS, 80, 791
Stone, J.M., & Norman, M.L. 1994, ApJ, 433, 746
Uchida, Y., & Shibata, K. 1985, PASJ, 37, 515
Yabe, T. et al 1991, Comp. Phys. Comm., 66, 233

SPIRAL STRUCTURE IN IP PEG: CONFRONTING THEORY AND OBSERVATIONS

TAKUYA MATSUDA, MAKOTO MAKITA,
HIROSHI YUKAWA
Department of Earth and Planetary Sciences, Kobe University, Kobe 657-8501, Japan

AND

HENRI M.J. BOFFIN
Department of Physics and Astronomy, University of Wales, Cardiff CF2 3YB, UK

Abstract. The first convincing piece of evidence of spiral structure in the accretion disc in IP Pegasi was found by Steeghs et al. (1997). We performed two kinds of 2D hydrodynamic simulations, a SFS finite volume scheme and a SPH scheme, with a mass ratio of 0.5. Both results agreed well with each other. We constructed Doppler maps and line flux-binary phase relations based on density distributions, the results agreeing well with those obtained by observation.

1. Introduction

The standard model of accretion discs is the α disc model proposed by Shakura and Sunyaev (1973). One of various alternative models is the spiral shock model, which was first proposed by one of the present authors (Sawada, Matsuda & Hachisu, 1986a, b; Sawada et al., 1987). The α disc model essentially predicts the axi-symmetric structure of the disc except for the stream from the L1 point, while the spiral shock model predicts a bi-symmetric structure except for the stream. This fact is very important in order to distinguish between the two models observationally.

Steeghs, Harlaftis & Horne (1997) found the first convincing piece of evidence of spiral structure in the accretion disc of the eclipsing dwarf nova

S. M. Miyama et al. (eds.), Numerical Astrophysics, 207–210.

binary IP Pegasi using the technique known as Doppler tomography. IP Pegasi consists of a 1.02 $M_\odot$ white dwarf and a 0.5 $M_\odot$ companion star.

2. Model and methods of calculation

We calculated two-dimensional flows in a compact binary system by two numerical methods: the Simplified Flux vector Splitting (SFS) scheme (Jyounouchi et al., 1993; Shima & Jyounouchi, 1994) and the SPH scheme used by Yukawa, Boffin and Matsuda (1997). We studied a case with a mass ratio equal to 0.5 in order to simulate the IP Peg.

Only the region surrounding the white dwarf was calculated. The origin of the coordinates was at the center of the white dwarf, and the computational region was $-0.57 \leq x,\ y \leq 0.57$. In the case of our SFS scheme, the region was divided into 228 × 228 grid points. On the other hand, the number of particles in the SPH scheme was about 20,000. We assumed an ideal gas which was characterized by a ratio of specific heats γ. The γ was assumed to be 1.2 in the present calculations. The gas was injected from a small hole at the L1 point, which was at $x = -0.57, y = 0$. After a few orbital periods, we obtained a nearly steady density pattern.

3. Comparison with the observation

Based on the density distributions obtained by the SFS and SPH schemes, we constructed two Doppler maps, which showed the density distributions in the velocity space. Steeghs et al. (1997) observed the line intensities of H_α and HeI. Therefore, strictly speaking, we had to calculate the line intensity distributions based on the temperature distribution. Since our calculations were two-dimensional, our temperature distribution might not be very relevant to construct realistic Doppler maps. Therefore, we used two-dimensional density distributions instead. Line flux-binary phase diagrams could also be produced from the calculated Doppler maps.

Figure 1 shows the observed line flux as a function of a binary phase with H_α and HeI obtained by Steeghs et al. (1997). Figure 2 shows our calculated line flux-binary phase relations (top row), calculated Doppler maps (middle row), and calculated density distribution (bottom row). We used two methods of calculation. The left column of Fig. 2 shows the results based on the SFS scheme, while the right column shows those obtained by the SPH scheme. In constructing calculated Doppler maps, a contribution was added due to the companion star. The contribution from the outer region of the accretion disc, $r > 0.22$, was omitted in order to avoid the effect of the stream from the L1 point. This contribution is not seen in observation, a point which has not been explained yet. As can be seen, the calculated line flux and the Doppler maps shown in Fig. 2 agree very well with observations found in Fig. 1.

4. Discussion

Godon, Livio, and Lubow (1998) recently carried out a work similar to ours, although they only showed the line of peak density in the Doppler map. They claim that the calculated disc, which best fits the observation, is very hot compared with that in other observations. This claim is true in our calculations as well, in the sense that $\gamma = 1.2$ produces a rather hot disc. We performed 3D calculations (see Makita and Matsuda in this volume), in which we obtained rather open spirals even for the case of $\gamma = 1.01$, which gives a cooler disc. Therefore, it might be speculated that the 3D disc is more promising than the 2D disc as a means to explain the observations.

References

Godon P., Livio M., Lubow S., 1998, MNRAS, 295, L11.
Jyounouchi T., Kitagawa I., Sakashita, Yasuhara M., 1993, *Proceedings of 7th CFD Symposium.*
Sawada K., Matsuda T., Hachisu I., 1986a, MNRAS, 219, 75.
Sawada K., Matsuda T., Hachisu I., 1986b, MNRAS, 221, 679.
Sawada K., Matsuda T., Inoue M., Hachisu I., 1987, MNRAS, 224, 307.
Shakura N.I., Sunyaev R.A., 1973, A&A, 24, 337.
Shima E., Jyounouchi T., 1994, *25th Annual Meeting of Space and Aeronautical Society of Japan,* pp.36-37.
Steeghs D., Harlaftis E.T., Horne K., 1997, MNRAS, 290, L28.
Yukawa H., Boffin H.M.J., Matsuda T., 1997, MNRAS, 292, 321.

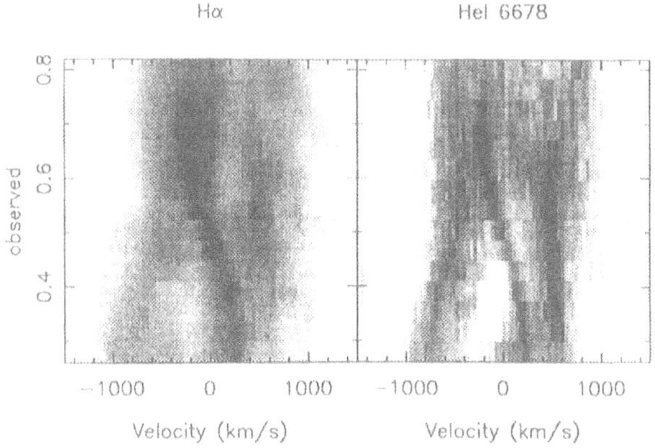

Figure 1. The observed line flux from IP Peg as a function of the binary phase with H_α on the left and HeI(6678) on the right; these figures are taken from Steeghs et al. (1997).

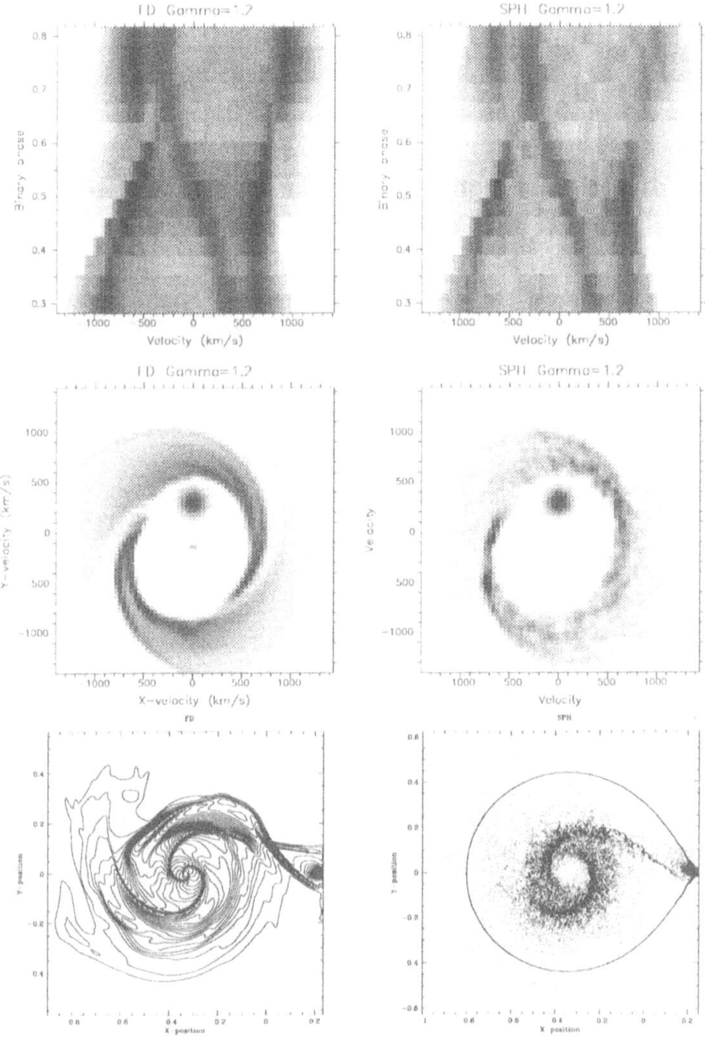

Figure 2. Top row: calculated line flux as a function of the binary phase (vertical axis). The horizontal axis represents the radial velocity. Middle row: calculated Doppler map. The horizontal axis and the vertical axis depict the horizontal velocity, v_x, and the vertical velocity, v_y, respectively. The bottom row shows the density distribution. The left column shows the results based on the SFS method, while the right column shows those based on the SPH method.

3-D SPH SIMULATIONS OF RADIATIVELY COOLING MAGNETIZED JETS

E.M. DE GOUVEIA DAL PINO AND A.H. CERQUEIRA
University of São Paulo, Instituto Astronômico e Geofísico
Av. Miguel Stéfano, 4200, 04301-904 São Paulo, SP, Brazil

Considerable amount of numerical work on magnetized, adiabatic, light jets has been done to study extragalactic jets (see, e.g., [1] for a review). Only recently, these MHD studies have been extended to heavy, radiatively cooling jets [e.g., 2, 3, 8, 9, 10].

In this work, we summarize the results of three-dimensional smooth particle magnetohydrodynamics numerical simulations of supermagnetosonic, radiatively cooling jets [3]. Two initial magnetic configurations (in $\sim$ equipartition with the gas) are considered: (i) a helical and (ii) a longitudinal field which permeate both the jet and the ambient medium [2]. These MHD simulations have been compared with a previous baseline non-magnetic radiative cooling and adiabatic calculation [e.g., 4, 5, 6, 7]. We find that magnetic fields have important effects on the dynamics and structure of radiative cooling jets, especially at the head. Both magnetic field geometries are able to improve jet collimation, although this effect is more pronounced in the helical case. In both magnetic configurations, the confining pressure of the cocoon is able to excite MHD Kelvin-Helmholtz pinch and kink modes that drive low-amplitude internal shocks and wiggling along the beam (see Fig. 1). These modes are however, inhibited by the presence of radiatively cooling and by increasing the density contrast, η, between the jet and the ambient medium [4, 5, 6, 7]. As a consequence, the weakness of the induced shocks makes it doubtful that they could produce by themselves the bright knots observed in the overdense, radiatively cooling protostellar jets. Also we find that the presence of a helical field (Fig. 1, bottom) suppresses the formation of the clumpy structure which is found to develop at the head of *purely hydrodynamical* [3] jets by fragmentation of the cold shell of shocked material. On the other hand, a cooling jet embedded in a longitudinal magnetic field (Fig. 1, top) retains the clumpy morphology at its head. This fragmented structure resembles the knotty pattern commonly observed in HH objects behind the bow shocks of protostellar jets.

S. M. Miyama et al. (eds.), Numerical Astrophysics, 211–212.

This suggests that a predominantly helical magnetic field configuration is unlikely at the jet head in those cases [2].

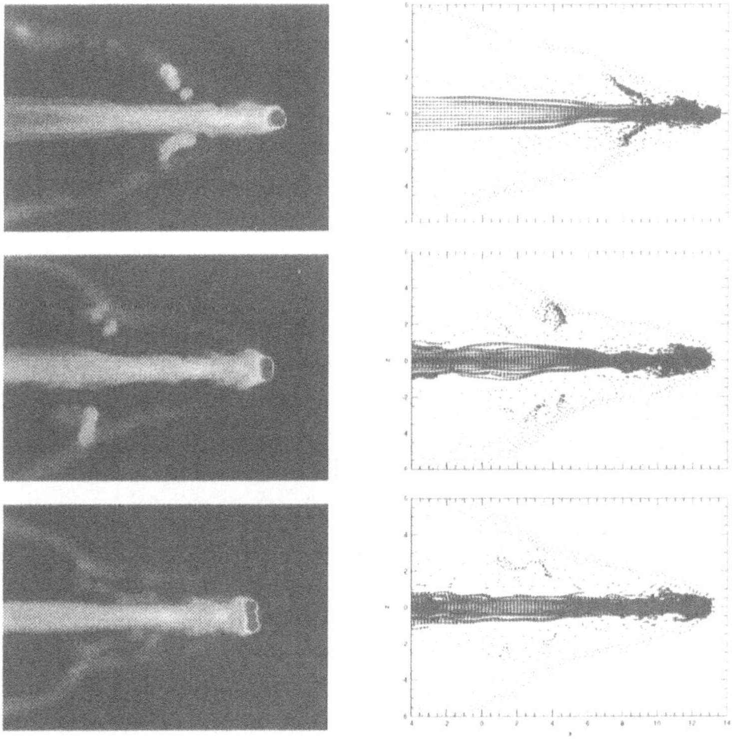

Figure 1. Figure 1: Left – gray-scale representation of the midplane density of the head of radiative cooling jets with $\eta = 3$ and ambient Mach number $M_a = 24$: a hydrodynamical jet (HD3r, top), an MHD jet with initial longitudinal B-field (ML3r, middle), and an MHD jet with initial helical B-field (MH3r, bottom) after they have propagated $\approx 30R_j$. Right – mid-plane velocity field distribution [2].

References

[1] Birkinshaw, M. (1997) in *Advanced Topics on Astrophysical and Space Plasmas*, Eds. E.M. de Gouveia Dal Pino et al., Kluwer Acad. Publs., Dordrecht, 17
[2] Cerqueira, A.H. & de Gouveia Dal Pino, E.M. (1998) (submitted)
[3] Cerqueira, A.H. de Gouveia Dal Pino, E.M. & Herant, M. (1997), *ApJL*, 489, L185
[4] de Gouveia Dal Pino, E.M. & Benz, W. (1993), *ApJ*, 410, 686
[5] de Gouveia Dal Pino, E.M. & Benz, W. (1994), *ApJ*, 435, 261
[6] de Gouveia Dal Pino, E.M., & Birkinshaw, M. (1996), *ApJ*, 471, 832
[7] de Gouveia Dal Pino, E.M., Birkinshaw, M. & Benz, W. (1996), *ApJL*, 460, L111
[8] de Gouveia Dal Pino, E.M. & Cerqueira, A.H. (1996), *Astro. Lett. & Comm.*, 34, 303
[9] Frank, A, Ryu, D., Jones, T.W. & Noriega-Crespo, A. (1998), *ApJL*, 494, L79
[10] Hardee, P.E. & Stone, J.M. (1997), *ApJ*, 483, 121

MHD SIMULATION OF ACTIVE GALACTIC NUCLEI JETS

SEIICHI KATO
Department of Astronomy, University of Tokyo
7-3-1 Hongo, Bunkyo-ku, Tokyo 113, Japan

AND

TAKAHIRO KUDOH AND KAZUNARI SHIBATA
National Astronomical Observatory of Japan
2-21-1 Osawa, Mitaka, Tokyo 181-8588, Japan

The formation mechanism of AGN jets is one of the most mysterious problems in astronomy. The most promising model at present is 'a magnetically driven jet model (Blandford and Payne, 1982; Uchida and Shibata, 1985; Shibata and Uchida, 1986; Kudoh and Shibata, 1995, 1997a,b; Matsumoto et al., 1996). Recently, Kudoh, Matsumoto, and Shibata (1998) investigated the acceleration mechanism of nonsteady jets from geometrically thick disks and found that a jet's ejection point (slow point) is determined by the effective gravitational potential made by the action of magneto-centrifugal force in spite of the fact that the model contains nonsteady accretion. Using a two-dimensional nonsteady MHD code (CIP-MOCCT), we investigated the relation between accretion and jets from geometrically *thin disks* that are not in dynamical equilibrium in the initial state. We confirmed that the dependence on B is approximately consistent with Kudoh and Shibata's theory (1997) $V_{\rm jet} \propto B^{1/3}$ (see TABLE 1), and found a knot-like structure in weak magnetic field cases. The results are shown in CD-ROM. Numerical computations were carried out on VPP300/16R at NAOJ.

References

Blandford, R. D. & Payne, D. G. (1982), *MNRAS*, **199**, pp.883–903
Kudoh, T., Matsumoto, R., & Shibata, K., (1998) *ApJ, in press.*
Kudoh, T., & Shibata, K. (1995), *ApJ*, **452**, L41-L44
Kudoh, T., & Shibata, K. (1997a), *ApJ*, **474**, pp.362 377
Matsumoto, R. et al. (1996), *ApJ*, **461**, pp.115–126
Shibata, K., & Uchida, Y. (1986), *PASJ*, **38**, pp.631–660
Uchida, Y., & Shibata, K. (1985), *PASJ*, **37**, pp.515–535

S. M. Miyama et al. (eds.), Numerical Astrophysics, 213–214.

TABLE 1. The relation between magnetic field strength (represented by $\beta = P_{\rm gas}/P_{\rm mag} \propto B^{-2}$) and jets' maximum speed.

$\beta_{\rm disk} = P_{\rm gas}/P_{\rm mag}$	Kepler case	Sub-kepler case ($V_\varphi/V_{\rm Kepler} = 0.8$)
0.5	1.7 $V_{\rm Kepler}$	2.5 $V_{\rm Kepler}$
50	0.8 $V_{\rm Kepler}$	1.0 $V_{\rm Kepler}$

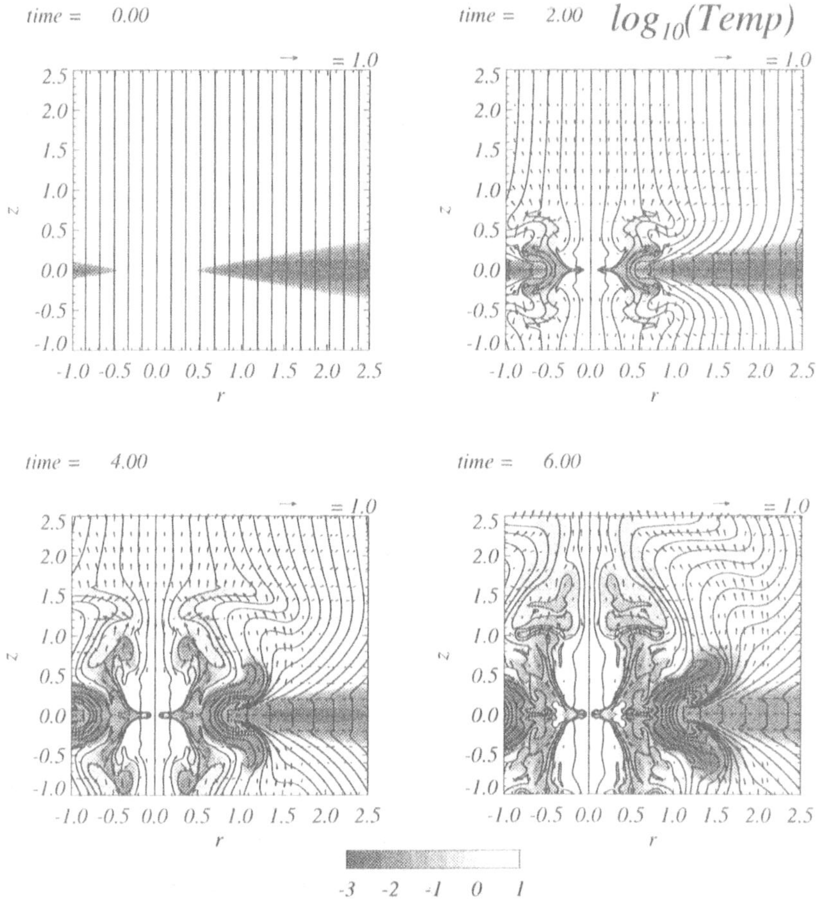

Figure 1. These figures show the time evolution of the temperature ($\log_{10} T$) when the magnetic field is weak ($\beta_{\rm disk} \simeq 50$) and the initial rotational velocity of the disk is smaller than Keplerian velocity ($V_\varphi/V_{\rm Kepler}$=0.8). The thick lines are magnetic field lines. Interestingly, the magnetic field lines show a highly bending shape corresponding to a knot-like structure.

GENERAL RELATIVISTIC JET FORMATION FROM MAGNETIZED ACCRETION DISK AROUND BLACK HOLE

S. KOIDE
Faculty of Engineering, Toyama University, Gofuku, Toyama 930, Japan

AND

K. SHIBATA AND T. KUDOH
National Astronomical Observatory, Mitaka, Tokyo 181, Japan

1. Introduction

Radio observations have revealed the compelling evidence of the existence of relativistic jets not only from active galactic nuclei (AGNs) but also from "microquasars" in our galaxy (Mirabel and Rodriguez 1994). It is believed that a black hole exists in the cores of these objects and that magnetic phenomena occur near the black hole to form these relativistic jets (Blandford and Payne 1982; Uchida and Shibata 1985; Shibata and Uchida 1986; Kudoh and Shibata 1997). To simulate jet formation in the magnetosphere, we recently developed the general relativistic magnetohydrodynamic (GRMHD) code (Koide, Shibata and Kudoh 1998). The numerical results of the jet formation are presented here.

2. Numerical Method

Our study is based on general relativistic equations of conservation laws about mass, momentum, and energy of conductive fluids and on Maxwell equations. We use the Schwarzschild metric, which provides the space-time around the black hole at rest. We employ a *simplified total variation diminishing* (TVD) method. We assume that there exist a free fall corona, a uniform magnetic field, and a relativistic Keplerian disk at the initial stage. The inner edge of the disk is assumed to be located at the last stable circular orbit.

S. M. Miyama et al. (eds.), Numerical Astrophysics, 215–216.

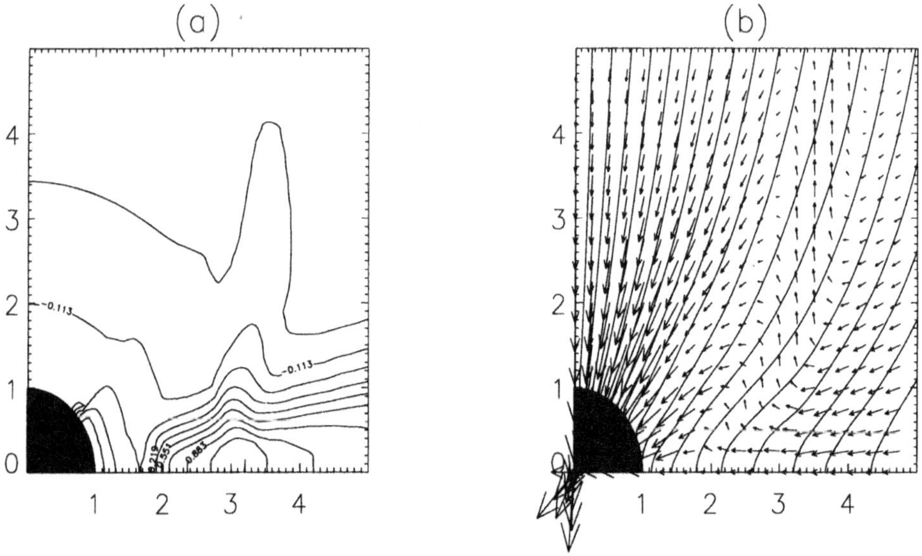

Figure 1. A snapshot of (a) proper mass density in the logarithmic scale, (b) velocity (vector) and magnetic field lines (solid lines) in the early stage of jet formation.

3. Results

Figure 1 shows the early stage of jet formation when the disk rotates almost one cycle. The black region shows the black hole (inside event horizon, $r \leq r_S$). The disk loses its angular momentum due to magnetic dragging and falls into the black hole. However, the centrifugal barrier stops the advection-dominated flow between $2r_S$ and $3r_S$ [Fig. 1 (a), vector in Fig. 1 (b)]. A high-pressure region is produced in front of the barrier, pushing the plasma of the disk upward. The plasma is accelerated upward by both the gas pressure and the magnetic force of the oblique magnetic field [Fig. 1 (b)]. In a nonrelativistic calculation, such a structure of jet formation is not found. Remarkably, a jet is formed from the disk very near the event horizon even in the free-fall corona of the non-rotating black hole.

We thank M. Koide for her valuable input on this study and the National Astronomical Observatory for permitting the use of their super-computers.

References

Blandford, R. D. and Payne, D. G. 1982, MNRAS, **199**, 883.
Koide, S., Shibata, K., and Kudoh, T., 1998, ApJ, **495**, L63.
Kudoh, T. and Shibata, K. 1997, ApJ, **476**, 632.
Mirabel, I. F. and Rodriguez, L. F. 1994, Nature, **371**, 46.
Shibata, K. and Uchida, Y. 1986, PASJ, **38**, 631.
Uchida, Y. and Shibata, K. 1985, PASJ, **37**, 515.

3-D RELATIVISTIC MHD SIMULATIONS OF EXTRAGALACTIC JETS

K.-I. NISHIKAWA, J. FRANK AND D. M. CHRISTODOULOU
Department of Physics and Astronomy,
Louisiana State University
Baton Rouge, LA 70803, USA

S. KOIDE AND J.-I. SAKAI
Laboratory of Plasma Physics and Fusion Science
Faculty of Engineering, Toyama University
Gofuku, Toyama 930, Japan

H. SOL
DARC, Observatoire de Paris-Meudon
5, place Jules Janssen
92195 Meudon Cedex, France

AND

R. L. MUTEL
Department of Physics and Astronomy
The University of Iowa
Iowa City, IA 52242, USA

We present the numerical simulations of relativistic jets propagating initially oblique to the field lines of a magnetized ambient medium. Our simulations incorporate relativistic MHD in a four-dimensional spacetime and clearly show that (a) relatively weak, oblique fields (at 1/16 of the equipartition value) have only a negligible influence on the propagating jet and they are passively pushed away by the relativistically moving head (Koide *et al.*, 1996; Nishikawa *et al.*, 1997); (b) oblique fields in equipartition with the ambient plasma provide more resistance and cause bending at the jet head (Nishikawa *et al.*, 1998), but the magnitude of this deflection and the associated backflow are small compared to those identified by previous studies with a 2-D slab model (Koide, 1997). The new results are understood as follows: Relativistic simulations have consistently shown that these jets are effectively heavy and so they do not suffer substantial mo-

S. M. Miyama et al. (eds.), Numerical Astrophysics, 217–218.

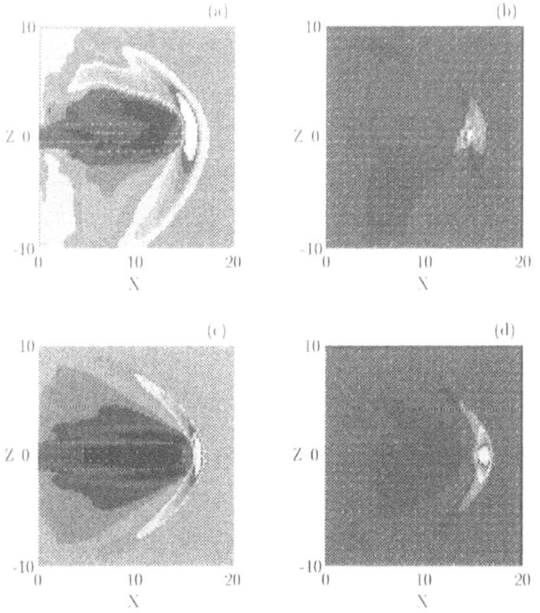

Figure 1. Run A (panels a,b) and Run B (panels c,d) with strong and weak magnetic field, respectively: XZ (vertical) cross-sections through $Y = 0$ at $t = 8.0\tau_s$, ($\tau_s = 1/v_s$). Panels a,c show contours of rest mass density and velocity vectors. Panels b,d show contours of thermal pressure and magnetic-field vectors. The maximum values represented by white color are: (a) $\rho_{max} = 1.68$; (b) $P_{max} = 3.56$; (c) $\rho_{max} = 2.17$; and (d) $P_{max} = 3.70$.

mentum losses and are not decelerated as efficiently as their nonrelativistic counterparts. In addition, the ambient magnetic field, however strong, can be pushed aside with relative ease by the beam, provided that the degrees of freedom associated with all three spatial dimensions are followed self-consistently during the simulations (see Figure 1). The effect is analogous to pushing Japanese "noren" or vertical Venetian blinds out of the way while the slats are allowed to bend in 3-D space rather than as a 2-D slab structure. Applied to relativistic extragalactic jets from blazars, the new results are encouraging since superluminal outflows exhibit bending near their sources and their environments are profoundly magnetized—but observations do not provide support for irregular kinematics such as large-scale vortical motions and pronounced reverse flows near the points of origin.

References

Koide, S., Nishikawa, K.-I. and Mutel, R. L. (1996) A two-dimensional Simulation of relativistic magnetized jet, ApJ, 463, pp. L71–L74.

Koide, S. (1997) A two-dimensional Simulation of a relativistic jet bent by an oblique magnetic field, ApJ, 478, pp. L66–L69.

Nishikawa, K.-I., Koide, S., Sakai, J.-I., Christodoulou, D. M., Sol, H. and Mutel, R. L. (1997) Three-dimensional magnetohydrodynamic simulations of relativistic jets injected along a magnetic field, ApJ, 483, pp. L45–L48.

Nishikawa, K.-I., Koide, S., Sakai, J.-I., Christodoulou, D. M., Sol, H. and Mutel, R. L. (1997) Three-dimensional magnetohydrodynamic simulations of relativistic jets injected oblique to a magnetic field, ApJ, 498, pp. 166–169.

JETS FROM TIME-DEPENDENT ACCRETION FLOWS ONTO A BLACK HOLE

KOJI NOBUTA AND TOMOYUKI HANAWA
Department of Astrophysics, School of Science,
Nagoya University, Chikusa-ku, Nagoya 464-8602, Japan

1. Introduction

Using numerical simulations, we investigate highly variable hydrodynamical accretion onto a black hole as a model of accretion disk and jets in AGNs. Numerical simulation of hydrodynamical accretion was studied by Hawley et al. (1984), Molteni et al. (1994), and others. In these simulations jets were not clearly formed, while our numerical simulations show clear semi-relativistic bipolar jets.

2. Models

We investigate axisymmetric accretion onto a black hole having the mass of $10^7 M_\odot$. For simplicity the gravity of the black hole is approximated by the pseudo-Newtonian potential. The computational domain covers from $r = 2\,r_g$ to $104\,r_g$.

The domain of computation is divided into eight concentric shells. In a given shell the grid spacing is uniform. The radial grid spacing, Δr, is twice larger in a given shell than in the inner adjacent shell. We applied the multi time step method of Chiang, van Leer, & Powell (1992) to integrate the hydrodynamical equations. In the multi time step method the time step, Δt, is set to be proportional to Δr.

3. Results

Figure 1 shows the jet formation mechanism schematically. Panel (a) shows the initial stage, at which a dense gas torus with high specific angular momentum ($\ell = 2.1\,cr_g$) is placed at $r = 20\,r_g$ in ambient gas flow of low specific angular momentum ($\ell \leq 1\,cr_g$). Panel (b) shows the bounce of the

S. M. Miyama et al. (eds.), Numerical Astrophysics, 219–220.

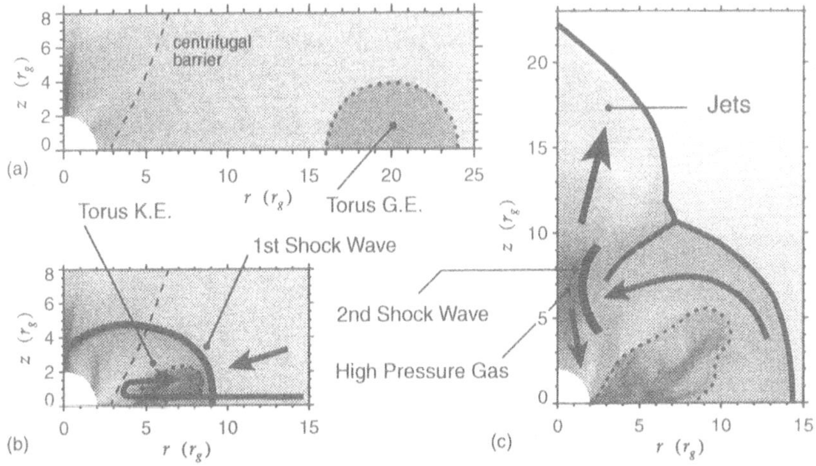

Figure 1. The jet formation mechanism is shown schematically. The grayness denotes the density distribution. The thick solid lines denote the shock fronts. The labels denote the form of energy. The abbreviations "G.E." and "K.E." mean "gravitational energy" and "kinetic energy", respectively.

torus at the centrifugal barrier. A shock wave forms at the head of the torus. Panel (c) shows the jet formation. The high pressure at the second shock pushes gas to outflow along the rotation axis. The outflowing gas evolves into bipolar semi-relativistic jets.

Summary:

(1) Bipolar jets emanate when the blobs have high specific angular momentum ($\ell \geq 2\, cr_g$). The mass ejection is largest when the specific angular momentum of the blobs is $\ell = 2.2\, cr_g$.
(2) When the blobs have low specific angular momentum ($\ell < 2\, cr_g$), they accrete smoothly onto the hole without producing shock waves and jets.
(3) The efficiency of the energy gain of the jets is $3 \sim 11\%$.
(4) When the ambient gas is colder, the jets are better collimated.

See Nobuta & Hanawa (1998) and the MPEG movie in the CD-ROM for more details. The numerical simulations were carried out on VPP300/16R at the National Astronomical Observatory, Japan.

References

Chiang, Y., van Leer, B., & Powell, K. G. 1992, AIAA Paper 92-0443, AIAA 30th Aerospace Sciences Meeting
Hawley, J. F., Smarr, L. L., & Wilson, J. R. 1984, ApJS, 55, 211
Molteni, D., Lanzafame, G., & Chakrabarti, S. K. 1994, ApJ, 425, 161
Nobuta, K. & Hanawa, T. 1998, ApJ, submitted

ON MAGNETODYNAMIC JET FORMATION FROM ACCRETION PROCESS OF MAGNETIZED GAS IN AGN CORE

Y.UCHIDA, M.NAKAMURA, S.HIROSE AND S.UEMURA
Department of Physics, Science University of Tokyo
1-3 Kagurazaka, Shinjyuku-ku, Tokyo 162-8601, Japan

In this paper, we discuss the extension of the Sweeping Magnetic Twist model to the situation of AGN jets from its accretion disk in the core. The model was first proposed for the star formation bipolar flow case by Uchida and Shibata (1985) with a genetic consideration. The gravitational contraction of the interstellar gas permeated by the large scale magnetic field ends up with an accretion disk whose angular momentum axis is parallel to the large scale field, and the field in the disk is stretched by the differential rotation into the azimuthal direction. The magnetic tension exerts braking to the rotation of the disk material and enhances the accretion, while the external part of the field lines is dragged, and the non-linear torsional Alfven waves are emitted along the large scale field. Those pinch the hourglass shaped field into the form of slender jets along the axis. The earlier phases in this scenario, providing the disk with an hourglass-shaped field as assumed in the initial work, have actually been verified recently by Tomisaka (1998) by using an adaptive mesh simulation.

We consider that this model is viable, at least in the star formation case, because the observations of L1551 bipolar flows actually gave some characteristic features predicted by the model, namely, (i) a hollow cylindrical structure, (ii) a long-ranged acceleration, and (iii) the spinning, of the flow (Uchida et al. 1987). Those are not explicable by hydrodynamic flow model with the central object as the source. Our model has a merit that it provides a genetic picture for the process.

Is the model applicable to the AGN situation with a tremendously different geometrical scales and in physical parameters? It is, as long as the situation of the system, namely, an accretion disk with magnetic field is rotating around the central gravitator, can be taken to be similar, and the sets of three non dimensional coefficients in the non-dimensionalized equations, namely, the ratios of acoustic, Alfvenic, and Keplerian velocities to the rotational velocity, all at a same typical point in the similar situations

S. M. Miyama et al. (eds.), Numerical Astrophysics, 221–222.

above, are close with each other (Uchida and Shibata 1986). Very rough estimates show that the latter holds at least conceptually, and the former, too, in a rough sense. Therefore, it is well expected that an extension is worthwhile. Actually, steps to this direction have been taken (Matsumoto et al. 1996), and it was shown that a similar but stronger actions takes place in the case of AGN accretion disk. It should be added that in this model the torsional Alfven wave train transports energy in the Poynting flux, pinching the magnetic flux into slender structures, and deposits the energy at the hub of the bunched field of the lobe, as a hotspot.

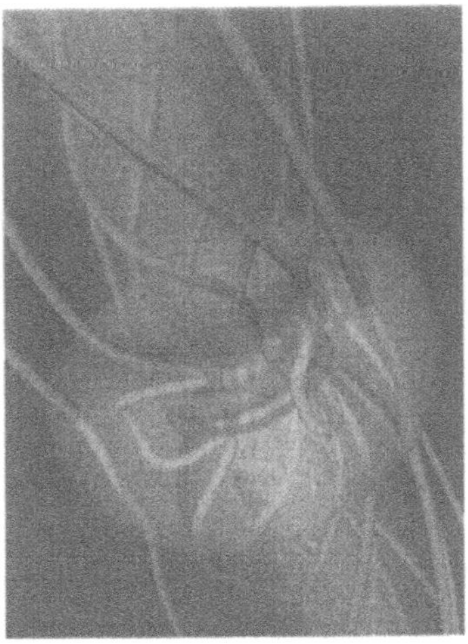

Figure 1. A result of simulation for AGN jetting. The accretion takes place preferentially near the surface of the disk like avalanches, by losing angular momentum most.

Questions that may be asked for this application are about the presence of primordial large scale field. We believe that the systematic field in the jets (stems) are brought in there by the pinch, and twisted in the above process. Presence of a strong magnetic field at the origin of the jets requires higher spatial resolution of observation to verify. If the gas jets are actively pulling out magnetic field, there is no reason for the systematic magnetic structure in the jets, and if the field is strong enough to hold such a structure, the reaction of pulling out the field may decelerate or destroy the jets.

References

Matsumoto, R., et al., 1996, Astrophys. J., 461, 115
Tomisaka, K., 1998, in preparation
Uchida, Y., and Shibata, K. 1985, Publ. Astron. Soc. Japan, 37, 515
Uchida, Y., and Shibata, K., 1986, Can. J. Phys. 64, 507
Uchida, Y., et al., 1987, Publ. Astron. Soc. Japan, 39, 907

GENERAL RELATIVISTIC MAGNETOHYDRODYNAMIC STRUCTURE AROUND A ROTATING BLACK HOLE

M. YOKOSAWA AND S. FUKAZAWA
Department of Natural Science
Ibaraki University, 2-1-1 Bunkyo, Mito 310-8512, Japan

1. Introduction

It is now well established that blazars are strong γ-ray emitters. The polarization of optical emission from these blazars indicates evidence of a strong magnetic field around a central engine. The rapid variation of polarization suggests the ejection of a beamed jet from a central object toward us. When a strong emission of γ-ray is generated at the front of a jet, the velocity of the jet must be relativistic. It is yet not clear how a relativistic jet is produced around a massive black hole at the center of a blazar. It may be necessary for the generating energy of a relativistic jet to be stored in the vicinity of the event horizon. One possible mechanism is due to the rotating effects of a black hole. The angular momentum of the accreting disk is increased by a rotating black hole, rotating energy being then accumulated in an inner disk around a black hole. We had calculated the dynamical evolution of magnetohydrodynamical (MHD) accretion in Kerr space-time by a numerical method. We show here the rates of transport of energy and angular momentum from a rotating black hole to accreting fluid by MHD interaction.

2. Angular momentum and rotational energy of accretion flow generated by a rotating black hole

We developed a new simulation method in which Maxwell's equations were solved using the 3+1 formalism, and momentum equations were solved using the four-dimensional form. By using a method involving the transformation law, the 3+1 components of the electro-magnetic field were transformed to the four-dimensional components (Yokosawa 1994). Calculations concerning the accretion of gas initially at rest were performed. The initial

S. M. Miyama et al. (eds.), Numerical Astrophysics, 223–224.

distributions of gas and magnetic field were set to be homogeneous, the directions of the magnetic field lines being aligned to the rotation axis of the black hole. We studied a case of inertia-dominance, i.e., the fact that the magnetic stress was weak compared with the inertia of the fluid. After a few dynamical times of the calculation domain, the MHD flow became quasi-stationary around a black hole.

We analyzed the angular momentum and rotational energy formed in the accretion around a black hole. The torsion of the magnetic field was produced over extensive space by means of the differential rotation of space-time. MHD interaction produced angular momenta in the opposite direction, i.e., the direction of the angular momentum of the magnetic field was counter-rotating against the hole's spin while the fluid rotated in the same direction as the hole. A region with negative energy-at-infinity for the magnetic field was formed near the horizon. Since magnetic torsion rotated the fluid, the specific angular momentum of fluid u_φ was increased in proportion to the square of the strength of magnetic field B^2 and to the angular momentum of the hole a, and inversely proportional to the density of fluid ρ, $u_\varphi \propto aB^2\rho^{-1}$. The density of the angular momentum of the fluid ϵ_{Lz} was distributed as $\epsilon_{Lz} = (\rho + P)\gamma v^{\hat{\varphi}}\varpi \propto r^{-3.5}$. When the flow fell freely, $\rho \propto r^{-1.5}$. Thus, the angular momentum for the fluid J on a shell with an inner radius of r_1 and outer radius r_2 approximated such that

$$J = \int_{r_1}^{r_2} \epsilon_{Lz} dV \approx \epsilon_{Lz}(r_1) r_1^3. \tag{1}$$

The quasi-radial MHD accretion in rotating space-time concentrated the angular momentum on the inner shell around the horizon. The quasi-radial accretion in flat space-time with a small constant value of the angular momentum concentrated it on the outer shell, since the fluid rotation was $v^{\hat{\varphi}} \propto r^{-1}$. The rotational component of the energy $E_{\rm rot}$ was $E_{\rm rot} = \int \omega \epsilon_{Lz} dV$, which was also concentrated on the inner shell. The rapidly rotating black hole efficiently supplied the angular momentum to an accretion disk by means of MHD interaction, allowing for the possibility that the inner structure of the disk might then expand with excess heat.

References

Yokosawa, M. (1994) Formation of Shock Waves in Magnetohydrodynamical Accretion onto a Black Hole, *PASJ*, **Vol. 46**, pp. 73-86

NUMERICAL COMPUTATION OF TWO-DIMENSIONAL WIND ACCRETION OF ISOTHERMAL GAS

EIJI SHIMA
Kawasaki Heavy Industries
Kakamigahara, Gifu 504, Japan

TAKUYA MATSUDA
Department of Earth and Planetary Science, Kobe University
Kobe 657, Japan

ULRICH ANZER AND GERHARD BÖRNER
Max-Planck Institut für Astrophysik
Postfach 1523, 85740 Garching, Germany

AND

HENRI M.J. BOFFIN
Department of Physics and Astronomy, University of Wales
Cardiff CF2 3YB, UK

1. Introduction, Method and Result

The hydrodynamics of two-dimensional plane symmetric accretion flows of supersonic isothermal gas in wind-fed sources is analyzed numerically. The Polar coordinates are used to ensure fine resolution near the object. The absorbing boundary condition is set on the surface of an accretor (central hole) and the typical value of the hole is $0.01Ra$ and the smallest mesh size is $0.001Ra$, where Ra is the Bondi-Hoyle accretion radius. The velocity outside the computational region is given by particle orbits. The Finite Volume Method is used and the conservation of angular momentum is adopted as a governing equation. This has an important effect on the formation of accretion disks.

It is found that a thin accretion column is formed which shows wave-like oscillations. This kind of motion was found also in a computation using a particle code (Boffin & Anzer 1994) and in a study of wind accretion in a binary system (Matsuda et al. 1987). A small accretion disk forms and disappears repeatedly around the object (Fig. 1). Accretion occurs

S. M. Miyama et al. (eds.), Numerical Astrophysics, 225–226.

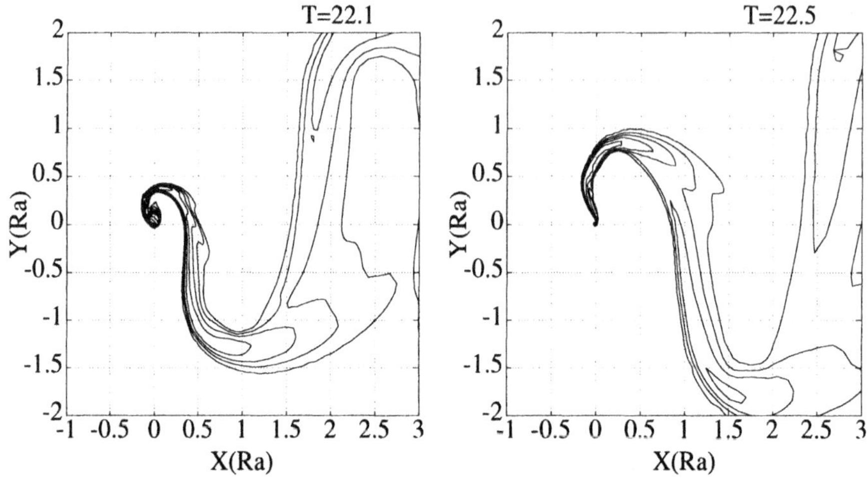

Figure 1. Density contours at two different times for the Mach 4 case. An accretion disk is formed (left), then collapses (right).

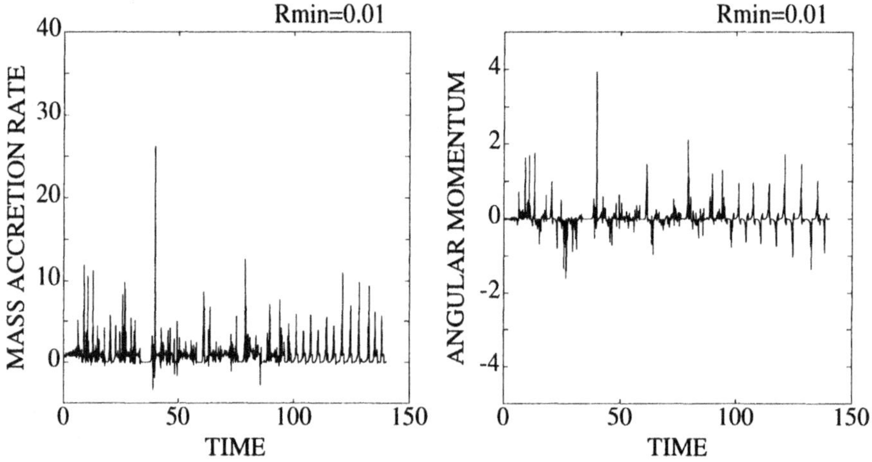

Figure 2. History of mass accretion rate (left) and angular momentum accretion rate (right), for the Mach 4 case. The mass accretion rate is normalized by the analytical value.

when the disk is destroyed by the infalling matter. Thus, mass and angular momentum accrete in a short period. The mass accretion rate and angular momentum accretion rate exhibit quasi-periodic oscillations (Fig. 2). The average accretion rates were close to analytical values.

References

Boffin, H.M.J., Anzer, U., 1994, A&A, 284, 1026
Matsuda, T., Inoue, M., Sawada, K., 1987, MNRAS, 226, 785

3D FINITE VOLUME SIMULATION OF ACCRETION DISCS WITH SPIRAL SHOCKS

MAKOTO MAKITA AND TAKUYA MATSUDA
Department of Earth and Planetary Sciences, Kobe University, Kobe 657-8501, Japan

1. Introduction

One of the models of accretion discs in a close binary system is the spiral shock model. This model was first proposed by one of the present authors (Sawada, Matsuda & Hachisu 1986a, b, Sawada et al. 1987). A number of authors have confirmed since then that spiral shocks appear in two-dimensional discs. In the case of 3D, Sawada & Matsuda (1992) used a Roe scheme to calculate the case of $\gamma = 1.2$ with mass ratio of unity. They found the existence of spiral shocks, but their calculation was done up to only half a revolution period. Yukawa, Boffin & Matsuda (1997), who made numerical simulations using SPH with mass ratio of unity, also demonstrated that spiral shocks existed in the case of $\gamma = 1.2$.

The purpose of the present paper is to perform 2D and 3D finite volume numerical calculations with higher resolution up to enough time to confirm the existence of spiral shocks.

2. Calculations and Conclusions

We considered a binary system composed of a mass-accreting primary star and a mass-losing secondary star. The mass ratio of the mass-losing star to the mass-accreting star was one. We set a numerical grid centered at the primary star and just touching the inner Lagrangian point L1. In addition, we used Cartesian coordinates having 200×200 grid points in 2D and $200 \times 200 \times 50$ grid points in 3D, respectively. Finally, we assumed a polytropic gas with constant specific heat ratio γ.

In this paper, we examined the cases of γ =1.2, 1.1, 1.05, and 1.01. Symmetry about the orbital plane was assumed in our 3D calculations. We used the Simplified Flux vector Splitting (SFS) scheme (Jyounouti et

S. M. Miyama et al. (eds.), Numerical Astrophysics, 227–228.

al. 1993; Shima & Jyounouti 1994). In 2D calculations, we found that the smaller γ was, the more tightly the spiral wound. The density contours in the orbital plane in the cases of $\gamma = 1.2$ and $\gamma = 1.01$ of our 3D simulations are presented in Fig.1. These figures show the existence of spiral shocks in the cases of $\gamma = 1.2$ as well as in all the other cases. However, they do not show as clear a difference in spiral structure as they do in our 2D calculations. (Other results can be seen in CD-ROM).

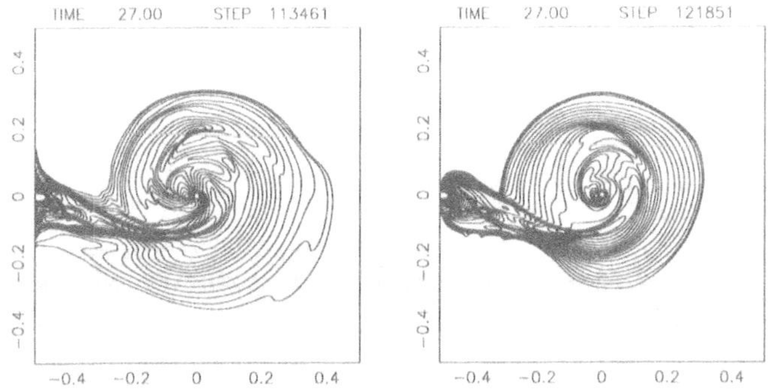

Figure 1. Density contours in the orbital plane at $t = 27$. Left: the case of $\gamma = 1.2$, Right: the case of $\gamma = 1.01$.

Steeghs, Harlaftis & Horne (1997) found the first convincing evidence of spiral structure in accretion discs observationally. We have now succeeded in reproducing their results using our numerical calculations. (Please see Matsuda et al. in this volume).

Acknowledgments

The 2D calculations were mainly performed on VX/1R at the NAOJ. The 3D calculations were performed on SX-4 at IPC at Kobe University.

References

Jyounouchi T., Kitagawa I., Sakashita, Yasuhara M., 1993, Proceedings of 7th CFD Symposium, in Japanese.
Sawada K., Matsuda T., 1992, MNRAS, 255, s17.
Sawada K., Matsuda T., Hachisu I., 1986a, MNRAS, 219, 75.
Sawada K., Matsuda T., Hachisu I., 1986b, MNRAS, 221, 679.
Sawada K., Matsuda T., Inoue M., Hachisu I., 1987, MNRAS, 224, 307.
Shima E., Jyounouchi T., 1994, 25th Annual Meeting of Space and Aeronautical Society of Japan, pp.36-37.
Steeghs D., Harlaftis E.T., Horne K., 1997, MNRAS, 290, L28.
Yukawa H., Boffin H.M.J., Matsuda T., 1997, MNRAS, 292, 321.

RESISTIVE MHD SIMULATIONS OF MAGNETICALLY DRIVEN MASS ACCRETION IN ACTIVE GALACTIC NUCLEI

TAKUHITO KUWABARA
Graduate School of Science and Technology, Chiba University
RYOJI MATSUMOTO
Department of Physics, Faculty of Science, Chiba University
AND
KAZUNARI SHIBATA
National Astronomical Observatory

The effects of resistivity/magnetic diffusivity on the magnetically driven mass accretion and jet formation in active galactic nuclei are studied by 2.5-dimensional resistive MHD simulations. The initial state of simulation is a rotating, constant angular momentum torus threaded by large scale vertical magnetic fields. Since the angular momentum is extracted by magnetic braking, the surface layer of the torus falls into the central region like an avalanche (Matsumoto et al. 1996). The infalling matter constitutes a disk accreting toward the central supermassive black hole. The magnetic twists generated by the infalling matter drive bipolar jets (Uchida & Shibata 1985).

Three dimensional MHD simulations of magnetized disks (e.g., Hawley, Gammie & Balbus 1995) indicate that accretion disks become turbulent due to the growth of the Balbus & Hawley instability. The effective magnetic diffusivity η in the turbulent flow may reduce the efficiency of magnetic braking. Figure 1 shows velocity and magnetic field distribution in non-diffusive model ($\eta = 0$) and diffusive model ($\eta = 0.01 v_K r_0$) where v_K is the Keplerian rotation speed at the reference radius r_0. In diffusive model (Figure 1b) the avalanching accretion is not so prominent as the non-diffusive model (Figure 1a). Figure 2a shows the mass accretion rate $\dot{M}_{\rm acc}$ through the surface $r = 0.5 r_0$ and $0 < z < 0.2 r_0$. The maximum accretion rate in diffusive model (dotted curve) is 30% that of the peak accretion rate in non-diffusive model. Figure 2b shows the mass outflow rate $\dot{M}_{\rm jet}$ through the surface $z = 1.5 r_0$ and $0 < r < 1.6 r_0$. In non-diffusive model, 10–30%

S. M. Miyama et al. (eds.), Numerical Astrophysics, 229–230.

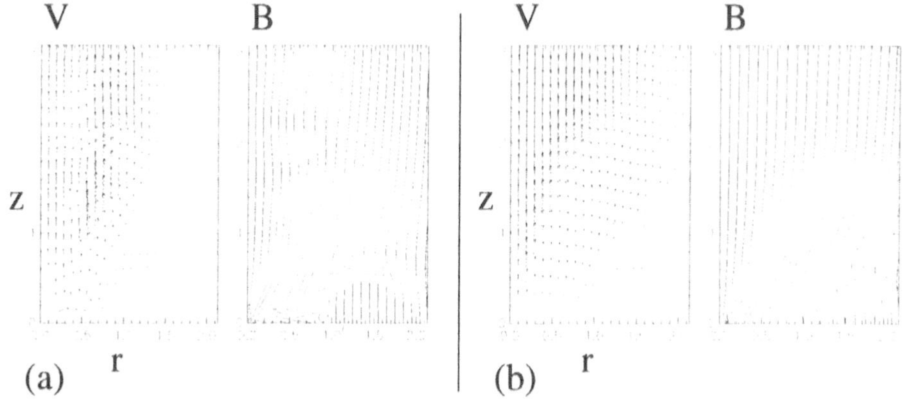

Figure 1. Velocity vectors and magnetic field lines for (a) non-diffusive model at t=6.28 (one rotation) and (b) diffusive model with $\eta = 0.01 r_0 v_{K0}$ at t=20

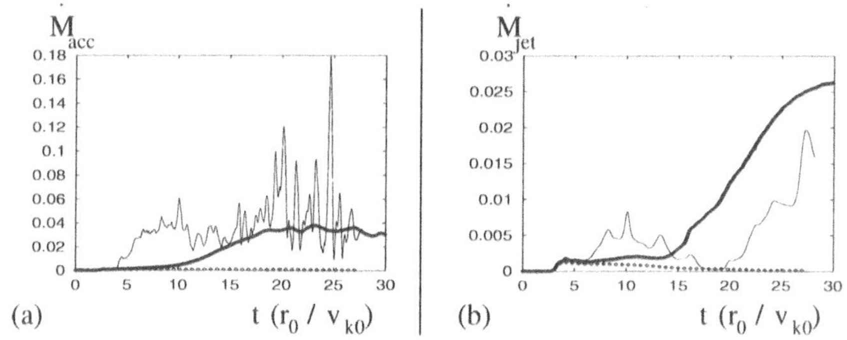

Figure 2. Mass accretion rate (a) and mass outflow rate (b) for non-diffusive model (solid curve), diffusive model with $\eta = 0.01 r_0 v_{K0}$ (wide-solid curve) and diffusive model with $\eta = 0.033 r_0 v_{K0}$ (dotted curve).

of the accreting gas is ejected. In diffusive model, 50% of the accreting gas is ejected. The accretion rate in diffusive model is $\dot{M}_{\rm acc} = 12 M_\odot/{\rm yr}$ when $M = 10^9 M_\odot$ and $M_{\rm torus} = 10^8 M_\odot$ and $r_0 = 100{\rm pc}$. We conclude that although the magnetic diffusivity reduces the magnetically driven mass accretion rate, magnetic braking can still supply matter sufficient to explain the activity of quasars.

References

Hawley, J. F., Gammie, C. F. & Balbus, S. A. (1995) *ApJ*, **440**, pp.742

Matsumoto, R. Uchida, Y., Hirose, S., Shibata, K., Hayashi, M.R., Ferrari, A., Bodo, G., & Norman, C. (1996) *ApJ*, **461**, pp.115

Uchida, Y. & Shibata, K. (1985) *PASJ*, **37**, pp.515

MAGNETOHYDRODYNAMIC SIMULATIONS OF RECURRENT X-RAY FLARES IN PROTOSTARS

Simulations of Recurrent X-ray Flares

MITSURU HAYASHI
Japan Atomic Energy Research Institute, 801-1 Mukouyama, Naka-machi, Naka-gun, Ibaraki 311-01, Japan

K. SHIBATA
National Astronomical Observatory, 2-21-1 Ohsawa, Mitaka, Tokyo 181-8588, Japan

AND

R. MATSUMOTO
Chiba University, 1-33 Yayoi-cho, Inage-ku, Chiba 263-8522, Japan

1. Introduction

By using ASCA, Koyama et al. (1994, 1996) carried out a systematic survey of hard X-ray sources in molecular clouds and revealed that protostars are strong hard X-ray emitting sources. Some of them show flare-like activities. Protostellar flares differ from solar flares in their total energy (10^{35-36} erg), size (several times the radius of protostar), and higher temperature (8keV). Protostellar flares are also observed in lower energy band by ROSAT in YLW15 (Grosso et al. 1997). By extending the model of solar flares associated with footpoint shearing motions, we proposed a model of protostellar flares in which the magnetic field lines connecting the protostar and the disk disrupt following the expansion driven by the twist injection from the rotating disk (Hayashi et al. 1996).

2. Numerical Models & Results

We solve the resistive MHD equations in cylindrical coordinate by applying a modified Lax-Wendroff scheme with artificial viscosity. The effects of radiative cooling are neglected. Magnetic field lines connecting the central star and the disk are twisted by the rotation of the disk. As the magnetic twist

S. M. Miyama et al. (eds.), Numerical Astrophysics, 231–232.

accumulates, the magnetic loops expand quasi-statically in the early stage but later, they expand dynamically. Magnetic reconnection takes place in the current sheet formed inside the expanding loops. Outgoing magnetic islands and post flare loops are formed after the reconnection (See Fig. 1 in Hayashi et al. 1996 and figures in CD-ROM).

We carried out longer time-scale simulations by including the rotation of the central star to investigate the evolution of the post flare loops created by the reconnection event. We found that magnetic reconnection and plasmoid ejection occur intermittently and that the time interval between the flares is about 15 hours when we assume 0.03 AU as the radius of the inner edge of the disk.

3. Discussion

We found that the dipole magnetic fields of the protostar can become partially open by imposed twists within one rotation period of the disk. We also showed that hot plasmoids are ejected in bipolar directions with velocity 2–5 times the Keplerian rotation speed (v_{K0}) around the inner edge of the disk. Plasma heating occurs both by Joule heating in the current sheet and by shock waves created by the magnetic reconnection. The plasma temperature corresponding to the reconnection flow speed ($\sim$ Alfvén speed of 2–5 v_{K0}, where v_{K0} is the Keplerian velocity at unit radius $r = r_0 = 1$), can be $10^7 - 10^8$ K, consistent with the observed spectrum which extends up to 10 keV. The size of the flaring loop is several times larger than the stellar radius.

In addition, recently ASCA detected X-ray flaring variability in YLW15 which repeated three times recursively in about 20 hours (Tsuboi et al. 1997). This observation indicates intermittent flaring variability in star forming region. Such recurrent activities can be explained by intermittent magnetic reconnection in magnetic loops connecting the disk and the protostar.

Numerical computations were carried out by using VPP300/16R at NAOJ.

References

Grosso, N., Montmerle, T., Feigelson, E.D., André, P., Cassanova, S., & Gregorio-Hetem, J. 1997, *Nature*, **385**, 56
Hayashi, M.R., Shibata, K. & Matsumoto, R. 1996, *ApJ*, **468**, L37
Koyama, K., Hamaguchi, K., Ueno, S., Kobayashi, N., & Feigelson, E. D. 1996, *PASJ*, **48**, L87
Koyama, K. et al. 1994, *PASJ*, **46**, L125
Tsuboi, Y. et al. 1997 IAU Symposium No.188, "The Hot Universe", p.83

DYNAMICS OF A MAGNETIC FLUX TUBE IN DIFFERENTIALLY ROTATING DISKS

ATSUSHI KUWABARA AND RYOJI MATSUMOTO
Faculty of Science, Chiba University, Inage-ku, Chiba 263-8522, Japan

KAZUNARI SHIBATA
National Astronomical Observatory, Mitaka, Tokyo 181-8588, Japan

AND

WENCHIEN CHOU
Institute for Fusion Studies, the University of Texas at Austin, Austin, TX 78712, U.S.A.

We carried out three-dimensional magnetohydrodynamic simulations of the dynamics of a magnetic flux tube in rotating disks. The effects of spatial variation of gravity and differential rotation were taken into account. We adopted a local shearing box model (Hawley, Gammie, & Balbus, 1995) to study the dynamical evolution of an isolated, initially azimuthal flux tube. In rigidly rotating disks, the flux tube is twisted due to the coupling of Parker instability and Coriolis force (Chou et al., 1998). In differentially rotating disks, the magnetic energy may increase because the differential rotation stretches the twisted flux tube. Furthermore, magnetorotational instability (Balbus & Hawley, 1991) also affects the dynamical evolution of the flux tube. Figure 1 shows the simulation box and the initial condition of the simulation. An isolated horizontal flux tube whose radius is r is embedded in the Keplerian disk. The magnetic field is assumed to be purely toroidal. The model parameters are r, plasma $\beta = (P_{\rm gas}/P_{\rm mag})$ at the center of the flux tube, and the height of the initial flux tube is h. We assume a periodic boundary in azimuthal direction and a sliding periodic boundary in radial direction. Figure 2 shows the time development of $\langle B^2/8\pi P_0\rangle$ and $\langle -B_xB_y/4\pi P_0\rangle$, where the bracket means spatial average. Figure 3 shows the isosurface of magnetic field strength (white surface) and magnetic field lines (gray) for a model with $r = 0.2H$, $\beta = 1$, and $h = 1.0H$, where H is the scale height. We found that, in differentially rotating disks, (1)

S. M. Miyama et al. (eds.), Numerical Astrophysics, 233–234.

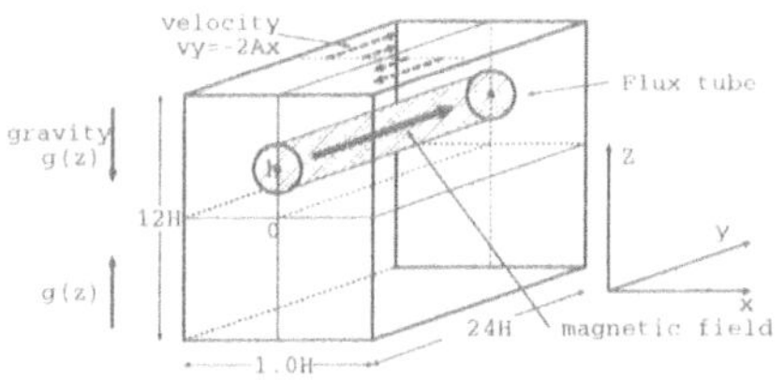

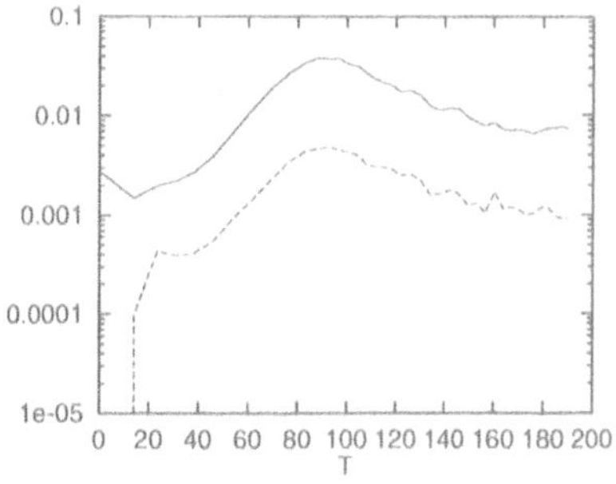

Figure 1. Simulation box and initial condition.
Figure 2. Time development of $\langle B^2/8\pi P_0 \rangle$ (upper) and $\langle -B_x B_y/4\pi P_0 \rangle$(lower).

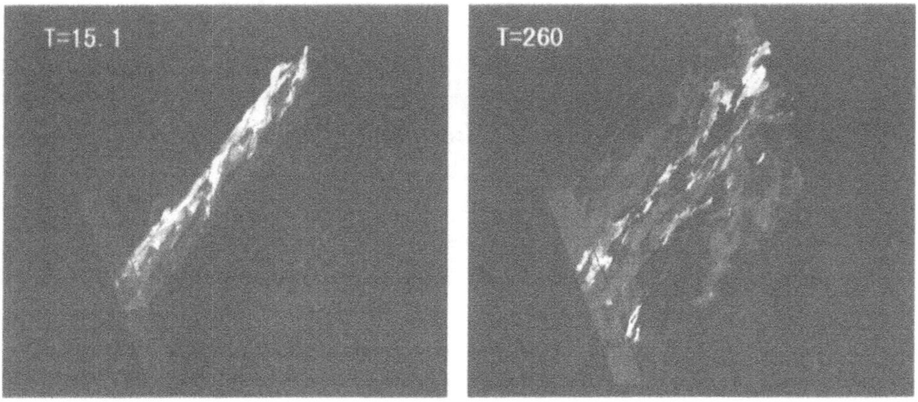

Figure 3. The isosurface of magnetic filed strength (white) and magnetic field lines (gray) for a model with $\beta = 1.0$ and $r = 0.2H$.

an isolated flux tube is destructed and generates turbulence due to the increase of magnetorotational instability; (2) intense azimuthal magnetic fields are intermittently created; and (3) after the initial amplification of magnetic energy saturates, the magnetic energy approaches a quasi-steady value. Numerical computations were carried out on VPP300/16R at NAOJ.

References

Chou, W., Tajima, T., Shibata, K., and Matsumoto, R., *submitted to ApJ* (1998)
Hawley, J.F., Gammie, C.F., & Balbus, S.A., *ApJ*, 440, 724 (1995)

MAGNETOHYDRODYNAMIC SIMULATIONS OF HIGH-BETA DISKS AND LOW-BETA DISKS

TAKAAKI MATSUZAKI
Graduate School of Science and Technology, Chiba University
1-33, Yayoi-Cho, Inage-ku, Chiba 263-8522, Japan

KAZUNARI SHIBATA
National Astronomical Observatory
Mitaka, Tokyo 181, Japan

TOSHIKI TAJIMA
Institute for Fusion Studies, the University of Texas at Austin
Austin, TX 78712, USA

AND

RYOJI MATSUMOTO
Department of Physics, Faculty of Science, Chiba University
1-33, Yayoi-Cho, Inage-ku, Chiba 263-8522, Japan

We present results of three-dimensional MHD simulations of Keplerian accretion disks which indicate that accretion disks have two distinct states; a gas pressure dominated quasi-steady state and a magnetic pressure dominated cataclysmic state.

In a frame corotating with the disk at $r = r_0$ from the gravitating center, we take a local Cartesian coordinate with x in the radial direction, y in the azimuthal direction, and z in the vertical direction. The size of the simulation box is $(L_x, L_y, L_z) = (H, 18H, 16H)$ where H is the scale height defined by $H = C_s/\Omega_k$. Here, C_s is the sound speed and $\Omega_k = (GM/r_0^3)^{1/2}$ is the Keplerian angular speed. Typical number of grid points is $(N_x, N_y, N_z) = (42, 62, 165)$. Under the shearing box approximation (Hawley, Gammie & Balbus, 1995), the equilibrium Keplerian shear flow is given by $v_y = -1.5\Omega_k x$. The vertical gravity is assumed to be $g_z = -GMz/(r_0^2 + z^2)^{3/2}$. The initial state is a gravitationally stratified, isothermal gas layer threaded by azimuthal magnetic field. We assume that at the initial state, plasma $\beta (= P_{\rm gas}/P_{\rm mag})$ is uniform. In the following, we use the normalizations $\rho(z=0) = C_s = \Omega_k = 1$. We set the thickness of

S. M. Miyama et al. (eds.), Numerical Astrophysics, 235–236.

the disk as $H/r_0 = 0.25$. The boundary conditions are periodic, sliding periodic (Hawley, Gammie & Balbus, 1995) and free, in the azimuthal, radial and vertical direction, respectively. For simplicity, we assume isothermal equation of state and neglect the effects of resistivity. In order to initiate the evolution, sinusoidal perturbation with amplitude $0.01C_s$ are imposed at $t = 0$ on azimuthal velocity.

In weakly magnetized disks, magnetic fields are amplified due to the growth of the Balbus & Hawley instability (Balbus & Hawley, 1991). In the nonlinear stage, the disk approaches to a quasi-steady state with $\beta = 20-30$ and $\alpha_B = -\langle B_x B_y/(4\pi P)\rangle \sim 0.01$, consistent with the results by Stone et al. (1996). When the initial magnetic energy is comparable to the thermal energy ($\beta \sim 1$), we found that the Balbus & Hawley instability coupled with the Parker instability further amplifies magnetic fields unless β is extremely low. The disk stays in the low-β state for time scale much longer than the rotation period. The value of α_B in the low-β state is $\alpha_B \sim 0.1$. When the magnetic energy in low-β disk is released, the disk will show large amplitude sporadic time variations observed in low-states of black hole candidates. High-β disks and low-β disks may correspond to high-state disks and low-state disks, respectively.

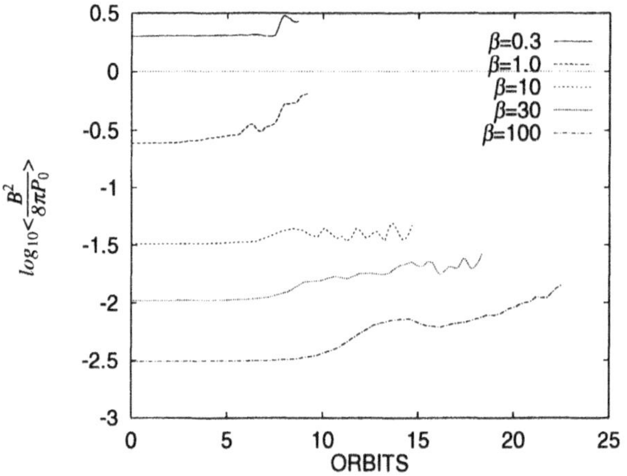

Figure 1. Time evolution of mean square magnetic field strength: $\log_{10}\left\langle B^2/(8\pi P_0)\right\rangle$. The unit of time is the rotation period: $2\pi/\Omega_k$.

References

Balbus, S. A., & Hawley, J. F. 1991, ApJ, 376, 214

Hawley, J. F., Gammie, C. F., & Balbus, S. A. 1995, ApJ, 440, 742

Stone, J. M., Hawley, J. F., Gammie, C. F., & Balbus, S. A. 1996, ApJ, 463, 656

3D MHD SIMULATIONS OF THE MAGNETO-ROTATIONAL INSTABILITY IN VERTICALLY STRATIFIED DISKS

K. A. MILLER AND J. M. STONE
Department of Astronomy, University of Maryland
College Park, MD 20742, USA

1. Introduction

The magneto-rotational instability has been shown to produce angular momentum transport via anisotropic turbulence in accretion disks (see Hawley et. al. 1995, 1996). It also shows evidence of nonhelical dynamo action. The work of Stone et. al. (1996) and Brandenburg et. al. (1995) examined the evolution of the instability in stratified disks. Their results confirm that the instability leads to nonhelical dynamo action and indicate that buoyancy may not be important in determining the saturated magnetic energies.

We present preliminary results of a stratified accretion disk simulation which extends the work of Stone et. al. (1996) by increasing the vertical domain to 10 scale heights. Our goal is to examine the nature of the magneto-rotational instability over many vertical scale heights. In particular we will study the transition between turbulence in the disk and laminar flow in the corona and determine its effect on the disk structure and field evolution.

2. Results

We use a shearing box model but allow outflow in $\hat{z}$. The grid is $32 \times 64 \times 120$ zones, covering $1H_z$ in $\hat{x}$ and $10H_z$ in $\hat{y}$ and $\hat{z}$ (H_z is a disk scale height). The disk is Gaussian in $\hat{z}$; the rotation is Keplerian. The field is purely azimuthal and confined to the disk, such that the plasma β ($= 25.0$) is constant within $2H_z$. We evolve the simulation with an isothermal equation of state.

We find that the action of the magnetorotational instability dominates the evolution, producing a turbulent disk core and a strongly magnetized corona. Figure 1 shows averaged β, and Maxwell (solid line) and Reynolds (dotted line) stresses (normalized by the pressure) after 20 orbital times.

S. M. Miyama et al. (eds.), Numerical Astrophysics, 237–238.

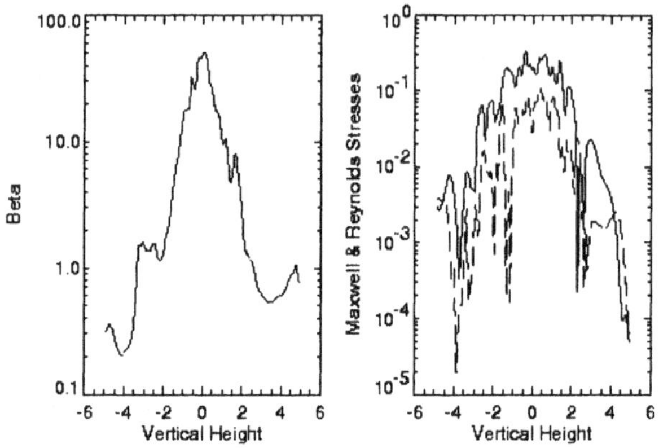

Note that β is two orders of magnitude higher in the disk midplane than in the corona. Both stresses are highest in the core, with the Maxwell stress dominating by a factor of about 3; the α parameter is of order 0.2–0.3 in the disk midplane. A color image of the time evolution of magnetic energy and Maxwell and Reynolds stresses is included in the CD Rom. This image clearly shows the bouyant rise of magnetic flux out of the disk and the formation of a magnetically dominated corona.

3. Conclusions

We have performed self consistent 3D MHD simulations of an accretion disk containing a purely azimuthal field, examining the evolution over 10 scale heights in the vertical direction. We find that an azimuthal field initially confined to the midplane of an accretion disk will rise bouyantly out of the disk, producing a magnetically dominated corona. The field generally maintains an ordered structure well above the disk. This suggests that a turbulent, magnetized disk can, through the action of the magnetorotational instability, produce a magnetized corona in laminar flow.

References

Brandenburg, A., Nordlund, A., Stein, R.F., and Torkelsson, U. (1995) Dynamo-Generated Turbulence and Large-Scale Magnetic Fields In A Keplerian Shear Flow, *Astrophysical Journal*, **Vol. no. 446**, pp. 741–754

Hawley, J.F., Gammie, C.F., and Balbus, S.A. (1995) Local Three-Dimensional Magnetohydrodynamic Simulations of Accretion Disks, *Astrophysical Journal*, **Vol. no. 440**, pp. 742–763

Hawley, J.F., Gammie, C.F., and Balbus, S.A. (1996) Local Three-Dimensional Magnetohydrodynamic Simulations of an Accretion Disk Hydromagnetic Dynamo, *Astrophysical Journal*, **Vol. no. 464**, pp. 690–703

Stone, J.M., Hawley, J.F., Gammie, C.F., and Balbus, S.A. (1996) Three-Dimensional Magnetohydrodynamic Simulations of Vertically Stratified Accretion Disks, *Astrophysical Journal*, **Vol. no. 463**, pp. 656–673

GENERATION OF CURRENT SYSTEM AND ENERGY DEPOSITION IN A MAGNETOHYDRODYNAMIC MODEL FOR THE PRODUCTION OF RADIO JETS FROM ACTIVE GALACTIC NUCLEI (AGN'S)

M.NAKAMURA, Y.UCHIDA, S.HIROSE AND S.UEMURA
Department of Physics, Science University of Tokyo
1-3 Kagurazaka, Shinjyuku-ku, Tokyo 162-8601, Japan

We have performed 2.5D MHD numerical simulations based on our model of the MHD jet formation (Uchida and Shibata 1985). We have already shown also in the case of AGN jets that a magnetically driven, well-collimated jets can be accelerated in the direction of the rotating axis of the accretion disk around the central object (Uchida and Shibata 1986, Matsumoto et al. 1996). In this model, it is our concept that the jets are formed by the interaction between the accretion disk and a large scale magnetic field. By this interaction, angular momentum of the disk gas is taken away, and the gas of the disk accretes towards the central massive blackhole, and the gravitational potential energy of the disk is continuously liberated in a much enhanced way. At the same time, a large scale magnetic field is helically twisted and this twist propagates as non-linear torsional Alfvén wave. We consider the generation, transfer, and depositon of the electromagnetic energy according to this MHD model of jets formation.

From the numerical results, we found that gravitational potential energy of the accretion disk is partly converted into the electromagnetic energy by the interaction between the accretion disk and a large scale magnetic field. In the evolution of this system, the toroidal magnetic field is generated, and the electromagnetic energy is carried out as the Poynting flux out of the disk. In all of the stage, the Poynting flux is dominant compared with other energy fluxes (i.e. kinetic energy flux, thermal energy flux) in the case of medium-β disk (magnetic pressure is comparable order to gas pressure). This tendency is the same as the case of the high-β disk (gas pressure dominant). As is shown in the following figures, at the front of the torsional Alfvén wave, a part of the energy is deposited through the work done by Lorentz force. Also, the configuration of the poloidal magnetic field is bunched towards the central axis by the pinch effect by the propagating toroidal magnetic field. Growing of toroidal magnetic field with time is, in

S. M. Miyama et al. (eds.), Numerical Astrophysics, 239–240.

other words, the growth of poloidal current. As a whole, the electromagnetic energy generated by the rotating disk flows in the form of the Poynting flux with the development of the poloidal current, and at its front, this energy is partly deposited through the work done by Lorentz force. Thus, the propagation of torsional Alfvén wave plays an essential role in the transportation of the energy, and this energy will be deposited through the interaction between this torsional Alfvén wave and the inter-protogalactic gas which is left in the surroundings of the cavity produced in the formation of AGNs (Nakamura et al. 1998).

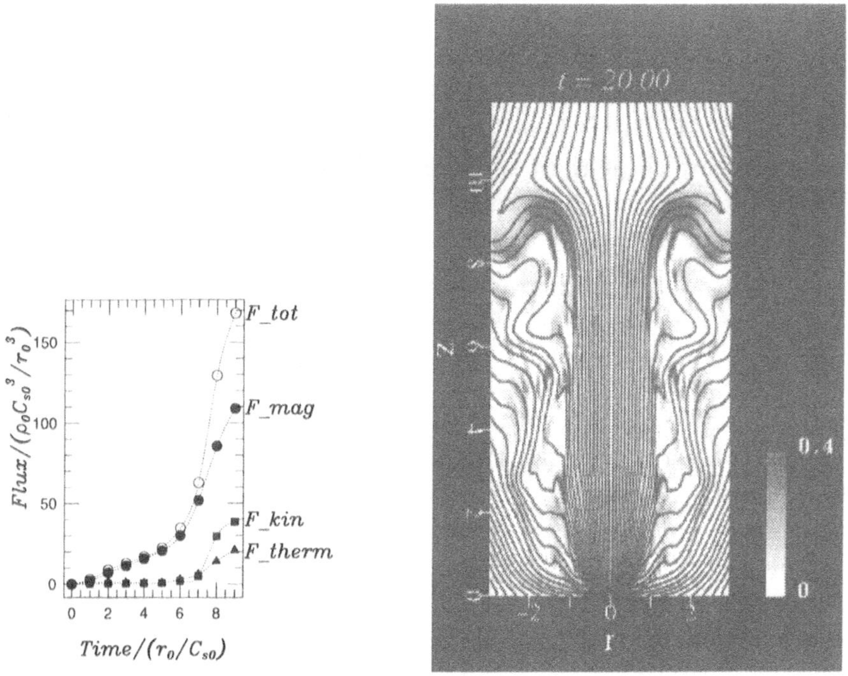

Figure 1. The figure on the left shows the time evolution of the Out-flow energy flux, and the figure on the right shows the snapshot of the time evolution of the poloidal magnetic field line and the specific powers exerted due to the Lorentz force by the poloidal current and the toroidal magnetic field. We normalize all of the physical quantities by the typical length r_0, the typical density ρ_0, and the typical velocity C_{s0} in the system.

References

Uchida, Y., and Shibata, K. 1985, Publ. Astron. Soc. Japan, 37, 515
Uchida, Y., and Shibata, K. 1986, Can. J. Phys., 64, 507
Matsumoto, R., Uchida, Y., Hirose, S., Shibata, K., Ferrari, A., Bodo, G., and Norman, C. A. 1996, Astrophys. J., 461, 115
Nakamura, N., Hirose, S., and Uchida, Y. 1998, in preparation.

ACCRETION DISKS AROUND COMPACT OBJECTS AND PROTOSTELLAR OBJECTS

TORU OKUDA
Hakodate College, Hokkaido University of Education
1-2 Hachiman-cho, Hakodate 040-8567, Japan

AND

MITSUTAKA FUJITA
Department of Physics, Faculty of Science, Hokkaido University, Kita-ku, Sapporo 060-0810, Japan

Abstract Two-dimensional accretion disks around a neutron star near Eddington luminosity and an FU Ori object during eruptions are examined. In both cases convection plays an important role in vertical energy transport and matter accretion onto the central star. The neutron star model is applicable to SS 433 and the protostar model suggests a new observational appearance to an FU Ori object.

1. Numerical Methods

Basic equations and numerical methods are the same as those described by Kley (1989) and Okuda et al. (1997). The equations include the full viscous stress tensor, heating and cooling of the gas, and radiation transport. The radiation transport is performed in the grey, flux-limited diffusion approximation (Levermore and Pomraning 1981). The methods are based on an Eulerian explicit-implicit finite difference scheme. The α-model for viscosity is used. We consider a non-rotating neutron star with mass $M_* = 1.4M_\odot$, radius $R_* = 10^7$ cm, and mass accretion rate $\dot{M} = 2\dot{M}_{\rm E}$, and a rapidly rotating FU Ori object with $M_* = 0.5M_\odot$, $R_* = 4R_\odot$, and $\dot{M} = 5 \times 10^{-5} M_\odot$ yr^{-1}. Details of the numerical methods and other conditions are referred to in Okuda et al. (1997).

S. M. Miyama et al. (eds.), Numerical Astrophysics, 241–242.

2. Results

In the neutron star model, after time integrations over a few hundred Keplerian orbital periods, the disk tends to settle into a nearly steady state. The stable disk belongs to a category of *slim accretion disk* (Abramowicz et al. 1988), where the convective energy flux dominates the integrated vertical radiative flux. The obtained disk is geometrically, optically thick and its thickness H is ~ 0.4–$0.5r$. The disk is highly convective and many convective cells are formed in the inner disk. The heat generated by viscous dissipation is transported vertically to the disk plane by convection, and convection produces a powerful acoustic energy flux which leads to a coronal heating of atmosphere above the disk ($T \sim 10^8$–10^{10} K), accompanying powerful winds with velocities of 6–8×10^3 km s^{-1}. On the other hand, gas near to the stellar surface and on the rotating axis mainly undergoes only radiation force, not a large mass-supply from the inner disk, and develops into a moderate outflow, which in turn, leads to lower densities there. Resultantly, the radiation force becomes to dominate more and more the gravitational force per unit volume. Finally we have a very rarefied, hot, and optically thin high-velocity jet region only along the rotational axis. The mass outflow rate from the system is comparable to the input accretion rate $2\dot{M}_{\rm E}(\sim 2 \times 10^{19}$ g s^{-1}). Some features of the unique X-ray source SS 433, such as the two oppositely directed relativistic jets moving with a velocity of $0.26c$, the jet acceleration and collimation, and the observed radiative energy flux, are likely to be explained in terms of this supercritical accretion disk onto the neutron star.

For the FU Ori object, a geometrically and optically thick disk is also obtained. Dominant outward force induces outflow constantly in the inner region of the disk. Interaction of the outflow with the incoming gas from the outer disk leads to the convective flow near the stellar surface and matter accretes onto the star through the circulatory flow. These processes circulate and the convective mass flux increases with time. As soon as the increasing convective mass flux near to the stellar surface attains a critical value, a high velocity jet is formed in the polar region and the luminosity increases abruptly. The bursting luminosity lasts for a month and recurs approximately every 8 months.

References

Abramowicz, M. A., Czerny, B., Lasota, J. P., and Szuszkiewicz, E. (1988), ApJ 332, 646
Kley, W. (1989), A&A 208, 98
Levermore, C. D. and Pomraning, G. C. (1981), ApJ 248, 321
Okuda, T., Fujita, M., and Sakashita, S. (1997), PASJ 49, 679

ON SUB-KEPLERIAN VELOCITY DISTRIBUTION OF THE MEGAMASER SOURCE IN NGC 1068

A.T. OKAZAKI
Faculty of Engineering, Hokkai-Gakuen University
Toyohira-ku, Sapporo 062-8605, Japan

NGC 1068 is considered to be the archetypal galaxy, harboring an active galactic nucleus heavily obscured by a dense molecular torus. In the nucleus of NGC 1068 lies a megamaser source which enables us to probe the structure and the velocity field of the innermost part of the torus. Recently, Greenhill et al. (1996) observed the water maser emission in NGC 1068 with the VLBA and obtained images with sub-milliarcsecond angular resolution. Their best fit rotation curve for the redshifted maser emission was sub-Keplerian; the line-of-sight velocity of the maser source decreased as $r^{-0.31}$ for $0.57\,\mathrm{pc} < r < 0.92\,\mathrm{pc}$.

In this paper, we examined the effect of global oscillation modes on the kinematics of the maser emission region of NGC 1068. For this purpose, we used a simplified disk model and solved the resulting eigenvalue problem in the three dimensions. We found that there exist global eccentric (z-symmetric $m = 1$) modes as well as warping (z-antisymmetric $m = 1$) modes in the masing disk of NGC 1068. Figure 1 shows the three-dimensional structure of the fundamental eccentric mode.

The global eccentric mode can explain the observed sub-Keplerian velocity distribution of the maser source of NGC 1068. Figure 2 shows the position-velocity diagram calculated under the assumption that the masing disk is perturbed by the fundamental eccentric mode. As shown in Figure 2, the eccentric mode flattens the velocity gradient of the redshifted emission and shifts the systemic velocity blueward by $\sim 70\mathrm{km\,s^{-1}}$. Our model predicts that future VLBA observation of the blueshifted maser emission of NGC 1068 will reveal that its line-of-sight velocity distribution is steeper than the Keplerian distribution.

References

Greenhill, L.J., et al. 1996, ApJ 472, L21.

S. M. Miyama et al. (eds.), Numerical Astrophysics, 243–244.

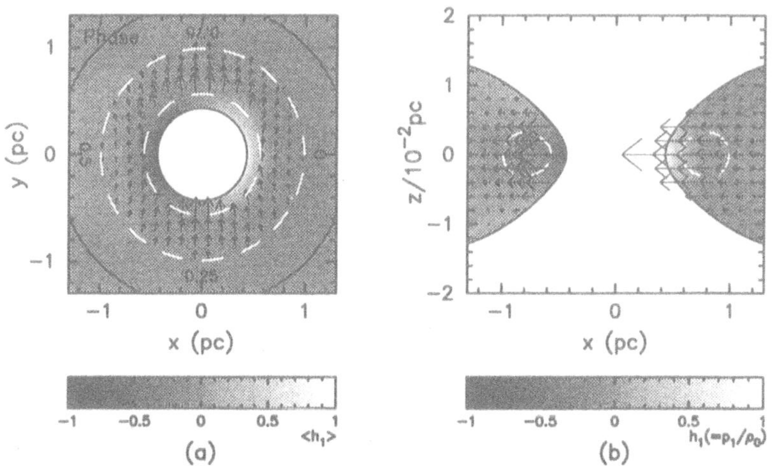

Figure 1. Fundamental eccentric mode in the masing disk of NGC 1068. (a) The (r, ϕ)-distribution of the vertically-averaged perturbations. At phase 0, the observer is in the x-direction. (b) The (r, z)-distribution of the perturbations. A gray-scale representation denotes the enthalpy perturbation h_1. Arrows superposed on the gray-scale plot are the perturbed velocity vectors in the maser emission region (surrounded by dashed lines).

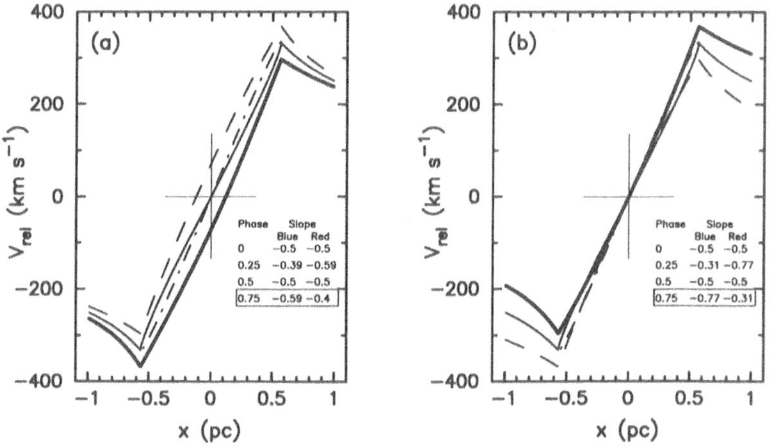

Figure 2. Position-velocity diagrams for the disk perturbed by the fundamental eccentric mode. In panel (a), the relative velocity is measured with respect to the systemic velocity, whereas in panel (b) it is measured with respect to the line-of-sight velocity of the systemic component at $x = 0$. Maser source is assumed to lie along the x-axis and along the inner radius of the maser emission region. The mass of the central black hole is $1.5 \times 10^7 \, M_\odot$ and the inclination angle of the disk is 80°. The amplitude of the mode is normalized so that the maximum value of the velocity perturbation is 60% of the local Keplerian velocity. In the bottom-right part of each panel, the slope of $V_{\rm rel}$ distribution at each phase is given. Note that at phase 0.75 (thick solid line) the model reproduces the observed sub-Keplerian distribution. A thin solid line denotes the PV diagram at phase 0. The cross indicates the origin of the diagram.

GAMMA-RAY EMISSION FROM GALACTIC BLACK HOLE CANDIDATES AND PARTICLE ACCELERATION BY PLASMA WAVES

MASAAKI KUSUNOSE
School of Science, Kwansei Gakuin University
Nishinomiya 662-8501, Japan

AND

HUI LI
Los Alamos National Laboratory
Los Alamos, NM 87545, U.S.A.

Abstract. Gamma-rays up to several MeV are observed from some Galactic black hole candidates, such as Cyg X-1, GRO J0422+32, and GRO J1655–40. These gamma-ray spectra are hard to explain by simple thermal emission. We introduce an acceleration mechanism of electrons, i.e., acceleration by fast-mode plasma waves owing to so called transit-time damping. We solve coupled kinetic equations for waves, particles, and photons, and we show the time-evolution of the spectra of particles, waves, and photons.

Some Galactic black hole candidates (GBHCs) are known to emit gamma-rays up to several MeV (e.g., Grove et al. 1998). As a model of radiation processes for such high energy emission, thermal Comptonization are often used, e.g., Sunyaev and Titarchuk (1980) and Titarchuk (1994). Although thermal models are simple to fit the data because there are only two parameters, i.e., electron temperature and Thomson scattering depth, it is uncertain whether electrons in the vicinity of black holes remain thermal or not. On the other hand, a power-law particle distribution function is adequate to fit gamma-ray data observed by OSSE from some of GBHCs such as GRO J1655–40. However, it is not well known how the power-law particles are generated.

Plasmas around a black hole, e.g., plasmas in accretion disk coronae, might be accelerated to relativistic energy by various processes such as shocks or resonant scatterings by plasma waves. Li, Kusunose, and Liang (1996) proposed a gyroresonant stochastic electron acceleration model to

S. M. Miyama et al. (eds.), Numerical Astrophysics, 245–246.

account for the MeV emissions from Cyg X-1 and GRO J0422. Li and Miller (1997) recently applied the theory of particle acceleration by obliquely propagating fast-mode waves to magnetically dominated accretion disk coronae.

In the present paper we study the emission spectra and particle distribution in the environment around black holes, by solving time-dependent kinetic equations for electrons, waves, and photons, including synchrotron radiation and Compton scattering as emission processes. Results are shown in Figure 1. The time-evolution is followed from $t = 0$ to $10t_{\rm dyn}$, where $t_{\rm dyn} = R/c$ with c the light speed, $R = 10GM/c^2$ (cloud radius), $M = 7M_\odot$ (black hole mass). Here plasma-β is 0.5, the Thomson scattering depth of the plasma cloud is 0.1, and electrons are assumed to obey the Maxwellian distribution with $T_e = 50$keV at $t = 0$.

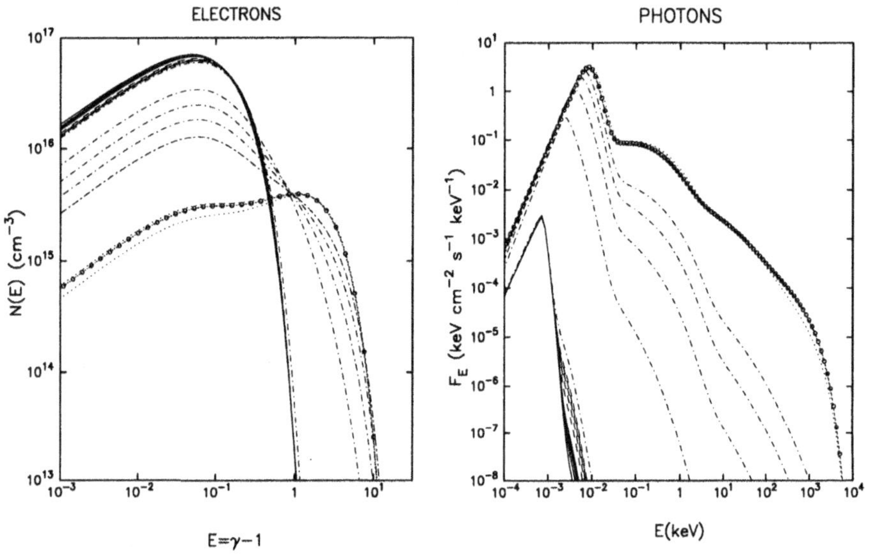

Figure 1. Left: Time-evolution of the electron number spectrum for $t/t_{\rm dyn} = 0 - 0.01$, $0.01 - 0.1$, $0.1 - 1$, and $1 - 10$. The steady state is denoted by open circles. *Right*: Time-evolution of the photon spectrum.

References

Grove, J.E., et al. (1998), to appear in ApJ (Astro-ph/9802242)
Li, H., Kusunose, M., and Liang, E. P. 1996, *ApJ*, **461**, L73
Li, H. and Miller, J. A. 1997, *ApJ*, **478**, L6
Sunyaev, R. A. and Titarchuk, L. 1980, *A&Ap*, **86**, 121
Titarchuk, L. 1994, *ApJ*, **434**, 570

A WAY TO 3D NUMERICAL RELATIVITY

— *Coalescing Binary Neutron Stars* —

TAKASHI NAKAMURA
Yukawa Institute for Theoretical Physics
Kyoto University, Kyoto 606-01, Japan

AND

KEN-ICHI OOHARA
Department of Physics, Niigata University Niigata, 950-2181, Japan

1. Introduction

Long baseline interferometers to detect gravitational waves such as TAMA-300 [1], GEO600 [2], VIRGO [3], LIGO [4] will be in operation by the end of this century or the beginning of the 21st century. One of the most important sources of gravitational waves for such detectors are coalescing binary neutron stars. PSR1913+16 is the first binary neutron stars found by Hulse and Taylor. The existence of the gravitational waves is confirmed by analyzing the arrival time of radio pulses from PSR1913+16. At present there are three systems like PSR1913+16 for which coalescing time due to the emission of gravitational waves is less than the age of the universe. From these observed systems the event rate of the coalescing binary neutron stars is estimated as $10^{-6} \sim 2 \times 10^{-5}$ events yr^{-1} galaxy^{-1} [5, 6, 7]. From the formation theory of binary neutron stars, the event rate is estimated as $2 \times 10^{-5} \sim 3 \times 10^{-4}$ events yr^{-1} galaxy^{-1} [8, 9, 10]. If the event rate is $\sim 2 \times 10^{-5}$, two different methods yield the agreement. Since the number density of galaxies is $\sim 10^{-2}$ Mpc^{-3}, we may expect several coalescing events yr^{-1} within 200Mpc ($\sim 2 \times 10^{-7}$events Mpc^{-3}). The coalescing binary neutron stars is a possible central engine of cosmological gamma ray bursts. Coalescing binary neutron stars is also a possible site of r-process element production.

The eccentricity e and the semimajor axis a of binary neutron stars decrease due to the emission of gravitational waves, and these quantities

S. M. Miyama et al. (eds.), Numerical Astrophysics, 247–256.

are related with each other as

$$e = e_0 \left(\frac{a}{a_0}\right)^{\frac{19}{12}} \tag{1}$$

where e_0 and a_0 are the initial eccentricity and the semimajor axis, respectively [11, 12]. If $a_0 \sim R_\odot$ like PSR1913+16, $e \sim 10^{-6}$ for $a \sim 500$km so that the orbit is almost circular. The merging time $t_{\rm mrg}$ due to the emission of the gravitational waves is given by

$$t_{\rm mrg} = 3{\rm min}\left(\frac{m_1}{1.4{\rm M}_\odot}\right)^{-1}\left(\frac{m_2}{1.4{\rm M}_\odot}\right)^{-1}\left(\frac{m_1+m_2}{2.8{\rm M}_\odot}\right)^{-1}\left(\frac{a}{470{\rm km}}\right)^{4}, \tag{2}$$

where m_1 and m_2 are each mass of neutron stars. The frequency of the gravitational waves $\nu_{\rm GW}$ is given by

$$\nu_{\rm GW} = 19{\rm Hz}\left(\frac{m_1+m_2}{2.8{\rm M}_\odot}\right)^{\frac{1}{2}}\left(\frac{a}{470{\rm km}}\right)^{-\frac{3}{2}} \tag{3}$$

This is just the frequency range ($\nu_{\rm GW} = 10{\rm Hz} \sim 1{\rm kHz}$) of long baseline interferometers to detect gravitational waves such as TAMA300, GEO600, VIRGO, LIGO. The number of rotation before the final merge is given by

$$N = 2635\left(\frac{m_1}{1.4{\rm M}_\odot}\right)^{-1}\left(\frac{m_2}{1.4{\rm M}_\odot}\right)^{-1}\left(\frac{m_1+m_2}{2.8{\rm M}_\odot}\right)^{-\frac{1}{2}}\left(\frac{a}{470{\rm km}}\right)^{\frac{5}{2}} \tag{4}$$

The merging phase of coalescing binary neutron stars is divided into two stages; 1)The last three minutes [13] and 2) The last three milliseconds. In the first stage a is much larger than the neutron star radius $R_{\rm NS}$ and the rotation velocity of the binary v_r is $\sim 0.05c$ so that each neutron star can be considered as a point particle. The post-Newtonian expansion will converge in the first stage. However the quite accurate theoretical calculations of the wave form are needed since only the uncertainty of $1/2635$ in the theoretical prediction of the energy loss will cause one rotation ambiguity at last. If the accurate theoretical template is obtained [14], by making a cross correlation with the observational data we may determine each mass, spin, inclination and the distance of binary neutron stars [15].

In the second stage a is comparable to $R_{\rm NS}$ and the gravity is so strong that the post-Newtonian expansion is not a good approximation and the finite size effect of neutron stars is important. In the final collision phase, the general relativistic hydrodynamics are also important. Only 3D numerical relativity can study this important final merging phase related to gravitational waves, gamma ray burst and production of r-process elements.

2. Newtonian Simulations

To clarify the phenomena in the last three milliseconds many numerical simulations have been done so far. They are classified into several categories.

A) Newtonian hydrodynamics without radiation reaction

In this category, using the Newtonian hydrodynamics, equilibria and stability of the binary neutron stars are studied for various initial parameters and equation of states [16, 17, 18, 19, 20, 21, 22].

B) A)+ simple radiation reaction up to contact of neutron stars

In this category, radiation reaction is included so that the binary spirals in due to the emission of the gravitational waves. However after the contact the radiation reaction is switched off [23, 24, 25].

C) A) + radiation reaction

In this category the radiation reaction is fully taken into account even after the contact. However the computation is time consuming because to estimate the radiation reaction potential two more poisson equations other than the poisson equation for the Newtonian gravitational potential should be solved [17, 26, 27, 28, 29, 30, 31, 33].

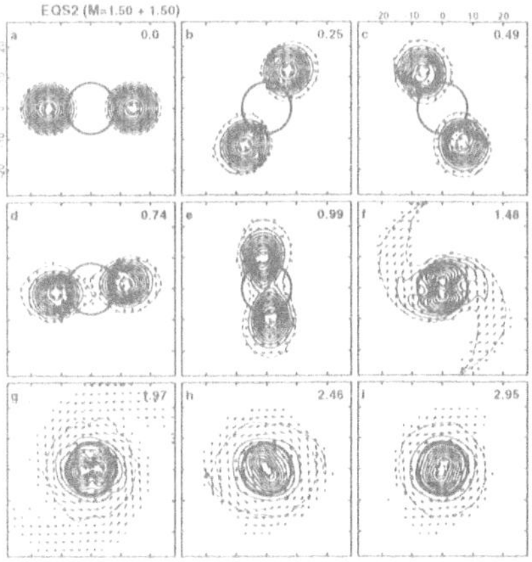

Figure 1. Density and velocity on the x-y plane for EQSP2. The time in units of milliseconds is shown. Arrows indicate the velocity vectors of the matter. A fat line shows a circle of radius $2GM_t/c^2$.

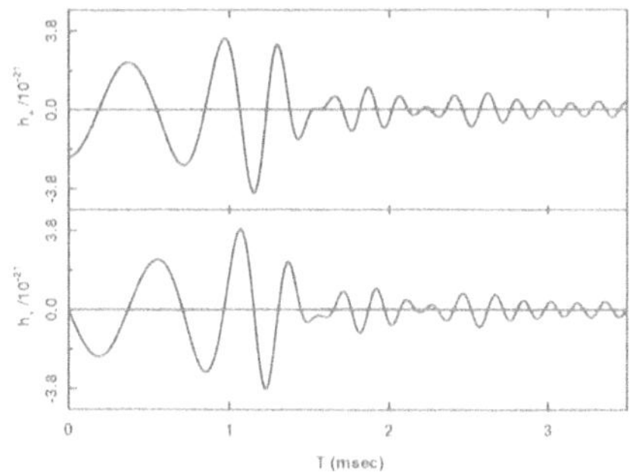

Figure 2. Wave forms of h_+ and $h_\times$ observed on the z-axis at 10Mpc for EQSP2.

In Fig.1 we show an typical example of such a simulation [26]. In this case each mass of the binary is the same. We start the simulation from the equilibrium obtained in category **A)**. Due to the loss of the angular momentum by gravitational waves the merging starts spontaneously. As merging proceeds, two arm spiral extends outward. The spiral arms become tightly bound later and becomes a disk. The final result is an almost axially symmetric central object and a disk around it. In Fig.1 the thick circle shows the Schwarzschild radius of the total mass. Although it is not possible to say the formation of a black hole in Newtonian simulations, we expect that the central object is a black hole. So the final destiny may be a black hole and a disk. This behavior is seen also in many other simulations in category **B)** and **C)**. In Fig.2 we show a typical wave pattern of the gravitational waves.

D) A)+1PN effect of Post Newtonian Hydrodynamics and Radiation Reaction

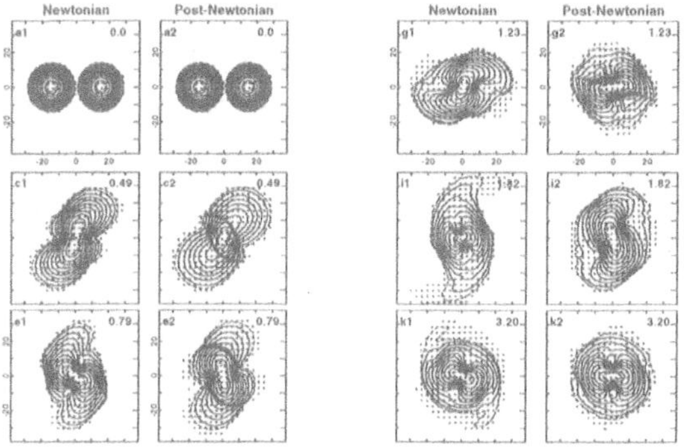

Figure 3. Density and velocity on the x-y plane. The left and right figures are for the Newtonian (N) and post-Newtonian (PN) calculations, respectively. Notations are the same as for Fig.1.

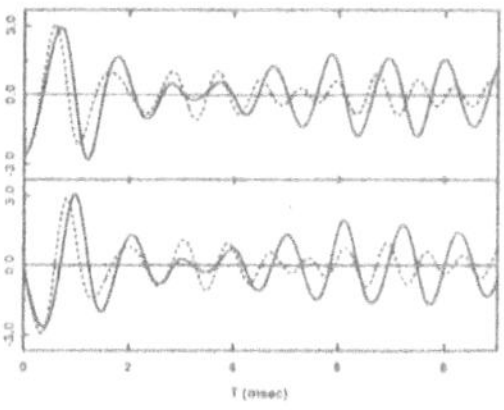

Figure 4. Wave forms of h_+ and $h_\times$ observed on the z-axis at 10Mpc. The solid and dashed lines are for PN and N, respectively.

Since neutron stars are general relativistic compact objects, the Newtonian hydrodynamics is zeroth approximation to the problem. One possible refinement is to include 1PN force [i.e. $O\left((v/c)^2\right)$ effect of the general relativistic gravity after the Newtonian gravity]. Oohara and Nakamura performed such a simulation [32]. They first include 1PN force to the simulation in Fig.1 and found that the 1PN effect is too strong. They decrease the mass of the neutron star to $0.62M_\odot$ from$1.5M_\odot$ keeping the radius of the neutron the same so that the gravity becomes weaker. To see the difference between Newtonian and 1PN cases, in Fig.4 two simulations from the same initial data are shown [32]. We clearly found that they are different even in this rather weak gravity case. In Fig.5 the wave forms are shown. The solid line (1PN) and the dashed line (Newton) are different. Due to the strong gravity effect of 1PN force, the strong shock in the central is formed and the coalescence is delayed. If the higher post Newtonian effects (2PN,...) are included, different results may be obtained. This suggests that fully general relativistic simulations are needed. This suggests also that all the results of coalescing binary neutron stars based on Newtonian hydrodynamics or Post-Newtonian hydrodynamics need to be viewed with caution.

There is another point to be considered. Suppose the point particle limit of each neutron star. In the Newtonian gravity there exists a circular orbit for any small separation a. However in general relativity a stable circular orbit does not exist for $a < r_{\rm ISCO}$ where ISCO stands for Innermost Stable Circular Orbit. For a test particle motion in the Schwarzschild black hole $r_{\rm ISCO} = 6GM/c^2$ where M is the mass of black hole. For equal mass point mass binary $r_{\rm ISCO}$ is estimated as $\sim 14Gm/c^2$ where m is each mass of the binary [34]. Since the radius of the neutron star is $\sim 5Gm/c^2$, when two neutron stars contact, the separation a is $\sim 10Gm/c^2$, which is comparable to $r_{\rm ISCO}$ in the point particle limit. Therefore in the last three milliseconds the general relativity is also important in the orbital motion. Recently ISCO for binary neutron stars including the effect of general relativity and the finite size of the neutron stars has been extensively studied. Shibata discusses ISCO problem in this volume.

3. General Relativistic Simulations

For full general relativistic 3D simulations of coalescing binary neutron stars, we need ISCO of binary neutron stars as initial data. However at present such initial data are not available. We therefore discuss the development of 3D general relativistic code and test simulations. We present only the summary of the basic equations. For details of formalisms, basic equations and numerical methods please refer [35].

3.1. BASIC EQUATIONS

3.1.1. *Initial Value Equations*

We adopt the (3+1)-formalism of the Einstein equation. Then the line element is written as

$$ds^2 = -\alpha^2 dt^2 + \gamma_{ij}(dx^i + \beta^i dt)(dx^j + \beta^j dt), \tag{5}$$

where α, β^i and γ_{ij} are the lapse function, the shift vector and the metric of 3-space, respectively. The lapse and shift represent the coordinate degree of freedom and its choice is essential for numerical simulations. γ_{ij} is dynamical variable including gravitational waves.

We assume the perfect fluid. The energy momentum tensor is given by

$$T_{\mu\nu} = (\rho + \rho\varepsilon + p)u_\mu u_\nu + pg_{\mu\nu}, \tag{6}$$

where ρ, ε and p are the proper mass density, the specific internal energy and the pressure, respectively, and u_μ is the four-velocity of the fluid. We define $\rho_H \equiv n^\mu n^\nu T_{\mu\nu}$, $J_i \equiv -h_i{}^\mu n^\nu T_{\mu\nu}$ and $S_{ij} \equiv h_i{}^\mu h_j{}^\nu T_{\mu\nu}$ where n_μ is the unit normal to the t=constant hypersurface and $h_{\mu\nu} = g_{\mu\nu} + n_\mu n_\nu$ is the projection tensor to the hypersurface.

The initial data should satisfy the Hamiltonian and momentum constraint equations given by

$$R + K^2 - K_{ij}K^{ij} = 16\pi\rho_H, \tag{7}$$

$$D_j\left(K^j{}_i - \delta^j{}_i K\right) = 8\pi J_i, \tag{8}$$

where R is the 3-dimensional Ricci scalar curvature, K_{ij} is the extrinsic curvature and D_i denotes the covariant derivative with respect to γ_{ij}.

3.1.2. *Relativistic Hydrodynamics*

In order to obtain equations similar to the Newtonian hydrodynamics equations, we define $\rho_N = \sqrt{\gamma}\alpha u^0 \rho$ and $u_i^N = J_i/\alpha u^0 \rho$ where $\gamma = \det(\gamma_{ij})$. Then the equation for the conservation of baryon number is expressed as

$$\partial_t \rho_N + \partial_\ell\left(\rho_N V^\ell\right) = 0, \tag{9}$$

where $V^\ell = u^\ell/u^0$. The equation for momentum is

$$\begin{aligned}\partial_t(\rho_N u_i^N) + \partial_\ell\left(\rho_N u_i^N V^\ell\right) &= -\sqrt{\gamma}\alpha\partial_i p - \sqrt{\gamma}(p+\rho_H)\partial_i\alpha \\ &\quad + \frac{\sqrt{\gamma}\alpha J^k J^\ell}{2(p+\rho_H)}\partial_i\gamma_{k\ell} + \sqrt{\gamma}J_\ell\partial_i\beta^\ell. \end{aligned}\tag{10}$$

The equation for internal energy becomes

$$\partial_t(\rho_N\varepsilon) + \partial_\ell\left(\rho_N\varepsilon V^\ell\right) = -p\partial_\nu\left(\sqrt{\gamma}\alpha u^\nu\right). \tag{11}$$

To complete hydrodynamics equations, we need an equation of state, $p = p(\varepsilon,\rho)$

3.1.3. *Time Evolution of the Metric Tensor*

The evolution equation for the extrinsic curvature which is essentially the time derivative of γ_{ij} is given by

$$\partial_t K_{ij} = \alpha\left\{R_{ij} - 8\pi\left[S_{ij} + \tfrac{1}{2}\gamma_{ij}\left(\rho_H - S^\ell{}_\ell\right)\right]\right\} - D_iD_j\alpha \tag{12}$$

$$+ \ \alpha\left(KK_{ij} - 2K_{i\ell}K^\ell{}_j\right), \tag{13}$$

$$+ \ K_{mi}\partial_j\beta^m + K_{mj}\partial_i\beta^m - K_{ij}\partial_m\beta^m, \tag{14}$$

where R_{ij} is the 3-dimensional Ricci tensor.

Now we define the conformal factor ϕ as $\phi \equiv (\det[\gamma_{ij}])^{1/12}$ and $\tilde{\gamma}_{ij}$ as $\tilde{\gamma}_{ij} = \gamma_{ij}/\phi^4$. The conformal factor ϕ is determined by

$$\tilde{\Delta}\phi = -\frac{\phi^5}{8}\left(16\pi\rho_H + K_{ij}K^{ij} - K^2 - \phi^{-4}\tilde{R}\right), \tag{15}$$

where $\tilde{\Delta}$ and $\tilde{R}$ is the Laplacian and the scalar curvature with respect to $\tilde{\gamma}_{ij}$, respectively. The time evolution of $\tilde{\gamma}_{ij}$ is given by

$$\begin{aligned} \partial_t\tilde{\gamma}_{ij} &= -2\alpha\left(\widetilde{K}_{ij} - \frac{1}{3}\tilde{\gamma}_{ij}K\right) + \tilde{D}_i\tilde{\beta}_j + \tilde{D}_j\tilde{\beta}_i - \frac{2}{3}\tilde{\gamma}_{ij}\tilde{D}_\ell\beta^\ell \\ &\equiv A^T_{ij}, \end{aligned} \tag{16}$$

where

$$\widetilde{K}_{ij} = \phi^{-4}K_{ij}, \quad \tilde{\beta}_i = \phi^{-4}\beta_i \tag{17}$$

and $\tilde{D}_i$ denotes the covariant derivative with respect to $\tilde{\gamma}_{ij}$.

3.1.4. *Coordinate Conditions*

On of the most serious technical problems in numerical relativity is the control of the space and time coordinates. Difficulties arise from (a) coordinate singularities caused by strong gravity and the dragging effect, where coordinate lines may be driven towards or away from each other, and (b) spacetime singularities which appear when a black hole is formed. We should demand coordinate conditions so that these singularities can be avoided.

Spatial coordinates — the shift vector β^i

We demand $\partial_i A^T_{ij} = 0$. Then A^T_{ij} becomes transverse-traceless. Equation (16) is reduced to the equation for the shift vector β^i as

$$\begin{aligned} &\nabla^2\beta^i + \frac{1}{3}\partial_i\left(\partial_\ell\beta^\ell\right) = \\ &\qquad \partial_j\left[2\alpha\left(\widetilde{K}_{ij} - \frac{1}{3}\tilde{\gamma}_{ij}K\right)\right] \\ &\qquad - \partial_j\left[\hat{\gamma}_{j\ell}\partial_i\beta^\ell + \hat{\gamma}_{i\ell}\partial_j\beta^\ell - \frac{2}{3}\hat{\gamma}_{ij}\partial_\ell\beta^\ell + \beta^\ell\partial_\ell\hat{\gamma}_{ij}\right], \end{aligned} \tag{18}$$

where ∇^2 is the simple flat-space Laplacian and $\hat{\gamma}_{ij} = \tilde{\gamma}_{ij} - \delta_{ij}$.

Time Slicing — the lapse function α

We require a time slice to have the following properties: (1) In the central region, spacetime singularities should be avoided. These singularities are regions of spacetime where curvature, density and other quantities are infinite. (2) In the exterior vacuum region, the metric should be stationary except for the wave parts to catch gravitational waves numerically.

The lapse function α is given by

$$\alpha = \exp\left[-2\left(\widehat{\phi} + \frac{\widehat{\phi}^3}{3} + \frac{\widehat{\phi}^5}{5}\right)\right], \tag{19}$$

where $\widehat{\phi} = \phi - 1$. In this slicing, the space outside the central matter quickly approaches the Schwarzschild metric. Since the non-wave parts decrease as $O(r^{-4})$ for large r, we can catch gravitational waves numerically with $O\left((M/r_c)^3\right)$ accuracy, where M is mass of the system and r_c is distance from the origin of the position where gravitational waves are evaluated.

4. Coalescing Binary Neutron Stars in 3D Numerical Relativity

We show our test simulations of coalescing binary neutron stars. We place two spherical neutron stars of mass $M = 1.0M_\odot$ and radius $r_0 = 6M$ with density distribution (ρ_N) of $\gamma = 2$ polytrope at $x = 7.2M$ and $x = -7.2M$. We add a rigid rotation with angular velocity Ω as well as an approaching velocity to this system such that

$$v_x^N = \begin{cases} -\Omega y - \Omega r_0 & \text{if } x > 0, \\ -\Omega y + \Omega r_0 & \text{if } x < 0, \end{cases} \tag{20}$$

$$v_y^N = \Omega x. \tag{21}$$

We adopt (x, y, z) coordinates with $201\times201\times201$ grid size (9GB) and typical cpu time being 10 hours on VPP300 (FUJITSU). The details of a numerical method is shown in [35].

We performed two simulations, BI1 and BI2, with different values of Ω. The total angular momentum divided by the square of the gravitational mass are 0.99 and 1.4 and for BI1 and BI2, respectively. Figure 6 shows the evolution of the density on the x-y, y-z and x-z planes of BI2. The binary starts to coalesce and rotates about 180 degree. In Fig.7 we show $flux \equiv r^2\dot{\gamma}_{ij}^2/32\pi$ which can be considered as the energy flux of the gravitational wave energy if we take appropriate averaging. $flux$ shows an interesting feature of generation and propagation of the waves. A spiral pattern appears on the x-y plane while different patterns with peaks around z-axis appear on the x-z and y-z planes. This can be explained by the quadrupole wave pattern. In Fig.8 we show the luminosity of the gravitational waves as a function of the retarded time $t - r$ calculated at various r. For large r it

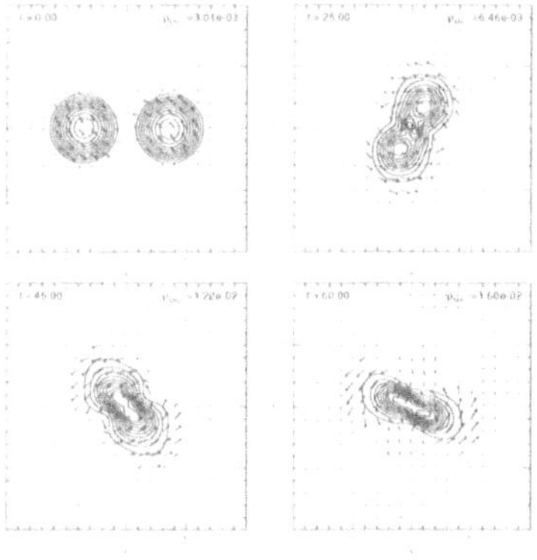

Figure 5. Mass density ρ_N on the $x - y$ plane. Arrows indicate the velocity vector V^i. Quantities are represented in units of $c = G = M_\odot = 1$.

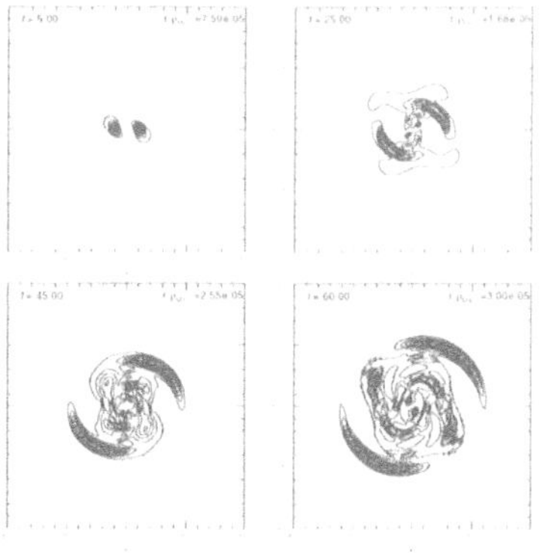

Figure 6. $r^2 \dot{\tilde{\gamma}}_{ij}^2 / 32\pi$, which is expected to represent "Energy density of the gravitational waves."

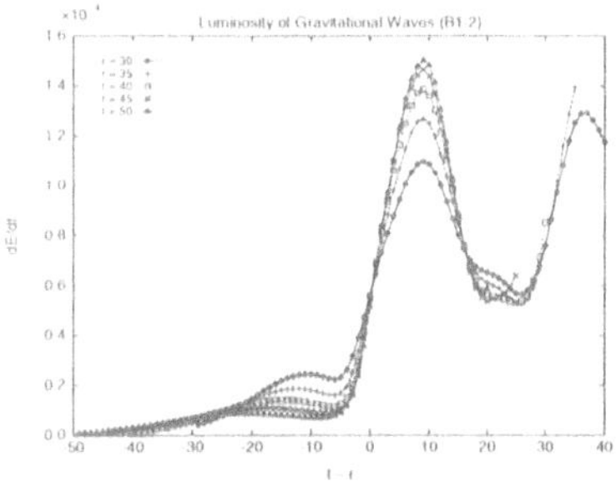

Figure 7. Luminosity of gravitational waves $L_{\rm GW}$ as a function of the retarded time $t - r$ calculated at $r = 30 \sim 50$.

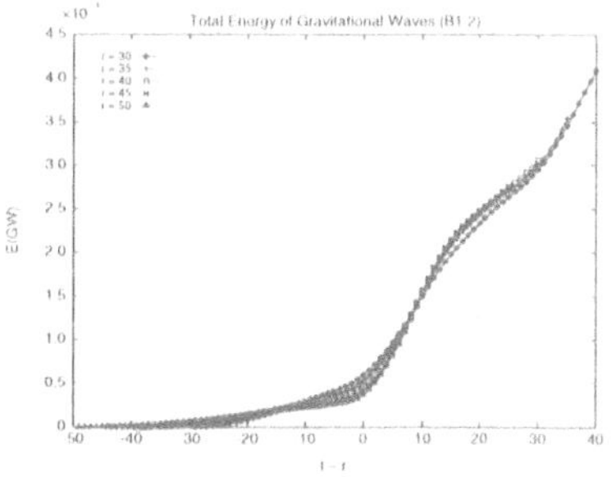

Figure 8. Energy of gravitational waves $E_{\rm GW}$ as a function of the retarded time $t - r$ calculated at $r = 30 \sim 50$.

seems that the luminosity converges, which suggests that the gravitational waves are calculated correctly in our simulations.

The total energy of the gravitational radiation as a function of the retarded time is shown in Fig.9. The total energy amounts to $4 \times 10^{-3} M$(0.2% of the total mass) and $1 \times 10^{-3} M$ (0.05% of the total mass) for BI2 and BI1, respectively.

In conclusion, the present test simulations suggest future fruitful development of 3D fully general relativistic simulations of coalescing binary neutron stars.

Numerical simulations were performed by VPP300 at NAO. This work was also supported by a Grant-in-Aid for Basic Research of the Ministry of Education, Culture, and Sports No.08NP0801.

References

1. Tsubono, K., in *Gravitational Wave Detection* eds, K. Tsubono, M.-K. Fujimoto and K. Kuroda, Universal Academic Press, INC.-TOKYO (November 12-14 1996 Saitama, Japan) pp.183-192
2. Hough, J., in *Gravitational Wave Detection* eds, K. Tsubono, M.-K. Fujimoto and K. Kuroda, Universal Academic Press, INC.-TOKYO (November 12-14 1996 Saitama, Japan) pp.175-182
3. Brillet, A., in *Gravitational Wave Detection* eds, K. Tsubono, M.-K. Fujimoto and K. Kuroda, Universal Academic Press, INC.-TOKYO (November 12-14 1996 Saitama, Japan) pp.163-174
4. Barish, B., in *Gravitational Wave Detection* eds, K. Tsubono, M.-K. Fujimoto and K. Kuroda, Universal Academic Press, INC.-TOKYO (November 12-14 1996 Saitama, Japan) pp.155-162
5. Phinney, E. S. 1991, *Astrophys. J*, 380, L17
6. Narayan, R., Piran, T., and Shemi, A. 1991, *Astrophys. J*, 379, L17
7. Van den Heuvel, E. P. J. and Lorimer, D. R. 1996, *Mon. Not. Roy. Astro. Soc.*, 283, L37
8. Tutukov, A. V. and Yungelson, L. R. 1994, *Mon. Not. Roy. Astro. Soc.*, 268, 871-879
9. Lipunov, V. M., Postnov, K. A., and Prokhorov, M. E. 1997, *Mon. Not. Roy. Astro. Soc.*, 288, 245-259
10. Yungelson, L. and Zwart, S. 1998, astro-ph/9801127
11. Peters, P. C. & Mathews, J. 1963, *Phys. Rev.*, 131, 435
12. Peters, P. C. 1964, *Phys. Rev.*, 136, B1224
13. Cutler, C. et al. 1993, *Phys. Rev. Letters*, 70, 2984
14. Blanchet, L. 1996, *Phys. Rev.*, D54, 1417-1438
15. Cutler, C. and Flangan, E. E. 1994, *Phys. Rev.*, 49, 2658-2697
16. Oohara, K. and Nakamura, T. 1989, *Prog. Theor. Phys.*, 82, 535
17. Nakamura, T. and Oohara, K. 1989, *Prog. Theor. Phys.*, 82, 1066
18. Rasio, F. A. and Shapiro, S. L. 1992, *Astrophys. J*, 401, 226-245
19. Centrella, J. M. and McMillan, S. L. W. 1993, *Astrophys. J*, 416, 719-732
20. Rasio, F. A. and Shapiro, S. L. 1994, *Astrophys. J*, 432, 242-261
21. Houser, J. L., Centrella, J. M., and Smith, S. G. 1989, *Phys. Rev. Letters*, 72, 1314-1317
22. New, K. C. B. and Tohline, J. E. 1997, *Astrophys. J*, 490, 311-327
23. Davies, M. B., Benz, W., Piran, T., and Thielemann F. K. 1994, *Astrophys. J*, 431, 742
24. Zhuge, Xing, Centrella, J. M., and McMillan, S. L. W. 1994, *Phys. Rev.*, D50, 6247
25. Zhuge, Xing, Centrella, J. M., and McMillan, S. L. W. 1996, *Phys. Rev.*, D54, 7261
26. Oohara, K. and Nakamura, T. 1990, *Prog. Theor. Phys.*, 83, 906
27. Nakamura, T. and Oohara, K. 1991, *Prog. Theor. Phys.*, 86, 73
28. Shibata, M., Nakamura, T., and Oohara, K. 1992, *Prog. Theor. Phys.*, 88, 1079-1095
29. Shibata, M., Nakamura, T., and Oohara, K. 1993, *Prog. Theor. Phys.*, 89, 809-819
30. Ruffert, M., Janka, H. T., and Schaefer, G. 1996, *Astron. Astrophys.*, 311, 532-566
31. Ruffert, M., Janka, H. T., Takahasi, K., and Schaefer, G. 1997, *Astron. Astrophys.*, 319, 122-153
32. Oohara, K. and Nakamura, T. 1992, *Prog. Theor. Phys.*, 88, 307
33. Ruffert, M., Rampp, M., and Janka, H. T. 1997, *Astron. Astrophys.*, 321, 991-1006
34. Kidder, L. E., Will, C. M., and Wiseman, A. G. 1992, *Class. Quantum Gravity*, 9, L125-31
35. Oohara, K., Nakamura, T., and Shibata, M. 1997, *Prog. Theor. Phys.* supplement, 128, 183-250

BLACK HOLES AND GRAVITATIONAL WAVES: SPACETIME ENGINEERING

STUART L. SHAPIRO
University of Illinois at Urbana-Champaign
1110 West Green Street, Urbana, IL 61801-3080, U.S.A.

1. Introduction

General relativity – Einstein's theory of relativistic gravitation – is the cornerstone of modern cosmology, the physics of neutron stars and black holes, the generation of gravitational radiation, and countless other cosmic phenomena in which strong-field gravitation is believed to play a dominant role. Yet the theory remains largely untested, except in the weak-field, slow-velocity regime. Moreover, solutions to Einstein's equations, except for a few idealized cases characterized by high degrees of symmetry, have not been obtained as yet for many of the important dynamical scenarios thought to occur in nature. Only now, with the advent of supercomputers, is it possible to tackle these highly nonlinear equations numerically and explore these scenarios in detail. That is the main goal of numerical relativity, the art and science of developing computer algorithms to solve Einstein's equations for physically realistic, high-velocity, strong-field systems. Numerical relativity also has a pressing goal – to calculate gravitational waveforms from promising astrophysical sources, in order to provide theoretical templates for the gravitational wave laser interferometers now under construction in the US, Europe and Japan.

The current focus of numerical relativity is handling two and three dimensional systems with strong dynamical fields. The underlying equations are multidimensional, highly nonlinear, coupled partial differential equations in space and time. They have in common with other areas of computational physics, like fluid dynamics and MHD, all of the usual problems associated with solving such nontrivial equations. However, solving Einstein's equations poses some additional complications that are unique to general relativity. The first complication concerns the choice of coordinates. In general relativity, coordinates are merely labels that distinguish

S. M. Miyama et al. (eds.), Numerical Astrophysics, 257–264.

points in spacetime; by themselves coordinate intervals have no physical significance. To use coordinate intervals to determine physically measureable (proper) distances and times requires the spacetime metric, but the metric is determined only after Einstein's equations have been solved. Moreover, as the integrations proceed, it often turns out that the original (arbitrary) choice of coordinates turns out to be bad, because, for example, singularities eventually are encountered in the equations. The gauge freedom inherent in general relativity – the ability to choose coordinates in an arbitrary way – is not always easy to exploit successfully in a numerical routine.

Another complication arising in numerical relativity involves the appearance of black holes. Black holes inevitably contain spacetime singularities – regions where the gravitational tidal field, the matter density and the spacetime curvature all become infinite. Encountering such singularities results in some of the terms in Einstein's equations becoming infinite, causing overflows in the computer output and premature termination of the numerical integration. Thus, when dealing with black holes, it is crucial to choose a technique which avoids the spacetime singularities inside.

Finally, one of the main goals of a numerical relativity simulation is to determine the gravitational radiation generated from a dynamical scenario. However, the gravitational wave components usually constitute small fractions of the background spacetime metric. Moreover, to extract the waves from the background requires that one probe the spacetime in the far-field or radiation zone, which is typically at large distance from the strong-field central source. Yet it is the strong-field region which usually consumes most the computational resources (e.g spatial grid) to guarantee accuracy. Furthermore, waiting for the wave to propagate to the far-field region usually takes nonnegligible integration time. Overcoming these difficulties to reliably measure the wave content thus requires that a code successfully cope with the problem of dynamic range inherent in such a simulation.

2. The Grand Challenge Efforts

The most outstanding unsolved problem in classical general relativity is the two-body problem. Consequently, significant computational effort is going into solving the coalescence of binary black holes and binary neutron stars. Obtaining the fully dynamical solutions to these scenarios is not only of theoretical interest, but it is also crucial to gravitational wave astronomy, since coalescing binaries are the most promising sources of gravitational waves detectable by the laser interferometers now being built. So important is this effort that in the US, both the NSF and NASA have funded Grand Challenge teams of computational physicists and computer scientists to work on these binary coalescence problems.

The 'holy grail' of these Grand Challenge efforts is the calculation of the complete gravitational waveform in each of the two polarization states, h_+ and $h_\times$, from binary inspiral and coalescence. Having the theoretical waveform as a template is crucial to the 'matched filter' technique of gravity wave detection by the laser interferometers under construction. It is also important for using the observed waveforms to probe the strong-field geometry around coalescing neutron stars and black holes.

Despite the importance of solving the coalescence problem and the intensive effort to date that has gone into obtaining the answer, *no computer code currently exists that can integrate two black holes or two neutron stars in binary orbit long enough to get a waveform out to an accuracy anywhere near 10%.* Nevertheless, considerable progress has been achieved. A number of computer modules crucial to solving the problem on parallel machines have been assembled and tested. Some new physical insights have emerged from simulations performed to date (see Section 3). Additionally, new formulations of Einstein's equations have emerged which recast them into a flux-conservative, first order, hyperbolic form where the only nonzero characteristic speed in that of light; (Choquet-Bruhat and York, 1995; Abrahams et. al., 1995). As a consequence, these equations may permit a more reliable "excision" of black holes and their nasty interior singularities from the numerical grid and the replacement of these regions by a well-behaved boundary condition ("horizon boundary conditions").

3. Applications: Topology of a Black Hole Event Horizon

Two definitive results that emerged from the early phase of the Binary Black Hole Grand Challenge work have to do with the topology of the black hole event horizon. The event horizon is the surface of a black hole; events which occur inside cannot be seen from the outside because nothing can escape the interior, including light signals. The first result of the simulations dealt with the event horizon formed from the merger of two identical, nonspinning black holes which collide head-on. The second dealt with the formation of the event horizon from the collapse of a rotating toroid. Both results were obtained by performing simulations with a mean-field, particle simulation code in axisymmetry (Shapiro and Teukolsky, 1992; Abrahams et. al., 1994). The code solved the collisionless Boltzmann equation for the matter by integrating particle geodesic equations in the mean gravitational field of the distribution. The matter provided the source terms for the gravitational field (ADM) equations, which were solved by finite differencing.

3.1. HEAD-ON COLLISION AND COALESCENCE OF TWO BLACK HOLES

Consider the the head-on collision of two identical clusters of collisionless particles (Shapiro and Teukolsky, 1992). These particles may represent stars in a cluster, or any other form of collisionless matter which interacts exclusively by gravity. At $t = 0$ the velocity dispersion in each cluster is zero, so that each cluster collapses on itself to form a black hole prior to merger. In this way we set up a simulation of a head-on collision of black holes.

In Figure 1 we show spatial snapshots of the collision at three different instants of time. The upper figure shows the initial configurations, which are hurling towards each other at velocity $v = 0.15c$. The middle frame shows the formation of two black holes from the collapse of each of the clusters. Note how the tidal field of the companion causes a distortion in the shape of the black hole, giving rise to the "hour glass" appearance of the merging holes when they first come into contact. The bottom frame shows the configuration at late times, when the two black holes have coalesced into a single, spherical Schwarzschild hole which encompasses all of the matter.

Figure 2 is a spacetime diagram of the collision and merger. The time axis is along the vertical direction, while spacelike hypersurfaces (with one of the spatial directions suppressed) are horizontal planes at any instant of time. The collision axis goes from left to right and the black hole horizon is shown as the dark shaded surface. Some of the light rays which generate the horizon ("null generators of the horizon") are shown. Their trajectories were traced by numerically propagating light rays in the background curved spacetime. The inset shows a closeup view of the formation and merger of the two horizons and how the rays enter the horizons at those early events.

Figure 2 is the famous "pair of pants" picture of the event horizon for coalescing black holes that was sketched in general relativity textbooks more than 25 years ago (Hawking and Ellis, 1973; Misner et. al., 1973). The figure shown here is the first real calculation of such a diagram. Many things were known about the the topology of the merging horizons and the null generators prior to these numerical simulations, but a few important details were not. It was well known that the black hole is spanned by null generators, which can intersect or cross each other only at those points at which they enter the horizon. Once on the horizon, a null generator can never propagate off, nor can it ever cross another null generator. These properties were all understood and nicely corroborated by the simulation. But in addition, the simulation revealed for the first time a line of crossover points for the null generators extending from the "crotch" on the "pair of pants" down along each inside trouser seam, around each bottom, and continuing a small distance up each outside "seam" (see Figure 2). The points on the outside seam at which the line of crossover terminates are caustics,

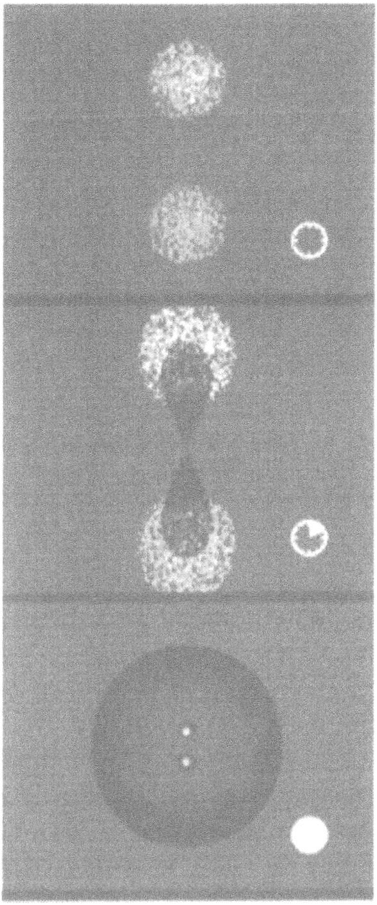

Figure 1. The collision and coalescence of two black holes formed by the collapse of two spheres of particles. The black hole event horizon is shown by the dark shaded surface. The clock in each snapshot indicates the fraction of time elapsed during the simulation.

where the intensity of the intersecting light rays becomes infinite. For a single, isolated cluster undergoing spherical collapse to a black hole, the caustic would arise at the base of the spacetime diagram at the point at which the horizon first forms, and there would be no crossovers. Here, the gravitational tidal field of the colliding black holes shifts the location of the caustic and produces the line of crossovers. An analysis of the simulation (Matzner et. al., 1995) shows that the line of crossovers is spacelike, which means that they cannot be traced by light rays or particles moving slower than c. As this line approaches the caustics on the sides, it becomes asymptotically light-like.

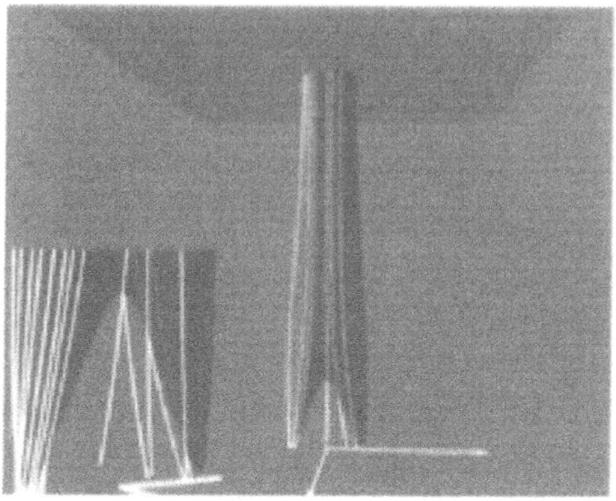

Figure 2. Spacetime diagram for the collision shown in Figure 1. Some of the light rays generating the horizon are drawn. The collision axis goes from left to right and the time axis is vertical. The inset is a blow up of the caustic and crossover structure at the birth of the horizon.

3.2. TOROIDAL BLACK HOLES AND TOPOLOGICAL CENSORSHIP

Consider the collapse of rotating toroidal clusters of collisionless particles to Kerr black holes. One of the interesting findings of our numerical simulations was a case in which the black hole event horizon emerges as a toroid (Abrahams et. al., 1994; Hughes et. al., 1994). The initial matter distribution is based on a solution for a relativistic toroidal cluster in dynamical equilibrium. Collapse was induced by reducing the angular velocity of each particle by a factor of two, resulting in a nonequilibrium cluster with a total angular momentum of $J/M^2 = 0.7$.

The spactime diagram for the collapse looks quite similar to the diagram plotted in Figure 2. The line of crossover points at which light rays enter the horizon, and the cusp formed by rays at the point at which the line of crossovers terminates, are all present as in Figure 2. The toroidal horizon first forms entirely within the vacuum, between the origin and the inner edge of the collapsing, toroidal cluster. The horizon then expands to fill up the "doughnut hole," becoming topologically spherical at about the instant when the outer edge of the horizon reaches the inner edge of the matter toroid.

When the results of the numerical simulation were first reported, it appeared that they might be in conflict with recent theorems regarding "topological censorship." These theorems permit a nonstationary black hole to have the topology of a two-sphere or a torus (Gannon, 1976), but a torus can persist only for a short time. According to topological censorship (Friedmann et. al., 1993; Jacobson and Venkataramani, 1995) the hole in a toroidal horizon must close up quickly, before a light ray can pass through. Moreover, the black hole must be two-sphere if no null generators enter the horizon at later times (Browdy and Galloway, 1995).

Once the numerical simulations were studied in detail and the geometry of the transient toroidal event horizon was analyzed by means of a spacetime diagram like Figure 2, the results were shown to be completely consistent with the theorems concerning the topology of black hole horizons (Shapiro et. al., 1995). At late times, when equilibrium has been reached, the topology is spherical, in agreement with the theorem of Hawking (1972). At early times, the topology is temporarily toroidal. However, the line of crossovers traced out by a point on the inner rim of the torus is *spacelike.* That implies that the "hole" in the torus indeed closes up faster than the speed of light, in compliance with topological censorship. Finally, at intermedidate times, when the horizon is spanned by its full complement of null generators, the rotating black hole has a spherical topology, in accord with the theorems of Browdy and Galloway (1995).

4. Conclusions

Numerical relativity is only in its infancy, or at most very early adolescence. It is still under development as a valuable scientific tool to solve Einstein's equations of general relativity for realistic, dynamical spacetimes involving strong-field sources like black holes. Already it has been used successfully to determine gravitational waveforms for simple cases characterized by high degrees of spatial symmetry, like head-on collisions and axisymmetric collapse. Its refinement is absolutely crucial for interpretating the waveforms likely to be observed by the new gravitational wave laser interferometers now under construction, like LIGO, VIRGO, GEO and TAMA. These waveforms will arise from the inspiral and coalescence of binary black holes and neutron stars.

Numerical relativity has also demonstrated that beyond the calculation of accurate numerical solutions, it can provide *qualitative* insight into Einstein's equations in those cases where uncertainty still prevails. It can even be helpful as a guide to proving (or disproving) theorems about strong-field spacetimes in those instances where analytic means alone have not proven adequate. Numerical relativity is an important component of modern computational astrophysics whose role is likely to mushroom in the not too distant future.

References

Abrahams, A., Anderson, A., Choquet-Bruhat, Y. and York, J. W., Jr. (1995) *Phys. Rev. Lett.*, **75**, p 3377.

Abrahams, A., Cook, G. B., Shapiro, S. L. and Teukolsky, S. A. (1994) *Phys. Rev. D*, **49**, p 5153.

Browdy, S. and Galloway, G. J. (1995) *J. Math. Phys.*, **36**, p 4952.

Choquet-Bruhat, Y. and York, J. W., Jr. (1995) *C. R. Acad. Sci., Ser. I: Math.*, **A321**, p 1089.

Friedmann, J. L., Schleich, K. and Witt, D. M. (1993) *Phys. Rev. Lett.*, **71**, p 1486.

Gannon, D. (1976) *Gen. Relativ. Gravit.*, **7**, p 219.

Hawking, S. W. (1972) *Commun. Math. Phys.*, **25**, p 152.

Hawking, S. W. and Ellis, G. F. R., *The Large Scale Structure of Spacetime*, (Cambride Univ. Press, Cambridge, 1973).

Hughes, S. A., Keeton, C. R., Walker, P., Walsh, K., Shapiro, S. L. and Teukolsky, S. A. (1994) *Phys. Rev. D*, **49**, p 4004.

Jacobson, T. and Venkataramani, S. (1995) *Class. Quantum Grav.*, **12**, p 1005.

Matzner, R. A., Seidel, H. E., Shapiro, S. L., Smarr, L., Suen, W.-M., Teukolsky, S. A., and Winicour, J. (1995) *Science*, **270**, p 941.

Misner, C. W., Thorne, K. S., and Wheeler, J. A., *Gravitation*, (Freeman, San Francisco, 1973).

Shapiro, S. L. and Teukolsky, S. A. (1992) *Phys. Rev. D*, **45**, p 2739.

Shapiro, S. L., Teukolsky, S. A. and Winicour, J. (1995) *Phys. Rev. D*, **52**, p 6982.

SUPERNOVA EXPLOSIONS

WOLFGANG HILLEBRANDT
Max-Planck-Institut für Astrophysik
D-85748 Garching, Germany

Abstract.
Because of their complexity models of supernova explosions have always been a challenge. In the case of core collapse models, which utilize gravitational binding for the explosion, the problem is not so much an energy but rather a momentum problem. There is plenty of energy available and only a small fraction of it has to be converted into outward momentum of the stellar envelope, but this is a non-trivial problem. Thermonuclear explosion models, on the other hand, suffer from the fact that once nuclear burning is ignited the progenitor star, presumably a white dwarf, it tends to expand and cool. It has been shown that an explosion results only if the burning front propagates with a velocity much larger than the laminar speed of nuclear flame in degenerate matter. Since for most supernova models hydrodynamic instabilities play a key role numerical experiments to prove or disprove certain ideas have to be 2- or 3-dimensional and, in fact, they have become feasible recently. Here we discuss the present status of 2- and 3-d numerical simulations of supernova explosions. In the case of explosions triggered by gravitational energy release, the crucial problem is the transport of neutrinos in convectively unstable matter, and for explosions powered by nuclear energy release, the problem of turbulent combustion has to be addressed.

1. Some observational facts

In general, supernovae are classified according to their maximum light spectra. Those showing Balmer lines of H are called Type II's, and all the others are of Type I. Those Type I's which show a strong Si absorption feature at maximum light are named Type Ia, and the others are Ib's or Ic's, depending on whether or no they have also He I features in their spectra (Wheeler

S. M. Miyama et al. (eds.), Numerical Astrophysics, 265–272.

& Harkness 1990). At later times, several months after the explosion, when the supernova ejecta become optically thin, Type II spectra are dominated by emission lines of H, O, and Ca, whereas Type Ia's have no O, but Fe and Co. Type Ib,c's, on the other hand side, show emission lines of O and Ca, just like the Type II's (Filippenko 1988).

In contrast to SN II which have rather different light curves, those of Type Ia's are quite similar, making them good candidates for standard candles to measure the cosmic distance scale. Moreover, since they are the brightest among all supernovae, they can be observed even at high red-shifts and, in fact, a Type Ia supernova at a red-shift of about 1 has recently been observed (Garnavich et al. 1998). Although their absolute peak luminosity may vary by about one magnitude, an observed correlation between the luminosity and light curve shape (the brighter ones have broader light curves) allows one to correct for the differences. So recently observations of Type Ia's have become a powerful tool in attempts to determine cosmological parameters (Garnavich et al. 1998).

Other observational information on supernovae is, in general, less solid. With the exception of SN 1987A, we do not have direct information on the properties of the progenitors, the energetics, or the masses of the ejecta and of the (compact) remnants, if there are any. Indirect information can be obtained from the light curves and the spectra leading, however, to an "inverse" problem as far as theoretical interpretations are concerned.

2. Theoretical classification

The theoretical classification of supernovae is usually done according to the suspected progenitor stars and the explosion mechanism, and it dates back to a classic paper of Fred Hoyle and Willy Fowler in 1960 (Hoyle & Fowler 1960). Based on very few observational facts available at hat time, they postulated that Type II supernovae are the consequence of an implosion of non-degenerate stars, whereas Type I's are the result of the ignition of nuclear fuel in degenerate stars, and today this is still believed to be true in general, if "Type I's" are substituted by "Type Ia's." It is now generally believed that Type II and Ib.c supernovae stem from collapsing massive stars with and without hydrogen envelopes, respectively, and that SN Ia originate from explosions of white dwarfs.

The classification with respect to the energetics is even more complex, and a variety of possible models is available with no unique and well accepted answer (see Fig.1). For example, in the case of thermonuclear explosions, it is not even clear what the mode of propagation of the burning front is, and observations supply poor constraints only. A (fast) deflagration wave in a Chandrasekhar-mass white dwarf can explain the spectra and light

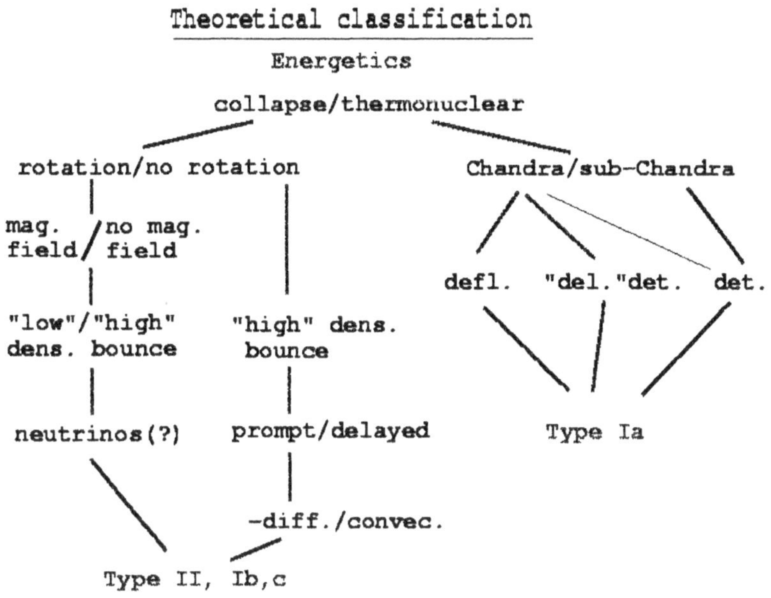

Figure 1. Theoretical classification scheme of supernovae according to their energy sources.

curves of Type Ia's, but so can deflagrations changing into detonations at low densities with or without pulsations (so-called delayed detonations), or even pure detonations in stars with low enough densities are not completely ruled out. As far as the physics of core collapse supernovae is concerned, the situation is not significantly better. Whether or not rotation is important, whether or not magnetic fields play any role, whether neutrinos are diffusively or convectively transported from the cooling proto-neutron star to the stellar mantle, and many other questions, are largely unanswered.

3. Numerical simulations

Of course, a simple way to try to improve the rather unsatisfactory situation outlined in the previous section is to call for more observations. Since computing realistic theoretical light curves and synthetic spectra for a given supernova model has become feasible recently (although computing reliable spectra for Type Ia's still requires novel techniques) high quality data might help to rule out certain possibilities. But as long as most theoretical models contain a large number of more or less free parameters one may doubt the success of this approach.

A few examples may serve as illustrations. As long as for a thermonuclear explosion the propagation velocity of the burning front and the degree of mixing can still be treated as free functions, it is easy to fit a given light curve and spectrum of a Type Ia supernova (see, e.g. Branch et al. 1995), but one does not prove that the model is correct. Similarly, in the core collapse scenario, neutrinos can transfer momentum to the mantle of a star and cause a supernova explosion, provided their luminosity is sufficiently high. If this luminosity results from low opacities or from convection does not matter. Also, only thousands of neutrino events from a single supernova seen in a neutrino detector could possibly tell the difference. Finally, similar momentum transfer can, in principle, be mediated by appropriate combinations of rotation and magnetic fields. Again, a lucky accident of a galactic supernova may supply enough data, including detections of gravity waves, to eliminate several models, but only for one special event and not for the entire class.

Therefore, it appears to be more promising to search for ways to reduce the freedom one still has in building theoretical models. An obvious possibility is to replace parameter studies by first principle calculations, whenever this is possible. The numerical methods that are widely used in this context are finite volume schemes based on Godunov-type algorithms or, more recently, also smoothed particle hydrodynamic codes, coupled to either a neutrino transport scheme (in the case of core collapse supernovae) or to the reaction kinetics (for thermonuclear ones).

3.1. CORE COLLAPSE MODELS

Again, a few examples are given to demonstrate that there is plenty of room for improvements, both in the physics and the numerics of core collapse supernova models. A first and obvious example is again neutrino transport. One will note that very primitive (and likely incorrect) approximations for neutrino interactions with dense nuclear matter are commonly used in all hydrodynamic simulations, and only very recently first attempts have been made to compute those cross sections on the basis of microscopic theories (Raffelt et al. 1995, Raffelt & Strobel 1997, Reddy et al. 1997). Similarly, well-developed methods to calculate the properties of dense nuclear matter are available, but the equations of state used in numerical studies are still computed ignoring nucleon-nucleon correlations, despite the fact that stiffness of the equation of state is very important ingredient.

As far as numerical simulations of stellar collapse and the subsequent supernova explosion are concerned, the most crucial phase seems to be when the newly born neutron star looses its leptons and thermal energy by neutrinos of all flavours. During this phase, both the proto-neutron star

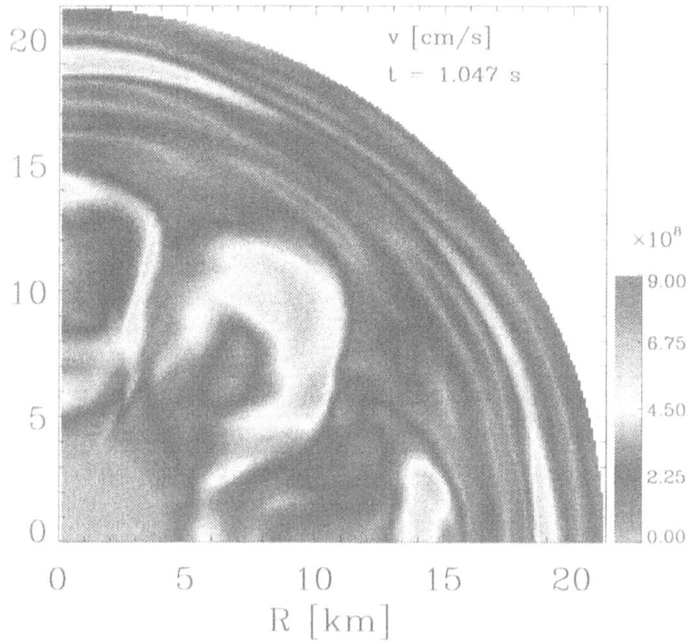

Figure 2. Flow patterns in a newly born neutron star about 1s after core bounce. Shown is the absolute value of the velocity field in order to indicate the vorticity of the convective motions.

itself and the matter behind the stalled shock in the mantle are found to be unstable to entropy and/or lepton number driven buoyancy instabilities which can increase the neutrino luminosity considerably [see, e.g. Janka & Keil (1997) for a recent review]. It is commonly believed that this increase in luminosity is required in order to revive the stalled shock by neutrino heating.

A typical example of a 2-d simulation of proto-neutron star convection is shown in Fig. 2 (from Keil et al., 1996). Here the matter is convectively unstable because of unstable lepton number gradients, the so-called Ledoux-criterion. Typically, convection is found to be very sub-sonic but it can boost the neutrino luminosity by up to almost a factor of 2. The problem is, however, that neutrino transport is treated in a rather approximate fashion in these calculations, i.e by a flux-limited equilibrium diffusion scheme. More elaborate methods, such as direct solutions of the Boltzmann transport equation or the multi-energy group flux-limited diffusion approximation are presently extended from the well-studied 1-d case to multi dimensions which is also necessary to settle the question of neutrino-driven convection outside the proto-neutron star (Messer et al. 1998). It is interesting to note in passing that due to neutrino viscosity motions on scales much smaller than those resolved by recent numerical simulations are heavily damped. So we are in a situation that is almost unique in astrophysics: We do not have to worry much about high Reynolds numbers and turbulence.

3.2. THERMONUCLEAR EXPLOSIONS

In this latter respect the situation is very different as far as thermonuclear supernova models are concerned. Here the main problem is the large scale space which makes direct numerical simulations nearly impossible. Formally, in a white dwarf, near the Chandrasekhar mass, the viscous dissipation scale as well as the width of a nuclear burning front are of the order of 10^{-3}cm, whereas the scale of convective motions can be as large as 10^7cm. Moreover, typical laminar velocities of nuclear flames inside white dwarfs ($\simeq$ 100km/s) are much less than the few % of the sound velocity ($\simeq$ several 1000km) which are needed to drive a thermonuclear explosion. Therefore, turbulent acceleration of the front is a key ingredient into this class of supernovae, and it has to be modelled in some way.

Numerical methods to handle turbulent combustion are presently developed for combustion in engines and for reactor safety (see, e.g., Smilianovski et al. 1997, and references therein), and there is no principle problem in applying them also to supernovae. Moreover, simple microscopic models for turbulence can be calibrated to laboratory combustion experiments, which again will help to reduce the number of parameters [see Kerstein (1991, 1997) for a recent approach]. The aim is always to model numerically unresolved scales and their impact on the resolved large scales.

Fortunately, this problem is not as complicated as it may seem to be. At high densities near the center of the white dwarf the thermal width of the thermonuclear combustion front is small and the turbulence intensity is moderate. The flame is said to be in the "flamelet"-regime and can be modelled by means of front capturing/tracking schemes (Smiljanovski et al., 1997), coupled to a turbulence model (Niemeyer & Hillebrandt, 1995). In this approach the front is tracked by a level set, and the thermodynamic properties of the matter before and after the front are reconstructed from the jump conditions for mass, energy, and momentum. This is possible if the normal velocity of the burning front is known which, in turn, can be obtained from a turbulence model, because in the limit of high propagation speed, the front propagates with the turbulent velocity and is independent of the laminar burning velocity.

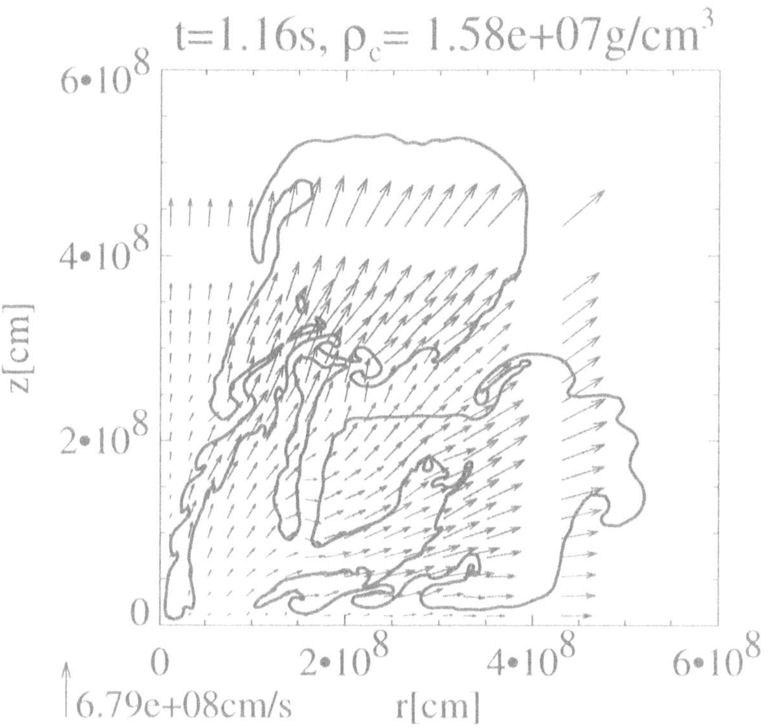

Figure 3. 2-d simulation of a thermonuclear deflagration front in a C+O white dwarf. Shown are the position of the front, the velocities, and the density contours about 1s after ignition. [From Reinecke et al. (1998)]

A first result of a simulation based on this approach is given in Fig. 3. Although the full reconstruction was not yet implemented the effects of a better resolution of the burning front can clearly be seen. In contrast to codes which capture fronts and smear them over several zones our new code can resolve rather complex geometries and, therefore, leads to a higher effective burning velocity because of a larger burning surface. However, even with the improved code we get at best a weak explosion. The ^{5}6Ni masses are in agreement with supernova observations but the velocities of the expanding white dwarf are too low. One can hope that the full implementation of the reaction kinetics together with the reconstruction of the thermodynamic properties will improve the models. One may also speculate that, due to a better modelling of turbulence, a transition to a detonation (which was not observed in any of our simulations) may result in the distributed regime of combustion at low densities ($\leq 10^7 \mathrm{g/cm}^3$). Finally, one of our basic assumptions, namely that the velocity spectrum of turbulence in the presence of combustion is still given by that of inertially driven free turbulence, might have to be modified.

4. Summary and conclusions

The physics of supernovae as well as the present status of numerical simulations of the explosion process have been discussed. By means of a few examples it was demonstrated that multi-dimensional models are needed and that, to a certain extend, they have been carried out already in 2-d with moderately good physics input. The bad news is that these calculations have not solved the supernova problem yet, neither for explosions driven by stellar collapse nor for thermonuclear ones. In both cases the limiting factor does not seem to be the numerical resolution of the simulations but rather still existing uncertainties in the physics.

In case of core collapse supernovae it is likely that with the next generation of supercomputers one will be able to carry out fully resolved simulations in 3-d, and most of the work has to go into improving the microphysics and the neutrino transport. In contrast, for thermonuclear explosions one will never be able to resolve all relevant scales numerically and, therefore, developing new and clever tools to overcome this difficulty is a must. However, these new tools will not make supercomputers obsolete because all simulations on whatever scale one is looking at will have to be done in three dimensions.

References

Branch D. et al., (1995) PASP **107** 1019
Filippenko A.V., (1988) ApJ. **96** 1941
Garnavich P.M. et al., (1998) ApJ. in press
Hoyle F. & Fowler W.A., (1960) ApJ. **132** 565
Janka H.-Th. & Keil W., (1997) preprint, astro-ph/970912
Keil W., Janka H.-Th., & Müller E., (1996) ApJ **473** L111
Kerstein A.R., (1991) J. Fluid Mech. **231** 361
Kerstein A.R., (1997), preprint, submitted to J. Fluid Mech.
Messer B. et al., (1998) preprint, astro-ph/980576, ApJ in press
Niemeyer J.C. & Hillebrandt W., (1995) ApJ.**452** 769
Raffelt G., Seckel D. & Sigl G., (1995) Phys. Rev. D **54** 2784
Raffelt G. & Strobel T., (1997) Phys. Rev. D **55** 523
Reddy S., Prakash M., & Lattimer J.M., (1997) preprint, astro-ph/9710115
Reinecke M., Hillebrandt W., & Niemeyer J.C., (1998) Proc. 16th Int. Conf on Numerical Methods in Fluid Dynamics; Arcachon, France, Springer, in press
Smiljanovski V., Moser V., & Klein R., (1997) Combust. Theory Modelling **1** 183
Wheeler J.C., & Harknesss R.P., (1990) Rep. Prog. Phys. **55** 1467

NUMERICAL ASPECTS OF SIMULATIONS OF MERGING NEUTRON STARS

M. RUFFERT
IoA, Madingley Road, Cambridge CB3 0HA, U.K.

AND

H.-TH. JANKA
MPI für Astrophysik, Postfach 1523, 85740 Garching, Germany

1. Introduction

In order to simulate the merging of two neutron stars with the aim to compute neutrino and gravitational-wave emission, the three-dimensional Newtonian equations of hydrodynamics are integrated by the 'Piecewise Parabolic Method'. We take into account the effects of the emission of gravitational waves and the corresponding backreaction on the hydrodynamics. The properties of neutron star matter are described by the equation of state of Lattimer and Swesty (1991). Energy loss and changes of the electron abundance due to the emission of neutrinos are taken into account by an elaborate "neutrino leakage scheme", which is based on a careful evaluation of the lepton number and energy source terms of all neutrino and antineutrino flavors. The results of various models with different parameters such as neutron star masses, mass ratios and impact parameters, shed light on the accessable physical conditions during the merger, on the conversion of energy into neutrino emission, on the timescales of mass accretion by black hole assumed to be formed after the merging, and on the amount of matter dynamically ejected during the merging.

We give a short summary of this topic which will be discussed in a broader context in Ruffert & Janka (1998) where our previously published calculations (Ruffert & Janka, 1997, and references therein) are extended with respect to the employed numerical procedures and the parameter space for the initial conditions. The former includes the use of nested grids to cover a larger computational volume, the latter concerns lower initial tem-

S. M. Miyama et al. (eds.), Numerical Astrophysics, 273–276.

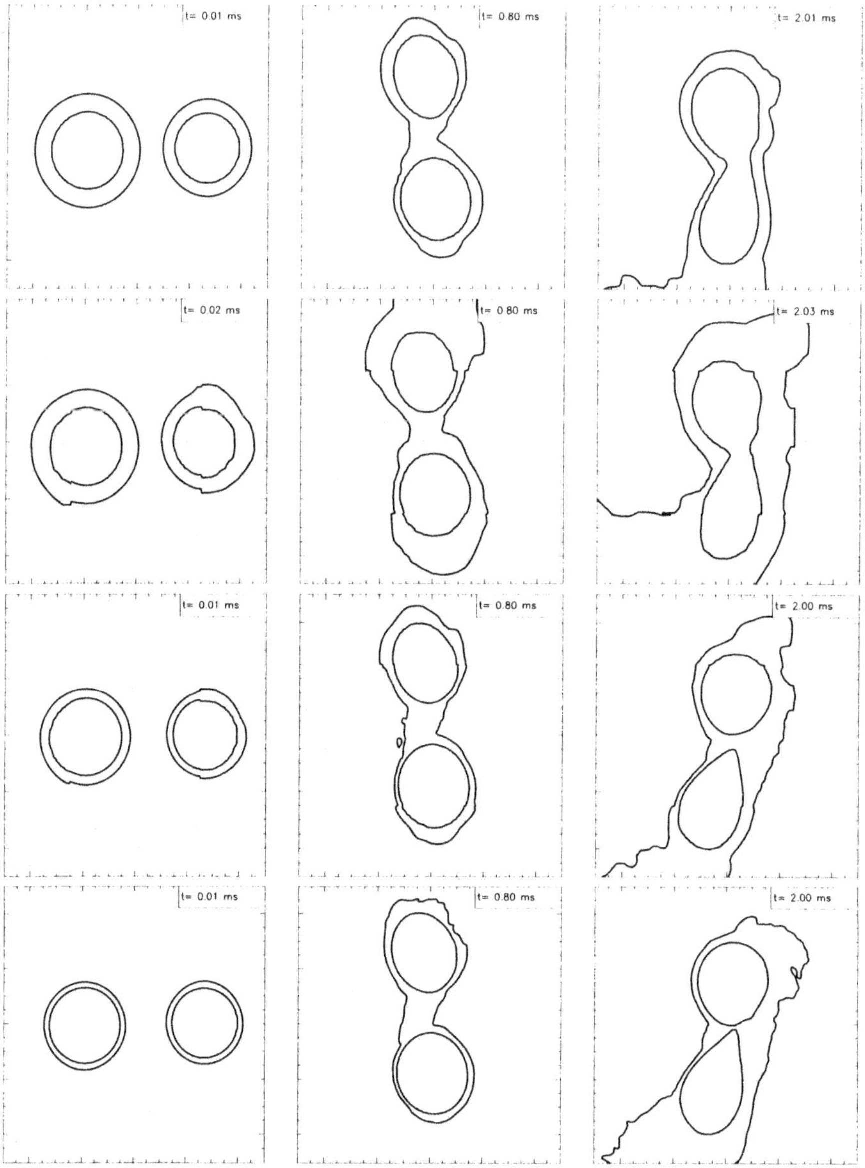

Figure 1. Contour plots of the mass-density distribution of the four Models D, A, B, C (from top to bottom). The outer density contour corresponds to 10^{10} g/cm^3, the inner to 10^{14} g/cm^3. In the top right corner of each panel one finds the time elapsed since the beginning of the simulation. The neutron stars orbit in counter-clockwise direction.

peratures in the neutron stars and lower minimum densities at their surfaces. Different masses and mass ratios of the two neutron stars are also explored and the dynamics of the dense gas torus around the central black hole formed by the collapse of the merged system is investigated.

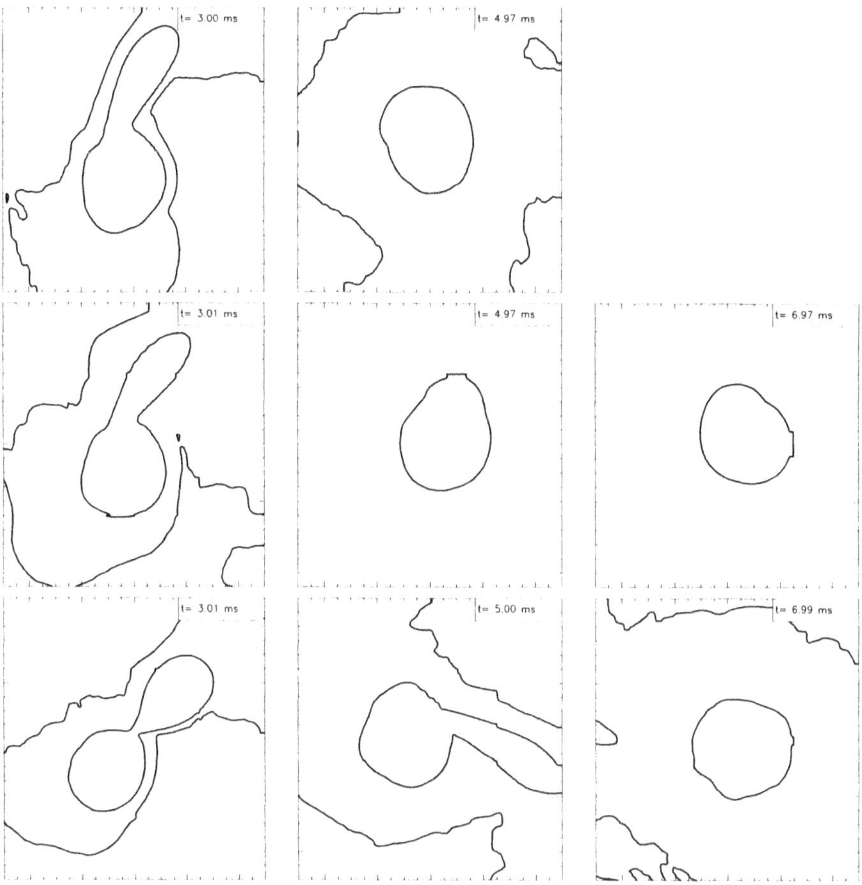

Figure 2. Contour plots of the mass-density distribution of the three Models D, A, B (from top to bottom). The outer density contour corresponds to 10^{10} g/cm^3, the inner to 10^{14} g/cm^3. In the top right corner of each panel one finds the time elapsed since the beginning of the simulation.

Only one aspect will be touched upon in this conference contribution: the extent of the influence of different numerical resolutions on the dynamical evolution of the merging stars.

From the many models shown and discussed in Ruffert & Janka (1998) we will concentrate here on the ones involving two neutron stars with masses of $1.2\,M_\odot$ and $1.8\,M_\odot$. They initially are in a circular orbit around each other at a distance of 46 km, are corotating and have a radius of approximately 15 km. The whole computational volume is a cube 400 km on a side. For the simulation with 4 nested grids, the size of one zone on the finest grid is 0.78 km.

2. Some Results

The four models we compare here differ in the number of zones per spatial dimension, n, and in the levels of gird nesting g: Models A, B, C all have g=4 but n=32, n=64, n=128, respectively, while Model D has g=3 and n = 64. Thus the finest zone of Model D has the same size as that of Model A. Figures 1 and 2 show the temporal evolutions of these models. In the left panels of Fig. 1, which give the initial situation, the more massive neutron star is on the left side.

Obviously, the higher resolution models exhibit a clearly steeper density gradient in the surface layers and the density contours shown in Fig. 1 are closer. After one quarter of an orbital revolution (0.8ms) a bridge of matter linking the two stars has formed and is well represented in all models. However, in Model A the surface regions inflate where the neutron star matter leaves the finest (g=4) level of the grid. The less massive neutron star is tidally strongly deformed after 3/4 of an orbital revolution. Models B and C show a stretched tip of the higher density contour of the secondary barely touching the primary, while in the less resolved Models D and A the cores are well into the process of merging. The very high resolution Model C had to be stopped at this point because of its computational costs. Since there is good agreement between this model and Model B, we conclude that the resolution of the latter is sufficient to follow at least the first 2 ms of evolution. However, at t=3ms (Fig. 2) it becomes even clearer that Models A and D do *not* have enough resolution to describe the tidal disruption of the secondary correctly. Compared with the better resolved Model B, the disruption proceeds too quickly in Models D and A, where the secondary has completely dissolved at t=5ms, whereas Model B is roughly 2ms behind! Although the final state of the mass distribution is again very similar in all the models, the different speed of the temporal evolution strongly affects the gravitational-wave signal from the merging phase and will show up in the thermodynamical conditions inside the remnant.

Ray cast images and movies showing different physical quantities can be found on the CD-ROM accompanying these proceedings.

Acknowledgements. MR is grateful for support by a PPARC Adv. Fellowship and by the LOC and Japanese Ministry of Science. HTJ acknowledges support by the "Sonderforschungsbereich 375-95 für Astro-Teilchenphysik" der DFG.

References

Lattimer, J.M., Swesty, F.D. (1991), *Nucl. Phys.* **A535**, 331.

Ruffert, M., Janka, H.-Th., Takahashi, K., Schäfer, G. (1997) Coalescing Neutron Stars – A Step Towards Physical Models II., *Astron. & Astrophys.*, **319**, 122.

Ruffert, M., Janka, H.-Th. (1998) Merging Neutron Stars, *Astron. & Astrophys.*, in prep.

HYDRODYNAMIC SIMULATIONS OF COALESCING BINARY STARS: STABILITY AGAINST GRAVITATIONAL COLLAPSE

MASARU SHIBATA
Department of Earth and Space Science, Graduate School of Science, Osaka University, Toyonaka, Osaka 560-0043, Japan

THOMAS. W. BAUMGARTE
Department of Physics, University of Illinois at Urbana-Champaign, Urbana, Il 61801, USA

AND

STUART. L. SHAPIRO
Departments of Physics and Astronomy and NCSA, University of Illinois at Urbana-Champaign, Urbana, Il 61801, USA

1. Introduction

Recently, Wilson, Mathews and Marronetti [1] (hereafter WMM) reported results obtained with a relativistic hydrodynamics code assuming several simplifying physical and mathematical approximations. Their results suggest that the central densities of the stars increase as the stars approach each other and that massive neutron stars, stable in isolation, individually collapse to black holes prior to merger. WMM therefore conclude that in general relativity, the presence of a companion star and its tidal field tend to destabilize the stars in a binary system. This conclusion is contrary to what is expected from Newtonian [3], post-Newtonian [4, 5, 6], perturbative [7] and matched asymptotic expansion [8, 9] treatments of the problem. Constructing self-consistent, fully relativistic initial data for two neutron stars in a circular, quasi-equilibrium orbit does not show any evidence of this "crushing effect" either [10]. Moreover, applying energy turning-point methods to sequences of these initial data suggests that inspiraling neutron star binaries are *secularly* stable all the way down to the innermost stable circular orbit [11]. To summarize, most researchers currently believe that the maximum allowed rest mass of neutron stars in close binaries is *larger*

S. M. Miyama et al. (eds.), Numerical Astrophysics, 277–280.

than in isolation, and that their central density is *smaller* than in isolation. If there exists any destabilizing, relativistic effect at high post-Newtonian order, then this effect is much smaller than the dominating stabilizing effect of the tidal field.

However, to date, the only fully *dynamical* treatment of the problem in general relativity – that of WMM – reports a star-crushing effect. In this paper, we perform a new, fully dynamical simulation for binary stars in post-Newtonian gravity. We use a formalism in which (1) all first post-Newtonian terms are taken into account, and (2) sufficient nonlinearity is retained, so that spherical, static stars satisfy the fully general relativistic equations exactly. This formalism is very suitable for studying binary neutron stars. We study relativistic effects in binary stars with $M/R \ll 1$, where M and R are typical values of the stellar mass and radius, so that a post-Newtonian treatment is completely adequate.

By performing a fully dynamical calculation, we can relax various constraints assumed in previous treatments. For example, Wiseman assumed the stars to remain spherically symmetric [6], Baumgarte *et al.* [2] assumed the binary stars to be corotating, and Thorne [9] assumed the stars' orbital separation to be much larger than the stars' radius. Here, we relax all these assumptions and study tidally deformed stars, both corotational and irrotational, at arbitrarily small separations. We still find that the presence of the tidal field of a companion star tends to stabilize neutron stars against catastrophic collapse.

2. Results

In our simulations, we assume the two stars are identical, and set up initial data so that they are in a circular binary orbit. We use a polytropic equation of state with $\Gamma = 1.4$, i.e., sufficiently a soft equation of state, so our post-Newtonian formalism will be accurate for the most massive stars and so that there is no innermost stable circular orbit [3]. Therefore, we can choose arbitrarily small binary separations, and the *orbit* of the binary will still be stable. We choose very close binaries, in which the separation of the surfaces of the two stars is much smaller than the orbital separation ($z_A \simeq 0$ in the terminology of Ref. [2]). For these binaries the tidal effects are strongest, and they are therefore the most suitable configuration to study the stability against gravitational collapse of the individual stars.

We evolve two different classes of binary initial data. The first class are corotational binaries. For these configurations, self-consistent equilibrium initial data can be constructed in the post-Newtonian approximation [13] and even in general relativity (where the stars are only in quasi-equilibrium, see [10, 2]). The other class are irrotational binaries. For this class, a nu-

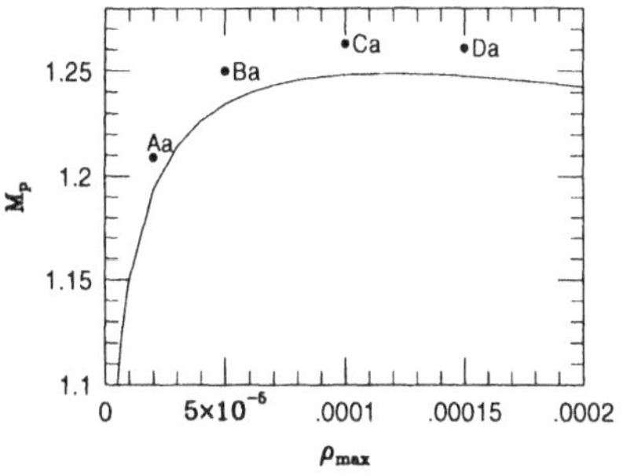

Figure 1. Proper mass as a function of the maximum density for each star in a corotational equilibrium binary (filled circles) and for an isolated spherical star (solid line). We use units in which G, c and K (adiabatic constant) are unity.

merical code for the construction of self-consistent equilibrium initial data is not available now, so that we put two spherical stars at a small separation and assign a (nearly) irrotational velocity (see [12] for details).

In Fig.1, we show the proper mass as a function of the maximum density for a sequence of corotating contact binaries in equilibrium as an example. Filled circles denote corotating binaries, and the solid line is for a spherical star in isolation. For each case, there exists a maximum mass for a critical density $\rho_{\rm crit}$. For the single star, this maximum marks the onset of radial instability and separates the stable branch ($\rho_{\rm max} < \rho_{\rm crit}$) from the unstable branch ($\rho_{\rm max} > \rho_{\rm crit}$). Here, our goal is to determine if all stars that are stable in isolation are also stable in a binary configuration.

We study the hydrodynamical evolution of models Ba and Da in Fig. 1. In Figs. 2, we show contour lines for the density and the velocity fields in the equatorial plane. The stars in model Ba have masses close to the maximum mass of an isolated star and are stable at large separation. It is obvious from the graphs that these stars remain stable in a close binary. Note also that the velocity field remains nearly corotational, and that we do not see any evidence of vorticity features which have been reported in [1]. The stars in model Da have higher mass and significantly higher density. As expected, the contours in Fig. 2 show that they contract and collapse. (Essentially the same results were found for the irrotational models [12].)

To summarize, we find the following results from our post-Newtonian simulations:

- we do not find any star crushing effect as reported by WMM,
- we find that stars in binaries can support more mass than in isolation,

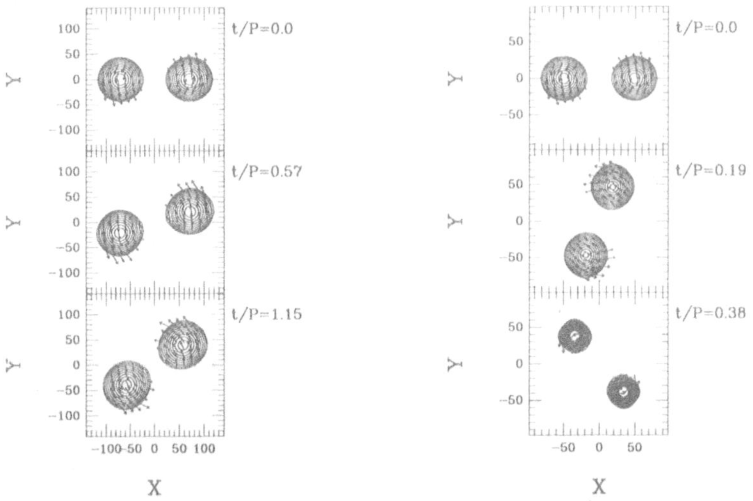

Figure 2. Snapshots of density contour lines and the velocity field in the equatorial plane for models Ba (left) and Da (right). Contour lines are drawn for $\rho_*/\rho_{*\,0} = 10^{-0.3j}$, where $\rho_{*\,0}$ denotes the maximum value at $t = 0$, for $j = 0, 1, 2 \cdots 10$. Vectors indicate the local velocity field. Time is shown in units of orbital period P.

– all stars that are stable in isolation will also stable in binaries.

Numerical computations were carried out on VX/4R in data processing center of National Astronomical Observatory of Japan.

References

1. J. R. Wilson and G. J. Mathews, Phys. Rev. Lett. **75**, 4161 (1995); J. R. Wilson, G. J. Mathews and P. Marronetti, Phys. Rev. D **54**, 1317 (1996); gr-qc/9710140 (WMM).
2. T. W. Baumgarte, S. L. Shapiro, G. B. Cook, M. A. Scheel, and S. A. Teukolsky, Phys. Rev. D, in press (1998) (gr-qc/9709026)
3. D. Lai, F. A. Rasio and S. L. Shapiro, Astrophys. J. Suppl. **88**, 205 (1993).
4. D. Lai, Phys. Rev. Lett. **76**, 4878 (1996).
5. J. C. Lombardi, F. A. Rasio and S. L. Shapiro, Phys. Rev. D **56**, 3416 (1997).
6. A. G. Wiseman, Phys. Rev. Lett. **79**, 1189 (1997).
7. P. R. Brady and S. A. hughes, Phys. Rev. Lett. **79**, 1186 (1997).
8. E. E. Flanagan, gr-qc/9706045.
9. K. S. Thorne, gr-qc/9706057.
10. T. W. Baumgarte, S. L. Shapiro, G. B. Cook, M. A. Scheel, and S. A. Teukolsky, Phys. Rev. Lett. **79**, 1182 (1997);
11. T. W. Baumgarte, S. L. Shapiro, G. B. Cook, M. A. Scheel, and S. A. Teukolsky, Phys. Rev. D, in press (1998) (gr-qc/9705023).
12. M. Shibata, T. W. Baumgarte, and S. L. Shapiro, submitted to Phys. Rev. D.
13. M. Shibata, Phys. Rev. D **55**, 6019 (1997); Prog. Theor. Phys. **96**, 317 (1996).

SUB-CHANDRASEKHAR MASS MODELS FOR TYPE IA SUPERNOVAE: SINGLE AND MULTIPOINT IGNITION

DOMINGO GARCíA-SENZ AND EDUARDO BRAVO
Dpt. de Física i Enginyeria Nuclear (UPC), and Institut d'Estudis, Espacials de Catalunya. Barcelona. Spain

1. What is meant by Sub-Chandrasekhar mass models

The name of Sub-Chandrasekhar mass models for Type Ia Supernovae has been recently coined to design those white dwarfs with masses well below $1.4M_{\odot}$ which through off-center explosive helium ignition and further induced carbon detonation, could become a bright Supernova belonging to the remarkably homogeneous class known as Type Ia Supernovae. The scenario usually invoked consists of a massively moderate carbon-oxygen white dwarf accreting matter from a companion. As a consequence of the accretion process a thick layer of helium of $0.15 - 0.25M_{\odot}$ settles and later ignites on top of a $0.6 - 0.9M_{\odot}$ C-O white dwarf. A hydrodynamical stage begins characterized by: 1) formation of a steady Chapman-Jouguet detonation, 2) transformation of the He layer in Fe-peak elements moving faster 10000 km s^{-1}, 3) induced ignition of the C-O core, and formation of a steady carbon detonation wave, 4) incineration of the core and formation of more Fe-peak and intermediate mass (Si-Ti) elements, 5) destruction of the star in the hydrodynamical time, giving rise to the optical display typical of Type Ia supernovae explosions.

Recent studies about this sort of models have been carried out by Woosley&Weaver (1994) (in 1D), and by Livne&Arnett (1995) (in 2D). Even though the process leading to the initial seeding of the ignition is insofar rather unknown it is highly probable for the ignition to begin in several hot spots spreaded near the edge of the C+O core. Thus the problem should be better handled in 3D for which the two aforementioned studies represent limiting cases. Previous attempts to modelize the explosion in 3D managed to reproduce well the propagation of the He detonation but failed to trigger a steady detonation in the underlying C-O core (Benz 1997, García-Senz, Bravo&Woosley 1997).

S. M. Miyama et al. (eds.), Numerical Astrophysics, 281–282.

2. Single point versus multipoint helium ignition

A SPH code has been used to carry out the simulations shown below. Our initial model was Model 3 of Woosley&Weaver (1994) consisting of a $0.8M_{\odot}$ C+O white dwarf plus a thick layer of $0.22M_{\odot}$ of He. Physics includes an accurate EOS with Coulomb corrections and electron-positron pairs, as well as a small network of 9 isotopes ranging from protons to nickel. Resolution in the low density helium layer was increased by putting less massive particles in that zone. A very important feature is the use of anisotropic interpolation kernels, wich was necessary in order to represent accurately the precursor shock waves associated to detonations. Table 1 sums up the main features of the three evolved models at the final time, once the nuclear reactions were quenched by the expansion. Models A, B and C refer to an ignition starting from a single point, from two simultaneous ignited points at opposite poles of the core and an asynchronized ignition starting at five points respectively.

TABLE 1. Models at final time.

Model	number of particles	M^f_{He} ($\odot$)	$M^f_{\mathrm{Si+Ni}}$ ($\odot$)	KE_{∞} (erg)
A	77712	0.045	0.819	1.40×10^{51}
B	35880	0.054	0.45	7.95×10^{50}
C	77712	0.052	0.905	1.51×10^{51}

As it can be seen the mass of ^{56}Ni synthezised and the kinetic energy are enough to power the light curve and to account for the energetics of a typical Type Ia Supernova. The lower values achieved by Model B were probably due to a lack of resolution, as a consequence many pockets of fuel remained unburned and the explosion was less energetic. Models A and C showed a healthy explosion. From the figures of Table 1 it becomes apparent that a multipoint asynchronized ignition is an optimal starting situation to obtain strong explosions. The complete history of the explosion for these three models is shown in the CD-ROM accompanying this proceedings.

References

Woosley, S.E. and Weaver, T.A. (1994), *Astrophysical Journal*, **Vol. 423**, pp. 371-379

Livne, E. and Arnett, D. (1995), *Astrophysical Journal*, **Vol. 452**, pp. 62-74

Benz, W. (1997), *Thermonuclear Supernovae*, pp. 457-474. Kluwer Academic Publishers, Dordrecht.

García-Senz, D., Bravo, E. and Woosley, S.E. (1997), *Thermonuclear Supernovae*, pp. 389-403. Kluwer Academic Publishers, Dordrecht.

NUCLEOSYNTHESIS IN TYPE II SUPERNOVAE AND ABUNDANCE TRENDS OF IRON-PEAK ELEMENTS IN METAL-POOR STARS

TAKAYOSHI NAKAMURA, H. UMEDA AND K. NOMOTO
Department of Astronomy, School of Sciences, Univ. of Tokyo

AND

FRIEDRICH-KARL THIELEMANN
Dept. für Physik und Astronomie, Univ. Basel, Switzerland

Recent high-resolution abundance surveys show interesting trends of [Cr/Fe], [Mn/Fe], and [Co/Fe] with [Fe/H]. Both [Cr/Fe] and [Mn/Fe] decrease as metallicity declines from [Fe/H] = −2.4 to −4.0, while [Co/Fe] increases (McWilliam et al., 1995; Ryan et al., 1996). We calculated nucleosynthesis in Type II supernovae (SNe II) for various parameters (explosion energy, mass cut, progenitor mass, and neutron excess) in order to explain these trends.

Hydrodynamical simulations of SNe II were carried out with a PPM code, solving a small nuclear reaction network. The progenitor models were taken from Nomoto & Hashimoto (1988). We exploded the star by depositing thermal energy below the mass that divides the ejecta and the collapsing core. Then, we solved a larger nuclear reaction network which contained 211 isotopes (Hix & Thielemann, 1996).

Figure 1 shows the isotopic composition of the ejecta of a typical SN II ($20M_\odot$). ^{52}Fe and ^{55}Co, which decay into ^{52}Cr and ^{55}Mn, respectively, are mostly synthesized in the outer region of incomplete Si burning ($1.66M_\odot \lesssim M_r \lesssim 1.74M_\odot$). ^{59}Cu, which decays into ^{59}Co, is produced in the inner region of the alpha-rich freeze-out ($M_r \lesssim 1.66M_\odot$). This fact suggests that the ratios [Cr/Fe], [Mn/Fe], and [Co/Fe] depend on the adopted mass cut. For a larger mass cut, the ratios of [Cr/Fe] and [Mn/Fe] are larger, while that of [Co/Fe] is smaller. On the contrary, these ratios show opposite trends for a smaller mass cut.

We applied the results of the nucleosynthesis calculation to two cases of material mixing in the early stages of galactic evolution: uniform mixing and no mixing. In Figure 2, the three curves for mixing and the three symbols

S. M. Miyama et al. (eds.), Numerical Astrophysics, 283–284.

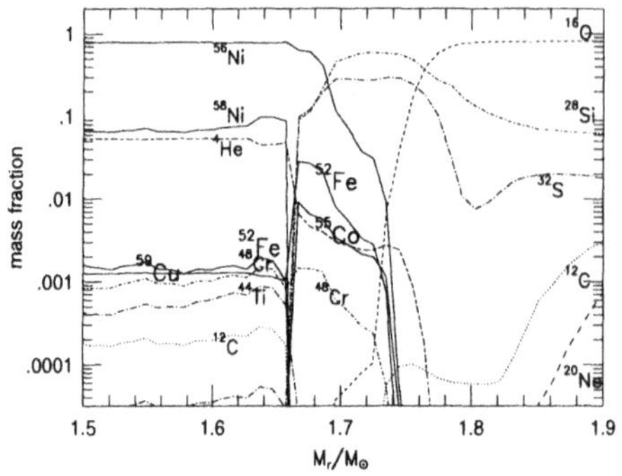

Figure 1. The isotopic composiotion.

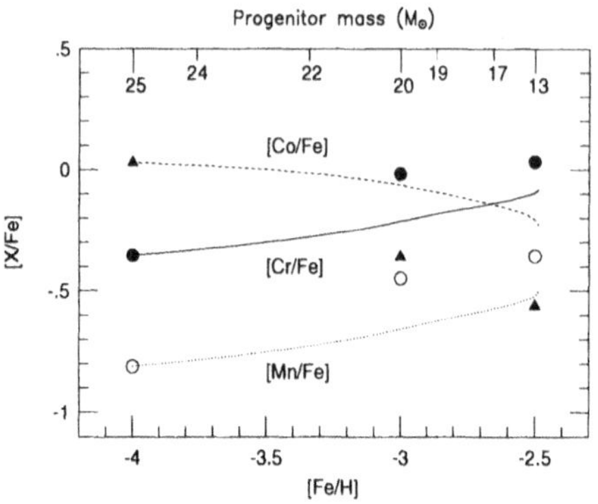

Figrure 2. The iron peak ratios.

for no mixing show changes in the iron-peak abundance ratios as functions of [Fe/H]. The filled circles stand for [Cr/Fe] in the individual supernova, the open circles for [Mn/Fe], and the filled triangles for [Co/Fe]. For each stellar model (25, 20, and 13 $M_\odot$), we took into account the observational relation between the ejected ^{56}Ni mass and the stellar mass, i.e., a larger ^{56}Ni yield for the $25 M_\odot$ star than for less massive stars (Iwamoto et al., 1998).

In uniform mixing, [Fe/H] declined monotonically towards the past. Lower [Fe/H] corresponded to more massive stars, because more massive stars have a shorter lifetime. The mass-[Fe/H] relation was obtained from a chemical evolution model for the galactic halo (Kobayashi, 1998). The Salpeter IMF of $\phi(M) \propto M^{-2.35}$ was adopted in this figure. The "unmixed" case showed similar trends but steeper slopes. If metal-poor stars were produced in a chemically inhomogeneous environment, i.e., stars were contaminated only by one or few supernovae and had the same heavy elemental abundance as the ejecta, a larger contrast would be expected in the abundance ratio.

Both models can explain the observed trends. The decline of the mixed model seems slightly shallow when compared with the observation, while the unmixed model can produce a steeper decline.

References

Hix, W. R. & Thielemann, F.-K. 1996, ApJ, **460**, 869

Iwamoto, K., Nakamura, T., Nomoto, K., Mazzali, P. A., Garnavich, P., Kirshner, R., Jha, S., Balam, D. 1998, ApJ, submitted

Kobayashi, C. 1998, private communication

McWilliam, A., Preston, G.W., Sneden, C., & Searle, L. 1995, AJ, **109**, 2757

Nakamura, T., Umeda, H., Nomoto, K., Thielemann, F.-K. & Burros, A. 1998, ApJ, submitted

Nomoto, K. & Hashimoto, M. 1988, Phys. Rep., **256**, 173

Ryan, S.G., Norris, J.E., & Beers, T.C. 1996, ApJ, **471**, 254

EXPLOSION OF A NEUTRON STAR BELOW THE MINIMUM MASS WITH IMPLICIT HYDRODYNAMICS

K. SUMIYOSHI
The Institute of Physical and Chemical Research (RIKEN), Hirosawa, Wako, Saitama 351-01, Japan

S. YAMADA
Department of Physics, The University of Tokyo, Hongo, Bunkyo, Tokyo 113, Japan

H. SUZUKI
High Energy Accelerator Research Organization (KEK), Oho, Tsukuba, Ibaraki 305, Japan

AND

W. HILLEBRANDT
Max-Planck-Institut für Astrophysik, Karl-Schwarzschild-Str. 1, 85740 Garching, Germany

Being a possible site for the origin of r-process elements, the explosion of a neutron star is extremely interesting. One of the possible origins of such an explosion can be the dynamical instability of a neutron star near the minimum mass (Colpi *et al.*, 1993). If the mass of a neutron star decreases below the minimum stable mass, it becomes dynamically unstable against adiabatic perturbations, which may lead to an explosion even if the perturbations do not conserve entropy. We numerically investigated the dynamical instability of such a neutron star, the thermal history of the dense matter, and its consequence for nucleosynthesis (Sumiyoshi *et al.*, 1998).

Numerical simulations of the evolution of neutron stars below the minimum mass were performed by solving the hydrodynamics and nuclear processes. The equations of general relativistic hydrodynamics in spherical symmetry were solved by an implicit method with Lagrange meshes adopting the code developed by Yamada (1997). Nuclear processes including β-decay, ν emission, neutron capture, and fission were coupled with hydrodynamics. As an initial model, we constructed a hydrostatic neutron star at the minimum mass and removed a small amount of mass from the

S. M. Miyama et al. (eds.), Numerical Astrophysics, 285–286.

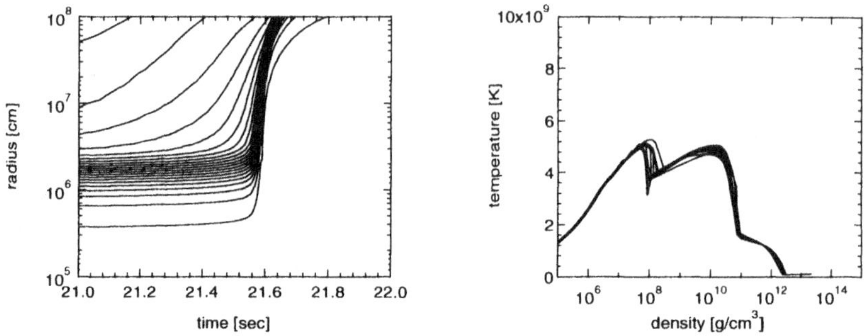

Figure 1. The trajectories of mass-elements as a function of time (left) and in the plane of temperature and density (right).

outer layer as an initial perturbation. We followed the evolution of such a neutron star for time scales much longer than the hydrodynamical time scale, taking advantage of our implicit hydrodynamics scheme in order to find out whether the neutron star expands secularly or explodes dynamically.

Even after the removal of a certain amount of mass, most parts of the neutron star remained almost hydrostatic, except that the surface layers slowly expanded as a result of β-decays having been suppressed in the central region. After a while, β-decays became fast enough to initiate dynamical instability. In the case of the removal of 22% of the total mass, the whole star blew off dynamically after 20 seconds (Fig. 1 left). The kinetic energy of the explosion was around 10^{49} erg, and electron anti-neutrinos were emitted in a burst. During the explosion, the matter was heated due to the β-decays and fissions of nuclei (Fig. 1 right). The peak temperature was found to be close to 5×10^9 K, which is high enough for nucleosynthesis but not high enough to reach nuclear statistical equilibrium. Therefore, nucleosynthesis proceeded as a kind of r-process reflecting the initial composition in the neutron star matter and producing heavy elements.

In summary, by adopting an implicit hydro code, we have shown that neutron stars with a mass somewhat below the minimum stable mass explode violently after a phase of quasi-static expansion caused by slow nuclear β-decays. Having a ν-burst, this phenomenon looks like a "mini- supernova" event, and the thermodynamic conditions of the expanding matter are favorable for r-process nucleosynthesis. Further quantitative studies are currently being carried out.

References

Colpi M., Shapiro S. L., Teukolsky S. A., 1993, ApJ 414, 717.
Sumiyoshi K., Yamada S., Suzuki H., Hillebrandt W., 1998, A&A 334, 159.
Yamada S., 1997, ApJ 475, 720.

MULTIGROUP FLUX LIMITED DIFFUSION SCHEME CODE FOR COOLING SIMULATIONS OF PROTONEUTRON STARS

HIDEYUKI SUZUKI
KEK: High Energy Accelerator Research Organization
Institute of Particle and Nuclear Studies
Oho 1-1, Tsukuba, Ibaraki, 305-0801, JAPAN

As is well known, supernova neutrinos are divided into two phases: the dynamical phase of the core and the quasi-static phase of the protoneutron star cooling. Here, we describe our numerical code for simulations of protoneutron star cooling. Starting from given initial models of protoneutron stars, we calculated the quasi-static evolutions of spherically symmetric protoneutron stars with neither additional mass accretion nor convection.

The prescription is similar to ordinary stellar evolution codes by the Henyey method. The general relativistic structure of the protoneutron star [radius $r(m)$, density $\rho(m)$ and metric function $\phi(m)$ as functions of enclosed baryon mass m] was calculated for every time step by solving the Oppenheimer-Volkoff equation and related equations. Equations of states for high density and finite temperature regime were also used. The evolution was driven by exchange of the entropy and the lepton number between the matter and neutrinos, that is, dS/dt and $dY_{\rm e}/dt$, due to various neutrino interactions.

Spatial transport of energy and lepton number was done by neutrinos. As for the neutrino transfer, we adopted the multigroup flux limited diffusion scheme (MGFLD). The neutrino number density per energy was calculated at every radial grid and at every time step, that is, $dn_\nu/d\omega(t, m, \omega)$. The energy dependence of neutrino transport coefficients could be handled with the multigroup scheme. A flux limiter was introduced in order to manage the transparent regime with the diffusion approximation. This was due to the fact that the diffusion approximation is unphysical in cases with very long (nearly infinite) mean free path: the diffusion flux exceeds the causal limit of cn_ν. In order to solve this inconvenience, various types of flux limiters were contrived and used. In addition, we included general relativistic effects such as time dilation and red shift of neutrino energy as

S. M. Miyama et al. (eds.), Numerical Astrophysics, 287–288.

well. The three neutrino species, ν_e, $\bar{\nu}_e$, and 'ν_μ', were treated separately, 'ν_μ' representing the average of ν_μ, $\bar{\nu}_\mu$, ν_τ, and $\bar{\nu}_\tau$.

The following neutrino interactions were included in our code.

$$\mathrm{p\ e^- \longleftrightarrow n\ \nu_e,\ n\ e^+ \longleftrightarrow p\ \bar{\nu}_e,\ A\ e^- \longleftrightarrow A'\ \nu_e}$$
$$\mathrm{e^-\ e^+ \longleftrightarrow \nu\ \bar{\nu},\ plasmon \longleftrightarrow \nu\ \bar{\nu}}$$
$$\mathrm{e^\pm\ \nu \longleftrightarrow e^\pm\ \nu,\ N\ \nu \longleftrightarrow N\ \nu,\ A\ \nu \longleftrightarrow A\ \nu}$$
$$\mathrm{nucleon\ bremsstrahlung:\ N\ N' \longleftrightarrow N\ N'\ \nu\ \bar{\nu}}$$

where ν represents all species of neutrinos, A is a representative heavy nucleus, and N is either a proton or a neutron. Most of the interaction rates were taken from Bruenn (1985) with some modifications. In addition to the standard neutrino interactions with electrons and nucleons, we included neutrino pair processes via nucleon bremsstrahlung. Neutrino interaction rates due to the nucleon bremsstrahlung process in formats suitable for our MGFLD code were obtained with approximations (Ishizuka: private communications, Suzuki, 1993) in which 1) nucleon interactions are described by one pion exchange model; 2) the transferred momentum is much smaller than the nucleon mass but much larger than the neutrino energy; and 3) only the global energy dependences are kept at integrations. Furthermore, at present, many body effects on neutrino opacity have not been included except for the multiple scattering suppression effects on the nucleon bremsstrahlung process pointed by Raffelt & Seckel (1991). It is important to note that neutrino emission will be suppressed if the third nucleon collides with the intermediate state of nucleon bremsstrahlung. We followed their method when managing this suppression.

We calculated the quasistatic time evolution of the protoneutron star described above with the so-called implicit scheme, which requires solving the linear equations with a coefficient matrix in the block tridiagonal form. Each block has $(NE+5) \times (NE+5)$ elements and the total matrix consists of $S \times S$ blocks, where N is the number of neutrino species (3 in the present case), E is the number of energy meshes, and S is the number of radial grids. We used the Feautrier method to solve these kinds of equations efficiently on the vector computer.

References

Bruenn, S.W. (1985) Stellar Core Collapse: Numerical Model and Infall Epoch, ApJS 58, pp. 771–841.

Raffelt, G. and Seckel, D. (1991) Multiple-Scattering Suppression of the Bremsstrahlung Emission of Neutrinos and Axions in Supernovae, PRL 67, pp. 2605–2608.

Suzuki, H. (1993) Supernova Neutrinos-Multigroup Simulations of Neutrinos from Protoneutron Star, in *Proc. of the International Symposium on Neutrino Astrophysics: Frontiers of Neutrino Astrophysics*, ed. Suzuki, A. and Nakamura, K., Universal Academy Press Inc., Tokyo, pp. 219–226.

DYNAMICAL EVOLUTION OF BOSON STARS

JAYASHREE BALAKRISHNA[1], G. L. COMER[1,2], ED SEIDEL[3], HISA-AKI SHINKAI[1,*] AND WAI-MO SUEN[1]

[1] *Dept. of Physics, Washington Univ., St. Louis, MO, USA*
[2] *Dept. of Sci. and Math., Saint Louis Univ., St. Louis, MO, USA*
[3] *Max-Planck-Ins. für Gravitationsphysik, Potsdam, Germany*
[*] *Email address:* shinkai@wurel.wustl.edu

Abstract.

We study the dynamical properties of self-gravitating complex scalar field configurations (boson stars) in general relativity (GR) and in scalar tensor (ST) gravity, for scalar fields both with and without self couplings. We compare the stability and scalar field emissions from this system in both gravity theories. We also simulate the formation of boson stars.

1. Introduction

Boson stars are self-gravitating objects made up of massive and complex scalar fields that are held together by the balance of forces between the attractive effects of gravity and the dispersive effects of the uncertainty principle (Kaup; Ruffini-Bonazzola). Many particle physics and cosmological models rely on the presence of scalar fields. These scalars have never been seen and are important dark matter candidates that could clump together by a Jeans instability to form boson stars. The scalar field exists in different states. The ground state consists of a radial field profile that has no nodes, whereas the n-th excited state has n nodes. In the ground state, the equal mass bosons making up the star form a zero-temperature condensate. Independently of their actual existence, boson stars do provide a good theoretical model for numerical studies of strong gravitational fields.

We report here on our results for the dynamical behavior of boson stars both in GR and ST. These results are an extension of (Seidel-Suen), in which the dynamical behavior of ground-state boson stars in GR is studied numerically.

S. M. Miyama et al. (eds.), Numerical Astrophysics, 289–290.

2. Boson Stars in General Relativity

We study dynamical boson stars with a self-coupling term in the ground and excited states. We examine the quasi-normal modes, the stability of the equilibrium configurations, and the scalar gravitational wave emissions during dynamical evolutions. The dynamical processes studied include (i) black hole formation of perturbed equilibrium configurations on the unstable branch; (ii) migration of perturbed equilibrium configurations from the unstable branch to the stable branch; (iii) transitions from an excited state to the ground state. We find that excited state boson star equilibrium configurations have branches similar to that of the ground state, but all branches are unstable. We also find that highly excited states spontaneously decay through a cascade of intermediate states similar to atomic transitions. The details are in (Balakrishna-Seidel-Suen).

3. Boson Stars in Brans-Dicke (BD) theory

We find that the dynamical behavior of boson stars in Brans-Dicke (BD) theory is quite similar to that of GR. The qualitative features of their stability are also extracted through use of catastrophe theory (Comer-Shinkai). The conclusions are consistent with dynamical numerical studies. These studies also show comparable scalar wave emissions to that of GR. This indicates that additional monopole effects from the scalar field of BD is not affecting much the dynamical evolutions.

We also simulate the formation of a stable boson star from a Gaussian scalar field packet with flat BD field initial data. This suggests that boson stars can be formed in BD in much the same way as in GR. The details are in (Balakrishna-Shinkai).

4. Boson Stars in Scalar-Tensor theory

Here we use a quadratic gravitational scalar coupling model (Damour-Nordtvedt), which is known to provide an attractor model to GR at late cosmological times. An application of catastrophe theory to this model indicates that boson stars formed during early cosmological eras are inherently unstable (Comer-Shinkai), which is very different from GR. The dynamical stability of this system is now under investigation.

References

Balakrishna, J., Seidel, E. and Suen, W-M. (1998) *gr-qc/9712064*, submitted to *Phys. Rev. D.*

Balakrishna, J. and Shinkai, H. (1998) gr-qc/9712065, submitted to *Phys. Rev. D.*

Brans, C.H. and Dicke, R.H. (1961) *Phys. Rev.* **124**, 925.

Comer, G.L. and Shinkai, H. (1998) *Class. Quant. Grav.* **15**, 669-688.

Damour, T. and Nordtvedt, K. (1993), *Phys. Rev. Lett.* **70**, 2217 ; *Phys. Rev.* **D48**, 3436.

Kaup, D. J. (1968), *Phys. Rev.***172**, 1331

Ruffini, R. and Bonazzola, S. (1969) *Phys. Rev.* **187**, 1767.

Seidel, E., Suen, W-M. (1990) *Phys. Rev.* **D 42**, 384.

STREAM-STREAM COLLISION HYDRODYNAMICS WITH RADIATIVE COOLING AROUND A MASSIVE BLACK HOLE

SUNGSOO S. KIM
Dept. of Physics, Korea Advanced Inst. of Sci. & Tech. Daejon 305-701, Korea

HYUNG MOK LEE
Dept. of Earth Sci., Pusan Natl. Univ. Pusan 609-735, Korea

AND

MYEONG-GU PARK
Dept. of Astron. & Meteorol., Kyungpook Natl. Univ. Taegu 702-701, Korea

1. Introduction

A star can be tidally disrupted around a massive black hole. The debris forms a precessing stream, which may collide with itself. The stream collision is a key process determining the subsequent evolution of the stellar debris. In this paper, we have studied such stream collision resulting from the encounter between a black hole and a normal star with TVD scheme.

The width ratios of incoming to outgoing streams are 5/3 for run 7, 12/3 for run 8, and 3/12 for run 9, respectively. The computational box consists of 101^3 cells. A simple treatment for radiative cooling has been adopted for both optically thick and thin regions. Readers are referred to Kim, Lee, & Park (1998) for detailed description for simulation parameters and radiative treatment.

2. Simulation Results

Figure 1 shows collision geometry of runs 8 and 9. The widths of the streams determine the collision geometry and therefore determine the efficiency of the collision in turing kinetic energy into thermal energy. However, the amount of radiation is less dependent on the stream parameters as shown in Figure 2, which is the evolution of thermal and radiative energy fractions.

S. M. Miyama et al. (eds.), Numerical Astrophysics, 291–292.

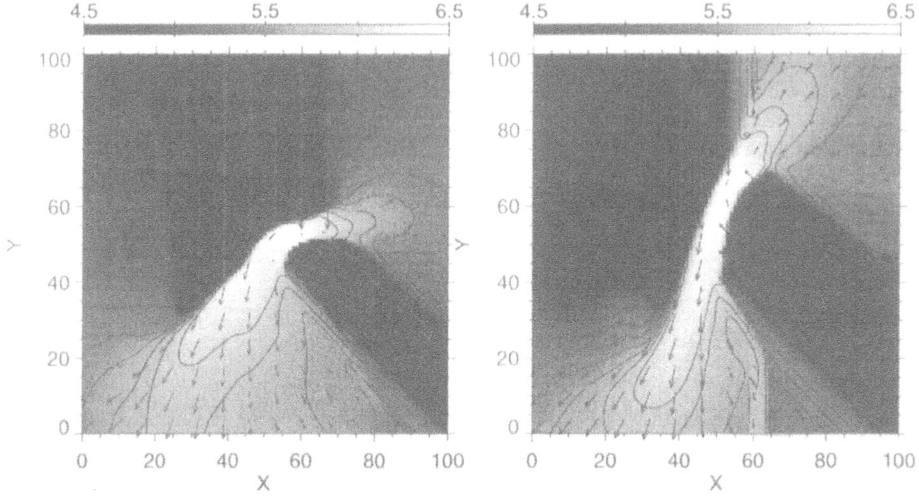

Figure 1. Density (contour), temperature (grey scale), and velocity vector (arrows) plot of z=50 plane for runs 8 (left) and 9 (right) at $t = 4 \times 10^4$ sec. Incoming stream is coming from top at $x = 50$ and outgoing stream 2 from lower-right corner. Grey-scale bar represents log T (K).

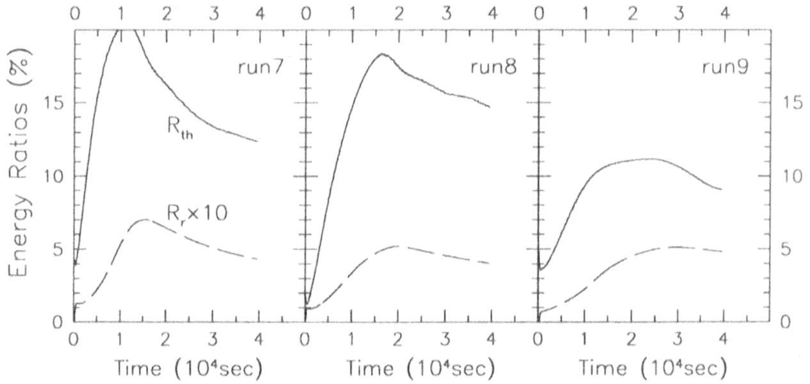

Figure 2. Evolution of the ratio of thermalized energy to input kinetic energy ($R_{\rm th}$; solid lines) and the ratio of cooling to input kinetic energy (R_r; dashed lines).

The collision increases the angular momentum of the streams by a factor of 2 to 3, and such increase, together with kinetic energy decrease, will reduce the circularization time scale of the gas debris. The cooling is dominated by diffusive photons from optically thick, almost adiabatically expanding, thermalized gas with effective temperature of $\sim 10^5$ K and bolometric luminosity of $\sim 10^{41}$ erg/sec.

References

Kim, S. S., Lee, H. M., & Park, M.-G. 1998, ApJ, submitted

SOLAR CONVECTION AND MHD

Å. NORDLUND
Theoretical Astrophysics Center, and Niels Bohr Institute for Astronomy, Physics, and Geophysics
Juliane Maries Vej 30, Dk-2100 Copenhagen Ø, Denmark

AND

R.F. STEIN
Dept. of Physics and Astronomy, Michigan State University, East Lansing, MI 48823, U.S.A.

abstract

We review the current status of efforts to model, in considerable detail, the entropy structure of the solar convection zone, in particular the layers near the surface of the Sun. Given an accurate, tabular equation of state, tables of continuous and binned spectral line opacities, suitable methods for handling radiative energy transport, and numerical methods that attempt to minimize viscous effects for given numerical resolution, while still capturing the shocks that develop, it is possible today to produce numerical models of the solar surface layers that have no arbitrary, free parameters other than those that specify the luminosity, surface acceleration of gravity, and the solar abundance. These models are reasonably converged at numerical resolutions that are affordable today, and they may be used as "test benches" for studying, e.g., the interaction of the turbulent solar convection with magnetic fields, and helioseismic oscillations. We also briefly review recent progress in understanding boundary driven magnetic dissipation and its consequences for the dissipation of magnetic energy in the solar corona.

1. Introduction

The Sun is an excellent "laboratory" for studying the ability of numerical models to accurately represent realistic (i.e. turbulent) astrophysical plasmas. Both direct observations of the solar surface (Berger & Title, 1996; Schrijver et al., 1997) and indirect observations of the solar interior through helioseismic methods (Christensen-Dalsgaard et al., 1996; Gough et al.,

S. M. Miyama et al. (eds.), Numerical Astrophysics, 293–301.

1996) are providing quantitative (sometimes exceedingly precise) information about the detailed structure of the Sun.

A number of current issues relate to the very surface layers of the Sun. These are characterized by intense (in places transsonic) convectively driven turbulent motions that significantly affect the shapes and strengths of photospheric spectral lines, generate the noise that excites the solar oscillations, and also interact strongly with the magnetic fields present at the solar surface. Small and large scale photospheric motions are able to exert work on the coronal plasma by the small and large scale horizontal motions of the footpoints of the coronal magnetic field.

Numerical simulations of stratified convection have proved very useful, both when attempting to understand these mechanisms qualitatively (Graham, 1977; Hossain & Mullan, 1991; Singh & Chan, 1993; Atroshchenko & Gadun, 1994; Porter & Woodward, 1994; Brummell et al., 1995), and when attempting to predict the average thermal structure of these layers quantitatively (Nordlund, 1982; Steffen et al., 1989; Stein & Nordlund, 1989; Nordlund & Dravins, 1990; Rast et al., 1993; Kim et al., 1996; Stein & Nordlund, 1998a).

These simulations have led to a shift in our paradigm for compressible convection, from that of a cascade of energy from large driving eddies to small dissipative eddies, to that of non-local driving by low entropy downdrafts produced on the intermediate scale of granulation by radiative cooling at the extremely thin surface thermal boundary layer, which drive both larger scale cellular flows and smaller scale turbulent motions (Nordlund, 1985; Spruit et al., 1990; Nordlund & Stein, 1997; Spruit, 1997).

In Section 2 we review current efforts to model solar and stellar surface layer convection, and in Section 3 we review some recent progress in the understanding of boundary driven dissipation in magnetically dominated plasmas. Color versions of the figures included here, and further illustrations and animations are available on the World Wide Web (Nordlund, 1998).

2. Convection

Solar convection is characterized by a range of scales, both in space and time. Fig. 1a illustrates the range of spatial scales involved qualitatively, while Fig. 2 shows the quantitative background; the mass density and gas pressure stratification cover about ten and fourteen orders of magnitude, respectively. The size ("outer scale") of convective motions scale roughly with the (density or pressure) scale height, that is again roughly proportional to the ratio of pressure to density. From the bottom of the convection zone to the photosphere, this ratio varies with about two and a half orders

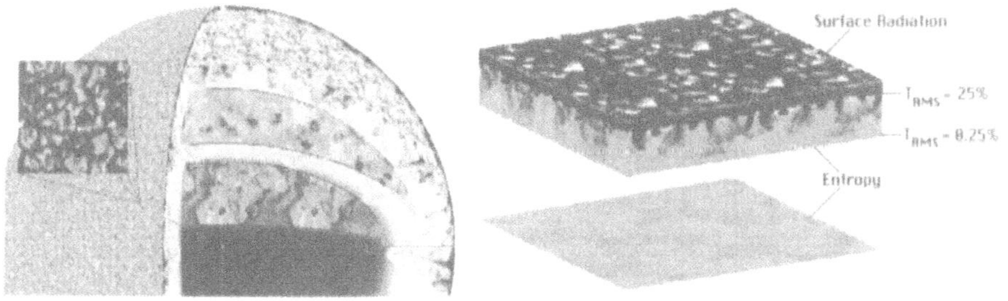

Figure 1. a) The huge range of convective scales of motion in the solar convection zone is illustrated in this "artists view" of the Sun. Images of entropy from detailed simulations of the solar surface layers and from "toy models" of the bulk of the convection zone have been appropriately scaled, and warped onto the spherical and annular surfaces (see also Nordlund, 1998). b) A perspective view of the solar surface layers, with surface radiation intensity shown on the top surface, and images of entropy shown on the sides and on the "drop-down" of the bottom surface.

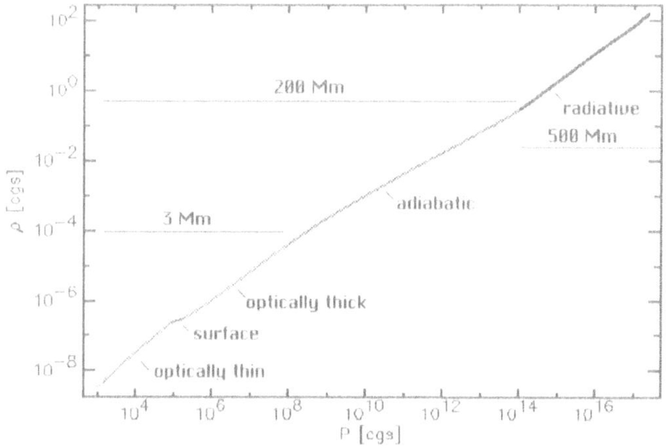

Figure 2. The stratification of mass density and gas pressure in the Sun, from the photospheric temperature minimum to the center of the Sun. The thin 'surface' layer is crucial for determining the depth at which the stratification changes from nearly adiabatic convection to radiative diffusion.

of magnitude. The time scales span an even larger range, since the velocities are over an order of magnitude smaller near the bottom of the convection zone than at the top.

Fortunately, in order to accurately compute the structure of stars similar to the Sun, one only needs to model a relatively thin surface layer in great detail; the rest of the structure can be computed with one dimensional stellar envelope codes that solve the equations for simultaneous hydrostatic and energy equilibrium (Rosenthal et al., 1998). In convectively unstable layers, such codes typically employ some variation of the "mixing length

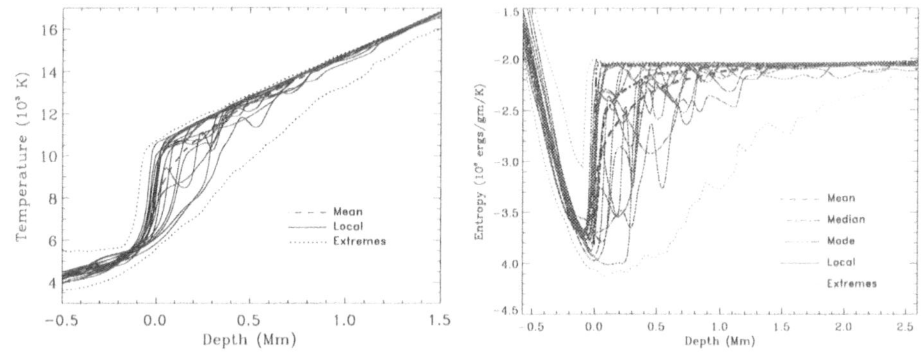

Figure 3. Samples of the run of temperature (a) and entropy (b) with depth, from a numerical simulation of the solar surface layers (Stein & Nordlund, 1998a). Also shown are the mean (dashed), mode (dashed-dotted) and extrema (dotted).

recipe" (Böhm-Vitense, 1958; Cox & Giuli, 1968) to describe convective energy transport. The resulting stratification is very close to adiabatic in the bulk of the convection zone, and hence the details of the convection treatment are of little importance there (Gough & Weiss, 1976).

The thin surface layer where convection is of crucial importance is illustrated in Fig. 1b. Even though the depth of this layer is only about 3 Mm (as compared to the total depth of about 200 Mm of the solar convection zone), the layer still covers about five orders of magnitude in pressure, because of the low temperatures (and hence small scale heights) of these surface layers.

The low temperatures, and the presence of the optical surface of the Sun in this layer imply that, in order to produce quantitative predictions, it is necessary to use elaborate equations of state, and to include methods that can accurately represent radiative transfer of energy in the presence of a large number of spectral lines. Spectral lines block about 12% of the luminosity of the Sun, as compared to a case with only continuum opacity. This means that a model that does not account for the spectral line blocking will have to be about 3% cooler in the optical surface layers, in order to have the right solar luminosity. That would again result in an interior with a lower temperature at a given pressure (a lower entropy), and ultimately to a convection zone that is too deep (Christensen-Dalsgaard, 1997).

As indicated in Fig. 1b, the root mean square temperature fluctuations are quite large near the top of the layer; at the optical surface of the Sun, temperatures in a horizontal layer vary from about 4,000 K to about 10,000 K. This is further illustrated in Fig. 3, that shows the temperature and entropy as functions of depth, for a sample of horizontal points in a numerical model (Stein & Nordlund, 1998a).

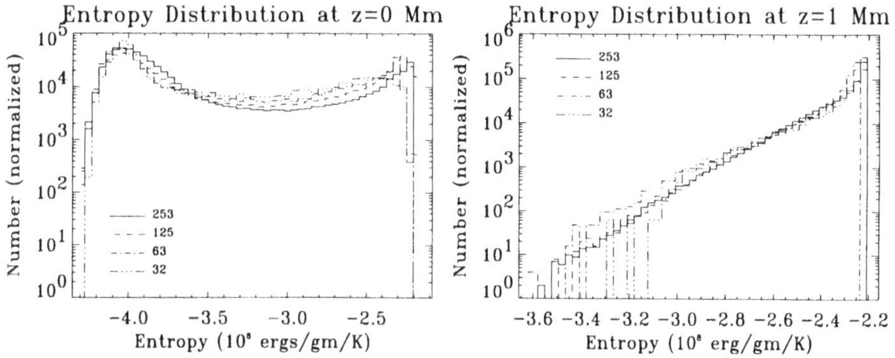

Figure 4. Histograms showing the distribution of entropy at the surface (a) and at a depth of 1 Mm (b), from the numerical simulations of Stein & Nordlund (1998a). The various curves illustrate the numerical convergence; the labels in the panels indicate the number of zones in the horizontal directions.

The statistical distribution of the entropy in the very surface layers is shown in Fig. 4a. It is somewhat bimodal, with a "hot" component at temperatures around 10,000 K and a "cold" component with temperatures around 5,000 K. The upper limit of the distribution is exceedingly sharp; this reflects the almost constant entropy of ascending material (Stein & Nordlund, 1989). At a depth of 1 Mm (less than the outer scale of convection cells but still several scale heights), there is still a sharp upper limit of the distribution at the same entropy, but the distribution towards lower entropy is now exponential, signalling that descending material is subject to a mixing process. Descending material is indeed mixed with overturning ascending material at a rate that is almost entirely determined by first principles; conservation of mass enforces overturning of ascending material, and the small scale heights imply that the rate is large, and thus dominates over turbulent mixing (Nordlund, 1975).

The turbulent and inhomogeneous surface layer is crucial for the solar p-mode oscillations. On the one hand it produces the main part of the noise that is responsible for the stochastic excitation of the modes (Goldreich & Keeley, 1977; Stein & Nordlund, 1991; Goldreich et al., 1994; Nordlund & Stein, 1998; Stein & Nordlund, 1998b), on the other hand it extends the acoustical cavity and hence lowers the frequency of the p-modes that extend into these layers (Rosenthal et al., 1998). In addition, the jump in average entropy across the turbulent surface convection layers determines the entropy of the bulk of the convection zone and hence ultimately, as mentioned above, the depth of the solar convection zone.

Stars over a wide range of parameters have surface layers that are qualitatively similar to the solar surface layers, and hence numerical models that can successfully predict the properties of the well observed solar oscillations may be used to predict the envelope structure and oscillation properties of other stars as well (Ludwig et al., 1997; Trampedach et al., 1997).

In summary, the requirements for accurate, quantitative modeling of stellar surface layers and predictions of convection zone depths are (Rosenthal et al., 1998):

- Continuum opacities good to the percent level.
- The inclusion accurate ionization and dissociation equilibria to an extent sufficient to account for all major electron sources and sink.
- Inclusion of the effects of spectral line opacities to an extent sufficient for predicting the emergent surface luminosity to within about a percent.
- Numerical resolution sufficient to utilize the above; in practice this means of the order of 200^3 mesh points, and a vertical resolution of about a tenth of the local scale height.

With these requirements fulfilled, it is possible to accurately predict the convection zone depth, to predict surface intensity statistics, to predict spectral line widths and shapes, to predict the excitation and damping of global oscillations, and to predict the influence of the surface layers on the frequency spectrum of the global oscillations, using as input parameters only the fundamental stellar parameters (effective temperature, surface acceleration of gravity, chemical abundances).

3. Magnetic Dissipation and Coronal Heating

Magnetically dominated astrophysical plasmas, such as the solar corona or the corona above a magnetized accretion disc, are likely to be continuously influenced by braiding motions on its boundaries. The solar photosphere, for example, is known to be threaded by magnetic flux concentrations that connect to the solar corona. Because the photosphere and the convection zone below are much denser than the solar corona, the photospheric flux concentrations are moved around more or less passively by the convective motions, which causes the coronal extension of the magnetic field lines to be continuously braided. A similar relation is likely to exist between accretion disk coronae and the much denser and turbulent layers near the equatorial plane of the disk (Matsumoto & Tajima, 1995; Brandenburg et al., 1995; Stone et al., 1996).

In such situations, the magnetically dominated ("low beta") part of the plasma is forced to dissipate the work done by the boundary ("high beta") plasma, since otherwise the angles of inclination (and hence the work) would increase indefinitely (Parker, 1972, 1983, 1987, 1988).

Analytical work and numerical experiments have clarified the mechanisms through which balance between the boundary work and the dissipa-

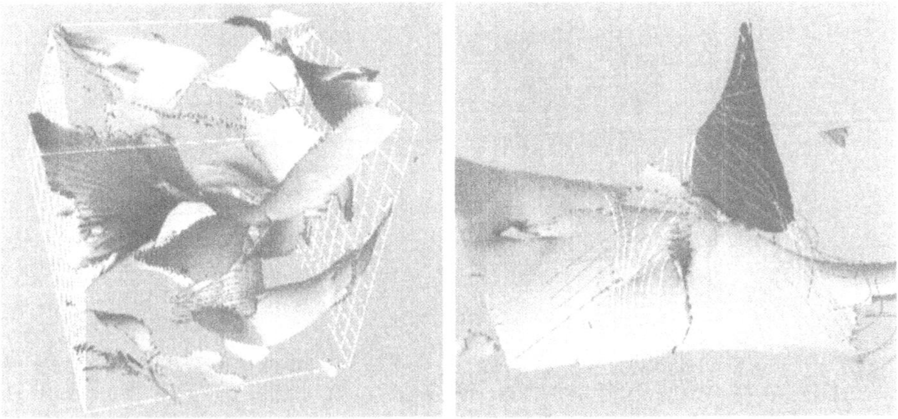

Figure 5. Isosurface visualizations of 3-D current sheets, from an experiment with boundary driven magnetic dissipation (Galsgaard & Nordlund, 1996). The right hand side panel shows a blow-up of the small box outlined near the bottom in the left hand side panel. The driving boundaries are to the left and right. The numerical resolution is 136^3. For color renderings, see Nordlund (1998).

tion in the interior is established (van Ballegooijen, 1986; Mikić et al., 1989; Strauss, 1991; Gómez & Fontán, 1992; Heyvaerts & Priest, 1992; Biskamp, 1993; Longcope & Sudan, 1994; Galsgaard & Nordlund, 1996, 1997; Ricca & Berger, 1996).

Figure 5 (from Galsgaard & Nordlund, 1996) shows an example of the complex, hierarchical current sheet structure that develops in boundary driven low beta plasmas, even when the driving only consists of large scale, incompressible shearing motions.

Such a hierarchical set of current sheets has dissipation properties that differ dramatically from those of the assumed, monolithic current sheets that have been at the center of attention for decades, in the quest for understanding "fast magnetic reconnection" (Petschek, 1964; Sonnerup, 1988; Priest & Forbes, 1992). Indeed, as the numerical resolution is increased in these experiments, the amount of resolved structure increases correspondingly, but the average level of dissipation stays approximately constant (Galsgaard & Nordlund, 1996).

This is a remarkable result, that may be taken as an indication that chaotic MHD phenomena obey scaling relations analogous to Kolmogorov scaling in turbulence (Kolmogorov, 1941; Hunt & Williams, 1991). A scaling relation between the average level of magnetic dissipation and the parameters that characterize the boundary driving is shown in Fig. 6 (from Galsgaard & Nordlund, 1996). The dissipation in a number of experiments, with varying driving amplitudes and aspect ratios, may all be fitted with a simple scaling relation. The scaling relation does not depend on the value

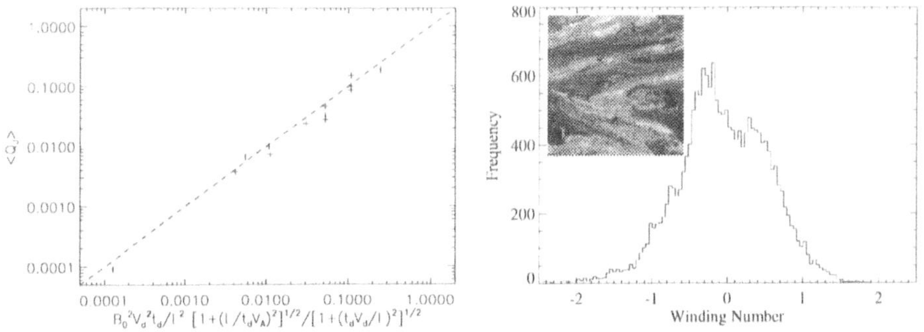

Figure 6. The left hand side panel shows the scaling law for magnetic dissipation derived by Galsgaard & Nordlund (1996). The mean magnetic dissipation is a function of the magnetic field strength B_0, the size of the box L, the characteristic velocity V_d and duration t_d of shearing events at the boundary, and the Alfvén speed V_A. The right hand side panel shows a histogram of the "local winding number" (Nordlund & Galsgaard, 1997), defined as the number of times two neighboring field lines wind around each other, as they pass from one boundary to another. The inset shows an image rendering of the winding number.

of the resistivity, just as the level of viscous dissipation in Kolmogorov turbulence does not depend on the value of the molecular viscosity.

This result is interesting also from a practical point of view, since it was obtained with relatively modest numerical resolutions (up to 136^3). We may surmise that, unless the geometry is such that a range of "outer" scales are involved, and/or extreme aspect ratios are required, one can expect to obtain reasonably realistic levels of dissipation in numerical MHD models that are affordable today, even on ordinary workstations.

4. Conclusions

Numerical models of the solar surface layers and the adjacent layers of the solar convection zone are now in many respects accurate enough for quantitative predictions, for example of the convection zone depth, and of the amount of excitation and damping of global oscillations. The models are "parameter-free," in the sense that the main physical parameters (effective temperature, surface acceleration of gravity, and chemical composition) are known from observations, and the main physical characteristics of the plasma (equation of state, continuous and spectral line opacities) are obtainable from standard packages. The remaining, numerical parameters (numerical resolution and viscosities, etc.) have been pushed into a regime where they only have a marginal effect, even quantitatively.

It is indeed fortunate that the asymptotic behavior of turbulent systems, both hydrodynamical and magnetohydrodynamical, is so benign that currently affordable numerical models are near, or already inside, the asymptotic regime where details of the numerical dissipation become unimportant.

Certainly, such models can be used with great benefit to increase our understanding of the hydrodynamic and magnetohydrodynamic mechanisms at work, in the surface layers of stars as well as in other astrophysical contexts. Apart from its intrinsic value, such an understanding is indeed also necessary when attempting to obtain similarly quantitative predictions in other circumstances; only if we understand the mechanisms at work can we assess if the models are providing fair numerical representations of the systems under study.

Acknowledgements

The work of Å.N was supported in part by the Danish Research Foundation, through its establishment of the Theoretical Astrophysics Center. RFS was supported by NASA grant NAG 5-4031 and NSF grant AST 9521785. Calculations were performed at the National Center for Supercomputer Applications, which is supported by the National Science Foundation, at Michigan State University and at UNI•C, Denmark.

References

Atroshchenko, I. N., Gadun, A. S. 1994, A&A, 291, 635
Berger, T. E., Title, A. M. 1996, ApJ, 463, 365
Biskamp, D. 1993, Nonlinear Magnetohydrodynamics, Cambridge Monograms on Plasma Physics
Böhm-Vitense, E. 1958, Zs. f. Ap., 46, 108
Brandenburg, A., Nordlund, Å., Stein, R. F., Torkelsson, U. 1995, ApJ, 446, 741
Brummell, N. H., Cattaneo, F., Toomre, J. 1995, Sci, 269, 1313
Christensen-Dalsgaard, J. 1997, in F. P. Pijpers, J.Christensen-Dalsgaard, C. S. Rosenthal (eds.), Solar Convection, Oscillations and their Relationship; SCORe'96, Kluwer Academic Press, Dordrecht, p. 3
Christensen-Dalsgaard, J., Däppen, W., et al. 1996, Science, 272, 1286
Cox, J., Giuli, R. 1968, Priciples of stellar structure I, Gordon and Breach, New York
Galsgaard, K., Nordlund, Å. 1996, Journal of Geophysical Research, 101(A6), 13445
Galsgaard, K., Nordlund, Å. 1997, Journal of Geophysical Research, 102, 231
Goldreich, P., Keeley, D. A. 1977, ApJ, 212, 243
Goldreich, P., Murray, N., Kumar, P. 1994, ApJ, 424, 466
Gómez, D., Fontán, C. 1992, ApJ, 394, 662
Gough, D. O., Kosovichev, A. G., Toomre, J., et al. 1996, Science, 272, 1296
Gough, D. O., Weiss, N. O. 1976, MNRAS, 176, 589
Graham, E. 1977, in Problems of Stellar Convection, Vol. 71 of *Lecture Notes in Physics*, Springer-Verlag, Berlin, p. 151
Heyvaerts, J., Priest, E. R. 1992, ApJ, 390, 297
Hossain, M., Mullan, D. J. 1991, ApJ, 380, 631
Hunt, J. C. R. Phillips, O. M., Williams, D. (eds.) 1991, Turbulence and stochatic processes: Kolmogorov's idaes 50 years on, Vol. 434, The Royal Society, University Press, Cambridge
Kim, Y.-C., Fox, P. A., Demarque, P., Sofia, S. 1996, ApJ, 461, 499
Kolmogorov, A. N. 1941, Dokl. Akad. Nauk. SSSR, 30, 301, Reprinted in Friedlander

S.K., Topper L., eds., 1961, Turbulence: Classic Papers on Statistical Theory, New York, Interscience
Longcope, D. W., Sudan, R. N. 1994, ApJ, 437, 491
Ludwig, H., Freytag, B., Steffen, M. 1997, in Solar Convection, Oscillations and their Relationship; SCORe'96, p. 59
Matsumoto, R., Tajima, T. 1995, ApJ, 445, 767
Mikić, Z., Schnack, D. D., van Hoven, G. 1989, ApJ, 338, 1148
Nordlund, Å. 1975, A&A, 50, 23
Nordlund, Å. 1982, A&A, 107, 1
Nordlund, Å. 1985, Solar Physics, 100, 209
Nordlund, Å. 1998, `http://www.astro.ku.dk/~aake/talks/NAP98`
Nordlund, Å., Dravins, D. 1990, A&A, 228, 155
Nordlund, Å., Galsgaard, K. 1997, in G. Simnett, C. E. Alissandrakis, L. Vlahos (eds.), Proc. 8th European Solar Physics Meeting, Vol. 489 of *Lecture Notes in Physics*, Springer, Heidelberg, p. 179
Nordlund, Å., Stein, R. F. 1997, in F. P. Pijpers, J.Christensen-Dalsgaard, C. S. Rosenthal (eds.), Solar Convection, Oscillations and their Relationship; SCORe'96, Kluwer Academic Press, Dordrecht, p. 79
Nordlund, Å., Stein, R. F. 1998, in F. L. Deubner, J. Christensen-Dalsgaard (eds.), Proc. IAU Symp. 185, Vol. 185 of *IAU Symp.*, Reidel, Dordrecht, (in press)
Parker, E. N. 1972, ApJ, 174, 499
Parker, E. N. 1983, ApJ, 264, 642
Parker, E. N. 1987, ApJ, 318, 876
Parker, E. N. 1988, ApJ, 330, 474
Petschek, H. E. 1964, in Physics of Solar Flares, NASA Spec. Publ. SP-50, 425
Porter, D. H., Woodward, P. R. 1994, ApJS, 93, 309
Priest, E. R., Forbes, T. G. 1992, J. Geophys. Res., 97, 16757
Rast, M. P., Nordlund, Å., Stein, R. F., Toomre, J. 1993, ApJ, 408, L53
Ricca, R. L., Berger, M. A. 1996, Physics Today, 49(12), 28
Rosenthal, C. S., Christensen-Dalsgaard, J., Nordlund, Å., Stein, R. F., Trampedach, R. 1998, A&A, (in press)
Schrijver, C. J., Hagenaar, H. J., Title, A. M. 1997, ApJ, 475, 328
Singh, H. P., Chan, K. L. 1993, A&A, 279, 107
Sonnerup, B. U. 'O. 1988, Comp. Phys. Commun., 49, 142
Spruit, H. C. 1997, Mem. Soc. Astron. It.
Spruit, H. C., Nordlund, Å., Title, A. 1990, ARA&A, 28, 263
Steffen, M., Ludwig, H.-G., Krüss, A. 1989, A&A, 213, 371
Stein, R. F., Nordlund, Å. 1989, ApJ, 342, L95
Stein, R. F., Nordlund, Å. 1991, in D. Gough, J. Toomre (eds.), Challenges to Theories of the Structure of Moderate Mass Stars, Vol. 388 of *Lecture Notes in Physics*, Springer, Heidelberg, p. 195
Stein, R. F., Nordlund, Å. 1998a, ApJ, 499, 914
Stein, R. F., Nordlund, Å. 1998b, ApJ, (in preparation)
Stone, J. M., Hawley, J. F., Gammie, C. F., Balbus, S. A. 1996, ApJ, 463, 656
Strauss, H. R. 1991, ApJ, 381, 508
Trampedach, R., Christensen-Dalsgaard, J., Nordlund, Å., Stein, R. F. 1997, in F. P. Pijpers, J.Christensen-Dalsgaard, C. S. Rosenthal (eds.), Solar Convection, Oscillations and their Relationship; SCORe'96, Kluwer Academic Press, Dordrecht, p. 73
van Ballegooijen, A. A. 1986, ApJ, 311, 1001

MHD SIMULATIONS OF MAGNETIC RECONNECTION AND SOLAR FLARES

K. SHIBATA AND T. YOKOYAMA
National Astronomical Observatory
2-21-1 Osawa, Mitaka, Tokyo 181-8588, Japan
shibata@solar.mtk.nao.ac.jp

Abstract: Recent development of numerical simulations of magnetic reconnection and solar flares is reviewed with emphasis on the numerical modeling of various evidence of reconnection in flares that are discovered by Yohkoh X-ray observations.

1. Introduction

Solar flares are explosions occurring in the solar atmosphere, releasing a huge amount of magnetic energy $10^{29} - 10^{32}$ erg in a relatively short time scale $10^2 - 10^3$ sec. Flares have spatial scales $10^4 - 10^5$ km, and emit almost all electromagnetic spectrum from radio waves to gamma rays. The energy release mechanism of solar flares has been one of the most challenging subjects in solar physics and plasma astrophysics in the latter half of this century.

What is the basic difficulty to understand solar flares? This is easily understood if we calculate the magnetic diffusion time (or Ohmic dissipation time) of the solar coronal magnetic loop with a spatial scale L and temperature T on the basis of the classical Spitzer conductivity;

$$t_D = \frac{L^2}{\eta_{\rm spitzer}} \simeq 10^{14}\left(\frac{L}{10^9{\rm cm}}\right)^2\left(\frac{T}{10^6{\rm K}}\right)^{3/2} \quad {\rm sec.}$$

This is nearly 3 million years (!), much longer than the observed time scale of solar flares. Hence the simple current dissipation mechanism cannot work for solar flares. Another way to see the difficulty is to calculate magnetic

S. M. Miyama et al. (eds.), Numerical Astrophysics, 303–310.

Reynolds number R_m;

$$R_m = \frac{t_D}{t_A} = \frac{V_A L}{\eta_{\rm spitzer}} \simeq 10^{13} \gg 1,$$

where t_A is the Alfven time

$$t_A = \frac{L}{V_A} \simeq 10\left(\frac{L}{10^9 {\rm cm}}\right)\left(\frac{B}{10{\rm G}}\right)^{-1}\left(\frac{n}{10^9 {\rm cm}^{-3}}\right)^{1/2} \quad {\rm sec.}$$

In general, $R_m \gg 1$ in astrophysical plasmas so that it is very difficult to dissipate magnetic energy in a short time scale ($\sim t_A$). Nevertheless, explosive energy release similar to solar flares often occur in astrophysical objects in a time scale comparable to Alfven time ($10 - 100 t_A$).

In order to overcome this difficulty, the theory of magnetic reconnection connecting microscopic physics (resistivity) and macroscopic dynamics (flow) has been proposed (Giovanelli 1947) and developed by Sweet (1958), Parker (1957), Petschek (1964) and others. Magnetic reconnection mechanism has been applied also to geomagnetic aurora-substorms by Dungey (1961) and disruption of magnetically confined fusion plasma (e.g., see Biskamp 1993). From late-70's, new era of reconnection study has begun; i.e., numerical simulations of reconnection have started, which greatly contributed to the development of not only reconnection theory but also realistic modeling of solar flares, geomagnetic substorms, and disruption of confined fusion plasma. In spite of the progress of both analytical and numerical studies (e.g., Forbes and Priest 1987, Biskamp 1993, Tajima and Shibata 1997), the magnetic reconnection theory has still not yet been established. The enormous gap between t_D and t_A is still a big obstacle for numerical simulations, laboratory experiment, and analytic theories.

In this article, we review basic reconnection theory, recent development of numerical simulations of reconnection, and new observations and numerical modeling of solar flares based on the reconnection model.

2. Basic Reconnection Theory

Sweet (1958) and Parker (1957) developed a simple theory of magnetic reconnection, taking into account of the effect of *flow*. This is now called *Sweet-Parker model.* In this model, the plasma inflow with a speed of V_i drives the reconnection in the current sheet with a length L. Since the magnetic energy is converted to plasma outflow, the speed of outflow (*reconnection jet*) becomes comparable to Alfven speed V_A just outside the current sheet. The time scale of reconnection (and energy release) is $t_{SP} \simeq L/V_i \simeq {R_m}^{1/2} t_A \sim 10^6 t_A$ when applied to solar flares, which is much faster than simple magnetic diffusion ($t_D \simeq R_m t_A$) but still much slower than the observed time scale of solar flares ($10 - 100t_A$).

Petschek (1964) proposed a remarkable idea that if the effect of *slow mode MHD shock (or wave)* is considered, it is possible to construct *fast reconnection* model. Namely, the slow shock enables the conversion of magnetic energy to plasma kinetic and internal energies, and the magnetic reconnection occurs only in a very small localized region (called *diffusion region*) where the *Sweet-Parker model* holds locally. The speed of outflow (reconnection jet) is again V_A, while the inflow speed becomes $V_i/V_A \sim (\pi/8)/\ln R_m \sim 0.01 - 0.1$, nearly independent of R_m. The time scale of reconnection (and energy release) is $10 - 100t_A$ which is exactly the observed time scale of solar flares. Although this model (called *Petschek model*) is very successful and has been considered to be the most promising model for solar flares, there are still some basic questions: (1) The Petschek model is not based on an exact solution of the resistive MHD equation, but on an approximate solution. Is the Petschek model a real solution of the resistive MHD equation? (2) If the origin of resistivity is Coulomb collisions (i.e., Spitzer resistivity), the length of diffusion region must be less than 1 cm to explain solar flares! (Note that the typical flare size is 10^9 cm.) Is it really true that such a small single diffusion region controls the entire flare process? In order to answer the first question, people had to await the development of computers.

3. Numerical Simulations of Magnetic Reconnection

Ugai and Tsuda (1977) first numerically solved the time dependent resistive MHD equation in 2D space. They assumed a *locally enhanced resistivity* (which is spatially and temporally fixed in a current sheet), and followed the subsequent evolution of magnetic field and plasma dynamics caused by the locally enhanced resistivity. They found that the solution is very similar to the Petschek model, i.e., they found appearance of reconnection jet with Alfven speed, a pair of slow shocks, and logarithmic dependence of reconnection rate on R_m.

Sato and Hayashi (1979) studied similar problem with different assumption; they assumed strong inflow from the side boundaries which drives the reconnection. They also assumed non-fixed resistivity that depends on current density such that $\eta = \eta_0 (j - j_c)^2$ (for $j > j_c$), and $\eta = 0$ (for $j < j_c$), mimicing the *anomalous resistivity* caused by plasma turbulence. They again found that the resulting solution is similar to the Petschek model, and proposed that *external driving* is essential to induce (Petschek type) fast reconnection. At this stage, it is clear that the Petschek model is a real solution of the resistive MHD equation under some special condition.

However, a fundamental question is remained; *what is the condition of the (Petschek type) fast reconnection?* Is it *external driving* or *locally enhanced resistivity such as anomalous resistivity*?

There were some controversies about this question. Biskamp (1986) studied the same problem but assuming *uniform resistivity*, and found that the external driving does *not* lead to the Petschek type fast reconnection. Instead, he found only the Sweet-Parker solution in his numerical simulations. On the basis of these simulation results, he criticized the Petschek model (Biskamp 1993). Priest and Forbes (1992), on the other hand, constructed a unified reconnection theory using analytical but approximate method, and demonstrated that the external boundary condition for inflow is essential for determining whether reconnection is fast or slow.

However, many self-consistent resistive MHD numerical simulations (Ugai 1986, Scholer 1989, Yan et al. 1992, Yokoyama and Shibata 1994) have shown that the fast reconnection (Petschek type reconnection) is realized only when the *locally enhanced resistivity* such as anomalous resistivity is assumed. Namely, if *uniform resistivity* is assumed, only the slow reconnection (Sweet-Parker reconnection) is realized irrespective of any external driving at the inflow boundary condition. [1]

This notion is important especially when the reconnection model is applied to actual solar flares and magnetospheric substorms using numerical simulations. If one assumes uniform resistivity, one would not find (global) slow shocks inherent to Petschek type reconnection. If such simulation results are applied to flares and substorms, one might predict that no slow shocks would be found in flares and substorms. But actually slow shocks have been found in the geotail current sheet during substorms (Saito et al. 1995).

It should be also mentioned that *ideal MHD numerical simulations* often lead to violent fast reconnection due to *numerical resistivity*. For example, according to our own experience, the "ideal MHD simulation" of reconnection using modified Lax-Wendroff scheme lead to (nonsteady) Petschek type reconnection. In this scheme, the numerical resistivity is spatially nonuniform and is similar to anomalous resistivity. It is interesting to note that the ideal MHD case (or the least resistive case in an exact sense) leads to the most violent magnetic reconnection (Yokoyama and Shibata 1994). Of course, the type of numerical resistivity strongly depends on the numerical scheme. Karpen et al. (1998) numerically simulated solar flares with FCT

[1] Of course, these are results of numerical simulations at the present stage, whose numerical magnetic Reynolds number is at most $10^3 - 10^4$ in a uniform grid. Biskamp (1998, private communication) conjectures that even a uniform resistivity case may generate effective turbulent anomalous resistivity leading to fast reconnection. However, this is only a conjecture at the present stage. No one knows at what numerical magnetic Reynolds number we can get such fast reconnection under a uniform resistivity.

scheme assuming ideal MHD equation, and obtained the results very similar to those found for uniform resistivity case. This is probably because of nearly uniform numerical resistivity in a dicontinuous layer (current sheet) in FCT scheme. MOCCT scheme often lead to explosive reconnection in a current sheet in ideal MHD numerical simulations (e.g., Hawley and Stone 1995, Kudoh et al. 1998). In such explosive case, however, the total energy is not conserved, so that numerical results cannot be applied to any physical problems. In summary, to use ideal MHD numerical simulations for reconnection problems is somewhat dangerous unless we should be well aware of the properties of numerical scheme. We recommend people to use resistive MHD equations with physical resistivity term (larger than numerical resistivity) if their MHD problems include reconnecting current sheets.

4. Numerical Modelings of Solar Flares and Their Role in Interpretating New Observations of Flares

Before launch of Yohkoh in 1991, solar physicists have sometimes doubted the reconnection model since there were not enough observational evidence of reconnection. Now the situation has dramatically changed because a wealth of observational evidence of reconnection has been discovered through Yohkoh X-ray observations (see the review by e.g. Shibata 1996). It must be stressed that during the course of analysis and interpretation of Yohkoh data, the numerical simulations of magnetic reconnection played a very important role as listed in the following examples of observational evidence of reconnection discovered by Yohkoh.

(1) **Cusp-shaped flare loops**: Although the cusp-shaped magnetic field configuration was predicted by *CSHKP model* (Carmichael-Sturrock-Hirayama-Kopp-Pneuman), the numerical simulations by Ugai (1987), Forbes and Malherbe (1991), Magara et al. (1996) were extremely useful to understand physics of reconnection. Tsuneta et al. (1992, 1996) discovered that outer regions of cusp-loops have systematically higher temperature, which have been predicted by the reconnection model. More recently, Yokoyama and Shibata (1998) succeeded to develop a model of flares including reconnection, heat conduction, and chromospheric evaporation, and reproduced observed temperature distribution in cusp-shaped loops.

(2) **Loop top hard X-ray sources**: Masuda et al. (1994) discovered a hard X-ray source well above the top of the soft X-ray flare loop, which indicated that energy release occurred outside of the soft X-ray flare loop. In interpreting this observation, numerical simulations of Ugai (1987) played a leading role, since the simulation clearly showed the appearance of the fast shock above the reconnected loop. Hence Masuda et al. (1994) interpreted that the loop top hard X-ray source may correspond to the very hot region

just behind the fast shock, and they indeed found that the observed temperature is consistent with the theoretically predicted temperature behind the fast shock.

(3) **X-ray plasmoid ejections**: On the basis of numerical simulations of reconnection, Shibata et al. (1995) conjectured that if the impuslive flares are caused by the reconnection, the plasmoid ejections would be found high above the main flare loop (reconnected loop). They indeed discovered plasmoid ejections in many compact impulsive flares. Magara et al. (1997) developed a numerical simulation model explaining the observed motion of X-ray plasmoids analyzed by Ohyama and Shibata (1997).

(4) **X-ray jets**: Yohkoh discovered many jet-like mass ejections in the solar corona (Shibata et al. 1992b, Shimojo et al. 1996). Shibata et al. (1992b) proposed the magnetic reconnection model for these X-ray jets, and Yokoyama and Shibata (1995, 1996) carried out extensive numerical simulations of reconnection occurring between emerging flux and coronal field to construct a model of X-ray jets (Fig. 1 and CD-ROM for movies). They successfully reproduced observed properties of X-ray jets, and predicted the coexistence of hot X-ray jets and cool Hα jets (surges). The latter has been confirmed observationally by Canfield et al. (1996).

5. Summary

MHD reconnection model successfully explained Yohkoh observations, such as cusp-shaped flares, loop top hard X-ray sources, X-ray plasmoid ejections, and X-ray jets. On the basis of these solar studies, the reconnection model has been applied also to protostellar flares (Hayashi et al. 1996) and "galactic flares" (Tanuma et al. 1998).

There remains some fundamental questions, i.e., origin of resistivity and particle acceleration mechanism, both of which are closely related microscopic physics of reconnection such as non-MHD effects (e.g., Hoshino 1998). The former is related also to macroscopic turbulence (e.g., Biskamp 1994, Nordlund 1998). Another important remaining question is the preflare energy build-up; how the magnetic energy is accumulated in the solar atmosphere? Is it due to photospheric shear flow (e.g., Kusano 1998) or emergence of twisted flux tubes (e.g., Matsumoto et al. 1998)?

Consequently, future subjects for numerical astrophysicists are: (1) 3D MHD modeling of flares (e.g., emergence of twisted flux tubes and resulting 3D reconnection). (2) To develop adaptive mesh MHD code. This is necessary to resolve thin current sheet in high R_m plasma which has probably a fractal structure (Tajima and Shibata 1997). Note again that $L_{flare} \simeq 10^9$ cm $\gg r_{ion,Larmor} \simeq 100$ cm, the latter being the typical scale of anomalous resistivity. (3) To develop numerical code including both MHD and non-

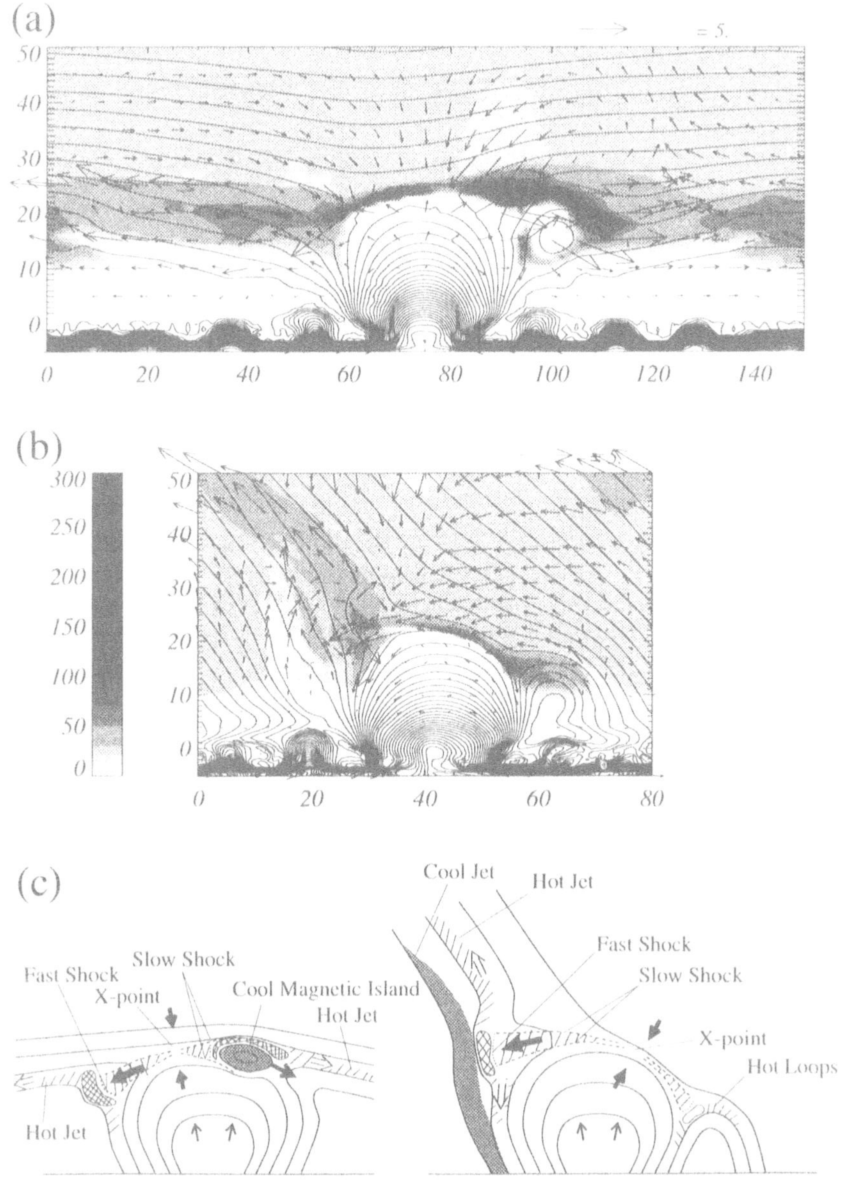

Figure 1. MHD numerical simulation of magnetic reconnection model for X-ray jets (from Yokoyama and Shibata 1995, 1996). This is an extension of the simulation by Shibata et al. (1992a). (a) *Two-sided-loop type*, in which the initial coronal field is horizontal. (b) *Anemone-jet type*, in which the initial coronal field is oblique. Both figures show temperature distribution (grey scale; darker region is hotter), magnetic field lines (lines), and velocity vectors. The unit of length is $\sim$ 200 km. The velocity of the hot jet is about 0.3 – 1.0 in unit of coronal Alfven speed $V_{A,cor}$ ($\sim$ 1000 km/s). (c) Schematic illustration of physical processes found from numerical simulations.

MHD (kinetic) effects for a spatial scale ranging from 1 cm to 10^9 cm to construct self-consistent flare model.

Numerical computations were carried out on VPP300/16R at NAOJ.

References

Biskamp, D., 1986, Phys. Fluids, 29, 1520.
Biskamp, D., 1993, Nonlinear Magnetohydrodynamics, Cambridge Univ. Press.
Canfield, R. C., Reardon, K. P., Leka, K. D., et al., 1996, ApJ, 464, 1016.
Dungey, J. W., 1961, Phys. Rev. Lett., 6, 47.
Forbes, T. G. and Priest, E. R. 1987, Rev. Geophys., 25, 1583.
Forbes, T., and Malherbe, J. M., 1991, Sol. Phys. 135, 361.
Hawley, J.F., & Stone, J. M., 1995, Comp. Phys. Comm. 89, 127.
Hayashi, M. R., Shibata, K., & Matsumoto, R. 1996, ApJ, 468, L37.
Hoshino, M., 1998, in these proceedings.
Giovanneli, R. G., 1947, MNRAS, 107, 338.
Karpen, J. T., Antiochos, S. K., DeVore, C. R., and Golub, L., 1998, ApJ, 495, 491.
Kudoh, T., Matsumoto, R., and Shibata, K., 1998, ApJ, Nov. issue, in press.
Kusano, K., 1998, in these proceedings.
Magara, T., Mineshige, S., Yokoyama, T., and Shibata, K., 1996, ApJ, 466, 1054.
Magara, T., Shibata, K., & Yokoyama, T. 1997, ApJ, 487, 437.
Matsumoto, R., Tajima, T., Chou, W., Okubo, A., and Shibata, K., 1998, ApJ, 493, L43.
Masuda, S., Kosugi, T., Hara, H., Tsuneta, S., and Ogawara, Y., 1994, Nature, 371, 495.
Nordlund, A., 1998, in these proceedings.
Ohyama, M., and Shibata, K. 1997, PASJ, 49, 249.
Parker, E. N., 1957, JGR, 62, 509.
Petschek, H. E., 1964, in Proc. of AAS-NASA Symp. on the Physics of Solar Flares, ed. W. N. Hess, NASA Sp. Pub. 50, p. 425.
Priest, E. R. and Forbes, G. T., 1992, JGR, 97, 16757.
Saito, Y., Mukai. T., Terasawa, T., et al. 1995, JGR, 100, 23567.
Sato, T., and Hayashi, T., 1979, Phys. Fluids, 22, 1189.
Scholer, M., 1989, JGR, 94, 8805.
Shibata, K., Nozawa, S., & Matsumoto, R. 1992a, PASJ, 44, 265.
Shibata, K., Ishido, Y., Acton, L., et al., 1992b, PASJ, 44, L173.
Shibata, K., Masuda, S., Shimojo, M., et al., 1995, ApJ, 451, L83.
Shibata, K., 1996, Adv. Space Res., 17, (4/5)9.
Shimojo, M., Hashimoto, S., Shibata, K., et al. 1996, PASJ, 48, 123.
Sweet, P. A., 1958, in *Electromagnetic Phenomena in Cosmic Physics, IAU Symp. No. 6*, ed. Lehnert, B. (Cambridge Univ. Press), p. 123.
Tajima, T., & Shibata, K. 1997, *Plasma Astrophysics*, Addison Wesley, Massachusetts.
Tanuma, T., Yokoyama, T., Kudoh, T., and Shibata, K., 1998, in these proceedings.
Tsuneta, S., et al., 1992, PASJ, 44, L211.
Tsuneta, S., 1996, ApJ, 456, 840.
Ugai, M. and Tsuda, T., 1977, J. Plasma Phys. 17, 337.
Ugai, M., 1986, Phys. Fluids, 29, 3659.
Ugai, M., 1987, Geophys. Res. Lett., 14, 103.
Yan, M., Lee, L. C., and Priest, E. R., 1992, JGR, 97, 8277.
Yokoyama, T., & Shibata, K. 1994, ApJ, 436, L197
Yokoyama, T., & Shibata, K. 1995, Nature, 375, 42.
Yokoyama, T., & Shibata, K. 1996, PASJ, 48, 353.
Yokoyama, T., & Shibata, K. 1997, ApJ, 474, L61.
Yokoyama, T., & Shibata, K. 1998, ApJ, 494, L113.

TOWARDS THE UNDERSTANDING OF MAGNETIC RECONNECTION: SIMULATION AND SATELLITE OBSERVATIONS

M. HOSHINO
The Institute of Space and Astronautical Science (ISAS)
3-1-1 Yoshinodai, Sagamihara, Kanagawa 229-8510, JAPAN

1. Introduction

Magnetic reconnection processes are thought to be important to understanding fundamental plasma phenomena in geophysical/astrophysical plasmas such as the earth's magnetosphere, solar and stellar flares, and the astrophysical accretion disks. By virtue of the recent intensive plasma and magnetic field observations of the earth's magneotail by the ISEE, AMPTE, and GEOTAIL satellites and the X ray observations of solar flares by the YOHKOH satellite, it is getting to be possible to discuss self-consistently detailed physical pictures of microscopic and macroscopic reconnection processes. The computer simulation study of magnetic reconnection also plays an important role on understanding of non-linear time evolutions of reconnection.

Magnetohydrodynamic (MHD) description of reconnection so far adequately accounts for the hot and high speed plasma flows and the plasmoid formation/ejection in both the solar corona and the Earth's magnetotail, though the difficulty is that the physics of the violation of the MHD approximation remains in doubt. From the recent satellite observations in the Earth's magnetotail, it is now recognized that non-Maxwellian, ion distribution functions are often observed [Lui et al., 1977; DeCoster and Frank, 1979; Sarris and Axford, 1979, Mukai et al., 1996]. Moreover, the distribution functions show sometimes non-gyrotropic behavior with several bunched structure in the velocity space [Frank et al., 1994, Hoshino et al., 1997; Tu et al., 1997; Hoshino et al., 1998a]. Since those non-MHD, kinetic features appear in the macroscopic scale length such as the thickness of the plasma sheet, the self-consistent understanding of both the macroscopic evolution of reconnection and the microscopic behavior seen in the

S. M. Miyama et al. (eds.), Numerical Astrophysics, 311–318.

distribution functions becomes important in the reconnection modeling of a sudden energy release mechanism.

We study the kinetic, collisionless reconnection processes beyond the MHD reconnection picture, and discuss the electron and ion dynamics in the magnetic reconnection region in terms of comparison between the particle-in-cell simulation and the Geotail observations.

2. Ion Dynamics in Magnetic Reconnection

An important step toward a collisionless reconnection model is, we believe, the study of ion distribution functions in a thin plasma. The motion of thermal particles near the X-type neutral line is described by the meandering motion [Sonnerup, 1971], and the unmagnetized particles are trapped in the distance of $\sqrt{\lambda r_L}$, where λ is the thickness of the plasma sheet and r_L is the Larmor radius. Those particles contribute to the energy diffusion by being accelerated by the dawn-dusk electric field during the finite transition in the weak magnetic field region [Laval and Pellat, 1968; Coppi et al., 1966]. On the other hand, the motion within a weakly magnetized plasma sheet outside the meandering particle region is described by the Speiser motion [Speiser, 1965]. Thermal particles remain within the current sheet for a time comparable to the gyroperiod based on the normal magnetic field B_z, and after the interaction of the particles with the dawn-dusk electric field, the particles are ejected from the plasma sheet along the strong magnetic field. This finite time interaction between particles and the electric field in the current sheet is also important as the origin of an effective conductivity [Lyons and Speiser, 1982].

The meandering/Speiser motion can produce a unique signature in the velocity space. If the thermalization time scale is slower than the dynamic time scale, the bunched ion structure as the "memory" of the meandering/Speiser motion could be observed in the velocity space. Since the lobe thermal energy is much smaller than the typical kinetic energy gain during the interaction of particles with the dawn-dusk electric field, the lobe origin plasma forms a coherent structure representing the meandering/Speiser motion, and multi-component plasma populations could be found in the velocity space. It is expected that the structure of the plasma velocity distribution functions can provide very important information on the dynamical evolution of magnetic reconnection. From this point of view, we study the structure of the plasma distribution functions seen in the typical GEOTAIL observations during the passage of the plasmoid, and then we compare the result with the numerical simulation result obtained by using a two-dimensional, full particle-in-cell simulation.

Figure 1 shows the GEOTAIL plasmoid observations when the space-

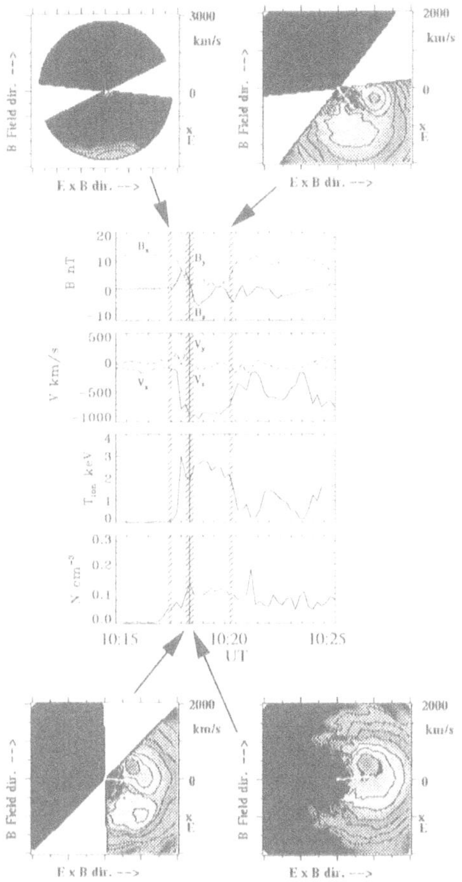

Figure 1. Typical plasmoid event observed by Geotail on September 18, 1993. Top and bottom panels show the ion distribution functions, and the magnetic field and the plasma moments data are shown in the middle panel. (color figure in CD-ROM)

craft was situated in the night side magnetotail, about 70 R_e from the Earth. A ground-based magnetometer at Kakioka station detected a Pi-2 onset at 1014UT. The top and bottom corners shows four typical distribution functions during the plasmoid passage, and the center figure shows time series of three components of the magnetic field, bulk flow velocities, ion temperature, and plasma density. The time series of magnetic variations showed a north-south bipolar signature in the B_z magnetic field around 1018UT. Associated with the bipolar signature, the tailward plasma flows are enhanced, and plasma density and temperature increases. Prior to the northward turning of the B_z magnetic field at 1017:47 UT, the tailward high speed ion of $\geq$ 2500km/s is observed in the plasma sheet boundary layer (PSBL). The vertical and horizontal axes are respectively aligned to

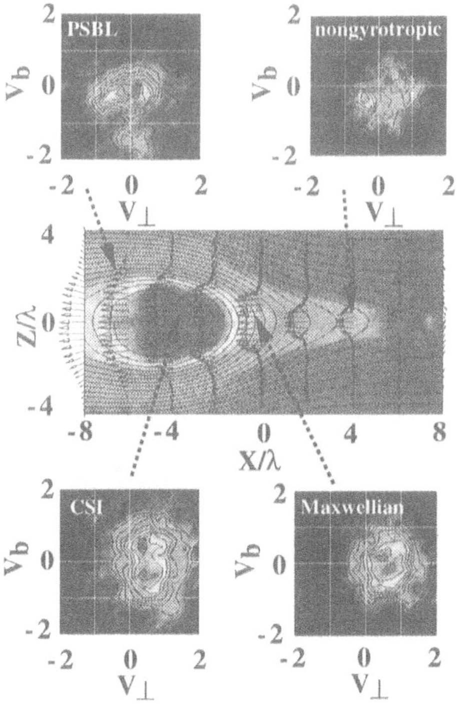

Figure 2. Particle-in-cell simulation result. Ion flow vectors, magnetic field lines, and plasma density in color contour are shown in the middle panel. Ion distribution functions obtained in the simulation are shown in top and bottom panels. (color figure in CD-ROM)

the directions of the magnetic field and the plasma convection. The small black spot situated in the center corresponds to the cold lobe ions, and the high-speed component flowing into the direction opposite that of the magnetic field is the PSBL ion beam. The ion distribution inside the plasmoid at the time interval of 1018:11-1018:23 UT is characterized by two cold ion components parallel to the magnetic field [Mukai et al., 1996]. Around the turning point of the B_z polarity, namely at the time interval of 1018:23-1018:35 UT, an almost thermal ion distribution function is observed. Just after the passage of the plasmoid at the time interval of 1020:11-1020:23UT, we observe two main ion components bunched perpendicular to the magnetic field. Together with another slice of the three-dimensional distribution function, we find that the ion distribution function is characterized by non-gyrotropic behavior [Hoshino et al., 1998a]. We found several other similar non-gyrotropic ion events when GEOTAIL is located in a weak B_x magnetic field region after the passage of plasmoids.

Figure 2 shows the particle simulation result. The center shows a snapshot of magnetic field lines and plasma density in a color contour of a reconnecting current sheet. The red and blue regions correspond to high and low density regions, respectively. Four typical ion distribution functions obtained in this simulation box are shown in the top and bottom four panels. The distribution shows a slice of a three-dimensional distribution function in a plane including the magnetic field and the convection vectors. The velocity scale is normalized by the Alfven velocity which is defined by the lobe magnetic field and the plasma sheet density. By comparing the behavior of those ion velocity distribution functions obtained by the kinetic reconnection simulation with the distributions observed by the Geotail satellite, we find that most of ion dynamics observed in the earth's magnetotail can be well reproduced by the two-dimensional kinetic simulation.

Hoshino [1998b] have discussed that the plasma mixing under the action of the meandering/Speiser particles accelerated in the current sheet play an important role on the formation of a wide variety of distribution functions. It has been discussed that those kinds of distributions are strongly coupled with the magnetic diffusion process around an X-type region, and the macroscopic evolution of the plasma sheet is also affected by the formation mechanism of non-Maxwellian distributions.

3. Electron Dynamics in Magnetic Reconnection

In the thin plasma sheet where the plasma sheet thickness becomes comparable to the ion inertia length (or ion Larmor radius), the electric field appears in the reconnection plane and affects on the dynamics of electrons as well as of ions.

The origin of the electric field can be understood as follows: In the lobe region outside of the plasma sheet, the characteristic scale length is much larger than the ion inertia length, the both ions and electrons follow the motion of magnetic field line. Inside the plasma sheet, however, the plasma sheet thickness becomes comparable to the ion inertia length, ions start to be unmagnetized and cannot follow the motion of the magnetic field line, while electrons are still magnetized until the scale size becomes less than the electron inertia length. Therefore the electric fields in the reconnection plane are excited in order to maintain the charge neutrality [Hoh 1966; Dobrowolny 1968; Hoshino, 1987], as long as the evolution of magnetic reconnection is slower than the Alfven transit time of the plasma sheet. In a linear growing stage of the tearing mode instability, the electric field parallel to the magnetic field is roughly given by $E_{\|} \sim ku_i/\gamma E_y$, where k is the wave number of the tearing mode, u_i is the ion drift velocity, and γ is the linear growth rate. The large parallel electric field may play an important role on the electron dynamics in the reconnection region, because of the

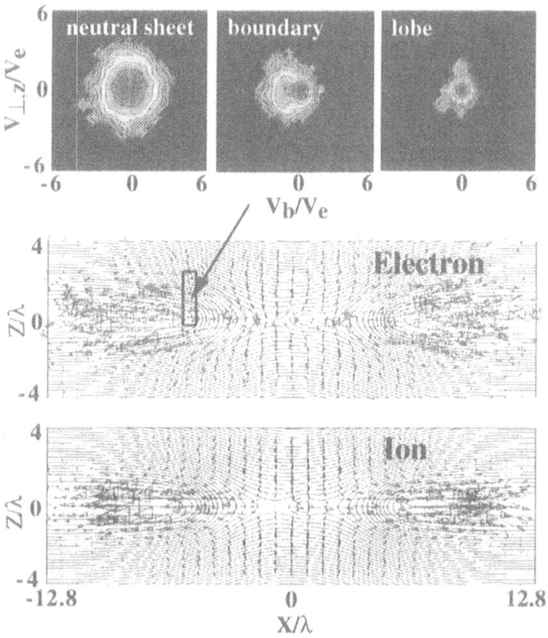

Figure 3. Particle-in-cell simulation result. From top to bottom, electron distribution functions, electron flow vectors superposed with the magnetic field lines, and ion flow vectors are shown. (color figure in CD-ROM)

magnitude of $ku_i/\gamma \leq 1$ in a thin plasma sheet.

Figure 3 shows the snapshot of the magnetic reconnection simulation. The plasma parameters used in this simulation are almost same as that shown in Figure 2, but the spatial resolution has been improved in order to study the electron dynamics. Three electron distribution functions across the boundary region are shown in the top panel. The middle and bottom panels are respectively the electron flow vectors with magnetic field lines and the ion flow vectors. The ion flow vectors direct from the lobe toward the plasma sheet, and then ions are ejected from an X-type region to an O-type region. On the other hand, we can find the electron flow vectors have the opposite directions along the boundary region between the lobe and the plasma sheet. The difference of ion and electron flows produces the Hall electric current in the reconnection plane [Sonnerup, 1979; Terasawa 1983]. The location of the electron distribution functions are indicated by the box in the electron flow vector panel. The distributions are a slice of a three-dimensional distribution function in a plane including the magnetic field and the convection flow vectors perpendicular to the magnetic field. The velocity scale is normalized by the initial electron thermal velocity. Inside the plasma sheet, a hot and isotropic Maxwellian distribution can be found, while in the lobe region the distribution is described as a cold Maxwellian distribution. The distributions obtained in the boundary region between the lobe and the plasma sheet, however, show the deformed distribution from the Maxwellian. The middle distribution shows the non-Maxwellian distribution in which a cold component flows toward the parallel to the magnetic field, that is, toward the X-type region, and a hot component flows away from the X-type region. By tracing the position of particles backward in time, we found that the hot electron component comes from the X-type region after the current sheet acceleration, which is the similar process to the ion beam population observed in PSBL. On the other hand, the cold component can be understood as a result of the Hall electric field acceleration given in the above equation.

The electron distributions similar to the above electron dynamics have been already observed by the Geotail satellite [Tu et al.,1997; Fujimoto et al.,1997]. The electron distributions composed with the cold core and the hot component escaping from an X-type region are often observed near the plasma sheet boundary. Although there remains uncertainty in the electric field measurement, we think that a strong parallel electric field is produced during the magnetic reconnection by virtue of the similarity of the distribution functions obtained by both the Geotail satellite and the numerical simulation study. The electric field produces the Hall electric current in the reconnection plane, and we find that the electrons are the main carrier of the field aligned electric current which could propagate toward the ionosphere in the form of the kinetic Alfven waves.

4. Discussion and Conclusions

We have discussed both ion and electron dynamics beyond ion inertia length in the magnetotail. A wide variety of non-Maxwellian ion distribution functions in the magnetotail have been observed in the thin plasma sheet. These observations suggest that the ion inertia length becomes comparable to the thickness of plasma sheet, namely, ion starts to be demagnetized and electron is still magnetized. These distributions are produced by non-adiabatic ion motions in such region as a result of the plasma mixing between the cold lobe ions and the hot plasma sheet plasmas. On the other hand, electron distributions observed by the Geotail satellite are almost gyrotropic with respect to the magnetic field, but the counter streaming electrons which components have the different temperature are sometimes observed near the boundary between the plasma sheet and the lobe. These ion and electron distributions have provided very important information about the microscopic transportation coefficients such as electric conductivity, viscosity, and thermal conduction.

The studies of electron dynamics beyond electron inertia length should be important to know the reconnection physics, because the final breakdown of the frozen-in condition is controlled by the electrons. It will be possible, in a near feature technology, to observe the velocity distribution function within the time resolution of the order of 10 mil second, and the electron inertia scale would be resolved. Kinetic electron behavior will provide variable information on plasma heating, mixing, and non-thermal particle acceleration. Our theoretical and observational study is only a first step to understanding more complete collisionless reconnection physics.

References

Coppi, B., G. Laval, and R. Pellat, Dynamics of geomagnetic tails, *Phys. Rev. Lett.*, *16*, 1207-1210, 1966.

DeCoster, R. J., and L. A. Frank, Observations pertaining to the dynamics of the plasma sheet, *J. Geophys. Res.*, 84, 5099-5124, 1979.

Dobrowolny, M., Instability of a neutral sheet, *Nuovo Cimento*, *B55*, 427, 1968.

Frank, L. A., W. R. Paterson, and M. G. Kivelson, Observations of nonadiabatic acceleration of ions in Earth's magnetotail, *J. Geophys. Res.*, *99*, 14,877-14,890, 1994.

Fujimoto, M., et al., Observations of earthward streaming electrons at trailing boundary of a plasmoid, *Geophys. Res. Lett.*, *24*, 2893-2896,1997.

Hoh, F. C., Stability of sheet pinch, *Phys. Fluids*, *9*, 277, 1966.

Hoshino, M., Electrostatic effect for the collisionless tearing mode, *J. Geophys. Res.*, *92*, 7368-7380, 1987.

Hoshino, M., A. Nishida, T. Mukai, S. Kokubun, and T. Yamamoto, Ion dynamics of magnetic reconnection in the magnetotail, *Adv. Space Res.*, *19*, 1827-1835, 1997.

Hoshino, M., et al., Ion dynamics in magnetic reconnection: Comparison between numerical simulation and Geotail observations, *J. Geophys. Res.*, *103*, 4509-4530, 1998a.

Hoshino, M., Kinetic ion behavior in magnetic reconnection region, *Geophysical Monograph*, AGU, in press, 1998b.

Laval, G. and R. Pellat, Stability of the plane neutral sheet for oblique propagation and anisotropic temperature, *ESRO SP-36*, 5, Euro. Space Agency, Paris, 1968.

Lui, A. T. Y., E. W. Hones Jr., F. Yasuhara, S.-I. Akasofu, and S. J. Bame, Magnetotail plasma flow during plasma sheet expansions: VELA 5 and 6 and IMP 6 observations, *J. Geophy. Res.*, *82*, 1235-1244, 1977.

Lyons, L. R. and T. W. Speiser, Evidence for current sheet acceleration in the geomagnetic tail, *J. Geophys. Res.*, *87*, 2276-2286, 1982.

Mukai, T., et al., Structure and kinetic properties of the plasmoid and its boundary region, *J. Geomagn. Geoelectr.*, *48*, 541-560, 1996.

Sarris, E. T., and W. I. Axford, Energetic protons near the plasma sheet boundary, *Nature*, *277*, 460, 1979.

Sonnerup, B. U. Ö., Adiabatic particle orbit in a magnetic null sheet, *J. Geophys. Res.*, *76*, 8211-8222, 1971.

Sonnerup, B. U. Ö., Magnetic field reconnection, *in Solar System Plasma Physics*, vol.3, pp. 45-108, North-Holland, New York, 1979.

Speiser, T. W., Particle trajectories in model current sheet, 1, Analytical solution, *J. Geophys. Res.*, *70*, 4219, 1965.

Terasawa, T. Hall current effect on tearing mode instability, *Geophys. Res. Lett.*, *10*, 457-478, 1983.

Tu, J.-N., et al., Geotail observations of ion velocity distributions with multi-beam structures in the post-plasmoid current sheet, *Geophy. Res. Lett.*, *24*, 2247-2250, 1997.

3D NONLINEAR MHD WAVE HEATING OF CORONAL LOOPS

S. POEDTS
Centre for Plasma Astrophyscis, KULeuven, Belgium
Research Associate of the FWO-Vlaanderen

R. KEPPENS
FOM Institute for Plasma Physics 'Rijnhuizen'
P.O.Box 1207, 3430 BE Nieuwegein, The Netherlands

AND

A.J.C. BELIËN
ESA Space Sci. Dept., SOHO Experimental Analysis Facility,
Goddard Space Flight Center, Greenbelt, MD 20771, U.S.A.

1. Numerical simulations of wave heating

The origin and dissipation mechanism of the energy required to heat the millions of hot magnetic loops contained in the solar corona is still unknown. Several candidate heating mechanisms have been proposed and subjected to extensive theoretical and observational studies. Alfvén waves can be *excited directly* by convective motions of the foot points of the loops and/or by reconnection events in the corona itself, or *indirectly* by sideways wave excitations of the loops. Their dissipation is considerably enhanced by the narrow resonances that occur where the incident wave frequencies match the local Alfvén frequency in the loops. Extensive parameter studies of the linear dissipative MHD steady state and time scales revealed that resonant dissipation is a viable candidate for coronal loop heating. However, it was shown that in the hot coronal plasma, the dynamics of the resonant layers is *highly nonlinear* (Ofman and Davila, 1995; Poedts and Goedbloed, 1997). Poedts et al. (1997) continued a linear steady state with the nonlinear code HERA and demonstrated that the periodic shear flow in the linear resonant layers is Kelvin-Helmholtz unstable. The simulations showed that the instability partly destroys the resonant layers, spreads the heat deposition over the whole loop volume, and results in a lower dissipation rate than that predicted by the linear calculations.

S. M. Miyama et al. (eds.), Numerical Astrophysics, 319–322.

In the present paper, we present fully nonlinear 3D results obtained with the new massively parallel processing (MPP) version of the HERA code for the same configuration so that these results can be compared to the earlier ones. The density update, previously neglected, affects the resonance positions and the dissipation rate, and the results are quite different from the earlier nonlinear continuations of linear steady states (saving a considerable amount of CPU time).

The results discussed below were obtained with an extended version of the HERA code, which solves the nonlinear resistive MHD equations in cylindrical geometry (Poedts and Goedbloed, 1997). The extensions involve a conversion to a pseudo-spectral method for the θ- and z-direction, a density update, and an MPP implementation based on high performance fortran (Keppens *et al.*, 1998). The pseudo-spectral method avoids the convolutions in nonlinear terms and is therefore much faster than the original spectral method, especially for high-resolution runs. It also enables taking the density update into account, which was previously neglected because of excessive CPU time requirements due to triple convolutions in the momentum equation. The MPP implementation yields a nearly linear scaling on different platforms (Keppens *et al.*, 1998).

2. Fully nonlinear results in 3D

We modeled a resonantly driven magnetized loop by a cylindrical plasma column with radius a (normalized to 1), excited periodically by a sideways incident wave with amplitude A, frequency ω_d, and azimuthal and axial wave numbers m and nk (with n an integer and k a quantization factor, due to the finite length L of the cylinder $k = 2\pi a/L$) determining the angle of incidence.

Figure 1 shows a snapshot of a moderate resolution run ($64 \times 16 \times 16$) in a fully nonlinear resistive MHD calculation, i.e. solving the nonlinear MHD equations starting from an ideal static configuration. The initial equilibrium was characterized by a constant density, $\rho(t = 0)$ normalized to 1, and a parabolic axial current density profile, i.e. the axial magnetic field was constant, $B_z(t = 0) = 1$, but the field was twisted. Time was normalized to the Alfvén transit time a/v_A. For Fourier mode $(m, n) = (2, 1)$ and $k = 0.05$, the dimensionless Alfvén frequency varied from 0.25 on the axis to 0.15 at the loop surface $r = a$. The wave excitation was simulated by imposing a small-amplitude ($A = 0.001\, v_A$, corresponding to 2 km/s in the corona) radial velocity perturbation of a single mode number $(2, 1)$ with a driving frequency of 0.205 matching the Alfvén frequency at radius $r_s = 0.67a$. At this resonant radius r_s the resistivity, taken to be $\sim 1/J_z(t = 0)$, was $\eta(r_s) = 10^{-5}$. In an initial linear phase, a resonant layer developed around

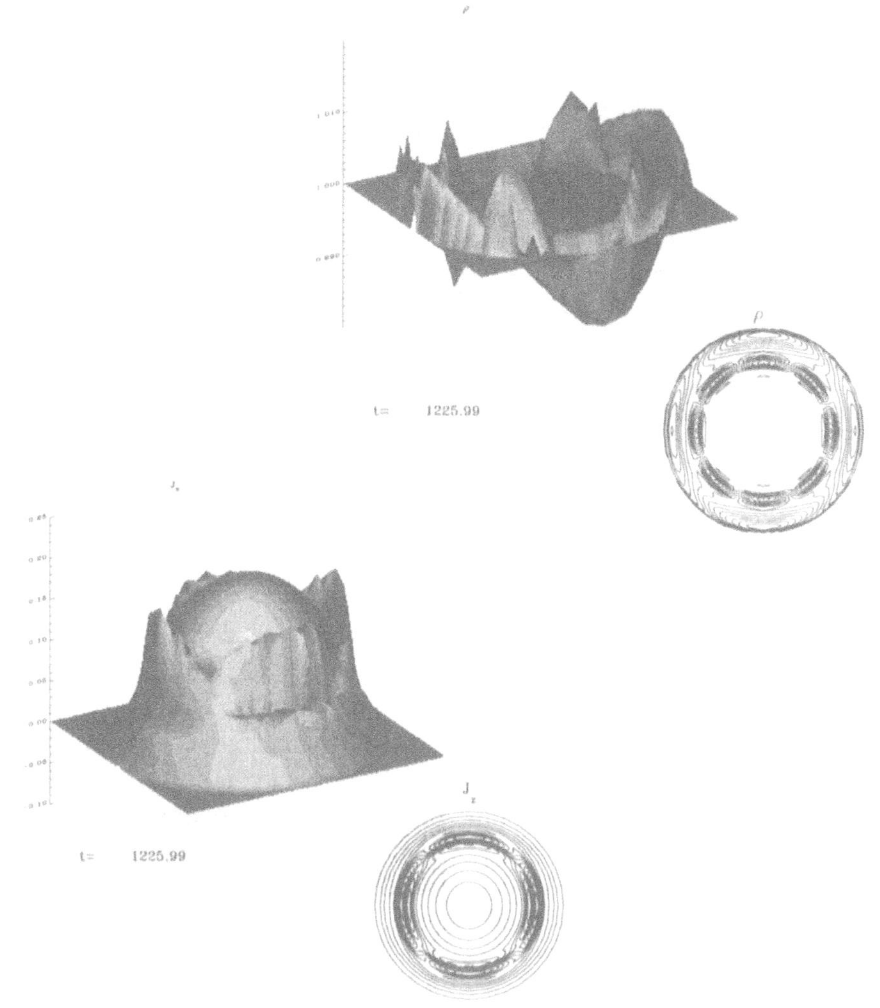

Figure 1. Surface plot and contour plot of the density ρ and current density J_z after 40 driving periods with $A = 0.001\,v_A$, corresponding to only 2 km/s in the solar corona.

r_s with a strong periodic shear flow. The high amplitudes of this flow induced nonlinear mode coupling restricted to modes with the same helicity as the incident wave mode, i.e. $m/n = 2$. In the nonlinear phase, the $m = 4$, $n = 2$ perturbation yielded a contribution to the density profile and the current density, as is clear from the snapshot in Figure 1. The velocity amplitude stayed relatively small ($\sim 0.005\,v_A$) in this highly dissipative case. The current density and density perturbations affected the local Alfvén frequency and, hence, the position of the resonance(s). The dynamics was less violent than for the nonlinear continuation of the linear steady state because the amplitudes of the periodic shear flow remained smaller than in the linear calculation, and the K-H instability did not develop for the considered (relatively high) resistivity. Also, in the simulated time interval, the energy dissipation was restricted to the resonant layer, which was wider than in linear calculations but much less disturbed than in the K-H unstable flow case. Moreover, the energy dissipation rate turned out to be much higher than in the linear case, but this may be due to substantial numerical dissipation resulting from the relatively low spatial resolution and needs to be confirmed by convergence studies.

3. Conclusions

We have discussed fully 3D nonlinear simulations of the resonant dissipation of MHD waves in cylindrical coronal loops. Due to their high magnetic Reynolds numbers, the dynamics of this heating process is nonlinear. The effect of heat deposition and of nonlinear mode coupling on background density and magnetic field yields lower wave amplitudes than in linear calculations, and the magnetic field can stabilize the Kelvin-Helmholtz instabilities that did develop in nonlinear continuations of more sheared linear steady state flows. Therefore, the nonlinear low-R_m results are qualitatively close to the linear results. High-resolution runs are required to resolve the resonant layers and to minimize numerical dissipation in order to obtain more quantitative information on the nonlinear MHD dissipation rates. Moreover, higher magnetic Reynolds numbers yield more sheared flows, which may result in completely different nonlinear behaviour.

Visualization can help to interpret the results of these numerical simulations in terms of UV and/or X-ray emission resulting from the resonant heating (Beliën *et al.*, 1996). Time sequences of such artificially generated pictures can be directly compared to similar observational data (cf. CD-ROM). Such a consistent comparison for different heating models and different filters may contain a clue for the identification of the heating mechanism(s) responisble for the hot solar corona.

This work was performed as part of the research programme of FOM with financial support from NWO and the Dutch National Computing Facilities Foundation for the use of supercomputer facilities. The work was carried out while S. Beliën held an ESA fellowship. ESA is not financially responsible for this publication.

References

L. Ofman and J.M. Davila. JGR, 100, 23427, 1995.

R. Keppens, S. Poedts, and J.P. Goedbloed. Proc. HPCN Europe 98, in press.

S. Poedts and J.P. Goedbloed. A&A, 321, 935-944, 1997.

S. Poedts, G. Tóth, A.J.C. Beliën, and J.P. Goedbloed. SPh, 172, 45-52, 1997.

A.J.C. Beliën, S. Poedts, H.J.W. Spoelder, R. Leenders, and J.P. Goedbloed: Comp. in Physics, 10, 573-583 (1996).

SIMULATION STUDY ON CATASTROPHIC INSTABILITY IN SOLAR CORONA

K. KUSANO
Graduate School of Advanced Sciences of Matter,
Hiroshima University, Higashi-Hiroshima 739-8626, Japan

1. Introduction

A variety of catastrophic events is observed in the solar corona. Solar flares are the largest and the most typical processes of them. In general, catastrophic processes are constituted by a slow build-up of free energy and the following quick liberation. Recent observations indicate that the energy liberation in solar flares must be caused by magnetic reconnection. However, our understanding of flares is not yet completed. Since the time scale of energy supply due to the photospheric activities is much longer than the time scale of flares, the energy liberation by reconnection must be prevented until a substantial free-energy is stored in the system. It is still open question why the free energy can be slowly built-up before flares.

Recently, Kusano et al. have proposed a new theory to solve the problem above based on the minimum energy principle (Kusano *et al.*, 1995). When the magnetic helicity becomes larger than a threshold for the bifurcation of minimum energy state, the mutual helicity which is coupled with the potential field has to be partially converted into the self-helicity which is decoupled from the potential field. In the case, the minimum energy state is given by a mixture of the coupled and the decoupled components, and the purely coupled solution becomes unstable even linearly. The instability may drive magnetic reconnection if it grows substantially. However, as long as the amplitude in the instability is much smaller compared to the equilibrium component, magnetic reconnection can be prevented and the free energy is not released even though the system is unstable. Namely, the free energy could be stored in the competitive process between the growth of instability and the helicity supply. In this article, we will briefly describe the results of numerical simulations which are performed to examine the theory above.

S. M. Miyama et al. (eds.), Numerical Astrophysics, 323–326.

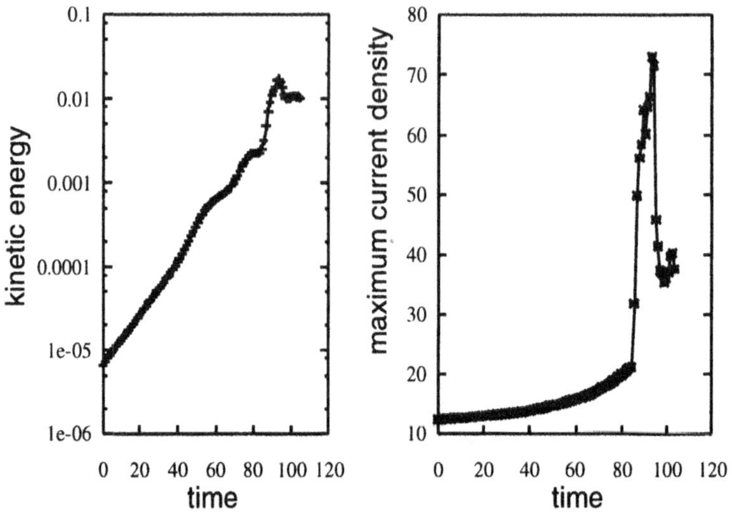

Figure 1. The total kinetic energy and the maximum current density in the simulation domain are plotted as a function of time.

2. Numerical Model

The geometry as well as the boundary condition is the same as those in the previous paper (Kusano *et al.*, 1996), except that the present model is restricted to a two dimensional system ($\partial/\partial x = 0$). The coronal region corresponds to a slab bounded by two horizontal planes, in which the bottom ($z = 0$) is the photospheric boundary and the top ($z = L_z$) is an artificial rigid boundary. Both boundaries electrically respond as infinitely conductive media. The lateral boundaries ($y = 0$ and $y = L_y$) satisfy a periodic condition. The aspect ratio of the simulation box is sufficiently large ($L_z/L_y = 10$), so that the top boundary does not influence the dynamics of the coronal loop connecting to the bottom.

The vertical components of the magnetic field at the bottom and the top are specified. The initial condition is composed of the purely coupled solution of Taylor's state ($\mathrm{rot}\vec{B}_0 = \alpha\vec{B}_0$) plus a small perturbation $\vec{b}_1$, which is given by the most unstable mode for $\vec{B}_0$. The basic equations to be solved are given by the MHD equations constituted by the continuous equation, the equation of motion, the induction equation, and the pressure equation. The viscosity ν, the resistivity η and the heat conductivity κ are fixed to 1.0×10^{-4}.

The numerical solver is composed of the finite difference scheme with the 2nd order accuracy of the spatial difference and the Runge-Kutta-Gill method with the 4th order accuracy of the temporal difference. The grid number ($N_y \times N_z$) in the simulation box ($L_y \times L_z$) is 256×400, and the

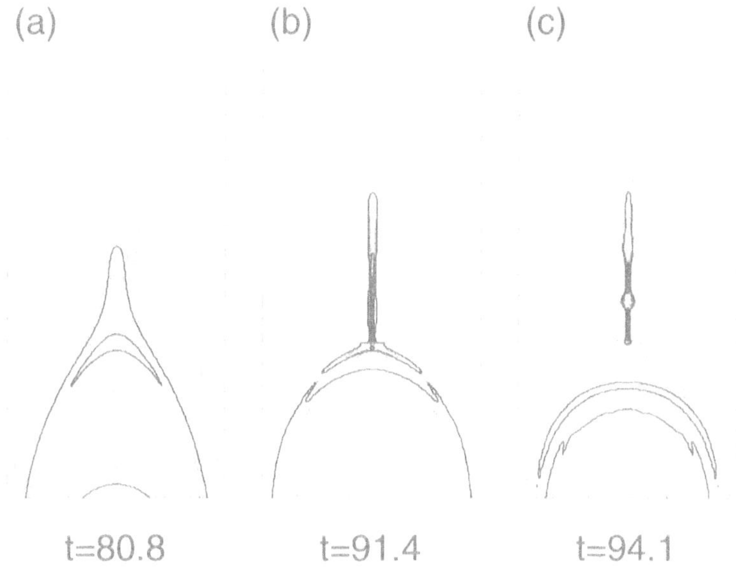

Figure 2. Contour plot of the current density.

grid size is spatially modulated in order to accurately capture the very thin current layers which appear in the center of magnetic loops. The mesh size for the y dimension, Δy, is minimized on the center of the magnetic arcades, where $\Delta y = 7.8 \times 10^{-4} L_y$.

3. Results

Figure 1(a) shows the evolution of the total kinetic energy in a logarithmic scale. We can see that the growth rate is enhanced in the late phase ($t > 80$) compared to the initial growth rate which well corresponds to the linear eigenvalue. This result indicates a self-catalyst effect in which the instability quickens the growth of itself. So, we may call that a catastrophic instability in a sense that it can grow faster than an exponential function of time.

The catastrophic feature is caused by a sequential dynamics of the initial linear phase and the subsequent nonlinear phase. After the initial linear phase, a current sheet is quickly created at the center of the coronal magnetic loop, as shown in Fig.2(b), and it works as the diffusion region in magnetic reconnection process. Figure 1(b) clearly indicates that the current sheet generation proceeds much faster than the growth of the original instability. Furthermore, it is also found that the current sheet becomes filaments if the width is satisfactorily thin, as shown in Fig.2(c). The filamentation and the ejection of the tiny plasmoids from the diffusion region

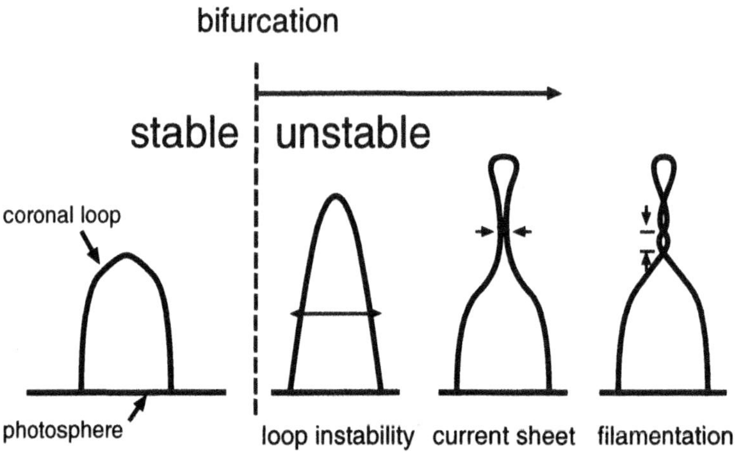

Figure 3. Conceptual scenario of the catastrophic instability in solar coronal loops.

are repeated several times until a substantial free energy is released.

4. Discussions

The simulation result implies that several classes of time and special scales are involved in the nonlinear dynamics of the coronal loop instability. As shown in Fig.3, once a larger helicity than the bifurcation criterion is supplied, a smaller size structure is successively created by the instability. First, the coronal loop becomes unstable against the loop size scale mode. Secondly, the large scale instability creates the thin current sheet. Thirdly, the thin current sheet becomes unstable against the local tearing mode, and the filamentation happens.

The creation of finer scale structure enhances the diffusion speed, and eventually the free energy is released catastrophically. However, if the growth of the original instability is faster than the energy supply, the free energy must be released before it is stored substantially, so that any catastrophe may disappear. Therefore, the simulation results suggest that the multiplicity in the time scale of instability is crucially important to form the explosive phenomena like solar flares.

References

Kusano, K., Suzuki, Y., and Nishikawa, K. (1995) A Solar Flare Triggering Mechanism Based on the Woltjer-Taylor Minimum Energy Principle, ApJ, 441, pp.942–951.

Kusano, K., and Nishikawa, K. (1996) Bifurcation and Stability of Coronal Magnetic Arcades in a Linear Force-Free Field, ApJ, 461, pp.415–423.

MHD SIMULATION OF A SOLAR FLARE BASED ON A MAGNETIC RECONNECTION MODEL

YOKOYAMA, T. AND SHIBATA, K.
National Astronomical Observatory
Mitaka, 181, Japan

Abstract. A two-dimensional simulation of a solar flare was performed using a newly developed magnetohydrodynamic (MHD) code including the nonlinear anisotropic heat conduction effect. We also derived a simple scaling law for the flare temperature described as $T_A \propto B^{6/7}$, where T_A and B are the temperature at the flare loop apex and the coronal magnetic field strength, respectively.

1. Introduction

Magnetic reconnection is now believed to be the energy release mechanism for a solar flare. The observation of cusp-shape loops in soft X-ray (Tsuneta et al. 1992) supports this model, which predicts the primary site of the energy release above the soft X-ray loops. The "chromospheric evaporation" (Neupert, 1968) is an ablation of chromospheric plasma by the energy released through reconnection and transported by heat conduction and/or by non-thermal electrons during a flare. We studied the reconnection and the chromospheric evaporation by means of a two-dimensional MHD simulation using a newly developed numerical code including the nonlinear anisotropic heat conduction effect (Yokoyama & Shibata, 1997, 1998).

2. Simulation models and results

We solved the two-dimensional, nonlinear, time-dependent, resistive, compressible MHD equations and ignored gravitational acceleration. Ohmic heating and the heat conduction effect were taken into account. The conduction coefficient was the Spitzer-type one, which was proportional to $T^{5/2}$, where T was the temperature, working only in the direction along

S. M. Miyama et al. (eds.), Numerical Astrophysics, 327–330.

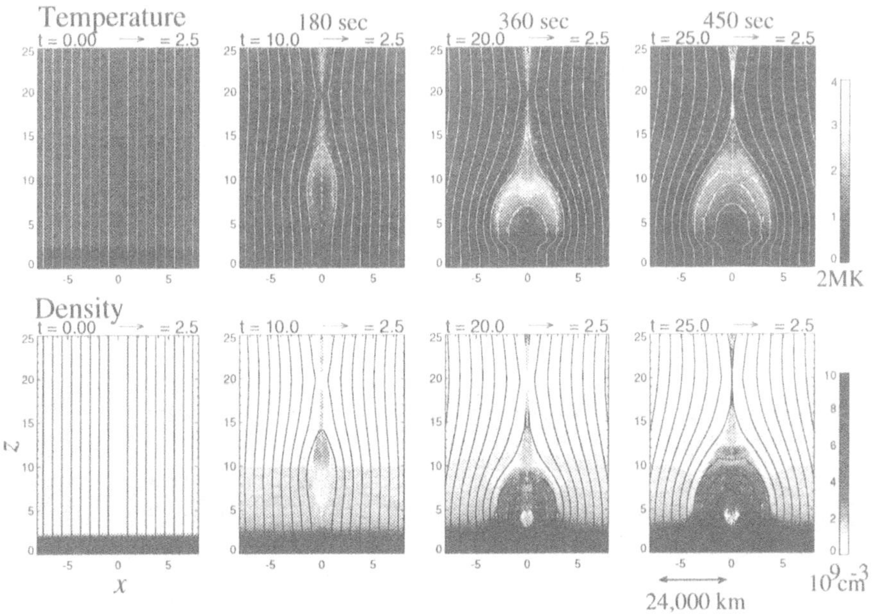

Figure 1. Results of the simulation. Upper and lower panels show the temporal evolution of temperature and density distribution, respectively. The arrows represent the velocity, and the lines show the magnetic field lines. The units of length, velocity, time, temperature, and density are 3000 km, 170 km s^{-1}, 18 s, 2×10^6 K, and 10^9 cm^{-3}, respectively. In the initial condition ($t = 0$) a dense region is located near the bottom of the simulation box, in which the density is about 10^5 times that of the other region.

the magnetic field line. The units of length, velocity, and time in the simulations were $\delta = 3000$ km, $C_{s0} = 170$ km s^{-1}, and $\tau \equiv \delta/C_{s0} \approx 18$ s, respectively, where δ and C_{s0} were the thickness of the initial current sheet and the sound speed of the initial coronal plasma, respectively. Density and temperature were normalized with the initial coronal values $n_0 = 10^9$ cm^{-3} and $T_0 = 2 \times 10^6$ K, respectively.

The initial plasma was in magnetohydrostatic equilibrium with antiparallel magnetic fields, between which a current sheet was situated ($t = 0$ of Fig. 1). The initial plasma beta $\beta = 0.2$ was uniform everywhere except for the current sheet region. We took a dense and low- temperature region located in $z < 1$ for modeling the chromosphere, whose density was 10^5 times the coronal one. The initial equilibrium was perturbed by a finite resistivity in $t \leq t_i = 2$ in a localized spot at $(x, z) = (0, 20)$ in the current sheet. For $t > t_i$, the resistivity was determined self-consistently from the local value of the current density for every time step. The sound-crossing time is $\tau \equiv \delta/C_{s0} \approx 18$ s, the Alfvén-crossing time is $\tau_A \equiv \delta/V_A \approx 7.4$ s, and

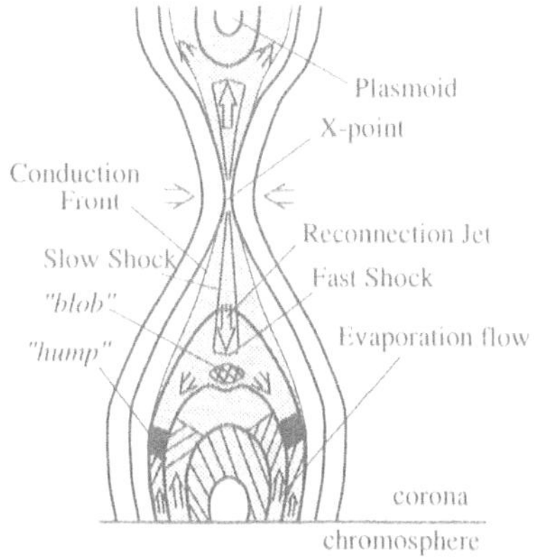

Figure 2. Schematic illustration of the simulation results. Thick solid lines show magnetic field lines.

the heat conduction time is $\tau_c \equiv \delta^2(n_0 k_B/\kappa) \approx 2.2$ s. Thus, the conduction time was shorter than the Alfvén-crossing time.

Figure 1 shows the results, and figure 2 describes them by means of a sketch. Because of the enhanced resistivity, the point $z = 20$ in the current sheet evolves to an X-point of the magnetic reconnection ($t = 10$). The reconnected field lines together with the frozen-in plasma are ejected as upward and downward jets from this X-point due to magnetic tension force. *Slow-mode MHD shocks* are formed at the boundary between the inflows and the outflows.

In the temperature distribution at $t = 20$ in Fig. 1, a clear cusp shape is seen. The outer edge of this cusp is a pair of *conduction fronts* propagating from the X-point. The conduction fronts transport the thermal energy along the field lines and form the cusp shape of hot plasma. Note that the global cusp shape of the hot region behind the conduction front is similar to the soft X-ray solar flare loops in long-duration (LDE) flares (Tsuneta et al., 1992). Since the conduction front propagates faster than the inflow of the shock, the temperature jump between the inflow and the outflow is smoothed away. The jumps in the density distribution at $t = 20$ in Fig. 1 are, therefore, *isothermal slow-mode MHD shocks*.

The reconnected magnetic field which goes down from the X-point finally forms a closed loop ($t = 25$ in Fig. 1). In the density distribution, a growing plasma mound can be seen in the panels of $t = 20$ and $t = 25$ of Fig. 1. This is a direct consequence of chromospheric evaporation. The velocity of the evaporation flow is $V_{\rm ev} = 0.2$–0.4, 0.2–0.3 times the local sound speed. The density of the evaporated plasma is ≈ 10 times the initial coronal density at maximum and ≈ 5 times in average.

What determines the flare temperature and its distribution? If we as-

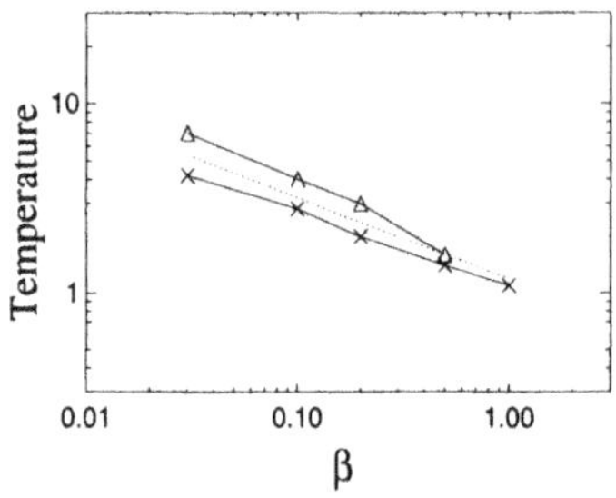

Figure 3. Dependence of the flare temperature on the initial plasma $\beta \equiv p/p_{\rm m}$, where p and $p_{\rm m}$ are gas and magnetic pressures, respectively. Triangles are the temperature at the loop top behind the fast-mode shock, and crosses are the averaged temperature in the loop. The dotted line is the theoretical result derived from the equation (1) in the text. When $\beta \geq 1$, no fast-mode shocks are formed, and no data point is thus shown.

sume that the input of energy to a loop balances with the conduction cooling rate, the temperature at the loop apex is $T_{\rm A} \approx (2QL^2/\kappa_0)^{2/7}$, where Q is the volumetric heating rate and L is the half-length of the loop (Fisher & Hawley, 1990). In our simulations, the heating mechanism used was magnetic reconnection, so the heating rate is described as $Q = B^2/(4\pi) \cdot V_{\rm in}/L \cdot 1/\sin\theta$, where $V_{\rm in}$ is the inflow velocity ($\approx 0.1 V_{\rm A}$ from our results), and θ is the angle between the slow-mode MHD shock and the loop and is approximately given by $\sin\theta \approx V_{\rm in}/V_{\rm A}$. By manipulating these, we found

$$T_{\rm A} \approx \left(\frac{B^3 L}{2\pi\kappa_0\sqrt{4\pi\rho}} \right)^{\frac{2}{7}} \propto B^{\frac{6}{7}} \propto \beta^{-\frac{3}{7}}. \tag{1}$$

The simulation results (Fig. 3) show very good agreement with this scaling law. When $\beta = 0.01$, which is the typical value in the real solar corona, this scaling law predicts the flare temperature $T \approx 10$ MK (see also Fig. 3), which is consistent with the observed flare temperature $T \approx 10$–20 MK (e.g. Tsuneta et al., 1992).

Computations were carried out on the supercomputers at NAOJ.

References

Fisher, G. H., & Hawley, S. L. (1990), *ApJ*, **357**, 243
Neupert, W. M. (1968) *ApJ*, **153**, L59
Tsuneta, S. et al. (1992) *PASJ*, **44**, L63
Yokoyama, T., & Shibata, K. (1997), *ApJ*, **474**, L61
Yokoyama, T., & Shibata, K. (1998), *ApJ*, **494**, L113

RISING MAGNETIC FLUX AND CONVECTION

A.D. MCLEOD
Communications Research Laboratory,
Koganei-shi, Tokyo, Japan

1. Background and Motivation

For several theoretical reasons (e.g. magnetoconvection and magnetic buoyancy) and from recent helioseismology measurements, it is widely believed that the large-scale toroidal magnetic field of the Sun resides in the convective overshoot zone (just beneath the convection zone proper). This field is believed to be generated by a dynamo mechanism and the breakup and subsequent escape of this field leads to surface manifestations such as sunspots and active regions. One of the main problems with solar dynamo models is explaining how the magnetic field is held in the dynamo region for a period of about 10 years in order to be amplified up to observed levels. Simple models using thin flux tube approximations and parametrised eddy viscosities to model the effects of convection found that even weak flux tubes rise too quickly to be able to maintain a large-scale internal field.

In this project our aims are twofold: firstly, to make a start at modeling the interaction of rising magnetic flux tubes and convection more realistically (i.e. in 3D) and secondly, to see whether the convection itself can act in such a way to keep the magnetic field preferentially held down (presumably via the topological pumping effect, also possibly via the geometrical pumping effect). This should then be of help in deciding what kind of dynamo models are feasible. (N.B. Because of the computational complexity of this problem, we limit ourselves to a simple conceptual model, viz. an electrically conducting perfect gas between two horizontal rigid planes with periodic sidewalls.)

2. Model and Strategy

We model our region of interest as a rotating F-plane of convecting perfect gas between two impenetrable, stress-free boundaries. The gas conducts

S. M. Miyama et al. (eds.), Numerical Astrophysics, 331–332.

and thus is also capable of interacting with the magnetic field. Only a simple model is considered because a truely realistic model would be too complicated and is currently unfeasible.

A statistically steady time-dependent cellular convecting state is evolved and into this a horizontal magnetic field is injected, in order to simulate the rise of a magnetic flux tube into the convecting region. Magnetic field reaching the top of the computational box is allowed to diffuse out in order to prevent an excessive buildup of magnetic field there. Different latitudes can be obtained by altering the angle of the rotation vector for the calculation of the Coriolis force.

The computational scheme used involves an explicit 2nd order Adams-Bashforth method. The spatial derivatives are calculated using FFT's in the horizontal directions and 4th order finite differences in the vertical. This makes the code ideally suited to smooth, periodic sidewall problems.

The code is parallel (about the z-direction) and has a projection method divergence cleaner to ensure the magnetic field remains solenoidal. It is based on a code originally written by Cattaneo with the initial magnetic field modifications by Matthews. The code currently runs on the Fujitsu VPP700E/17 parallel supercomputer at CRL.

3. Conclusions

Two-dimensional work is very suggestive of the possibility of the magnetic field being pinned down by convection. However, in two-dimensions an aspect of the buoyancy of the tube is lost and so three-dimensional simulations are necessary in order to be able to understand fully the interaction.

The code has only recently been completed and so an extensive set of runs and comparisons has yet to be carried out. However, a small sortie into parameter space has been made. We carried out two short simulations injecting magnetic field [one purely horizontal (i.e. pure toroidal field), and one with a small "poloidal" component present as must be the case within the Sun]. The "poloidal" field can be expected to behave differently from the purely "toroidal" case as the field entering the layer should be more rigid in some sense. However, no significant difference was observed between the simulations, though that is probably more the fault of the parameters chosen than the "poloidal" field in general having no effect.

A visual comparison with the equivalent static problem suggests that at least in the early stages, the magnetic field's rise through the layer is slightly slower than pure diffusion. However, it should be re-iterated that this investigation into these effects and also those too of rotation is still in its infancy.

THE EXTRAPOLATED 3-D SOLAR MAGNETIC FIELDS IN AR 7321

HUANING WANG[1,2], TAKASHI SAKURAI[1]
[1]*NAOJ, 2-21-1 Osawa Mitaka, Tokyo 181-8588, Japan*
YIHUA YAN
[2]*Beijing Astronomical Observatory, Datun Rd. #20A, Chaoyang District, Beijing 100012, China*
AND
YANG LIU[2,3]
[3]*CRL, Hiraiso Solar Terrestrial Research Centre, 3601 Isozaki, Nakaminato, Ibaraki 311-12, Japan*

1. Introduction

The solar magnetic field plays an important role in the solar atmosphere. Since it is only on the photospheric surface that the magnetic field can be measured directly, a reconstruction of the 3-D solar magnetic field above this surface is necessary to explore the structure of solar magnetic fields. Several techniques for this reconstruction have been proposed (Sakurai, 1989; McClymont et al, 1997; Yan and Sakurai, 1997). We here employ a new numerical method suggested by Yan and Sakurai (1997).

2. Numerical computation

The 3-D nonlinear force-free magnetic fields in AR 7321 were computed on October 26, 1992 by means of a new numerical method (Yan and Sakurai, 1997) in which force-free fields are represented by a boundary integral equation based on a specific Green's function. The photospheric vector magnetic fields measured in AR 7321 with the Solar Flare Telescope at Mitaka in Tokyo were taken as the boundary conditions of this equation. In this numerical extrapolation, the following three points were emphasized: (1) A package of IDL software suitable for large data arrays was developed, such software being able to draw 800 magnetic field lines on a boundary

S. M. Miyama et al. (eds.), Numerical Astrophysics, 333–334.

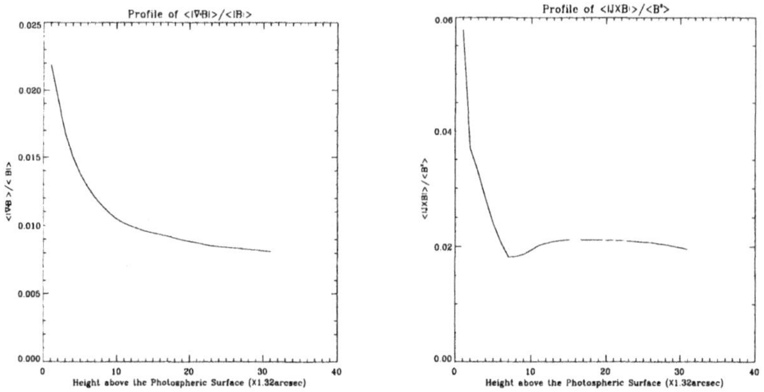

Figure 1. Examination of divergency and residual force of the extrapolated field

surface with 150×150 grid points in only 3 hours by means of SUN Sparc 10 workstation. This package, as is clear, was useful to deal with a large amount of observational data. (2) The extrapolated 3-D fields satisfied the divergence-free and force-free conditions with high precision (Figure 1). (3) A new method for data reduction was proposed, this method making extrapolated fields be in agreement with structures in H_α and Yohkoh soft X-ray images (Wang, Sakurai, and Yan, 1998).

3. Conclusions

The following conclusions can be drawn from our results: (a) The boundary integral method is a promising method in numerical computations of a 3-D static solar magnetic field; (b) Structures in the extrapolated 3-D magnetic fields are in agreement with those in H_α and soft X-ray images. This fact means that the boundary integral equation describes coronal magnetic fields very well; and (c) The reduction of boundary data is very important for the reconstruction of 3-D coronal magnetic fields due to uncertainties in these data. Our new method for data reduction makes the results of numerical computation reliable.

Numerical computations were carried out on VPP300/16R at NAOJ. Chinese authors are supported by NSFC grants.

References

Fletcher, C. 1984, *Computational Galerkin Methods*, Springer- Verlag, Tokyo.
McClymont A. N., Jiao, L. and Mikic Z. 1997 *Solar Physics*, **174**, 191
Sakurai, T. 1989, *Space Sci. Rev.*, **512**, 11
Yan, Y. and Sakurai, T. 1997, *Solar Physics*, **174**, 65
Wang, H. N., Sakurai, T. and Yan, Y. 1998, in preparation

THE MAGNETIC FIELD STRUCTURE OF NOAA 8100 ON 4 NOVEMBER 1997 FROM THE 3-D BOUNDARY ELEMENT SOLUTION OF THE FORCE-FREE FIELD

YIHUA YAN[1,3] AND YANG LIU[1,2]
[1]*Beijing Astronomical Observatory, Chinese Academy of Sciences, Beijing 100012, China*
MAKI AKIOKA[2]
[2]*CRL, Hiraiso Solar Terrestrial Research Center, 3601 Isozaki, Hitachinaka Ibraki 311-12 Japan*
AND
FENGSI WEI[3]
[3]*Laboratory of Heliospheric Physics, CSSAR, Chinese Academy of Sciences, Beijing 100080, China*

1. Instruments and event observations

The active region NOAA 8100 was the first event in the new 23rd solar cycle that produced GOES X-ray X-class flares during its pass on the Sun in November, 1997. A 2B/X2 flare occurred on November 4, 1997 in this active region, with the peak X-ray flux at 05:59UT (beginning: 05:54UT; ending: 07:13UT). Photospheric magnetograms taken at Huairou (Ai 1987), Hα images taken at Hiraiso (Akioka et al. 1994; Akioka 1996), and soft X-ray images observed by SXT on board Yohkoh (Tsuneta et al. 1991) have been used in this study.

2. Method and analysis

The boundary element method (Yan 1995) was used to solve the finite energy force-free magnetic field above the Sun (Yan & Sakurai 1996; Yan & Sakurai 1997). Thus, we obtained an approximation of the non-linear force-free field. Figure 1 shows the calculated field lines projected onto the photospheric magnetogram. The 2B/X2 flare occurred near the small N-polarity emerging flux region (EFR) in the north part of the region,

S. M. Miyama et al. (eds.), Numerical Astrophysics, 335–336.

and at the common footpoints of several magnetic loops that had different orientations. Thus, the flare may have been released due to the interaction of several interacting loops which have similar features to those of the 3B/X12 flare in NOAA 6659 in June 1991 (Yan & Wang 1995).

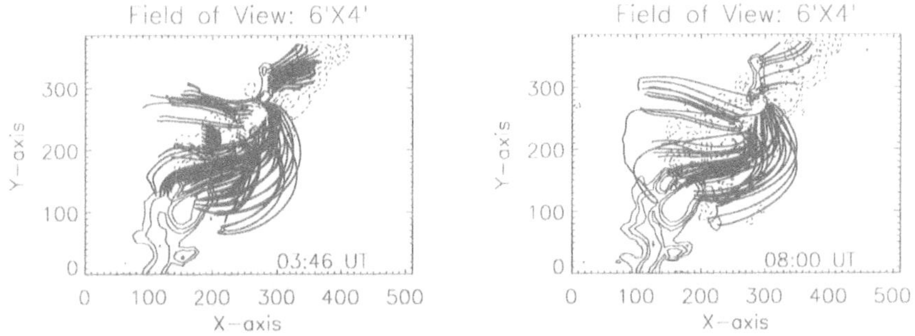

Figure 1. The calculated field lines before (a) and after (b) the flare. Solid (dash) contours indicate N(S)-polarity and thick solid lines are projected field lines

We found significant changes of magnetic structures at the EFR near the 2B/X2 flare site for both the pre- and post-flare cases, as shown in Fig. 1, which may indicate that the magnetic reconnection happened as a result of the energetic flare. However, the magnetic field lines also demonstrated a similar magnetic connectivity globally in both cases, as shown in Fig. 1. It should be noted that the twist situation did change, and the calculated magnetic field lines agreed well with the observed Yohkoh soft X-ray loops (e.g., attached CD-ROM). Other interesting features such as the structure of a soft X-ray bright point (Hanaoka 1996) were found as well. Detailed analyses of the present work will be published elsewhere.

Acknowledgments: This work was supported by the NNSF of China and partly by the SIGMA Weather Project. Y.Y. also thanks the CAS Education Bureau for its initial support. The Huairou, Hiraiso, and Yohkoh team members are acknowledged for providing the observational data.

References

Ai, G. (1987), Publ. Beijing Astron. Obs., **9**, 27
Akioka, M., Nishikawa, J., Isobe, T., and Igi, S. (1994), J. Commun. Res. Lab., **41**, 3
Akioka, M. (1996), J. Geomag. Geoelectr. , **48**, 5
Hanaoka, Y. (1996), Solar Physics , **165**, 275
Tsuneta, S., et al (1991), Solar Physics, **136**, 37
Yan, Y. and Wang, J. (1995), A&A, **298**, 277
Yan, Y. (1995), Solar Physics, **159**, 97
Yan, Y. and Sakurai, T. (1996), ApJ., submitted
Yan, Y. and Sakurai, T. (1997), Solar Physics, **174**, 65

MAGNETIC RECONNECTION AND FLARE LOOP MOTION

P. F. CHEN, C. FANG, Y. H. TANG AND M. D. DING
Department of Astronomy, Nanjing University, Nanjing, China

1. Introduction

The observation of two-ribbon flares show that, during a flare eruption, a bright loop rises, and the two H_α ribbons separate. To account for this phenomenon, a magnetic reconnection model was developed and generally confirmed by the high-resolution observations of *Yohkoh*. However, such a model needs further quantitative analysis.

We simulated a magnetic reconnection numerically, and the results showed loop motion. In particular, the effect of several physical quantities on the loop motion were studied. Section 2 describes the numerical method, the results being addressed in Section 3. Discussions and conclusions are made in Section 4.

2. Numerical method

The 2.5-dimensional MHD equations with dimensionless form as shown in Hu et al. (1997) were numerically solved to study magnetic reconnection with the inclusion of gravity. Heat conduction and radiation were not considered in the present study.

The domain of our simulation was taken to be $-3 \leq x \leq 3$, $0 \leq y \leq 8$. The initial magnetic configuration was a force-free field (a current sheet) surrounded by a potential field, similar to that of Hu et al. (1997). The anomalous resistivity was $\eta = \eta_0 \cos(5\pi x) \cos\{2(y-4)\pi\}$.

Note that in the dimensionless adiabatic MHD equations, all coefficients depend only on β_0 (the initial ratio of gas to magnetic pressures), T_0, L_0, and η_0 (characteristic values of temperature, length, and resistivity). These four quantities, as well as the half-width of the current sheet (w), serve as free parameters in this paper.

S. M. Miyama et al. (eds.), Numerical Astrophysics, 337–338.

3. Numerical results

The typical evolution can be described as follows: after resistivity is introduced, magnetic lines begin to reconnect. Two horizontal, convergent inflows set in and push field lines to form an 'X' shape. Two jets appear above and below the neutral point. The upward jet, along with the frozen-in field lines, flows out of the top boundary, which is an open boundary. Being line-tied with the bottom, the closed field lines below the point cease their decrease near the base to form an almost static magnetic loop. The downward jet collides with the magnetic loop, and a termination shock (Forbes & Priest 1983) is formed at the loop top. Besides, the plasma near the loop is also heated. As the newly-closed field lines pile up, the bright loop becomes higher and broader, giving the impression that the bright loop rises and its two footpoints separate. The loop rise is similar to the results of Magara et al. (1996). The rise speed of the loop (v_r) and the separation speed (v_s) are within the range of observational values (Nolte et al. 1979).

The results showed that: 1). $v_r \sim \beta_0^{-0.69}$, $v_s \sim \beta_0^{-0.7}$; 2). $v_r \sim T_0^{0.3}$, $v_s \sim T_0^{0.3}$; 3). $v_r \sim L_0^{0.06}$, $v_s \sim L_0^{0.01}$; 4). $v_r \sim w^{-0.05}$, $v_s \sim w^{-0.03}$; and 5). $v_r \sim \eta_0^{0.007}$, $v_s \sim \eta_0^{-0.002}$.

4. Discussions and conclusions

It was shown that resistivity (η_0) has no effect on loop motion speeds. Since there may be considerable numerical resistivity in our present simulation, such results should be checked in future simulations.

The numerical results in this paper support the so-called CSHKP model for two-ribbon flares. Besides, it is shown that: 1) The speeds of the loop motion are strongly dependent on β_0, and less dependent on T_0. However, L_0 and w (half-width of the current sheet) have a very weak effect on the speeds. This may mean that loop motion speeds have geometrical similarity. 2) Apart from the apparent loop motions, the reconnected field lines shrink with an amplitude about 10%, which qualitatively agrees with previous observations (e.g., Forbes & Acton 1996).

This work is supported by a fund from the National Natural Science Foundation of China and by a key project from the Chinese National Science and Technology Committee.

References

Forbes, T. G., & Acton, L. W. (1996) ApJ, Vol.459, p.330.
Forbes, T. G., & Priest, E. R. (1983), Solar Phys., Vol.84, p.169.
Hu, Y. Q. et al. (1997), Solar Phys., Vol.170, p.283.
Magara, T. et al. (1996), ApJ, Vol.466, p.1054.
Nolte, J. T. et al. (1979), Solar Phys., Vol.62, p.123.

A MAGNETOHYDRODYNAMIC MODEL TO DARK FILAMENT ERUPTION AND ARCADE FLARING

A Model with Quadruple Magnetic Source

S. HIROSE, Y. UCHIDA, S. UEMURA AND T. YAMAGUCHI
Department of Physics, Science University of Tokyo
1-3 Kagurazaka, Shinjuku-ku, Tokyo 162-8601, Japan

AND

S. CABLE
Department of Physics, Auburn University
206 Allison Laboratory, Auburn University, AL 36849, USA

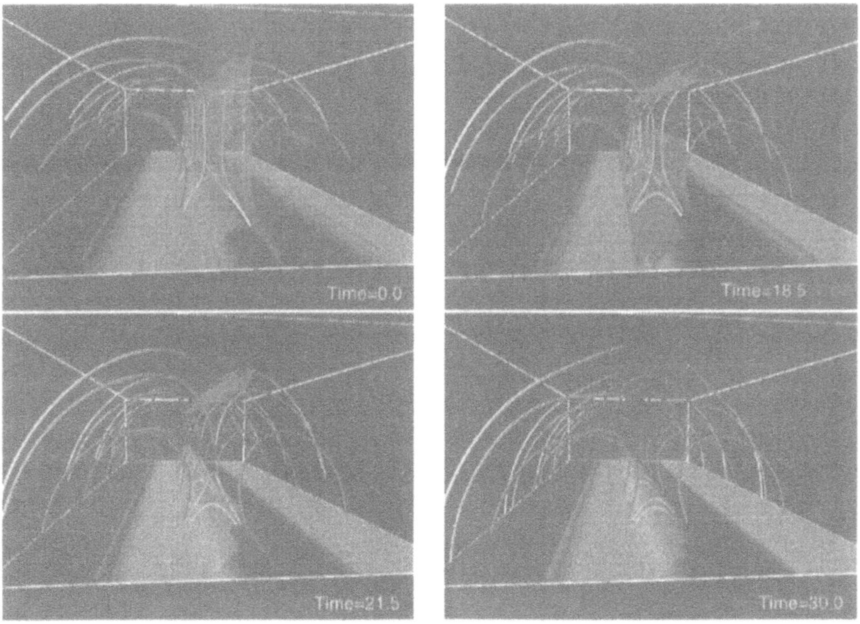

Figure 1. 3-dimensional "arcade" structure of the system and its time evolution. Dark filament gas (iso-density surface) and magnetic lines of force (tubes) are drawn.

We performed a 2.5-dimensional magnetohydrodynamic simulation for dark filament eruption and arcade flaring in the solar corona. The simula-

S. M. Miyama et al. (eds.), Numerical Astrophysics, 339–340.

tion was based on the quadruple magnetic source model created by Uchida (1980). The results of the simulation are summarized as follows (See CD-ROM for details.). In a quadruple source field energized by the photospheric motion, a certain amount of gas can be supported quasi-stably in the current sheet region. This "dark filament" gas contains a helical magnetic island and prevents magnetic reconnection (Figure 1: Time=0.0). After the dark filament gas is squeezed out both upwards (dark filament rise) and downwards (down-pressed S-shaped X-ray filament) by external triggers, magnetic reconnection can occur, releasing the stored magnetic stress energy (Figure 1: Time=18.5−21.5). (The erupted gas oscillates above the arcade and does not go away in this case, which may be due to the 2.5-dimensional approximation.) Thus, it is shown that this model can explain several key characteristic features of dark filament eruption as well as the arcade flaring found by Yohkoh.

Numerical computations were carried out on VX/4R at NAOJ.

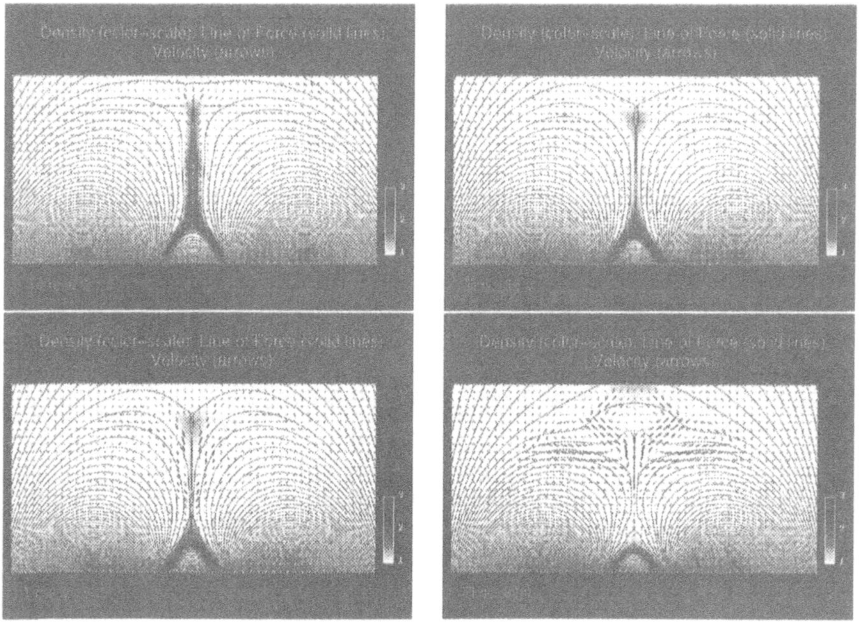

Figure 2. Time evolution of density (gray-scale), velocity field (arrows) and magnetic field (solid lines) in the section of the 3-dimensional "arcade" structure. Time sequence corresponds to that of Figure 1.

References

Uchida, Y., 1980, in *Skylab Workshop, Solar Flares*, ed. P.A.Sturrock (University of Colorado Press), p67 and p110.

THE 3-DIMENSIONAL NUMERICAL SIMULATION OF SOLAR FLARES

T. MAGARA
Department of Astronomy, Faculty of Science,
Kyoto University, Sakyo-ku, Kyoto 606-01, Japan

AND

K. SHIBATA
National Astronomical Observatory
Mitaka, Tokyo 181, Japan

In order to understand various 3-dimensional dynamical processes observed in solar flares, we have performed 3-dimensional MHD numerical simulations and obtained some preliminary results. In the following, we show these results and make some comments on them.

Figure 1 is a 3-dimensional view showing how the magnetic field evolves under a locally enhanced resistivity. Initially, physical variables are set to be in the state of force-free configuration. As the initial perturbation, we impose a locally enhanced resistivity around the central part of the calculation region. Lines represent the magnetic field lines and a color map represents the temperature. Elapsed times are $t = 0$ (top-left), 4 (top-right), 8 (bottom-left), and 12 (bottom-right).

This figure clearly indicate some important features of this simulation. First, as a result of the magnetic reconnection, sheared arcades are formed near the base of the calculation region. Moreover, by looking at the temperature map, we find that two hot sources are formed near the footpoints of these arcades. Those sources are produced by the hot plasma flows originating from the region where the resistivity is imposed (see the panels at $t = 8$, and 12 in both figures). These sources develop in both sides of the neutral line with time, which could be observed as Hα ribbons in solar flares. (Actually, the thermal conduction is considered to play a dominant role in the formation of Hα ribbons rather than such hot plasma flows.) Casting our eyes to the upper part of the calculation region, we find that a twisted magnetic-field line appears (see a blue line), which corresponds to a helical rope, i.e., a 3-dimensional version of magnetic islands seen frequently in

S. M. Miyama et al. (eds.), Numerical Astrophysics, 341–342.

2-dimensional MHD numerical simulations. (In the present study, an open-field configuration is assumed as the initial state so that a complete helical structure is not reproduced.)

Data analyses were carried out on VPP300/16R at NAOJ.

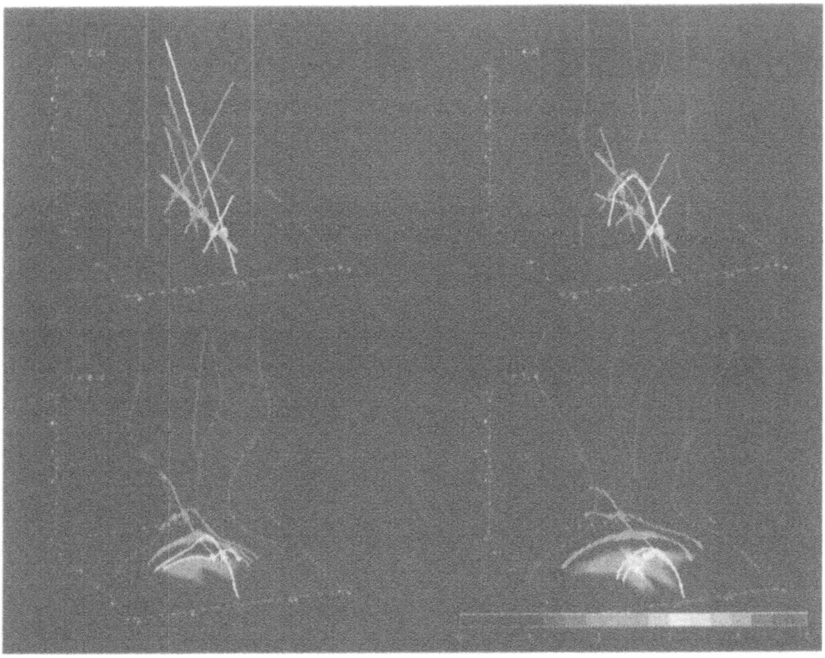

Figure 1. 3-dimensional view showing how the magnetic field evolves under a locally enhanced resistivity. Lines represent the magnetic field lines and a color map represents the temperature. Elapsed times are $t = 0$ (top-left), 4 (top-right), 8 (bottom-left), and 12 (bottom-right), which are normalized by the sound crossing time.

References

Magara, T., Mineshige, S., Yokoyama, T., and Shibata, K., 1996, *ApJ.*, **466**, 1054.
Magara, T., Shibata, K., and Yokoyama, T., 1997, *ApJ.*, **487**, 437.
Magara, T., 1998, Ph. D. Thesis, Kyoto Univ.
Shibata, K. 1996, *Adv. Space Res.*, **17**, (4/5) 9.

THREE-DIMENSIONAL MHD SIMULATIONS OF THE EMERGENCE OF TWISTED MAGNETIC FLUX TUBES IN THE SOLAR ATMOSPHERE

H. TONOOKA
Graduate School of Science and Technology, Chiba University, Inage-Ku, Chiba 263, Japan

R. MATSUMOTO
Faculty of Science, Chiba University, Inage-ku, Chiba 263, Japan

W. CHOU AND T. TAJIMA
Institute for Fusion Studies, the University of Texas at Austin, Austin, TX 78712, USA

AND

K. SHIBATA
National Astronomical Observatory, Mitaka, Tokyo 181, Japan

The emergence of twisted magnetic flux tubes in the Solar atmosphere is studied by three-dimensional magnetohydrodynamic (MHD) simulations. The initial state is a plane parallel, hydrostatic atmosphere with convectively unstable layer, isothermal cool photosphere/chromosphere and hot corona. A horizontal, twisted flux tube with Gold-Hoyle magnetic field distribution is imbedded in the convection zone. Small amplitude perturbation is imposed on the initial state. Typical number of grid points used in the simulations is $125 \times 125 \times 208$.

When the twist angle q over the length of the flux tube L is sufficiently large ($qL \gg 2\pi$), the flux tube is deformed into a helical shape due to the kink instability. On the other hand, for sufficiently long perturbations ($\lambda_p > 10H$) the undular mode of the magnetic buoyancy instability (Parker instability) undulates the magnetic flux tube. Figure 1 shows the result of three-dimensional simulation for a slim flux tube with $R = 0.25H$, $qH = 16$, $L = 12.5H$ and $\beta = P_{\rm gas}/P_{\rm mag} = 2$ where H is the scale height at $z = 0$. (see Matsumoto et al. 1998 for details) When the top portions of such helically deformed, undulating flux tube appear in the solar atmosphere,

S. M. Miyama et al. (eds.), Numerical Astrophysics, 343–344.

the emergent portions may show a series of S-shaped, kinked alignment of X-ray emitting regions as observed by the Yohkoh satellite. Figure 2 shows magnetic field lines projected on the photosphere for a model with $R = H$, $qH = 4$, $L = 10H$ and $\beta = 2$. Intense vertical magnetic fields corresponding to sunspots are created at the footpoints of the loops. Numerical results also reproduced typical motions of p- and f- spots.

Numerical computations were carried out on Fujitsu VPP300/16R at the Astronomical Data Analysis Center of NAOJ.

References

Matsumoto R., Tajima T., Chou W., Okubo A., and Shibata K., 1998, ApJ, **Vol.493**, L43-L46

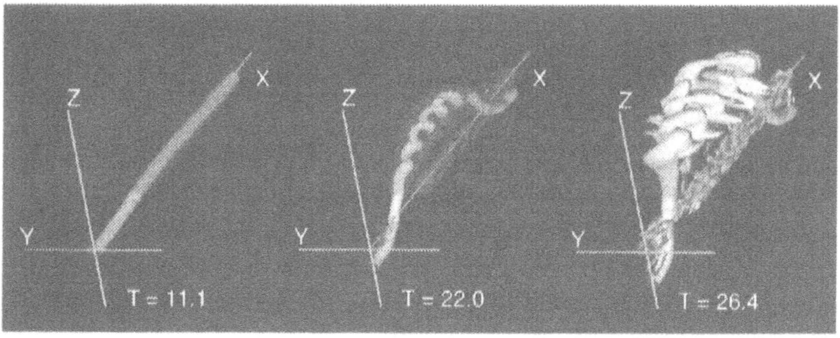

Figure 1. Isosurfaces of magnetic field strength for slim flux tube simulation. $(R = 0.25H)$

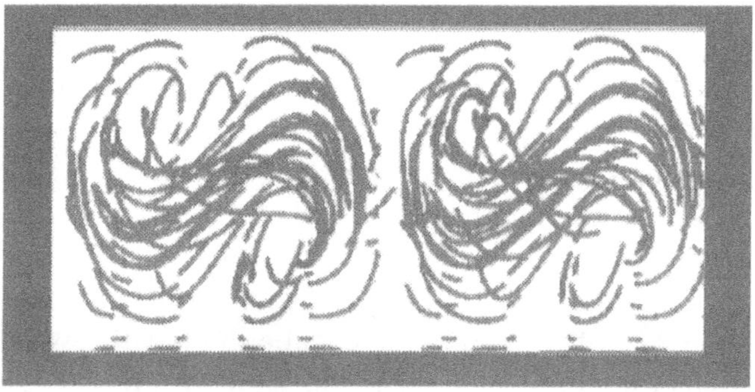

Figure 2. The projection of magnetic field lines onto the $x - y$ plane. The magnetic field lines below the photosphere are masked.

A RADIATIVE WIND MODEL FOR ACTIVE HOT STARS

PH. STEE, D. BONNEAU, D. MOURARD AND F. VAKILI
Equipe GI2T
Observatoire de la Côte d'Azur
2130 route de l'Observatoire
Caussols
06460 St Vallier de Thiey, France
email: stee@obs-nice.fr

Abstract. Radiative wind from active hot stars has been studied for more than two decades. During this period, numerous models which try to fit a small number of observables as closely as possible have been developed, such observables being Hα line profile and continuum energy distribution or polarization data. Rapid progress has been made in high angular resolution technics; however, little has been done to include these high angular resolution observations in simulations in spite of the fact that such data can strongly constrain radiative wind models. In this paper, the way in which long baseline interferometry can be used to constrain different models will be illustrated by means of a radiative wind model for active hot stars that we have developed at Observatoire de la Côte d'Azur, in France.

1. What are Be stars?

Be stars are a typical class of active hot stars which possess a B spectral type and, at one time in their lives, have exhibited emission lines in their spectra, most often Balmer lines. Be stars are not super-giant stars and are presumed to be fast rotators, spinning at 0.5 to 0.9 of their critical velocity. In addition, they have a large stellar wind and high mass loss rate. This stellar wind seems to be at the origin of a *two-component* envelope, which is characterized by an equatorial plane with high density and low expansion where the Balmer emission lines are formed. The polar region, where the UV absorption line profiles are Doppler-shifted up to 2000 km/s, is characterized by low density and high expansion.

S. M. Miyama et al. (eds.), Numerical Astrophysics, 345–346.

2. A radiative wind model for active hot stars

We have built a model which reproduces both spectroscopic, photometric, and interferometric synthetic data that have been compared to observations from the Grand Interféromètre à 2 Télescopes (GI2T) and other instruments. This model is a latitude-dependent radiative wind model for Be stars which clearly shows that the morphology of the circumstellar envelope depends strongly on the central observational wavelength and bandwidth (see the animated picture of γ Cas by Stee et al. in the attached CD-ROM). The line force in our model is the same as that used by Friend & Abbott (1986), but we introduce a varying contribution of thin and thick lines from pole to equator by adopting latitude-dependent radiative parameters. The velocity fields and density relationships derived from the hydrodynamic equations are then used for solving the statistical equilibrium equations. By adopting the Sobolev approximation, we could easily obtain a good estimate of both electron density and hydrogen level populations throughout the envelope. A numerical calculation was performed for parameters characteristic of the Be star γ Cassiopeiae. This model indicates that a radiative wind, driven mainly by optically thin lines at the equator, is a likely scenario for γ Cas. This is discussed in greater depth in Stee et al. (1995).

We also studied the rotational component of the wind in the envelope of γ Cas (Stee 1996), showing that only technics which can follow the photocenter displacement with wavelength, such as Michelson's spectro-interferometry or differential speckle interferometry, could infer the actual rotational law in the envelope of γ Cas. Finally, in a recent paper by Stee et al. (1998), we reported the first spectrally-resolved observations of γ Cas in the He I λ6678, and Hβ emission lines using the GI2T. We concluded that the emitting region must be smaller than 8.5 stellar radius in Hβ and close to 2.3 stellar radius in He I λ6678. This is, for He I λ6678, smaller than the nearby continuum extent. These results confirmed the γ Cas basic parameters for this star obtained by Stee et al. (1995).

References

Friend, D.B., Abbott, D.C. 1986, ApJ, **311**, 701
Stee, Ph., Araújo, F.X., Vakili, F., Mourard, D., Arnold, L., Bonneau, D., Morand, F. and Tallon-Bosc, I. 1995, A&A, **300**, 219
Stee, Ph. 1996, A&A, **311**, 945
Stee, Ph., Vakili, F., Mourard, D., Bonneau, D. 1998, A&A, **332**, 268

FORMATION OF THE 3-RING STRUCTURE IN THE SN 1987A NEBULA DUE TO THE MAGNETIC PRESSURE FORCE

H. WASHIMI
Shonan Institute of Technology
Tsujido, Fujisawa 251-8511, Japan

T. TANAKA
Communications Research Laboratory
Koganei, Tokyo 184-8795, Japan

AND

S. SHIBATA
Faculty of Sciences, Yamagata University
Yamagata 990, Japan

In this paper, we present a simulation for the formation of a 3-ring structure in the SN 1987A nebula obtained by HST observations. The progenitor of the SN 1987A is believed to be a red supergiant (RSG) which then evolved into a blue supergiant (BSG) just prior to the explosion. In such a case, the circumstellar material surrounding the SN should be composed of two types of stellar wind: a dense and slow RSG wind far removed from the SN, and a dilute, fast BSG wind near it.

We assumed that both the RSG and BSG had a dipole magnetic field around their rotation axes. This magnetic field then extended radially in the circumstellar space by stellar wind expansion. When the star rotated with angular speed Ω, a toroidal magnetic field B_ϕ was formed in the circumstellar space, i.e., $B_\phi = (R\cdot\Omega/V)\cdot B_r \cdot \sin\theta$. Here, R is the distance from the star, V the speed of the stellar wind in radial direction, B_r the radial field, and θ the colatitude angle. We also assumed that the wind speed in high and middle latitudes was faster than that in low latitudes, as observed by the Ulysses mission in the solar wind. Under the above conditions, the toroidal magnetic field was enhanced in the wind-wind interaction region, and the magnetic pressure effect began to work, resulting in the formation of the equatorial ring. The formation of the single, equatorial ring by the above MHD process has been discussed before (Washimi et al., 1996).

S. M. Miyama et al. (eds.), Numerical Astrophysics, 347–348.

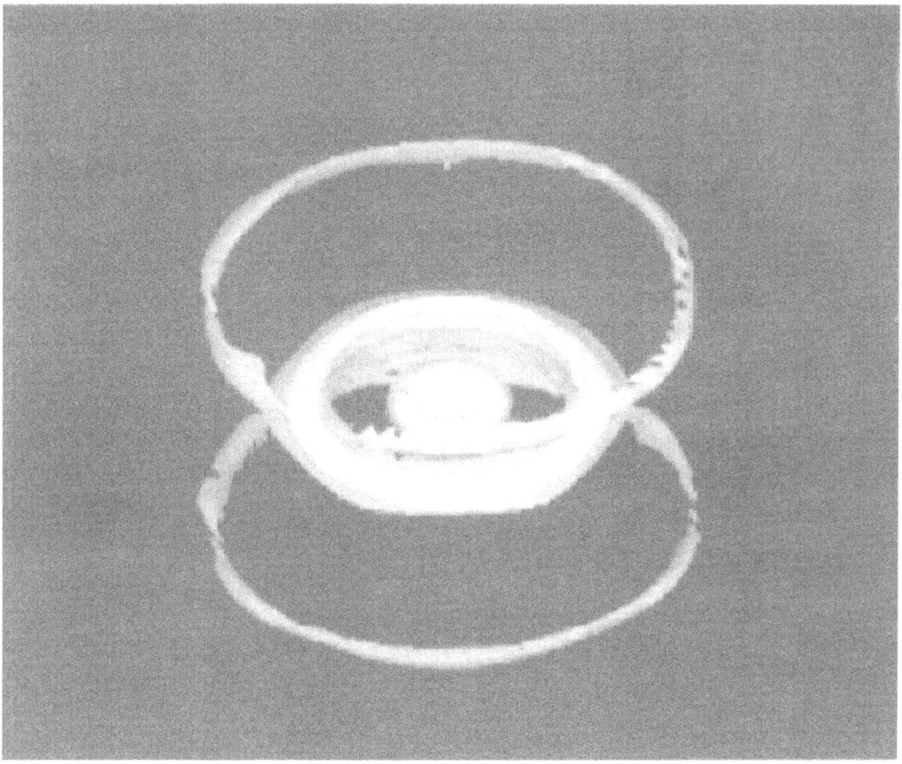

Figure 1. The 3-ring structure of the SN 1987A obtained by MHD simulation.

We now show that, in addition to the equatorial ring, the outer rings at a midlatitude in both the northern and southern hemispheres can also be reproduced in our MHD simulation. For this 3-D MHD simulation, we used the TVD finite-volume method (Tanaka, 1994). Our ring structure is shown in Fig. 1. This figure shows that the outer rings are formed by the balance of the poleward magnetic pressure force and the equatoward ram-pressure force at the middle latitudes.

References

Washimi, H., Shibata, S. and Mori, M. (1996) Formation of the ring-like structure in the SN1987A nebula due to the magnetic pressure of the toroidal field, PASJ, **vol. 276**, pp. 23–28.

Tanaka, T. (1994) Finite volume TVD scheme on an unstructured grid system for three-dimensional simulation of inhomogeneous systems including strong background potential field, J. Comp. Phys., **vol. 112**, pp. 381–389.

A DREAM TO SOLVE ALL PHASES OF MATTER, SOLID, LIQUID, AND GAS, SIMULTANEOUSLY

T. YABE, Y. OGATA, T. NAKAMURA AND Y. ZHANG

Department of Energy Sciences, Tokyo Institute of Technology
4259 Nagatsuta, Midori-ku, Yokohama 226-8502, Japan

1. Introduction

Various innovations in basic technology have made the developing speed of computer hardware faster than that of software technology. Similar innovations are required in order to accelerate the developing speed of software. A universal solver for computational engineering problems would be such an innovation. Recently developed high technology requires new tools for the combined analysis of materials in the different phase states, i.e., solid, liquid, and gas. A universal treatment of all phases by one simple algorithm is essential, and we are at a turning point to achieve this goal. We need to treat the topology and phase changes of the structure in processes like welding and cutting. In processes such as freezing, condensation, melting, and evaporation, the grid system initially aligned to the solid or liquid surface does not provide correct information of the surface at later time, and the mesh is sometimes distorted or even broken up.

Toward this goal, we have taken the Eulerian-approach based on the CIP (Takewaki *et al.*, 1985; Yabe *et al.*, 1991; Wang *et al.*, 1993; Yabe and Xiao, 1995; Ida and Yabe, 1995; Xiao *et al.*, 1996a, b) method developed by the author, which does not need the adaptive grid system and therefore eliminates the problems of grid distortion caused by structural break-up and topology change. Almost all the material surface can be captured by one grid throughout the computation. Furthermore, the code can treat all the phases of materials from solid through liquid state and two-phase state to gas without restrictions on the time step from high-sound speed. In this paper, a historical review of the CIP method and its strategy is given alongside with some examples.

S. M. Miyama et al. (eds.), Numerical Astrophysics, 349–356.

2. CIP Method

In order to attempt to solve the problems mentioned above, we must first find a method to solve the interaction of compressible gas with incompressible liquid or solid. For compressible fluid, elaborate schemes like TVD or ENO have proved to be quite effective in capturing shock waves. However, since these schemes employ a conservative form of fluid equations, divergence of velocity which becomes zero in the incompressible limit cannot be treated independently of the advection part. On the contrary, incompressible schemes like QUICK or higher-order upwind schemes combined with the improved MAC (Marker and Cell) procedure can treat divergence-free fluid vorticity and turbulence. However, these schemes cannot always treat a shock wave as a sharp discontinuity.

A scheme to treat both compressible and incompressible fluids simultaneously in one program is needed to simulate the interaction of gas with liquid or solid. Fully implicit solvers can treat this procedure, but the convergence of iteration in highly distorted states remains a problem. Recently, a new type of scheme CIP was proposed treat shock waves by a non-conservative scheme. By a simple extension, the CIP can be used for both compressible and incompressible fluids simultaneously. This code, called the CCUP (CIP Combined and Unified Procedure, Yabe and Wang, 1991; Xiao *et al.*, 1996b) can treat incompressible fluid with full hydrodynamic equations. In combination with the surface capturing scheme (called "digitizer" hereafter) presented in a previous paper (Yabe and Xiao, 1993; Xiao *et al.*, 1996b), this scheme provides a useful tool to describe various physical processes which had never been examined before.

In this section, we review the CIP method and illustrate its underlying principles. The key issues of the CIP method are the representation of the advection term and the separate treatment of other terms. By this separation, the code can be extensible to compressible and incompressible fluids. Let us first start with a one-dimensional linear advection equation.

$$\frac{\partial f}{\partial t} + u\frac{\partial f}{\partial x} = 0, \tag{1}$$

The solution of Eq.(1) gives a simple translational motion of wave with a velocity u. The initial profile [solid line of Fig.1(a)] moves like a dashed line in a continuous representation. At this time, the solution at grid points is denoted by closed circles and is the same as the exact solution. However, if we eliminate the dashed line as in Fig.1(b), it is difficult to visualize the original profile, and one may incline to retrieve the original profile like that shown by a solid line in (c). Thus, numerical diffusion arises when we construct the profile by linear interpolation even with the exact solution as shown in Fig.1(c). This process is the first-order upwind scheme. On the other hand, if we use quadratic polynomial for interpolation, it suffers from overshooting. This other process is the Lax-Wendroff scheme or the Leith scheme.

What made this solution worse ? For this reason, we have neglected the behavior of the solution inside the grid cell and tried to achieve a smooth solution. To this end, it is necessary to consider how to incorporate the real solution into the profile within a grid cell. We propose to approximate the profile as shown below. We differentiate Eq.(1) with spatial variable x; then, we get

$$\frac{\partial g}{\partial t} + u\frac{\partial g}{\partial x} = -g\frac{\partial u}{\partial x}, \tag{2}$$

where g stands for the spatial derivative of f, $\partial f/\partial x$. In the simplest case, where the velocity u is constant, Eq.(2) coincides with Eq.(1) and represents the propagation of spatial derivative with a velocity u. By this equation, we can trace the time evolution of f and g on the basis of Eq.(1). If g propagates as shown by the arrows in Fig.1(d), the profile after one step is limited to a specific profile. The solution is likely to become very close to the initial profile by this limitation. If two values of f and g are given at two grid points, the profile between these points can be described by cubic polynomial $F(x) = ax^3 + bx^2 + cx + d$. Thus, the profile at $n+1$ step can be obtained by shifting the profile by $u\Delta t$ like $f^{n+1} = F(x - u\Delta t)$, $g^{n+1} = dF(x - u\Delta t)/dx$.

3. Compressible and Incompressible Fluid

Since we are treating hydrodynamic equations in a non-conservative form, it is easy to extend them to include both incompressible and compressible fluids. In Fig.2, we plot the iso-temperature contour of aluminum automatically generated from the semi-analytical formula. In the gas phase where density is sufficiently low, the pressure is in proportion to the density. Therefore, we may solve the density first. Then, after the temperature is obtained, we use EOS (equation of state) in Fig.2. However, near the solid density of 2.7 g cm^{-3}, the pressure rises very sharply. If we use the same procedure there, the pressure can easily change by 3-4 orders of magnitude even with a small error of density of about a few tens of percent. Therefore, the strategy to solve the density first is not applicable in this area. This reinforces the idea of the difficulty in finding a universal treatment for solid, liquid, and gas matter. An interesting technique was devised by physicists in incompressible fluids to try to solve this problem. We will translate the strategy they used from a different viewpoint and reconsider the technique. If the pressure is very sensitive to the density, pressure has to be solved first. This means that we should rotate Fig.2 by 90 degrees. If we have a way

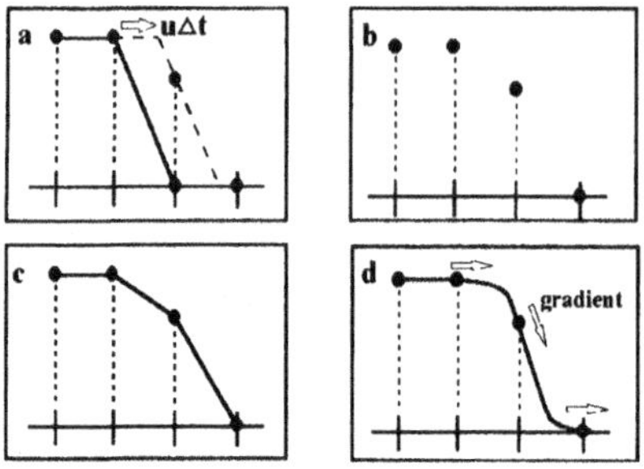

Figure 1

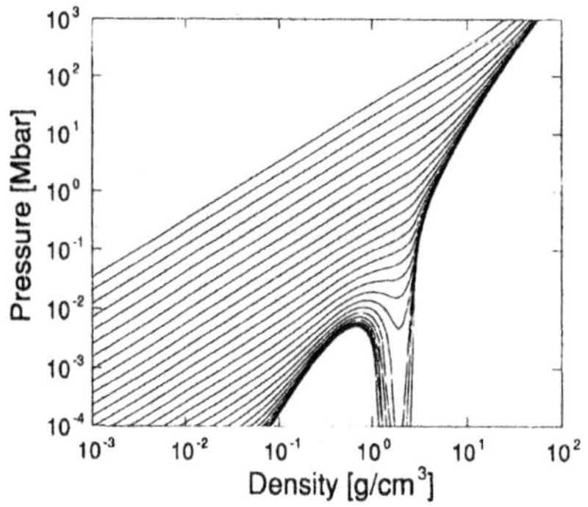

Figure 2

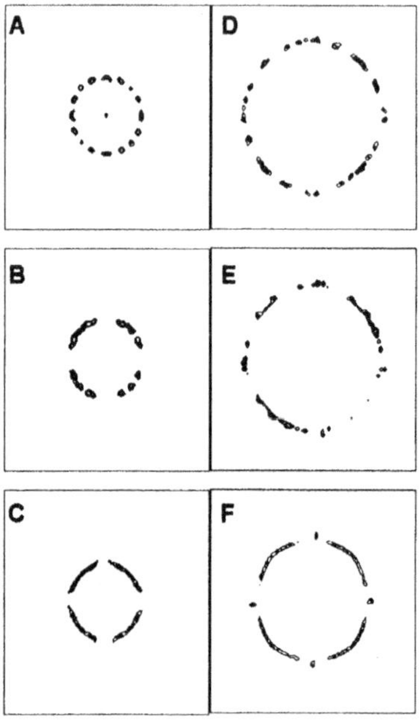

Figure 3

Figure 1. The principle of the CIP method. (a) the solid line is the initial profile, and the dashed line is an exact solution after advection, whose solution (note: "whose solution" is an incomplete phrase) (b) at discretized points (note: "at discretized points is also a fragment.) (c) When (b) is linearly interpolated, numerical diffusion appears. (d) In the CIP, the spatial derivative also propagates and the profile inside a grid cell is retrieved.
Figure 2. Equation of state of aluminum. Each line represents isotherms.
Figure 3. Density contour of the rim for the gas density of (A and D) 0.001, (B and E) 0.005 and (C and F) 0.01 g cm^{-3}. The left three figures (A, B, C) show the contours at 80 ms, and the right three figures (D, E, F) show those at 250 ms.

to solve pressure first, then we obtain a very accurate density at the solid density. Since the pressure is proportional to the density in the gas phase, this strategy does not jeopardize the solution there either. Our method to realize this strategy starts with the thermodynamic relation:

$$\Delta p = \left(\frac{\partial p}{\partial \rho}\right)_T \Delta\rho + \left(\frac{\partial p}{\partial T}\right)_\rho \Delta T, \tag{3}$$

where $\Delta p = p^{n+1} - p^*$ and $*$ represents a profile after advection. The same expression is used for ρ and T. Therefore, if $\Delta\rho$ and ΔT are predicted, Δp can be obtained since $(\partial p/\partial\rho)_T$ and $(\partial p/\partial T)_\rho$ are already given from EOS. The advantage of the CIP is the separate treatment of the non-advection term. Therefore, we can limit our discussion here to the non-advection term only. We should note that this advantage is crucial to obtain the final result. Then, we finally get

$$\nabla\left(\frac{\nabla p^{n+1}}{\rho^*}\right)=\frac{p^{n+1}-p^*}{\Delta t^2\left(\rho^* C_s^2+\frac{P_{TH}^2}{\rho C_v T}\right)}+\frac{\nabla\cdot\mathbf{u}^*}{\Delta t}, \tag{4}$$

where C_v is the specific heat for constant volume and $P_{TH} = T(\partial p/\partial T)_\rho$, $(C_s)^2 = (\partial p/\partial \rho)_T$. It is very important to note that in Eq.(4), ρ is inside the derivative on the left-hand side. At the interface between materials having large density differences, the continuity of acceleration $\nabla p/\rho$ is very important because the denominator ρ can change by several orders of magnitude in one grid. Equation (4) guarantees the continuity of $\nabla p/\rho$ at the discontinuity. By this procedure, we can treat all the material at once by simply changing its equation of state. Again, this property is a consequence of the separate treatment of advection and non-advection terms; otherwise, the continuity of $\nabla p/\rho$ is not guaranteed and large densities can not be traced.

4. Application

Since the pioneering work by Worthington (1908), the coronet or the so-called milk-crown has been investigated mostly by experiments that revealed many interesting features. However, simulation of the coronet has long been a dream in the field of computational physics because it can not only demonstrate the power of the scheme but also investigate the physics of coronet formation by choosing a situation which can be hardly realized by experiments. Although the present-day technique is already close to this goal, nobody has reported the three-dimensional coronet formation before except for several two-dimensional works pioneered by Harlow and Shannon (1967), in which instability and therefore break-up of rim leading to the coronet were not simulated. This is because the coronet is not merely a consequence of a free surface problem; ambient gas must be treated as well. This requires a special treatment at the complex boundary of the coronet where density ρ changes by 1000 times. In solving pressure p, it is quite a difficult task to guarantee the continuity of force $(\nabla p)/\rho$ across the complex interface, which has a very large density ratio. If this is not done, however, quite a large acceleration can occur because of the change of denominator from 1 to 0.001 g cm^{-3} across the interface.

The CIP has the advantage of being capable of solving all the phases of matter simultaneously from solid through liquid to gas because it can treat very large density changes while keeping the continuity of $(\nabla p)/\rho$ at the interface without any special boundary condition. These advantages of the

CIP method have been demonstrated in the simulation of the melting and evaporation dynamics of metal irradiated by laser light (Yabe and Xiao, 1995; Yabe *et al.*, 1996). This capability is essential for application to the instability during the dynamics of coronet formation.

We used 100×100(horizontal)$\times 34$ (vertical) fixed, equally spaced Cartesian grids with 0.1 cm grid spacing. A thin water film of 0.4 cm thick was placed on a solid plate, and a water drop of 0.8 cm radius impacted from the top at a speed of 50 cm s^{-1}. We solved gas as well as water. The density ratio at the interface was almost 1000, as already mentioned. At $t =$ 52 and 88 ms, the irregular ring, which we call "ornament" of the coronet, appeared at first, a laminar belt developing below the ornament later on.

If the break-up of the rim had been caused by the effect of finite-sized rectangular grids, its effect would have already appeared at an early stage ($t \leq 44$ ms). However, break-up abruptly occurred just at the time when deceleration started, and thus some instability might have taken place. Furthermore, as will be shown later, the wavelength of the irregularity along the rim changed depending on the density of ambient gas. This proves that the breakup observed here is not an artifact of finite grid size although it might have provided a reason for the instability.

From the simulation results, we can estimate the deceleration of the motion to be $a = 2.3 \times 10^3$ cm s^{-2}. From typical wavelength 0.67 cm ($k = 10$ cm^{-1}) of the irregularity along the rim during deceleration, we get the growth rate of $\gamma_{RT} = (ak)^{1/2} = 151$ s^{-1} for the R-T (Rayleigh-Taylor) instability. This growth rate seems to be sufficient to account for the evolution of the irregularity because of $\gamma_{RT}\Delta t = 3.93$ during the deceleration time $\Delta t = 26$ ms.

Since computer capability is limited, the validity of the theory cannot be checked by largely changing the parameters. The easiest and most interesting way to alter the situation is to change the gas density ρ_{gas}. Although denser gas seems to be unrealistic, it might be helpful to check the physics of situations that can only be realized by computer simulation. Since gas density is 1000 times less than that of water, it may not contribute to the evolution of the coronet. Surprisingly, however, it sensitively changed its evolution, as shown in this paper.

Figure 3 shows the density contour of water sliced horizontally in the middle of the rim. Interestingly, unstable wavelength became longer when gas density was increased, and finally at the gas density of 0.01 g cm^{-3}, the mode number was reduced to 4. Figures 3A-C show the density contour at $t = 80$ ms, and Figures 3D-F show that at $t = 250$ ms. As shown in the figure, the different nature of the breakup is already seen at this early stage of $t = 80$ ms when the deceleration has just stopped.

We have confirmed that the time evolution was similar for any density of gas, and the deceleration time was slightly longer for denser gas. If this deceleration had caused the breakup, then denser gas and hence long deceleration time would have made the irregularity be slightly larger. This contradicts the observation in Fig.3. Although the gas density comes into the growth rate as the Attwood number $= [(\rho_{\text{water}} - \rho_{\text{gas}})/(\rho_{\text{water}} + \rho_{\text{gas}})]^{1/2}$, it is already too small to change the number even for denser gas of 0.01 g cm^{-3} compared with the water density $\rho_{\text{water}} = 1$ g cm^{-3}.

If the instability is caused by the K-H (Kelvin-Helmholtz) instability, the growth rate is $\gamma_{\text{KH}} = k|U|(\rho_{\text{water}}\rho_{\text{gas}})^{1/2}/(\rho_{\text{water}} + \rho_{\text{gas}})$, where U is the relative velocity between gas and water and can be affected by the ambient gas density by a factor of 3 for 10 times difference of density. However, the growth rate is larger for denser gas, which contradicts the simulation result.

One possible explanation for this result is that the K-H instability drives the swelling-up of the rim to the vertical direction rather than to break-up, while the R-T instability owing to radial acceleration causes the break-up of the thin film of the rim along the azimuthal direction, thus causing the "ornament." If $\gamma_{\text{KH}} \geq \gamma_{\text{RT}}$, then the break-up becomes slower than the swelling-up, and the coronet is thus not formed. Therefore, the criterion for the break-up should be

$$\gamma_{\text{RT}} \geq \gamma_{\text{KH}} \Longrightarrow \frac{k\rho_{\text{water}}\rho_{\text{gas}}}{(\rho_{\text{water}} - \rho_{\text{gas}})(\rho_{\text{water}} + \rho_{\text{gas}})} \leq \frac{a}{U^2}, \tag{5}$$

If this relation is correct, the maximum wave number is inversely proportional to gas density. In fact, the wavelength of the ornament was approximately proportional to gas density. This justifies the relation (5) because a and U were similar regardless of gas density.

5. Summary

We have proposed a new tool to achieve the simultaneous solution of all the phases of matter. The success of the code is due to its high ability to trace sharp interfaces even with fixed grids and flexibility of extension to various materials and physics. The code's scope of application is quite large. It has already been applied, for example, to the break-up of comet Shoemaker-Levy 9 (Yabe *et al.*, 1995). A list of other applications includes bubbly flow, combustion and chemical reaction, volcanic explosion, and cavitation (Yabe and Xiao, 1995). Recently, the CIP method has been extended to many kinds of equations, such as diffusion equations, Poisson equations, and KdV equations. One of the most interesting applications will be the 6-dimensional phase space solver by the CIP method. Vlasov solution in 6-dimension is now possible on a DEC Alpha machine with 256 MB memory. Some of the above applications are not yet available in open literature, but they will soon be.

References

Harlow, F.H. and Shannon, J.P. (1967) The Splash of a Liquid Drop *J.Appl.Phys.* **Vol.38**, pp.3855–3866

Ida, M. and Yabe, T. (1995) Implicit CIP (Cubic-Interpolated Propagation) Method in One Dimension, *Comput.Phys.Commn.* **Vol.92**, pp.21–26

Takewaki, H., Nishiguchi, A. and Yabe, T. (1985) The Cubic-Interpolated Pseudo-Particle (CIP) Method for Solving Hyperbolic-Type Equations, *J. Comput. Phys.* **Vol.61**, pp.261–268

Wang, P.Y., Yabe, T. and Aoki, T. (1993) A General Hyperbolic Solver - the CIP Method - Applied to Curvilinear Coordinates, *J.Phys.Soc.Japan* **Vol.62**, pp.1865–1871

Worthington, A.M. (1908) A Study of Splashes, *(Longman and Green, London 1908)*

Xiao, F., Yabe, T. and Ito, T. (1996a) Constructing Oscillation Preventing Scheme for the Advection Equation by a Rational Function, *Comput.Phys.Commn.* **Vol.93**, pp.1–12

Xiao, F. *et al.* (1996b) An Efficient Model for Driven Flow and Application to GCB, *Comput. Model. & Sim. Eng.* **Vol.1**, pp.235–249

Yabe, T. *et al.* (1991) A Universal Solver for Hyperbolic Equations by Cubic-Polynomial Interpolation, *Comput.Phys.Commun.* **Vol.66**, pp.219–242

Yabe, T. and Wang, P.Y. (1991) Unified Numerical Procedure for Compressible and Incompressible Fluid, *J.Phys.Soc.Japan* **Vol.60**, pp.2105–2108

Yabe, T. and Xiao, F. (1993) Description of Complex and Sharp Interface during Shock Wave Interaction with Liquid Drop, *J.Phys.Soc.Japan* **Vol.62**, pp.2537–2540

Yabe, T. *et al.* (1995) Possible Explanation of the Secondary Flash and Strong Flare on IR Lightcurves upon Impact of Shoemaker-Levy 9, *Geophys. Res. Lett.* **Vol.22**, pp.2429–2432

Yabe, T. and Xiao, F. (1995) Simulation Technique for Dynamical Evaporation Processes, *Nucl Eng.& Design* **Vol.155**, pp.45–53

Yabe, T. *et al.* (1996) Anomalous Crater Formation in Pulsed-Laser-Illuminated Aluminum Slab and Debris Distribution, *Research Report NIFS (National Institute for Fusion Science)* **NIFS-417**

SMOOTHED PARTICLE HYDRODYNAMICS

J.J. MONAGHAN
Department of Mathematics, Monash University, Clayton, Victoria 3168, Australia

Abstract. In this review of SPH we describe advances made in the last few years. SPH is now seen as one of the simplest of the meshless numerical methods. Despite its simplicity it is a proven, robust tool for investigating fluid dynamics with or without gravity. One of the desirable features of SPH is that the resolution can vary with both space and time. This has now been extended to include variation with direction. The dissipation terms have been improved and their relation to Riemann solutions clarified. For relativistic blast waves and similar problems SPH gives results comparable to the best Riemann and FCT methods. New applications include impact and fragmentation, dusty gases, and nearly incompressible fluids. A wide range of problems in planetesimal formation and collision, engineering and geophysics can now be studied using SPH.

1. Introduction

SPH is a Lagrangian particle method which has now been used for many problems in astrophysics. Its main attraction is its simplicity, the ease with which complicated physics can be introduced, and the fact that it is very forgiving.

In this review I will report on some of the improvements in the formulation of SPH, on new applications and make some comments about the use of new hardware such as GRAPE and BEOWULF configurations. The improvements in formulation come from simulations of relativistic gas dynamics (Chow and Monaghan 1997) where Riemann solvers have led the way, and from work on the automatic adjustment of the resolution (Shapiro et al. 1996). The new applications range from the ground breaking work of Benz and Asphaug (1994) on fracture and fragmentation, and to dusty gas dynamics (Monaghan 1997a). There has also been a great deal of in-

S. M. Miyama et al. (eds.), Numerical Astrophysics, 357–366.

teresting work on problems in engineering and geology which, while not directly applicable to astrophysics, may suggest solutions to astrophysical problems.

I have noticed that many of the people that use SPH are often intimidated by the wonderful results achieved by finely tuned FCT and Riemann solvers applied to simple gas dynamics. I thought I might start this review by commenting on some of the problems with these finite difference methods. It is already clear that SPH has an obvious advantage in the ease with which resolution in space and time can be introduced. A further consideration is that some Riemann methods can fail. For example linear Riemann solvers fail where large regions of low density are suddenly created, as in the Sjörgreen test (Einfeldt et al. 1991). When a shock moves slowly through the finite difference mesh (Roberts 1990) significant down stream noise can be generated. Colella and Woodward (1984) spend one third of their paper discussing an artificial viscosity to get rid of this problem. Exact, or good, non-linear Riemann solvers do not have these problems, but you then need to solve the Riemann problem and, in physically complicated problems, this is a substantial undertaking. No Riemann solutions have yet been obtained for the important class of problems where chemical reactions occur and more than one phase exists. Other defects of Riemann methods are discussed by Quirk (1994) in a fascinating paper.

2. Advances in Formulation

2.1. DISSIPATION, SHOCKS AND RELATIVITY

The standard form of viscous dissipation in SPH was constructed by analogy with the dissipation in gases and then a quadratic term was included to prevent interpenetration of gas streams colliding with high relative velocity. While this works satisfactorily it gives no clue as to how a dissipation term should be constructed for relativistic gas dynamics. The solution of this problem was indirectly due to the Riemann solver finite difference solution simulation of relativistic gas dynamics by Marti and Müller (1996).

It is simpler to talk about the non-relativistic case. The Riemann solver formulation involves a dissipative term associated with eigenvalues and jumps in the momentum, energy and density across a cell boundary. The natural way to mimic these terms is to replace the eigenvalue by a signal velocity v_{sig} and the jumps in a variable by the difference of the variable between the two SPH particles being considered (Monaghan 1997a). Rather than use the thermal energy equation an equation for the energy/unit mass (kinetic and thermal energy) was used. In non-relativistic calculations there is an advantage in using the thermal energy equation because if the kinetic energy dominates small errors in the velocity can have severe effects on

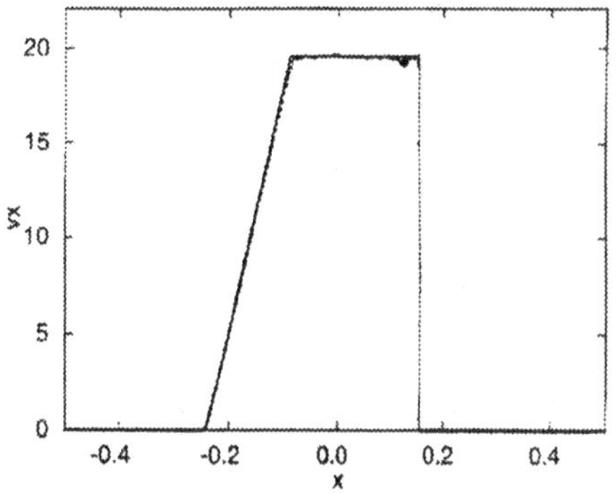

Figure 1. Velocity against distance for the non-relativistic blast wave using the new dissipation and the momentum and energy equations. Initial particle spacing is 0.001.

the thermal energy. However, if you wish to simulate relativistic gases, the natural variables are the momentum and energy.

Following the Riemann ideas the acceleration equation takes the form

$$\frac{d\mathbf{v_a}}{dt} = -\sum_b m_b \left(\frac{P_a}{\rho_a^2} + \frac{P_b}{\rho_b^2} - \frac{K v_{\rm sig}(a,b)}{\bar{\rho}_{ab}} \mathbf{v}_{ab} \cdot \mathbf{j} \right) \nabla_a W_{ab}, \qquad (2.1)$$

Where the non-standard SPH quantities are K a constant ~ 1, $v_{\rm sig}$ is a signal velocity replacing the Riemann eigenvalues, and $\mathbf{j}$ is a unit vector from b to a. The signal velocity is the velocity (measured in the computing frame) with which sound waves from a and b approach each other. In the simplest approximation this is

$$v_{\rm sig}(a,b) = c_a + c_b - \mathbf{v}_{ab} \cdot \mathbf{j}. \qquad (2.2)$$

If you now look at the viscous dissipation you will see that we have automatically included the quadratic term which was obtained previously by an argument about stopping interpenetration using a kinetic pressure term.

By combining these equations with the continuity equation very good results have been achieved for the blast wave, Roberts and Sjögreen problems (Monaghan 1997a). The only example we show here is the one dimensional blast wave problems with zero initial velocity, initial pressure to the left of the interface 1000 and on the right 0.01, density 1.0 and $\gamma = 1.4$. Typical results are shown in Figures 1 and 2.

While the results for the non-relativistic problem are similar to those with standard SPH dissipation terms the new formulation comes into its own for the relativistic case. We will not give the details but confine ourselves to one example, the relativistic blast wave. We assume the gas is composed of one type of baryon with rest mass m_0, the energy per unit

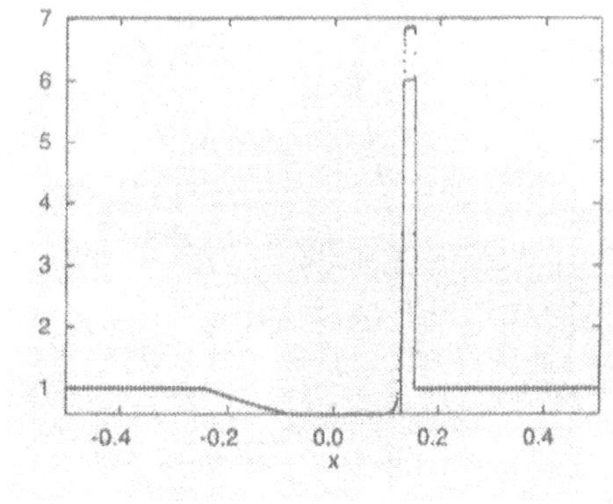

Figure 2. Density against distance for the non-relativistic blast wave.

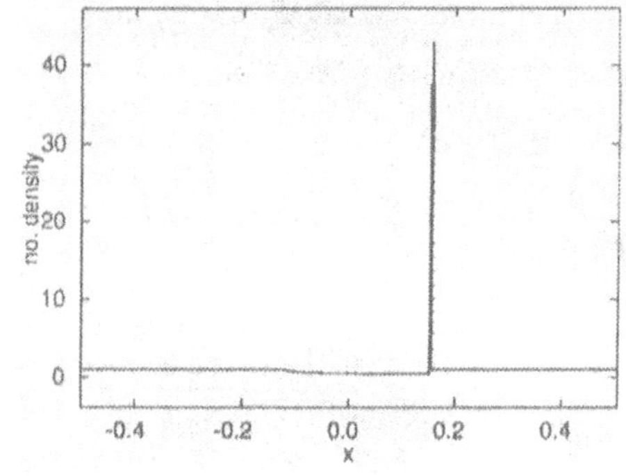

Figure 3. Baryon number density for the relativistic blast wave. Particle spacing 0.001.

mass is scaled with c^2, and the velocity with c. The initial conditions are $P = 1000$ to the left and $P = 0.01$ to the right with baryon number density $N = 1$ everywhere. These are the same scaled values as for the non-relativistic problem. The simulation uses a resolution varying in space. The results for the baryon number density is shown in Figure 3. These results are comparable in accuracy those obtained with the best Riemann solvers, LCPFCT, and RHHLE and substantially better than those obtained using SHASTA.

2.2. ADAPTIVE RESOLUTION

SPH with h varying in space and time has been used for some time. However, it would clearly be desirable to have a different resolution in some directions than in others e.g. when there is a shock along one of the axes. Significant advances in achieving this goal have been made by Shapiro et

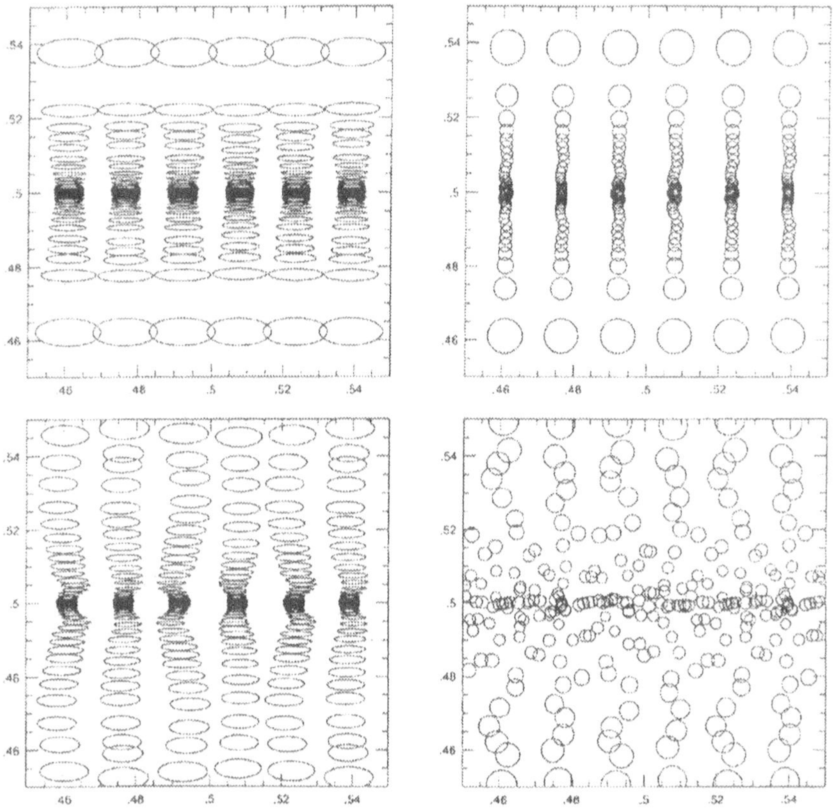

Figure 4. ASPH calculation of a one dimensional Zeldovich Pancake on the left, and a standard SPH calculation on the right.

al. (1996) and Owen et al. (1998) and incorporated in an ASPH code. The basic idea is to work with ellipsoidal kernels where the principal axes of the ellipsoid are orientated according to the motion of neighbouring particles. In the case of the spherically symmetric kernel, with single smoothing length h, it is common to determine h from the equation

$$\frac{d}{dt}\left(\frac{1}{h}\right) = \left(\frac{1}{h}\right)\frac{\nabla \cdot \mathbf{v}}{d}, \tag{2.3}$$

where d is the number of dimensions where I have used an equation for $1/h$ rather than for h. The $\nabla \cdot \mathbf{v}$ takes into account the average effect of the relative motion of the particles. In ASPH a matrix $\mathbf{G}$ replaces h. $\mathbf{G}$ has eigenvalues $1/h_k$ where h_k, with $k = 1, 2, 3$ are the smoothing lengths along the principal axis of an ellipsoid.

In test calculations involving shock tubes, blast waves, Zeldovich pancakes and rotating disks the ASPH code works extremely well. In these problems it is much more efficient than a standard SPH code. In Figure 4 a Zeldovich-pancake simulation. The figure (and those that follow are taken from Owen et al. 1998) shows the ellipsoidal contours of the ASPH simulation adjusting to the compression along the vertical axis.

Because the gradient of an ellipsoidal kernel is not along the line joining the particles the angular momentum is not conserved exactly. In problems involving disks the loss of conservation of total angular momentum is small (3 percent over the time for one rotation period of the outer layers and several rotation periods of the core). However if the material changes density rapidly the velocity gradients are not calculated accurately and **G** has very large errors.

3. New Applications

3.1. IMPACT AND FRAGMENTATION

When planetesimals collide they may fragment. The same physics occurs in rotating ore mixers and grinders. It is straightforward to include strength into the SPH formulation of a perfectly elastic solid and that has been done by a number of researchers (see for example Libersky and Petschek 1991). This only requires including terms which enable the rate of change of the elastic stress to be calculated in terms of the strain tensor. The fluid stress tensor is given by

$$\sigma_{ij} = -P\delta_{ij} + s_{ij}, \tag{3.4}$$

where P is the pressure and s_{ij} is the traceless deviatoric stress tensor. In the Hookian approximation

$$\frac{ds_{ij}}{dt} = 2\mu(\dot{\epsilon}_{ij} - \frac{1}{3}\delta_{ij}\dot{\epsilon}_{\gamma\gamma}) + \text{rotation terms} \tag{3.5}$$

where μ is the shear modulus and the strain tensor is defined by

$$\dot{\epsilon}_{ij} = \partial v_i/\partial x_j + \partial v_j/\partial x_i. \tag{3.6}$$

Benz and Asphaug (1994) begin with the elastic model and use a quantitative model of fracture damage due to Grady and Kipp (1980). The formulation depends on a damage factor D which enters the definition of the stress tensor in the form

$$\sigma_{ij} = -P^*\delta_{ij} + (1-D)s_{ij} \tag{3.7}$$

where $P^* = P$ if $P > 0$ and $P^* = (1-D)P$ if $P < 0$. The damage factor changes with time according to the local strain. Because the damage is a lagrangian quantity it is ideally suited to an SPH simulation.

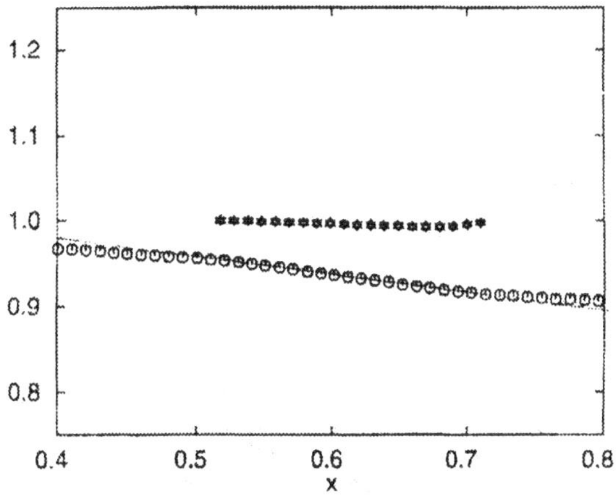

Figure 5. Density (vertical axis) at the SPH gas and dust particles for a layer of dust falling through air plotted against vertical distance x (horizontal axis). Upper symbols show the dust layer, lower symbols the gas atmosphere. In agreement with theory there is a slight change in the slope of the air density due to the dust layer falling through it. Further details in (Monaghan 1997b).

In their ground breaking paper Benz and Asphaug (1994) first showed that SPH gives good results for the ideal elastic material and for crack propagation When applied to the fragmentation case, SPH gives the best results ever achieved for these problems (see also Asphaug et al. 1998).

3.2. CLUMPING AND THE TENSILE INSTABILITY

In impact problems involving solids, the pressure can become negative. This is the physical effect of the forces between the atoms. When the solid is pulled (tensile stress) the atoms move further apart and the elastic contribution to the pressure becomes negative. When this happens the standard forms of the pressure gradient terms in SPH become negative and the particles tend to clump (Swegle et al. 1995). This was already seen in early MHD calculations (Phillips and Monaghan 1985). The clumping produces errors in the estimates of the fragments in an impact problem so it would be better to remove the problem. It also means that particles forced by some means to be within a distance r satisfying $r/h < 1/3$ in an ideal gas will tend to clump. There has been no entirely satisfactory solution to this problem for the fragmentation of solids (but see Libersky and Randles (1996) who use dissipation and Thomas and Couchman (1992) who truncate the gradient of the kernel so that for $r/h < 1/3$ the gradient of the kernel has the same value as at $r/h = 1/3$.

The same problem in MHD simulations can be evaded by working with the $\mathbf{J} \times \mathbf{B}$ force although momentum is not then conserved exactly. It would be clearly be useful to be able to solve this tensile instability for both the MHD and the impact problems.

3.3. DUSTY GAS

Once heavy elements have been made and dust can form we need to take into account the effect of dust on the gas. In the early and middle stages of star formation the effect of dust is primarily through the radiation attempting to drive the dust from the system. The drag between the dust and the gas allows the force to be transferred to the gas.

In many astrophysical problems the drag is enormous so it is natural to use implicit methods. Implicit methods have been tried with SPH for gas flow but they are cumbersome and so inefficient they are not worth using. Because the drag involves pair interactions it turns out that by working with each pair a very efficient implicit scheme can be worked out (Monaghan 1997b).

This technique has been tested for two basic problems (Monaghan 1997b, all in one dimension) wave propagation and dust layers falling under gravity through air. In Figure 6 we show the density of the dust and gas particles in a one dimensional simulation of a layer of dust falling through gas in hydrostatic equilibrium under gravity. The SPH simulation accurately predicts the slight change in gas density gradient within the moving dust layer, and the motion of the dust layer.

3.4. RADIATIVE TRANSFER AND CONDUCTION

Radiative transfer in the conduction form involves second derivative terms which require some care in an SPH formulation. The appropriate formulation has been known for some time. What isn't so well known is how to handle the case where there is a rapid jump in the effective thermal conductivity. The answer is surprisingly simple.

When the thermal conductivities changes discontinuously the flux of heat must still be continuous. By taking this into account the appropriate change is to replace the sum of the thermal conductivities according to

$$\kappa_a + \kappa_b \quad \text{by} \quad \frac{4\kappa_a \kappa_b}{\kappa_a + \kappa_b} \tag{3.8}$$

Amazingly this gives excellent results even when the thermal conductivity jumps by a factor 1000 from one particle to the next. An extensive series of calculations have been made by Cleary and Monaghan (1998).

3.5. ACCRETION, STAR FORMATION AND DISKS

Standard SPH, combined perhaps with some of the improvements discussed earlier, can be used to study the formation of stars. When a fragment in a cloud begins to collapse the density and scale eventually requires such small time steps that the calculation is paralysed. Bate et. al (1995) have worked out a way to continue the calculation by replacing the dense region by a single particle with appropriate boundary conditions. This allows the evolution of the cloud to be followed as binary and other n-body systems form.

SPH continues to be used as a tool to study the dynamics of disks. The work of Artymowicz and Lubow (1994), Murray (1997) and Nelson et al. (1998) are examples of the widespread application of SPH to these problems.

4. New Hardware

The GRAPE board was developed in Tokyo (Sugimoto et al. 1990) originally to calculate the gravitational forces on a set of particles by direct summation. Its advantage is cheapness (a board plus a work station). The summation was extended to work with a tree code (Makino and Funato 1993). They then extended to calculate SPH terms. Steinmetz (1996) has given a detailed analysis of version of GRAPE for cosmological SPH simulations (GRAPESPH). He shows that (given the technology of 3 years ago) it is possible to emulate results previously achieved by super computers. The situation is now more difficult to assess because the technology is changing so rapidly. For example the introduction of cheap BEOWULF configurations (http://qso.lanl.gov/~msw).

5. Summary

While there have been many applications of standard forms of SPH over the last few years there have been generalizations which increase the efficiency and extensions to deal with more complex problems. There has been a big interest in SPH by computational engineers and computational geophysicists because of the ease with which SPH can be applied to complicated geometry and complicated flow problems. Further developments can be expected in these areas as well as in astrophysical problems associated with relativistic flows, mixtures of solid particles and gas, radiative transfer and MHD problems. As always, new computers, in this case probably the BEOWULF configurations, will be a major resource.

References

Artymowicz, P., and Lubow, S.H., Ap.J. **421** (1994), 651.

Asphaug, E., Ostro, S.J., Hudson, R.S., Scheeres, D.J. and Benz, W., Nature, **393** (1998), 437.

Bate, M.R., Bonnell, I.A., and Price, N.M., Mon. Not. Roy. Astr. Soc. **277** (1995), 362.

Benz, W., and Asphaug, E., Icarus **107** (1994), 98.

Chow, E., and Monaghan, J.J., J. Computat. Phys. **134** (1997), 296.

Cleary, P., and Monaghan, J.J., J. Computat. Phys. (to appear 1998).

Colella, P., and Woodward. P., J. Computat. Phys., **54** (1984), 174.

Einfeldt, B., Munz, C.D., Roe, P.L., and Sjörgreen, B.J. J. Computat. Phys., **92** (1991), 273.

Grady, D.E., and Kipp, M.E. Int. J. Rock. Mech. Sci. Geomech. Abst., **17** (1980), 147.

Libersky, L.D., and Petscheck, A.G., Advances in the Free-Lagrange method (Trease and Cowley, eds), Springer Verlag, 1991 pp. 248.

Makino, J., and Funato, Y., PASJ **42** (1993), 279.

Martí, J.$M^{\underline{a}}$, and Müller, E., J. Computat. Phys. **123** (1996), 1.

Monaghan, J.J., Ann.Rev.Astron.Astrophys. **30** (1992), 543.

Monaghan, J.J., J. Computat. Phys. **136** (1997a), 298.

Monaghan, J.J., J. Computat. Phys. **138** (1997b), 801.

Murray, J., Mon. Not. R. astr. Soc. **297** (1998), 323.

Nelson, A.F., Benz, W., Adams, F.C., and Arnett, D., Dynamics of Circumstellar Disks, Preprint of the Steward Observatory, No. 1424, Univ. of Arizona, USA.

Owen, J.M., Villumsen, J.V., Shapiro, P.R., and Martel, H., Adaptive Smoothed Particle Hydrodynamics, II, Ap. J. (1998).

Phillips, G.J., and Monaghan, J.J., Mon. Not. R. astr. Soc. **216** (1985), 883.

Quirk, J.J., Int. J. for Num. Methods in Fluids, **18** (1994), 141.

Randles, P.W., and Libersky, L.D., Computer Methods in Applied Mechanics and Engineering (1996).

Roberts, T.W., J. Computat. Phys. **90** (1990), 141.

Shapiro., P.R., Martel, H., Villumsen, J.V., and Owen, J.M., Ap. J. Supp. **103** (1996), 269.

Steinmetz, M., *GRAPESPH: Cosmological SPH simulations with special purpose hardware GRAPE*, Max Planck Institute f. Astrophysik, Garching.

Sugimoto, D., Chikada, Y., Makino, J., Ito, T., Ebisuzaki, T.,Umemura, M., Nature **345** (1990), 33.

Swegle, J.W., Hicks, D.L., and Attaway, S.W., J. Comp. Phys. **116** (1995), 123.

Thomas, P.A. and Couchman, H.M.P., M.N.R.A.S. **257** (1992), 11.

ADVANCED METHODS IN PARTICLE HYDRODYNAMICS

SHU-ICHIRO INUTSUKA
National Astronomical Observatory
Mitaka, Tokyo 181-8588, Japan

Abstract. We have developed numerical methods that use particles to solve hydrodynamics and radiation hydrodynamics and reformulated Smoothed Particle Hydrodynamics using a new action principle. The force acting on each particle was determined by solving the Riemann problem. Use of the Riemann Solver strengthened the method, making it accurate for the study of strong shock phenomena. These techniques were implemented in conservation form. The method was also extended to multi-dimensional radiation hydrodynamics without invoking any diffusion approximation, being equally applicable to optically thin and thick regimes. The results of test problems in 1D, 2D, and 3D are shown.

1. Introduction

Since its invention by Lucy (1977) and Gingold & Monaghan (1977), smoothed particle hydrodynamics (SPH) has been used to study a variety of astrophysical problems. A broad discussion of the method can be found in a review by Monaghan (1992). However, some of its disadvantages remain to be overcome. SPH is less accurate with strong shocks than state-of-the-art finite difference schemes such as the Piecewise Parabolic Method (PPM, Colella & Woodward, 1984). Especially in the three-dimensional calculation of colliding gases, particles often penetrate into the opposite side, and large post-shock oscillations often appear, contaminating the numerical results.

In this paper, a new method to handle shocks in particle hydrodynamics is constructed (cf., Inutsuka, 1994, 1995). The force acting on each particle is determined by solving the Riemann problem (RP). Use of the Riemann Solver, the main ingredient of Godunov's method (1959), strengthens the accuracy of the method for the study of strong shock phenomena. The

S. M. Miyama et al. (eds.), Numerical Astrophysics, 367–374.

method may thus be regarded as a multi-dimensional Lagrangian sequel to Godunov's method.

Section 2 provides a description of the method. A few numerical examples involving a strong shock are presented in Section 3. In Section 4, we extend the application of the method to radiation hydrodynamics.

2. The method

In particle methods, each particle has its own mass, and its density is estimated by

$$\rho_i = \sum_j m_j W(\boldsymbol{x}_i - \boldsymbol{x}_j, h_1), \tag{1}$$

where subscripts denote particle labels, m_j is the mass of the j-th particle, and $W(\boldsymbol{x}, h_1)$ is a spherically symmetric kernel function. Although there are many possible forms for $W(\boldsymbol{x}, h_1)$, a typical form is the following:

$$W(\boldsymbol{x}, h_1) = \left(\frac{1}{h_1\sqrt{\pi}}\right)^d e^{-\boldsymbol{x}^2/h_1^2}. \tag{2}$$

In the above equation, d is the number of dimensions, and h_1 is a smoothing length which can vary. In this brief paper, we will concentrate on a case where the smoothing length is constant.

The standard procedure to derive sets of evolution equations for SPH has been previously described (e.g., Monaghan, 1992). In this paper, we will show how evolution equations are derived from action principles.

2.1. NEW ACTION PRINCIPLE

Barotropic fluid has a Lagrangian density $\mathcal{L}$ expressed in Lagrangian coordinates:

$$L = \int \mathcal{L} d\boldsymbol{x} = \int \left(\frac{1}{2}\rho v^2 - \rho u\right) d\boldsymbol{x}. \tag{3}$$

The equation of motion for the fluid can be derived by minimizing the action. By analogy, a new Lagrangian for the system of particles can be defined by the following equation:

$$L_{\text{new}} = \sum_k m_k \left[\frac{1}{2}{v_k}^2 - \int u(\boldsymbol{x}) W(\boldsymbol{x} - \boldsymbol{x}_k, h_2) d\boldsymbol{x}\right]. \tag{4}$$

Using the definition of density,

$$\rho(\boldsymbol{x}) = \sum_j m_j W(\boldsymbol{x} - \boldsymbol{x}_j, h_1), \tag{5}$$

the Euler-Lagrange equation,

$$\frac{d}{dt}\frac{\partial L_{\rm new}}{\partial \dot{\boldsymbol{x}}_i} - \frac{\partial L_{\rm new}}{\partial \boldsymbol{x}_i} = 0, \tag{6}$$

gives

$$\begin{aligned} \ddot{\boldsymbol{x}}_i &= -\sum_j m_j \frac{\partial}{\partial \boldsymbol{x}_i} \int \frac{P}{\rho^2} W(\boldsymbol{x}-\boldsymbol{x}_i, h_1) W(\boldsymbol{x}-\boldsymbol{x}_j, h_2) d\boldsymbol{x} \\ &\quad + \sum_j m_j \frac{\partial}{\partial \boldsymbol{x}_j} \int \frac{P}{\rho^2} W(\boldsymbol{x}-\boldsymbol{x}_i, h_2) W(\boldsymbol{x}-\boldsymbol{x}_j, h_1) d\boldsymbol{x}. \end{aligned} \tag{7}$$

The space-symmetry of $L_{\rm new}$ guarantees the conservation of linear momentum and angular momentum. In eq.(7), h_1 and h_2 are not necessarily the same. The choice of h_1 or h_2 gives different equations of motion.

Case 1 If we take $h_2 = 0$, then $W(\boldsymbol{x}, h_2) = \delta(\boldsymbol{x})$, and the above equation reduces to the standard SPH equation:

$$\ddot{\boldsymbol{x}}_i = -\sum_j m_j \left(\frac{P_i}{\rho_i^2} + \frac{P_j}{\rho_j^2} \right) \frac{\partial}{\partial \boldsymbol{x}_i} W(\boldsymbol{x}_i - \boldsymbol{x}_j, h_1). \tag{8}$$

Case 2 If we take $h_2 = h_1 = h$, then the above equation provides the following equation:

$$\ddot{\boldsymbol{x}}_i = -\sum_j m_j \left(\frac{\partial}{\partial \boldsymbol{x}_i} - \frac{\partial}{\partial \boldsymbol{x}_j} \right) \int \frac{P(\boldsymbol{x})}{\rho^2(\boldsymbol{x})} W(\boldsymbol{x}-\boldsymbol{x}_i, h) W(\boldsymbol{x}-\boldsymbol{x}_j, h) d\boldsymbol{x} \tag{9}$$

This equation has several good properties compared with equation (8). For example, the right-hand side vanishes identically when $P(\boldsymbol{x})$ is constant in space, while in equation (8) it does not.

2.2. THE INTRODUCTION OF THE RIEMANN SOLVER

At this point, the Riemann Solver is introduced into the particle method. The (finite difference) Godunov scheme uses the result of RP at each cell interface (van Leer, 1979). Here, we also use the result of RP at the middle point of the i-th particle and the j-th particle in the following:

$$\ddot{\boldsymbol{x}}_i = -\sum_j m_j P^* \left(\frac{\partial}{\partial \boldsymbol{x}_i} - \frac{\partial}{\partial \boldsymbol{x}_j} \right) \int \frac{1}{\rho^2(\boldsymbol{x})} W(\boldsymbol{x}-\boldsymbol{x}_i, h) W(\boldsymbol{x}-\boldsymbol{x}_j, h) d\boldsymbol{x} \tag{10}$$

where P^* is the result of RP between the i-th particle and the j-th particle. Similarly, the equation of energy becomes the following:

$$\ddot{u}_i = -\sum_j m_j P^*(\boldsymbol{v}_i - \boldsymbol{v}_j)\left(\frac{\partial}{\partial \boldsymbol{x}_i} - \frac{\partial}{\partial \boldsymbol{x}_j}\right)\int \frac{W(\boldsymbol{x}-\boldsymbol{x}_i,h)W(\boldsymbol{x}-\boldsymbol{x}_j,h)}{\rho^2(\boldsymbol{x})}d\boldsymbol{x}. \tag{11}$$

3. Numerical examples

The present method was tested on a variety of 1D, 2D, and 3D problems, a few of which are described below. In all calculations, a variable smoothing length was used. The computation time needed was quite similar to that needed by standard SPH.

3.1. SHOCK TUBES

3.1.1. *Sod's problem*

In Fig. 1, the results of 1D Sod's shock tube (Sod, 1978) are plotted. Solid curves correspond to the analytic solution. The number of equal-mass particles in the x-direction is only 40 on the right-hand side of the initial discontinuity. Shock surfaces are resolved by small number of particles.

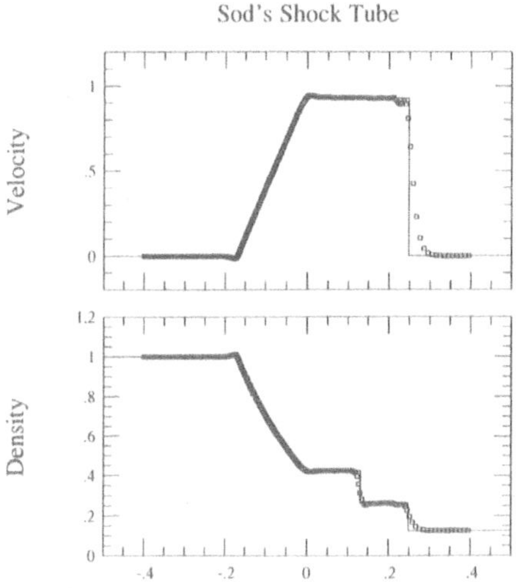

Fig. 1

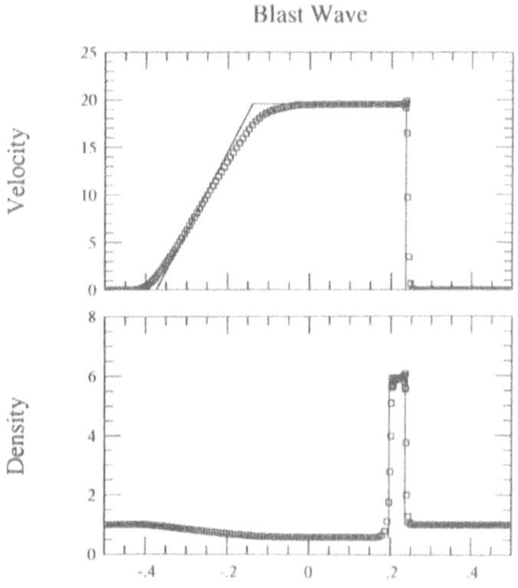

Fig. 2

3.1.2. *Blast wave*

Fig. 2 plots the result of an extremely strong shock-tube problem (Woodward & Colella, 1984) in which the Mach number was 200. Only 100 particles were used in the x-direction on each side of the initial discontinuity. Initial pressure of the fluid on the left-hand side was 10^5 times that of right-hand side. Even in this kind of severe problem, the present method gave stable and accurate results and was free from penetration problems.

3.2. FIFTEEN-DEGREE WEDGE CHANNEL FLOW

This problem, evaluated by Levy, Powell, & van Leer (1993), was a useful test to calibrate the method. The geometry was a two-dimensional channel with a 15-degree wedge on the lower wall. A 15-degree expansion corner was also included. The inflow Mach number was two. It was noticed, however, that this kind of problem is poorly suited for the particle method because (1) a rigid-wall boundary condition must be set up at the wall of the tunnel; and (2) in-flowing particles have to be prepared at all times at the left-side boundary.

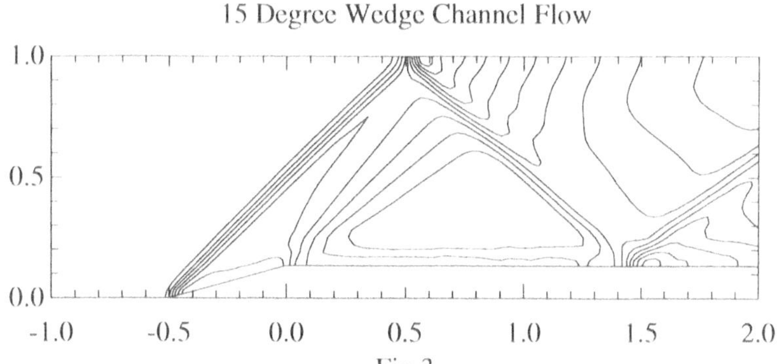

Fig.3

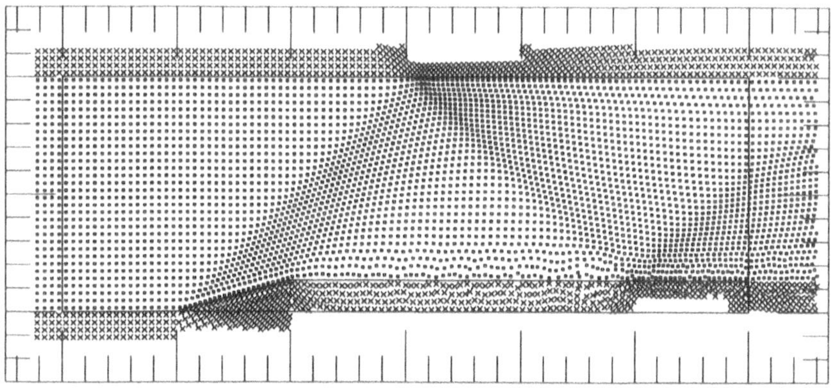

Fig.4

Figure 3 shows our method's result. Mach-number contours from 1.0 to 2.0 are plotted. Figure 4 shows particle positions. In this calculation, we realized the rigid-wall boundary condition by placing "ghost particles" in the wall. Initially, 32×96 particles were flowing inside the channel, which must be compared to the 32×96 grids finite difference calculation by Levy et al. Our results were satisfactory. The main difference between the results obtained by Levy et al. and ours was the thickness of the numerical boundary layer just above the lower wall. This was due to the fact that ghost particles inside the expansion corner do not realize reflecting boundary conditions completely, which causes a relatively thick numerical boundary layer. It is clear that a finite difference method must be used in the actual study of this kind of rigid-wall boundary problem. This result is presented only to demonstrate the capability of this particle scheme.

4. Extension to radiation hydrodynamics

Multi-dimensional calculations of radiation hydrodynamics (RHD) for modeling astrophysical phenomena constitute a frontier in theoretical astrophysics. Most RHD calculations presented to date invoke diffusion approximation at some level to simplify calculations. In such calculations, the direction of radiative energy flow is restricted to the direction of the temperature gradient of the media. However, that kind of treatment is inadequate in optically thin media. In this paper, a self-consistent description of radiation and fluid is presented by a Lagrangian particle scheme. This method is equally applicable to optically thin and thick media in three-dimensional calculations.

4.1. METHOD

In this section, we consider the following equations of RHD in Lagrangian coordinates:

$$\frac{d\rho}{dt} + \rho \nabla \cdot \boldsymbol{v} = 0, \tag{12}$$

$$\frac{d\boldsymbol{v}}{dt} = -\frac{1}{\rho}\nabla P - \nabla\Phi + \frac{1}{c}\int \kappa_\nu \boldsymbol{F}_\nu d\nu \tag{13}$$

$$\frac{du}{dt} = -\frac{P}{\rho}\nabla \cdot \boldsymbol{v} - 4\pi \int \kappa_\nu S_\nu d\nu + c\int \kappa_\nu E_\nu d\nu \tag{14}$$

$$P = (\gamma - 1)\rho u, \tag{15}$$

where c, u, κ_ν, S_ν, E_ν, and $\boldsymbol{F}_\nu$ are the speed of light, specific energy, mass opacity, source function, radiation energy density, and radiative flux, respectively. Other symbols have their usual meanings. The transfer equation is

$$\frac{dI_\nu}{ds} = \kappa_\nu \rho (S_\nu - I_\nu). \tag{16}$$

Hereafter, we suppress explicit mention of ν-dependence for the sake of brevity. Note that the present scheme is not restricted to gray problems.

4.1.1. *Algorithm*

From the formal solution of eq.(16), we can construct the radiation energy density at the position of the i-th particle as

$$E_i = \frac{1}{c}\sum_j \kappa_j m_j S_j \int \frac{e^{-\tau(x,x_i)}}{R(x,x_i)^2} W(\boldsymbol{x}-\boldsymbol{x}_j)d\boldsymbol{x}, \tag{17}$$

where $\tau(x,x_i)$ and $R(x,x_i)$ are the optical depth and geometrical distance from the position $\boldsymbol{x}$ to the j-th particle. Similarly, the following is found for radiative flux:

$$\boldsymbol{F}_i = \sum_j \kappa_j m_j S_j \int \frac{e^{-\tau(x,x_i)}}{R(x,x_i)^3}(\boldsymbol{x}_i-\boldsymbol{x})W(\boldsymbol{x}-\boldsymbol{x}_j)d\boldsymbol{x}. \tag{18}$$

4.1.2. *1D Analogue*

Considering a medium which is uniform in the y- and z-directions, one-dimensional analogues of eqs.(17) and (18) reduce to

$$E_i = \frac{2\pi}{c}\sum_j \kappa_j m_j S_j \int_{-\infty}^{\infty} W(\boldsymbol{x}-\boldsymbol{x}_j)\left[\int_{\tau(x,x_i)}^{\infty}\frac{e^{-\tau}}{\tau}d\tau\right]dx, \tag{19}$$

$$F_i^x = 2\pi\sum_j \kappa_j m_j S_j \int_{-\infty}^{\infty} \text{sign}(x_i-x)\, W(\boldsymbol{x}-\boldsymbol{x}_j)\left[\int_{\tau(x,x_i)}^{\infty}\frac{e^{-\tau}}{\tau^2}d\tau\right]dx. \tag{20}$$

In uniform optically thick medium, where S_ν (= B_ν: black body radiation), κ_ν, and ρ are constant in space, E_i reduces to the energy density of black body radiation:

$$E_i = \frac{4\pi}{c}B\int_0^{\infty}\left[\int_{\tau}^{\infty}\frac{e^{-\tau}}{\tau}d\tau\right]d\tau = \frac{4\pi}{c}B. \tag{21}$$

4.2. TEST ON THERMAL RELAXATION MODES

To test the ability of the present scheme, we considered temperature perturbations imposed on a static, gray, LTE ambient medium, initially in radiative equilibrium. The governing equation for gas energy reduced to

$$C_V\frac{dT}{dt} = -4\pi\kappa S + \kappa c E. \tag{22}$$

The standard normal mode analysis ($T = T_0 + T_1 = T_0 + \phi e^{ikx+nt}$) provided the dispersion relation for the linear fluctuation (Spiegel, 1957):

$$n(k) = -\frac{16\pi\kappa\sigma_{\rm SB}T_0^3}{C_V}\left[1 - \frac{\kappa\rho}{k}\cot^{-1}\left(\frac{\kappa\rho}{k}\right)\right], \tag{23}$$

which was the exact solution of the problem. Likewise, the Eddington approximation provided the following results (Unno & Spiegel, 1966):

$$n(k) = -\frac{16\pi\kappa\sigma_{\rm SB}T_0^3}{C_{\rm V}}\left[1+3\left(\frac{\kappa\rho}{k}\right)^2\right]^{-1} . \tag{24}$$

Figures 5 and 6 plot the result obtained in our scheme. The left panel shows a 1D case in which $N_p/\tau = 100$, and the right panel shows a 3D case in which $N_p/\tau = 3.25$. The solid curve denotes the exact solution (eq.[23]), while the dashed curve denotes the Eddington approximation (eq.[24]). Circles correspond to the numerical result of our scheme, which shows excellent agreement with the exact solution. Note that the Eddington approximation produces errors at $k/\kappa\rho > 1$.

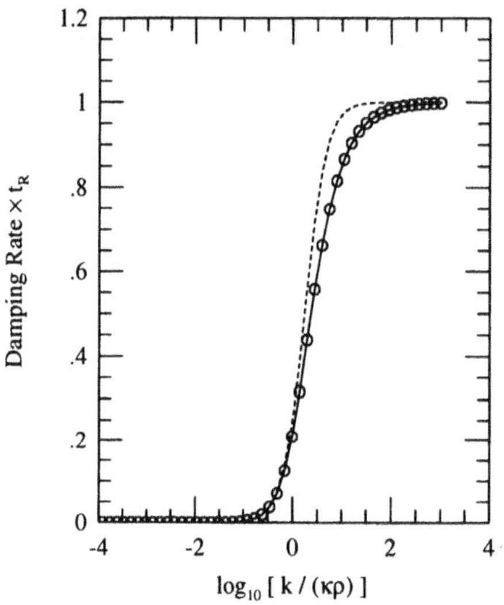

Fig.5 Thermal Relaxation Modes (1D)

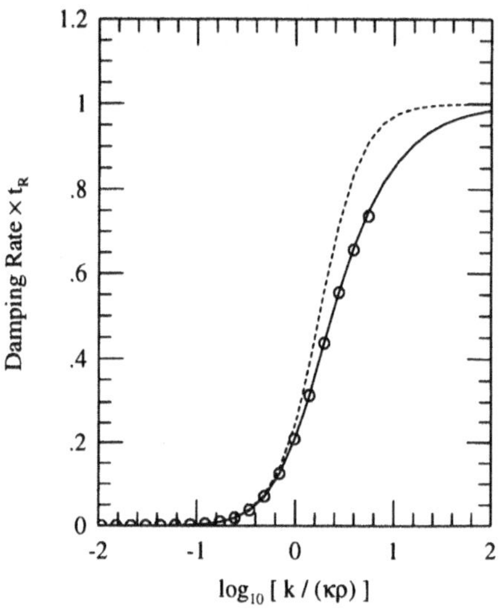

Fig.6 Thermal Relaxation Modes (3D)

References

Colella, P., & Woodward, P. (1984) JCP, 54, 174.
Godunov, S. K. (1959) MSB, 47, 271.
Gingold, R. A., & Monaghan, J. J. (1977) MNRAS, 181, 375.
Inutsuka, S. (1994) Mem. S. A. It., 65, 1027.
Inutsuka, S. (1995) in *Dynamical Evolution of Star Clusters, IAU Symp. 174*. ed. J. Makino & P. Hut, (Tokyo: Kluwer)
Levy, D. W., Powell, K. G., & van Leer, B. (1993) JCP, 106, 201.
Lucy, L. (1977) AJ, 82,1013.
Monaghan, J. J. (1992) ARAAP, 30, 543.
Sod, G. A. (1978) JCP, 27, 1.
Spiegel, E. A. (1957) ApJ, 126, 202.
Unno, W., & Spiegel, E. A. (1966) PASJ, 18, 85.
van Leer, B. (1979) JCP, 32, 101.
Woodward, P., & Colella, P. (1984) JCP, 54, 115.

A 3-D RADIATIVE TRANSFER SOLVER USING A MASSIVELY PARALLEL COMPUTER

TAISHI NAKAMOTO

Center for Computational Physics, University of Tsukuba, Tsukuba, Ibaraki 305-8577, Japan

1. Introduction

Radiation is one of the most fundamental and important physical processes in the astrophysics. It pushes matter, transfers energy, and provides observable information. Therefore, if we are to understand astrophysical phenomena, we have to take into account the radiative transfer appropriately.

It is difficult, however, to solve the radiative transfer equation in general situations without assuming any geometrical symmetry nor optical thickness in advance because it requires huge amount of calculations. The radiation field is described by the specific intensity $I(x, y, z, \theta, \varphi, \nu)$. It depends on six independent variables except the time. Therefore, the amount of calculations become more than of the order of $N_x N_y N_z N_\theta N_\varphi N_\nu$, where N_x, N_y, N_z are numbers of grid points in each space dimension, N_θ, N_φ are numbers of angle grid points, and N_ν is the number of the frequency grids.

The goal of this study is to look for an efficient way to use large computers, especially massively parallel computers, to carry out the radiative transfer calculations. It seems more difficult to use massively parallel computers with distributed local memory than to use shared memory parallel computers because a calculation sequence between processor units (PUs) has to be managed adequately. Particularly, the radiative transfer is a global problem in principle, we have to carefully carry out the calculation on massively parallel computers. In this article, we propose a new way to use massively parallel computers without breaking a calculation sequence restriction. The method is implemented on a massively parallel computer and found to be useful. Due to this method and a fast massively parallel computer, we can calculate very large problems such as $N_x N_y N_z \times N_\theta N_\varphi \times N_\nu = 128^3 \times 128^2 \times 6$ in a reasonable time.

S. M. Miyama et al. (eds.), Numerical Astrophysics, 375–378.

2. Radiative Transfer Equation

The integral form of steady radiative transfer equation is given by

$$I_A = I_O e^{-\tau} + \int_0^\tau S(t) e^{(t-\tau)} dt, \quad \tau = \int_O^A \chi \, dl, \tag{1}$$

where l is the one-dimensional coordinate along a line of ray, χ is the extinction coefficient, S is the source function, and τ is the optical thickness between two points O and A. When χ and S are known functions, the radiative transfer equations with different angles or frequencies are independent of each other.

3. Numerical Scheme

We discretize the equation (1) to solve it numerically. Note that finite difference formula derived from the integral form eq.(1) is applicable to any optical thickness regime. There are various ways to discretize eq.(1) for spatial dimensions. Following is a simple example;

$$I_A = I_O e^{-\tau} + \frac{1-(1+\tau)e^{-\tau}}{\tau} S_O + \frac{\tau - 1 + e^{-\tau}}{\tau} S_A, \quad \tau = \frac{\chi_O + \chi_A}{2} L, \tag{2}$$

where L is the distance between two points O and A.

Discretization of angle variables is simple and straightforward. We put as many angle grid points as spatial grid points on the boundary surface [1] in order to resolve all grid points on the boundary appropriately. Therefore, $N_\theta N_\varphi$ becomes of the order of $N_x N_y$. This large number of angle grids enables us to solve problems in which optically thick and geometrically small clumps are embedded in an optically thin medium; if there are not enough angle grids, such small clumps are not resolved adequately.

The frequency space is also discretized simply. Number of grids depends on problems under consideration; in some cases 6 grids or more may be needed, while in other cases only one grid may be enough.

Numerical calculation of eq.(2) at all spatial grid points for all direction angle and frequency grid points is carried out using the so called Short Characteristics method (Kunasz & Auer 1988). Although there is some extent of numerical diffusion, the Short Characteristics method demands rather small amount of calculations. Moreover the calculation procedure is simple and suitable for parallel computation. For these practical reasons, we adopt the Short Characteristics method to solve the equation.

[1]We employ Cartesian coordinates and a cubic computational domain. Therefore, the computational domain is surrounded by six plane boundary surfaces.

We should notice that a calculation sequence is restricted when we calculate the eq.(1) numerically. Calculation along a line of ray with a fixed direction and frequency has to be done from upstream side to downstream side, though the eq.(1) with different angle direction or frequency can be calculates independently. This calculation sequence restriction makes it difficult to use parallel computers. We have to find out an efficient way to use parallel computers without breaking the calculation sequence restriction.

4. Multiple Wave Front method

For an efficient use of large parallel computers we have to care about how to decompose the computational domain into sub domains on which computations are carried out parallelly. Since usual massively parallel computers do not have enough size of local memory to assign all space variables to each PU, we decompose the space dimensions into many sub-spaces and assign each sub-space to each PU. PUs are ordered in a 3-dimensional manner in 3-D space problems. In other words, all PUs form a 3-D PU space.

We manage the computation sequence among PUs obeying the calculation sequence restriction as follows. First, we classify the angle direction grid points into 8 groups according to whether it points to back or front, right or left, and up or down. Next we start the computation of eq.(2) from one corner of the PU space, say, front-left-bottom corner PU, when we calculate the intensity with a fixed direction grid in back-right-upward direction group. Computations in one PU is of course carried out following the calculation sequence restriction. After completion of the computation in the first PU, PU boundary intensities are transferred to next PUs and computations for the given direction are started in the secondary PUs. At the same time, in the first PU, computations of the intensity with an other direction grid in the back-right-upward direction group are started. Notice that computations in the secondary PUs for the former direction grid and in the first PU for the next direction grid can be done simultaneously. And this procedure is continued until all intensities with the direction grid in the back-right-upward direction group are obtained. Then the next direction group, say, back-left-downward direction group, will be calculated starting from front-right-up corner PU.

How efficiently does the method work? PUs on which intensities in the same direction are calculated at the same time form a wave front in the PU space. And the wave front propagates in the PU space. Moreover many wave fronts cover the PU space. So we call this method the *Multiple Wave Front* (MWF) method. By virtue of the covering of the PU space by multiple wave fronts, the MWF method can attain a high efficiency of parallelization. The theoretical parallelization efficiency f is given by

$$f = \frac{N_\theta N_\varphi N_\nu/8}{N_\theta N_\varphi N_\nu/8 + (P_x + P_y + P_z - 3)}, \tag{3}$$

where P_x, P_y, P_z are numbers of PUs in each PU-space dimension. The number $N_\theta N_\varphi N_\nu/8$ is a number of grid points in one direction group and $P_x + P_y + P_z - 3$ is a number of steps with which the first wave front crosses the 3-D PU space. When we use $N_\theta N_\varphi = 128^2$, $N_\nu = 1$, and $P_x = P_y = P_z = 8$, we have $f = 0.990$. It means that loss by non-active PUs in prologue and epilogue phases of the MWF method is negligible in practice. Actually, our test calculations showed that the computation time decreases almost linearly as the number of PU increases.

It may be worthy to notice that the MWF method is applicable to any message passing type parallel computers, not only massively parallel computers with distributed local memory but even shared memory type parallel computers. The MWF method is quite portable.

5. Applications

Our numerical code is applicable to various types of problems in a 3-D space; it can solve problems with any geometrical distributions and any optical conditions of matter. One interesting problem we have already applied is a re-ionization of an early universe. In the problem, photoionization of clumpy intergalactic matter is calculated consistently with the UV radiation field using our code. Details of the study will be shown in elsewhere.

6. Summary

We have developed a new method, the *Multiple Wave Front* method, to solve the radiative transfer equation employing massively parallel computers with distributed local memory. Though the calculation sequence restriction arises from the numerical scheme based on the formal solution, our MWF method manages the calculation sequence among PUs and attains very high efficiencies of parallelization in practical problems. Moreover, the MWF method is portable; it is applicable to any message passing type parallel computers.

Acknowledgements The author thanks Dr. Umemura for his fruitful discussions. A massively parallel computer the CP-PACS at the Center for Computational Physics, University of Tsukuba, was used for this work.

Reference

Kunasz, P. & Auer, L. H. 1988, JQSRT 39, 67

MULTI3D, 3D NON-LTE RADIATIVE TRANSFER

A Computer Program for Solving Multi-Level non-LTE Radiative Transfer Problems in 3D Atmospheres

ANDREAS BOTNEN
USIT, P.O. Box 1059 Blindern, N-0315 Oslo, Norway

AND

MATS CARLSSON
Institute of Theoretical Astrophysics, P.O. Box 1029 Blindern, N-0315 Oslo, Norway

Abstract. We have developed a computer program, Multi3D, for solving non-LTE radiative transfer problems in three dimensional geometry. Boundaries can be periodic or the radiation field may be prescribed. The atomic model may contain many energy levels. A short characteristic method is used for the radiative transfer and the rate equations are linearized such that a generalization to radiation hydrodynamics is straightforward. The code can be run on parallel computer architectures using MPI. The parallel speed-up is linear up to the number of nodes equal to the number of bound-bound transitions in the atom. Examples are given for the formation of the resonance lines of neutral sodium, NaI D, and singly ionized calcium, CaII K, in a simulation of the solar convection using 3, 6 and 12 level model atoms and 32 × 32 × 100 points in the atmosphere.

1. Formulation of the problem

If Local Thermodynamic Equilibrium (LTE) is assumed the population numbers of the energy levels in an atom can be found from the Boltzmann and Saha equations. In the outer parts of stellar atmospheres and whenever densities are low enough that collisional processes do not dominate over radiative processes LTE is a poor approximation. One will instead have to solve the equations of statistical equilibrium together with the radiative transfer equations to find the population numbers. The radiative transition probability from one energy level to another depends on the mean intensity and thus on conditions throughout the atmosphere through the transfer of

S. M. Miyama et al. (eds.), Numerical Astrophysics, 379–382.

energy with photons. With atomic models containing several energy levels one has to solve a very non-linear and non-local system of equations.

In one-dimensional geometry (plane-parallel or spherical) efficient codes to treat this problem have existed for a long time. Until recently the computational effort in 3D has been regarded as prohibitive but some implementations have started to appear. We will here describe a code that uses linearization instead of preconditioning to treat the multi-level non-linearities which makes adding more equations (like a momentum and an energy equation) straight-forward.

2. Implementation

Multi3D is implemented in Fortran90, using MPI for explicit message passing. The theory behind the implementation is outlined below.

2.1. THE SCHARMER AND CARLSSON SCHEME

Multi3D uses linearization similar to Scharmer & Carlsson (1985), as implemented in Multi by Carlsson (1986). The rate equations are also preconditioned using an expression that contains the diagonal of the familiar lambda operator to eliminate passive scattering.

2.2. OBTAINING A FORMAL SOLUTION

The process of obtaining the mean intensity, once the source function is known, is called a formal solution. We have examined several methods for performing this operation, and we decided to use a three point short-characteristic (SC) method. This method is described in Kunasz & Auer (1988), and Auer & Paletou (1994). The theory behind the method is very simple, in order to find the intensity at a certain point one only has to evaluate the following integral

$$I_\nu(\tau_\nu) = I_\nu(0)e^{-\tau_\nu} + \int_0^{\tau_\nu} S_\nu(t)e^{-(\tau_\nu - t_\nu)}\, dt_\nu. \tag{1}$$

In order to evaluate the integral numerically, the expression for the source function is replaced by a polynomial. The polynomial is then analytically integrated from 0 to τ_ν,

$$\int_0^{\Delta\tau_u} S(\tau)e^{-\tau} = S_0\omega_0 + c_U\omega_U + c_D\omega_D.$$

Using a second order expansion for the source function means that we must know the value of the source function, and the opacities at three points. An example of a three point ray is shown in Figure 1.

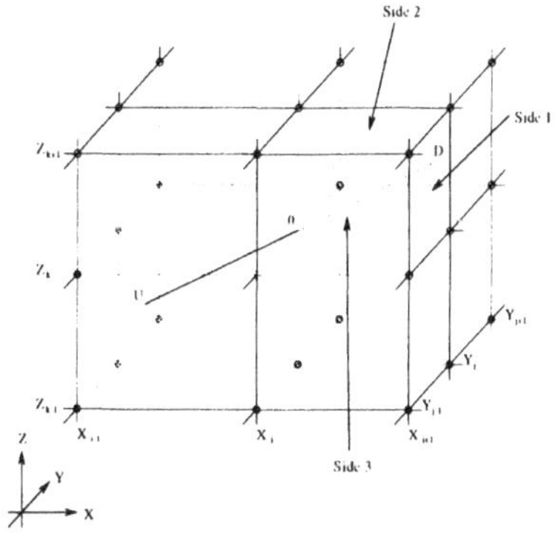

Figure 1. An example of a ray with a quadratic expansion for the source function. The grid is traversed with increasing z_k, and x_i, y_j is decreasing. The actual integration is only performed between the upwind point, where the intensity is known, and the 0 point. A solid black or gray circle indicates that the intensity has been computed at that point. When computing the intensity at (x_i, y_j, z_k) the downwind point can be at side 1, side 2, or side 3. The correlation shown between the location of the upwind point and the downwind point cannot in general be assumed.

3. Results

Modeling realistic physical problems in 3D is both CPU and memory demanding. Below we show some results from realistic use of Multi3D. These results are obtained using an RS6000 SP, with 120MHz POWER2 nodes.

3.1. MEMORY AND CPU USAGE

The memory consumption and the time to compute the formal solution for a 6 level Ca atom is given in Figure 2. Computation of the formal solution dominates the total iteration time. We have examined several ways of utilizing more than one processor; distributing over angle, over frequency, and over transition. For a limited number of processors, typically 1-20, the speed up will be linear. This also means that the time to compute the formal solution for the bound-bound transition containing the largest number of frequency points is the limit for the total execution time.

3.2. CONVERGENCE

Classical test problems caused no problems for the early implementation of the methods. It turned out that realistic snap-shots from convection simulations were much harder and checks had to be introduced in order to eliminate the problem of parabolic overshoot. Figure 3 shows the convergence, in this case the maximal correction to the population densities, for some combinations of atmospheres and atomic models. All cases investigated have converged to better than 10^{-3}.

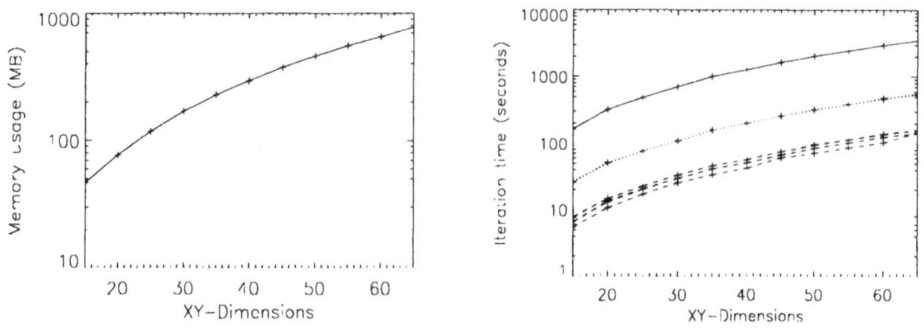

Figure 2. **Left panel :** The memory consumption for a 6 level Ca atom. **Right panel :** Time to compute the various transitions, total time (solid), the 5 bound-bound transitions (dotted), and the 5 bound-free transitions (dashed). Computing the formal solution of a bound-free transition requires less time as they contain fewer frequency points, and because they do not require the computation of an absorption profile. In both panels the number of points in the xy-direction is given by the x-axis, the number of points in the z-direction is 100.

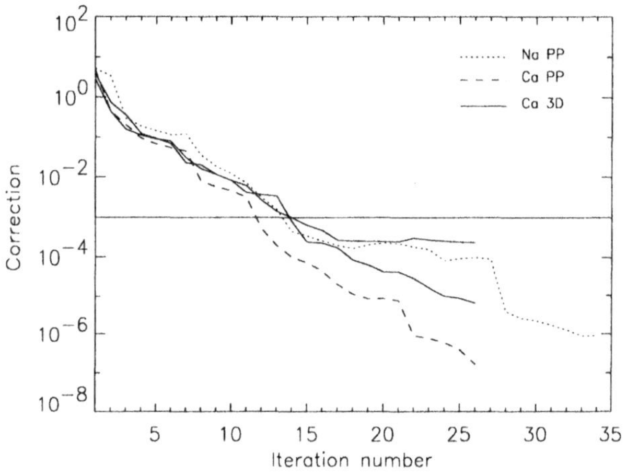

Figure 3. The two plan-parallel (PP) models, 12 level Na and 6 level Ca, both converge to 10^{-6} and beyond. However, one of the full 3D atmospheres will not give convergence below $2.3 \cdot 10^{-4}$ due to interpolation inaccuracies.

References

Auer, L. H., Paletou, F. 1994, A&A, 285, 675
Carlsson, M. 1986, Report No. 33, Uppsala Astronomical Observatory
Kunasz, P., Auer, L. H. 1988, J. Quant. Spectrosc. Radiat. Transfer, 39, 67
Scharmer, G.B., Carlsson, M. 1985 J. Comput. Phys., 59, 56

A HIGHER-ORDER GODUNOV SCHEME FOR NON-IDEAL MAGNETOHYDRODYNAMICS

T. SANO, S. INUTSUKA AND S. M. MIYAMA
National Astronomical Observatory, Mitaka, Tokyo 181-8588, Japan
E-mail: sano@th.nao.ac.jp

1. NUMERICAL METHOD

We propose a new Godunov-type numerical scheme for magnetohydrodynamics including the effect of ohmic dissipation. Several investigators have worked on the development of accurate numerical methods for magnetohydrodynamics (MHD). Among them, Godunov-type schemes are considered to be efficient for many astrophysical problems where flows are super-sonic. Brio & Wu (1988) and Hanawa *et al.* (1994) have applied Roe's method to MHD equations. Zachary & Colella (1992) and Dai & Woodward (1994) have extended PPM to MHD. Although their methods are successful in shock tube problems with moderate shock strength, neither of their methods maintains the divergence-free ($\nabla \cdot B = 0$) character of magnetic fields.

Our approach is to incorporate the MoC-CT method (Stone & Norman, 1992) into MUSCL scheme (van Leer, 1979) which is the second-order accurate Godunov-type scheme. MUSCL uses the exact solutions of Riemann problems and describes strong shocks accurately without any artificial viscosity. The constrained transport (CT) method proposed by Evans & Hawley (1988) guarantees the divergence-free constraint of magnetic fields at all time steps. Stone & Norman (1992) developed this CT algorithm using the method of characteristics (MoC) for accurate description of the Alfvén waves. Our scheme inherits all these advantages.

The method is divided into three parts; Lagrangian step, transport step, and magnetic field evolution. We adopt the second-order accurate van Leer advection in the transport step (van Leer, 1977). The update of magnetic fields is done by the MoC-CT algorithm described in Stone & Norman (1992). The MoC-CT algorithm is originally constructed for ideal MHD. We include the terms of ohmic dissipation into the CT algorithm.

S. M. Miyama et al. (eds.), Numerical Astrophysics, 383–386.

Our original point is mostly in the Lagrangian step. Therefore we describe the algorithm of the Lagrangian step in this short paper.

1.1. CENTERING OF VARIABLES

Before writing down finite-difference equations, we must decide how to center the variables. In our scheme, hydrodynamical variables (ρ, $\boldsymbol{v}$, and the specific total energy E) are defined at the center of each cell. On the other hand, magnetic fields are defined at the cell boundary. For the MoC-CT method, we need the velocity defined at the cell boundary. The solution of the Riemann problem is defined at each cell boundary in the Lagrangian step. Therefore we can obtain the velocity defined at the cell boundary. We use the solution of the Riemann problem for the magnetic field evolution. This is an important technique of our scheme.

1.2. LAGRANGIAN STEP

The heart of our scheme is the one-dimensional Lagrangian scheme, the results of which are remapped onto the desired Eulerian grids at each time step. Following the operator split method, the Lagrangian step itself is divided into two substeps.

First, we compute mainly hydrodynamical terms in the fundamental equations,

$$\frac{dV}{dt} = \frac{\partial v_x}{\partial m} , \tag{1}$$

$$\frac{dv_x}{dt} = -\frac{\partial P_{\mathrm{T}}}{\partial m} , \tag{2}$$

$$\frac{dE}{dt} = -\frac{\partial}{\partial m}(v_x P_{\mathrm{T}}) , \tag{3}$$

where $\partial m = \rho \partial x$, $P_{\mathrm{T}} \equiv P + (B_y^2 + B_z^2)/2$, and $V \equiv 1/\rho$ denotes the specific volume. The first partial update is performed in the Lagrangian step using the finite-difference equations;

$$\frac{V_i^{n+a} - V_i^n}{\Delta t} = \frac{v^*_{x,i+1/2} - v^*_{x,i-1/2}}{\Delta m} , \tag{4}$$

$$\frac{v_{x,i}^{n+a} - v_{x,i}^n}{\Delta t} = -\frac{P^*_{T,i+1/2} - P^*_{T,i-1/2}}{\Delta m} , \tag{5}$$

$$\frac{E_i^{n+a} - E_i^n}{\Delta t} = -\frac{v^*_{x,i+1/2}P^*_{T,i+1/2} - v^*_{x,i-1/2}P^*_{T,i-1/2}}{\Delta m} . \tag{6}$$

The time advanced ($+\Delta t/2$) variables v_x^* and P_{T}^* are calculated by solving a Riemann problem at each cell boundary. Here we solve the modified Riemann problem that includes only magnetic pressure terms as contributions from the magnetic field, instead of solving a full set of nonlinear MHD Riemann problem. The magnetic tension terms are considered in the second substep. This decomposition of the problem plays a key role in our scheme for the sake of simplicity and accuracy.

The second partial update is performed for the rest of the Lorentz force terms,

$$\frac{dv_x}{dt} = \frac{\partial}{\partial m}\left(\frac{B_x^2}{2}\right) , \tag{7}$$

$$\frac{dv_y}{dt} = -\frac{\partial}{\partial m}\left(B_x B_y\right) , \tag{8}$$

$$\frac{dv_z}{dt} = -\frac{\partial}{\partial m}\left(B_x B_z\right) , \tag{9}$$

$$\frac{dE}{dt} = -\frac{\partial}{\partial m}\left(v_x \frac{B_x^2}{2} + v_y B_x B_y + v_z B_x B_z\right) . \tag{10}$$

These equations utilize the time advanced values of v_y^*, v_z^*, B_y^*, and B_z^* calculated with the MoC in a Lagrangian frame of reference. An important simplification at this step is made by considering only the incompressible Alfvén wave characteristics to calculate the time advanced values. The validity of this simplification is confirmed by various test calculations.

The numerical result of the MHD Riemann problem is shown in *Figure 1*. Our scheme provides the accurate evolution of all modes of MHD wave families. This result shows excellent quantitative agreement with those of the other state-of-the-art methods (Brio & Wu, 1988; Dai & Woodward, 1994), while our method is considerably simpler than the others.

Numerical computations were partially carried out at the National Astronomical Observatory, Japan. Part of computations also has been carried out at the Hilo Facility of the National Astronomical Observatory, Japan.

References

Brio, M., & Wu, C. C. 1988, J. Comp. Phys., 75, 400
Dai, W., & Woodward, P. R. 1994, J. Comp. Phys., 115, 485
Evans, C. R., & Hawley, J. F. 1988, ApJ, 332, 659
Hanawa, T., Nakajima, Y., & Nobuta, K. 1994, Dept. of Astrophysics Nagoya Univ. Preprint No. 94-34
Stone, J. M., & Norman, M. L. 1992, ApJS, 80, 791
van Leer, B. 1977, J. Comp. Phys., 23, 276
van Leer, B. 1979, J. Comp. Phys., 32, 101
Zachary, A., & Colella, P. 1992, J. Comp. Phys., 99, 341

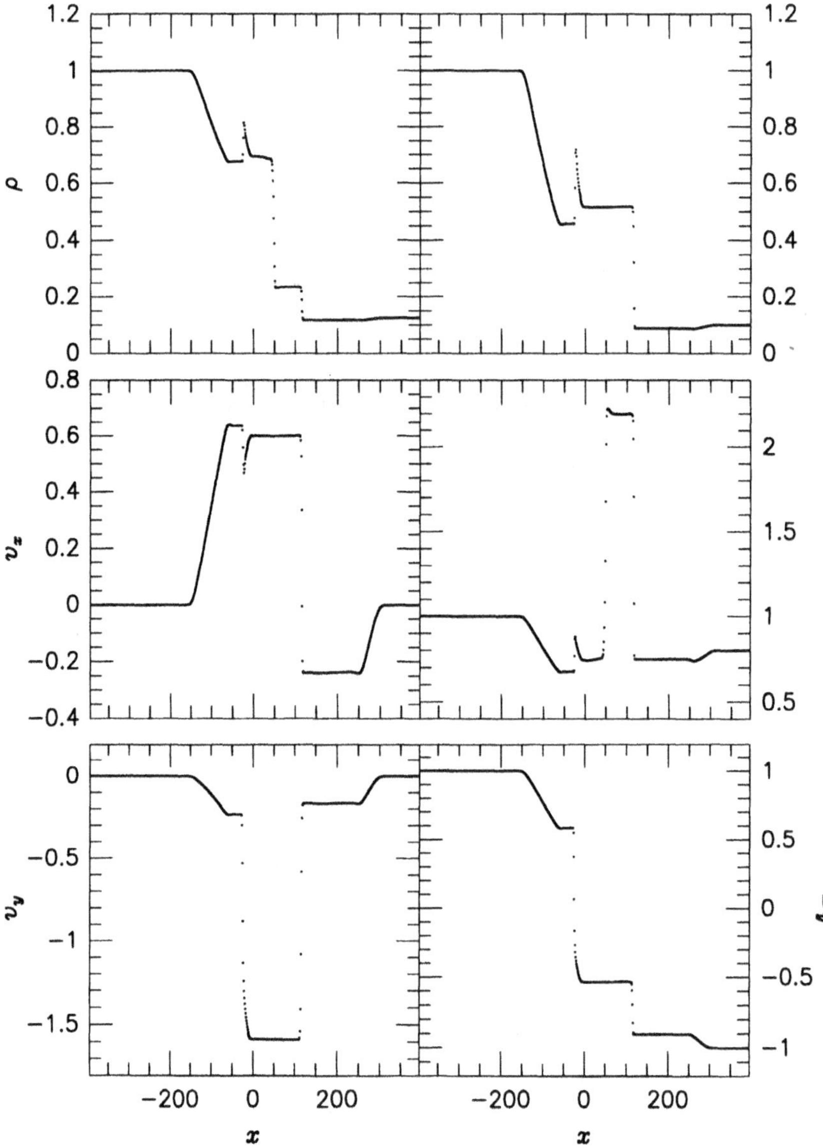

Figure 1. Results of the MHD shock tube problem with $\gamma = 2$ and $B_x = 0.75$. Initial conditions: $(\rho, P, u_x, u_y, B_y) = (1, 1, 0, 0, 1)$ for $x < 0$ and $(0.125, 0.1, 0, 0, -1)$ for $x > 0$. 800 uniformly separated Eulerian grid zones are used. Dots are the results at $t = 80$.

A SECOND ORDER ACCURATE NUMERICAL SCHEME FOR MHD EQUATIONS

NAOYA FUKUDA AND TOMOYUKI HANAWA
Department of Astrophysics, Nagoya University
Chikusa-ku, Nagoya 464-8602, Japan

Abstract. We present a second order accurate numerical scheme for ideal magnetohydrodynamical equations. Our numerical scheme is an upwind scheme based on the flux vector splitting and an extension of Roe's scheme. The second order accuracy is achieved by MUSCL, the Monotone Upstream-centered Scheme for Conservation Laws approach. Our numerical scheme ensures the total variation diminishing (TVD) as far as Courant number is smaller than unity.

1. A Second Order Accurate Numerical Scheme

In this paper we restrict ourselves to the one-dimensional problem to show the essence of our scheme. The ideal MHD equations are described as

$$\frac{\partial \boldsymbol{U}}{\partial t} + \frac{\partial \boldsymbol{F}}{\partial x} = 0 \,, \tag{1}$$

where $\boldsymbol{U}$ and $\boldsymbol{F}$ denote the conservative variable vector and the flux vector. The vectors are evaluated on the grid points and denoted as $\boldsymbol{U}_i^n \equiv \boldsymbol{U}(x = i\Delta x, t = n\Delta t)$, $\boldsymbol{F}_i^n \equiv \boldsymbol{F}(\boldsymbol{U}_i^n)$, where Δx and Δt denote the grid spacing in t and x, respectively.

First we compute the solution at $t = (n + 1/2)\,\Delta t$ using

$$\bar{\boldsymbol{U}}_i^{n+1} = \boldsymbol{U}_i^n - \frac{\Delta t}{2\Delta x}(\boldsymbol{F}_{i+\frac{1}{2}}^n - \boldsymbol{F}_{i-\frac{1}{2}}^n) \,, \tag{2}$$

and

$$\boldsymbol{F}_{i+\frac{1}{2}}^n = \frac{1}{2}[(\boldsymbol{F}_{i+1} + \boldsymbol{F}_i) - \boldsymbol{R}|\boldsymbol{\Lambda}|\boldsymbol{L}(\boldsymbol{U}_{i+1} - \boldsymbol{U}_i)] \,, \tag{3}$$

where $\boldsymbol{R}$, $\boldsymbol{L}$, and $\boldsymbol{\Lambda}$ denote the right eigenvectors, the left eigenvectors, and the phase velocities of the magnetohydrodynamical waves, respectively.

S. M. Miyama et al. (eds.), Numerical Astrophysics, 387–388.

Their forms are given as a function of $\boldsymbol{U}_i$ and $\boldsymbol{U}_{i+1}$ in Nakajima & Hanawa (1996). The second order accurate solution, $\boldsymbol{U}_i^{n+1}$, is given by

$$\boldsymbol{U}_i^{n+1} = \boldsymbol{U}_i^n - \frac{\Delta t}{\Delta x}(\boldsymbol{F}_{i+\frac{1}{2}}^{n,(2)} - \boldsymbol{F}_{i-\frac{1}{2}}^{n,(2)}) , \quad (4)$$

and

$$\begin{aligned} \boldsymbol{F}_{i+\frac{1}{2}}^{n,(2)} &= \frac{1}{2}\{(\bar{\boldsymbol{F}}_{i+1} + \bar{\boldsymbol{F}}_i) - \bar{\boldsymbol{R}}|\bar{\boldsymbol{\Lambda}}|\bar{\boldsymbol{L}}(\bar{\boldsymbol{U}}_{i+1} - \bar{\boldsymbol{U}}_i) \\ &+ \bar{\boldsymbol{R}}\left(\frac{|\bar{\boldsymbol{\Lambda}}| - \bar{\boldsymbol{\Lambda}}}{2}\right)\Psi[\bar{\boldsymbol{L}}(\bar{\boldsymbol{U}}_{i+2} - \bar{\boldsymbol{U}}_{i+1}), \bar{\boldsymbol{L}}(\bar{\boldsymbol{U}}_{i+1} - \bar{\boldsymbol{U}}_i)] \\ &+ \bar{\boldsymbol{R}}\left(\frac{|\bar{\boldsymbol{\Lambda}}| + \bar{\boldsymbol{\Lambda}}}{2}\right)\Psi[\bar{\boldsymbol{L}}(\bar{\boldsymbol{U}}_i - \bar{\boldsymbol{U}}_{i-1}), \bar{\boldsymbol{L}}(\bar{\boldsymbol{U}}_{i+1} - \bar{\boldsymbol{U}}_i)]\} , \end{aligned} \quad (5)$$

where Ψ denotes the limiting function. The barred symbols denote the values evaluated from $\bar{\boldsymbol{U}}_i$ and $\bar{\boldsymbol{U}}_{i+1}$. Note that the limiter operates the wave amplitude, $\bar{\boldsymbol{L}}(\bar{\boldsymbol{U}}_{i+1} - \bar{\boldsymbol{U}}_i)$. When the limiter operates on the conservative variable, the numerical damping of the Alfvén wave is serious. Our scheme can calculate the parametric instability of the Alfvén wave with large amplitude.

2. Test Problem

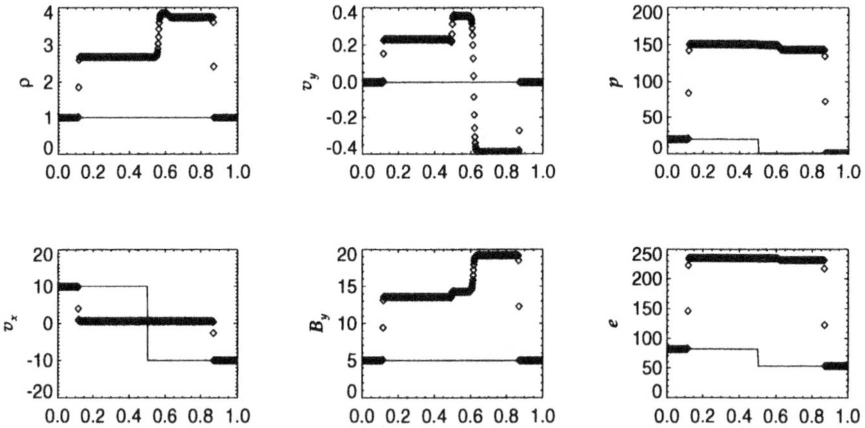

Figure 1. A result of the shock tube problem. The solid and dotted curves denote the condition at $t = 0.0$ and 0.08. The Courant number is 0.9. We applied the minmod limiter (see, e.g., Hirsch 1990).

References

Hirsch, C. (1990) *Numerical Computation of Internal and External Flows.* Wiley, Chichester.

Nakajima, Y., and Hanawa, T. (1996) Formation and evolution of filamentary molecular clouds with oblique magnetic field, *ApJ*, **Vol. 467**, pp. 321–333

NUMERICAL SIMULATIONS OF THE ASTROPHYSICAL MHD FLOWS

ALEXANDER E. DUDOROV, ANDREY G. ZHILKIN
Chelyabinsk State University, Russia

AND

OLEG A. KUZNETSOV
Institute of Applied Mathematics, RAS, Russia

Magnetic fields can play an important role in many astrophysical problems. At the present time, sufficient observational data have been gathered about the magnetic field of the interstellar medium, molecular clouds, star formation regions, Bok globules, and young stellar objects (Heiles et all, 1993; Dudorov, 1995; Vallee, 1997). This shows that star formation takes place in rotating magnetized interstellar clouds.

We have created a new 2D numerical code in order to investigate various astrophysical MHD problems and prove basic conclusions of the theory of Fossil Magnetic Fields (Dudorov 1995). Our code is based on a simple TVD-scheme for the solution of MHD problems. This scheme does not require calculating the eigenvectors of the hyperbolity matrix but only using the maximum modulus of the eigenvalues. It is a monotonic scheme which has high resolution in the smoothing region of the solution.

The rationale behind the construction of such a different scheme for a set of one-dimensional hyperbolical equations consists of the following (Vyaznikov et. all, 1989):

1) selection of a monotonic basic scheme of the first order of approximation;

2) transformation of the basic scheme to a high resolution scheme by adding antidiffusional terms;

3) correction of antidiffusional coefficients to guarantee monotonicity.

On the basis of this scheme, the code "Moon," which can simulate MHD flows both in the 1D and 2D statement of the problem, was developed. It has a module to solve the Poisson equation for gravitational potential, permitting the computation of the gravitational collapse of protostellar clouds. The Poisson equation is solved by the Douglas-Rachford ADI method. The code is realized by the programming language C++ and

S. M. Miyama et al. (eds.), Numerical Astrophysics, 389–390.

is completely object-oriented. Our code was tested on several problems with known exact or approximate analytical solutions, demonstrating its capability for solving most MHD-problems.

Numerical computations of the collapse of the magnetized protostellar cloud have confirmed earlier results in the frame of one and a half dimensional (1.5D) approximation (Dudorov 1990). The geometry of a magnetic field in a cloud with an initially small field becomes quasiradial. Magnetic fields lead to a flattening of collapsing clouds on the late stages of contraction. When the initial magnetic field is strong, the collapse is transferred to a quasistatic contraction. Such types of clouds must evolve on a diffusional time scale.

Numerical simulations show that for sufficiently large values of the initial magnetic field, the degree of flattening is proportional to $\varepsilon_m^{-1/2}$, where ε_m is the initial ratio of the magnetic energy to the gravitational one.

For sufficiently strong magnetic fields, the central density at the free-fall time $t_{\rm ff}$ is inversely proportional to ε_m dependence.

References

Dudorov, A.E. 1990, Itogy Nauki i Techniky. Astronomia, 39, 77-158. (in Russian)

Dudorov A.E. 1995, Astronomical Journal, 77, 884. (in Russian)

Heiles, C., Goodman, A.A., McKee, C.F., Zweibel, E.G. 1993, Protostars and Planets III. Edited by E.N. Levy and J. I. Lunine Tucson: Univ. Arizona Press, p.327

Vallee, J.P. 1997, Fundamentals of Cosmic Physics, 19, 1

Vyaznikov C.V., Tishkin V.F., Favorsky A.P. 1989, Matematicheskoe Modelirovanie, 1, 95 (in Russian)

HOW TO DETERMINE THE SMOOTHING LENGTH IN SPH?

J. KODA AND Y. SOFUE
Institute of Astronomy, University of Tokyo, Mitaka, Tokyo 181, Japan

AND

K. WADA
National Astronomical Observatory, Mitaka, Tokyo 181, Japan

1. Introduction

In Smoothed Particle Hydrodynamics (SPH), many authors have recently used a spatially and temporally varying smoothing length, h_i, to achieve high spatial resolution and a wide dynamic range. Although there are many ways to determine h_i, many of them are based on an approach introduced by either [1] Hernquist & Katz (1989) or [2] Benz (1990). In the former method, h_i is determined so that each particle interacts with a roughly constant number of neighbors, $N_{\rm nb}$, which helps achieving the same level of accuracy at all points (Hernquist & Katz 1989). This method, however, is time-consuming, and as pointed out by Steinmetz (1996), can lead to unphysical change in h_i, especially in a clumpy medium, since the selection criteria are computational rather than physical. In method [2], on the other hand, h_i is determined by the continuity equation and $h_i \propto \rho_i^{-1/3}$ in the method [2], where ρ_i is the density at the location of particle i. Although this is a physically natural approach, there is no guarantee that all particles have nearly constant $N_{\rm nb}$. In this paper we compare the above two methods for a test problem: adiabatic collapse of an initially isothermal gas sphere, and investigate the reliability of these two methods for determining h_i. We also propose here a modified method [3]: after h_i is calculated by the method [2], it is corrected to h'_i by using $h'_i = \frac{1}{2}h_i[1 + (N_{\rm min}/N_{\rm nb})^{1/3}]_i$, if $N_{\rm nb} < N_{\rm min}$, and $h'_i = \frac{1}{2}h_i[1 + (N_{\rm max}/N_{\rm nb})^{1/3}]_i$, if $N_{\rm nb} > N_{\rm max}$.

Our SPH code is basically the same as GRAPE-SPH used in Steinmetz (1996). However, we use the W_4-spline kernel, and search neighbors in a host computer instead of using the GRAPE neighbor list. The number of particles is 12,000. We set $N_{\rm min} = 54$, and $N_{\rm max} = 74$.

S. M. Miyama et al. (eds.), Numerical Astrophysics, 391–392.

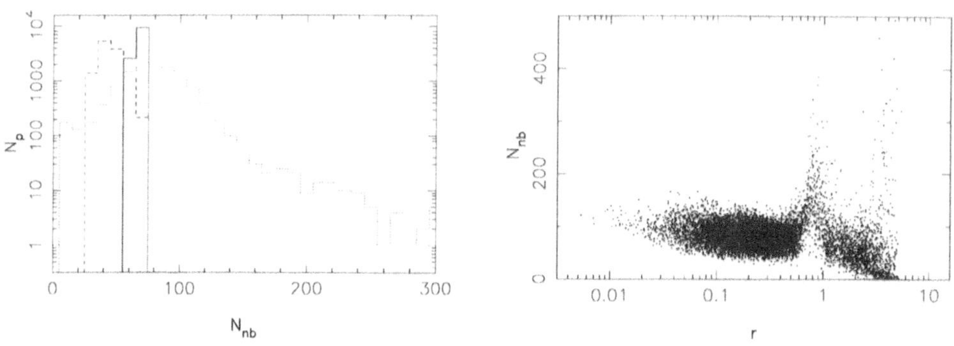

Figure 1. Left: $N_{\rm nb}$ vs. $N_{\rm p}$ at $t = 3t_{\rm ff}$. N_p is the number of particles with $N_{\rm nb}$ neighbors. The solid, dotted and dashed lines are N_p with the method [1], [2], and [3], respectively. Right: radial distribution of $N_{\rm nb}$ at $t = 3t_{\rm ff}$ with the method [2].

2. Results

A histogram of $N_{\rm nb}$ at $t = 3t_{\rm ff}$ (Fig.1: left), where $t_{\rm ff}$ is the free fall time, shows clearly that $N_{\rm nb}$ stays in a narrow range for methods [1] and [3], while it extends widely for method [2]. The right panel of Fig. 1 is the radial distribution of the number of neighbors in simulations using method [2]. $N_{\rm nb}$ is extremely dispersed in two regions: $r \simeq 0.9$ and 4.0. The former is a shock front between a central virialized gas ($r < 0.6$) and an outer expanding envelope ($r > 0.6$). This dispersion of $N_{\rm nb}$ causes unphysical fluctuations in the density distribution. The wide spread in $N_{\rm nb}$ is caused by a 'delayed' response of h_i, that is, when some particles with a large h_i move from a pre-shock region into the shock region, their h_i remains a large value. The region at $r \sim 4.0$ is the forefront of the expanding envelope. This result shows that method [2] fails in a non-stationary region with a high-density gradient.

By using method [3], on the other hand, $N_{\rm nb}$ stays in a desired range as with the method [1]. However, since h_i of most particles is determined by the local density in method [3], unphysical behavior of h_i that could happen in method [1] can be avoided. We conclude, therefore, that method [3] should be used instead of the two standard methods.

References

Benz, W. (1990) *in "The Numerical Modeling of Nonlinear Stellar Pulsations, Problems and Prospects", ed. J.R. Buchler, Kluwer Academic Publishers, Dordrecht, p.269-288*

Hernquist, L., and Katz, N. (1989) *ApJS*, **70**, 419.

Steinmetz, M., and Müller, E. (1993) *A&A*, **268**, 391.

Steinmetz, M. (1996) *MNRAS*, **278**, 1005.

PARALLEL TREESPH

CESARIO LIA
SISSA
Via Beirut 2, I-34013, Trieste, Italy
liac@sissa.it

AND

GIOVANNI CARRARO
Padova Department of Astronomy
Vicolo dell'Osservatorio 5, I-35122, Padova, Italy
carraro@pd.astro.it

1. Introduction

A new parallel implementation of a TreeSPH code is here presented. The code is written in F90 and the communications are handled using SHMEM libraries in the Cray T3E context. The original scalar code is described in details in Carraro et al. (1998), whereas the parallel algorithm is presented in Lia et al. (1998), where a comparison is also made with PTreeSPH (Davè et al. 1997).

2. A 3D test and something more

In this short paper we present the results of the adiabatic collapse of an initially isothermal gas sphere [see Davé et al. (1997) for any details] at the time of the maximum compression (see Fig. 1, where left panels refer to the scalar test, and right panels to the parallel one). The scalar test has been performed using 2176 particles and took 5400 secs, while the parallel one uses 6000 particles, and took 60 secs. All the qualitative and quantitative details are given in Lia et al. (1998). Fig. 2 shows a simulation of the formation of a galactic disk in a dissipative collapse of a rotating ($\lambda = 0.08$) spherical gas plus Dark Matter proto-galaxy. Gas settles down in a thin rotating disk due to cooling. Here we show gas evolution in the $x - z$ (right panels) and $x - y$ (left panels) planes, respectively.

S. M. Miyama et al. (eds.), Numerical Astrophysics, 393–394.

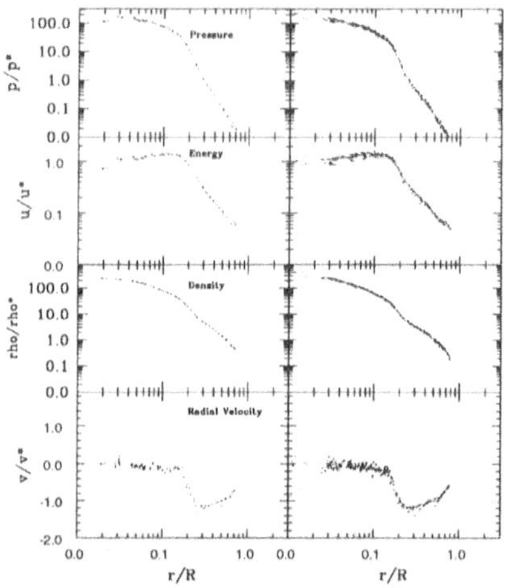

Figure 1. The adiabatic collapse test.

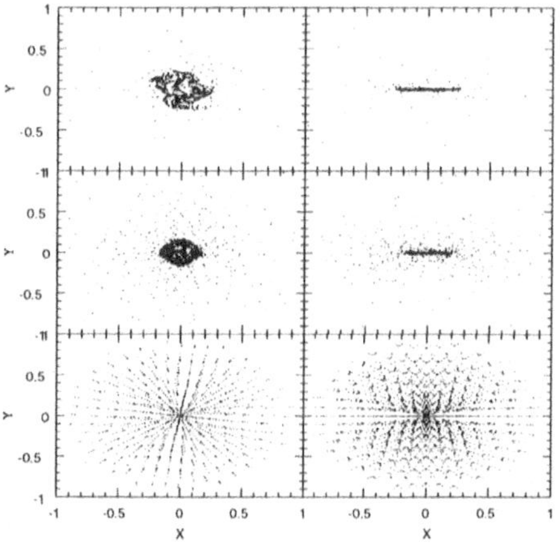

Figure 2. The formation of a galactic disk.

References

Carraro, G., Lia, C., and Chiosi, C. (1998) Galaxy Formation and Evolution. I. A new treeSPH code (PD–SPH), *MNRAS*, 297, 1021

Davè, R., Dubinski, C., and Hernquist, L. (1997) Parallel TreeSPH, *New Astronomy*, **Vol. no. 2(3)**, pp. 277–297

Lia, C., Carraro, G., Chiosi, C., and Voli, M. (1998) A new parallel TreeSPH code, *Science and Supercomputing at CINECA-Report 1998*, in press

PARALLEL PARTICLE SIMULATIONS

N. NAKASATO
Department of Astronomy, University of Tokyo,
Bunkyo-ku, Tokyo, 113-0033, Japan

AND

H. YAHAGI AND M. MORI
Institute of Astronomy, University of Tokyo,
Mitaka, Tokyo 181-8588, Japan

1. Introduction

In these days, demands of more powerful computers in astrophysics is getting harder and harder. One inexpensive way to satisfy such demands is using many computers in parallel. In this paper, we present our parallel particle simulation system, which is divided in two parts, e.g. Remote-GRAPE system and PC cluster system, and parallel particle simulation codes as following :

- "Parallel GRAPE-SPH" code using Remote-GRAPE system
- "Parallel TREE-SPH" code for general parallel computers including our PC cluster

2. Remote-GRAPE system and GRAPE-SPH code

The original ideas of the special purpose computer "GRAPE" was partly based on such a concept as "using many computers in parallel." The GRAPE-3AF, which is one of the GRAPE family, can efficiently calculate gravitational force between particles, and construct neighbor lists. All other calculations are performed on the host workstation which is directly connected to GRAPE. "Remote-GRAPE" is a subroutine library to use the special purpose computer GRAPE-3AF. The Remote-GRAPE allows us to use GRAPE-3AF with the computer which is not directly connected to GRAPE. Thus, we can use the state-of-the-art computer as a host computer. For detail description about "Remote-GRAPE", please see our paper (Nakasato et al. 1997).

S. M. Miyama et al. (eds.), Numerical Astrophysics, 395–396.

TABLE 1. SPH Benchmark result (N = 50k)

PE	total (sec)	sph (sec)	gravity (sec)
4	11.5	6.9	4.6
8	4.5	2.5	2.0
15	2.8	1.6	1.2

"PE" represents the number of processors. "total" shows the CPU time per one step and "sph" and "gravity" show the CPU time for the sph part and gravity part, respectively.

3. PC cluster system and Parallel TREE-SPH code

In a development stage of parallel numerical codes, much many compile and run trials are required till we can use the codes for real applications. Using a super computer located at some computer center for developing the codes may be very time consuming work due to its limitations of usage of the super computer. PC cluster system is a poor man's parallel computer, or a desk side parallel computer. We develop a PC cluster system, which consists of two PC, and have used it as a development machine of our parallel TREE-SPH code.

Detail description of our parallel TREE-SPH code will be presented elsewhere. In this paper, we present one SPH benchmark result (Table 1.), which is carried out on VPP300/16R and VX/4R at the Astronomical Data Analysis Center of the National Astronomical Observatory, Japan. In this benchmark we compute the adiabatic collapse of an isothermal gas sphere, which is the one of the standard test of SPH code done by many authors like Hernquist & Katz (1989) etc.. We set $\theta = 1.0$ in gravity parts and the mean number of neighbor particles ~ 50 in SPH parts. We have developed our parallel TREE-SPH code by using the Message Passing Interface (MPI) so that we can use it on not only VPP series but also a general parallel computer system including our PC cluster system. Although our code is not highly optimized for VPP architecture, we can obtain an almost linear scaling of performance.

References

Hernquist, L., & Katz, N., 1989, ApJS, 70, 419.
Nakasato, N., Mori, M., Nomoto, N., 1997, ApJ, 484, 608.

THE FOREST METHOD AS A NEW PARALLEL TREE METHOD

H. YAHAGI AND M. MORI
Institute of Astronomy, University of Tokyo, Mitaka, Tokyo 181-8588, Japan

N. NAKASATO
Department of Astronomy, University of Tokyo, Bunkyo-ku, Tokyo 113-0033, Japan

AND

Y. YOSHII
Institute of Astronomy, University of Tokyo, Mitaka, Tokyo 181-8588, Japan
Research Center for the Early Universe, University of Tokyo, Tokyo 113-0033, Japan

1. Introduction

Dark matter and stars are collisionless in galaxies and larger systems, and the evolution of a galaxy system consisting of such particles can be studied using an N-body simulation. There are many algorithms proposed for calculating the gravity force exerted on the particles. The particle-particle method, among others, calculates a direct sum of interactions between all pairs of N particles in a system, so that the computational time scales as N^2. In order to carry out simulations with a great number of particles, the tree method was developed with a more efficient algorithm whose time-complexity scales as $N \log N$ (Appel 1985; Barnes and Hut 1986). This tree method, if paralleled, can handle more particles faster compared to the serial tree method and is most appropriate for the study of galaxy formation and large-scale structures in the universe.

2. The Forest Method

We have developed a new parallel tree method and hereafter call it the forest method. This new method uses the sectional Voronoi tessellation (SVT) for the domain decomposition. SVT decomposes a space into polyhedra by changing their volumes according to the assigned weights (Okabe *et al.* 1992). The forest method determines these weights as follows based on the

S. M. Miyama et al. (eds.), Numerical Astrophysics, 397–398.

Figure 1. An example of the sectional Voronoi tessellation (SVT) with the over-load diffusion (OLD), applied to a truncated Mestel disk. Different colors are used only to indicate the location of the boundaries.

load balancing by the over-load diffusion (OLD). That is, each processor first checks individual loads of all ambient processors and is given a weight by which its own load, if higher than others, is lowered to the global average level. The load balancing among the processors is achieved by iterating this procedure. Figure 1 shows an example of SVT with OLD, applied to a truncated Mestel disk whose surface mass density is inversely proportional to the radius away from the disk center.

3. Performance and Summary

Elapsed time of the simulation code in each step is measured on the VPP300/16R, and the result is shown in Table 1 for the case of a homogeneous sphere consisting of $N = 128^3$ particles. The error tolerance is set to be $\theta = 1.0$ and the quadrupole terms are included.

Table 1: Timing Analysis

No. of processors	1	8	15
Elapsed time per step (sec)	336.2	39.8	20.9

Table 1 indicates that a good linearity of the code is maintained in the forest method. With this code, we are able to carry out the simulation using as large as 10^7 or more particles with a wide dynamic range.

Acknowledgements All calculations have been performed on the VPP300/16R of the National Astronomical Observatory, Japan. This work has been supported in part by the Research Fellowship of the Japan Society for the Promotion of Science for Young Scientists (6867) and the Grant-in-Aid for COE research (07CE2002) of the Ministry of Education, Science, and Culture in Japan.

References

Appel, A. W., 1985, SIAM J. Sci. Stat. Comp., 6, 85
Barnes, J. & Hut, P., 1986, Nature, 324, 446
Okabe A., Boots, B. & Sugihara, K., 1992, Spatial Tessellations Concepts and Applications of Voronoi Diagrams, Chichester, John Wiley & Sons Ltd

A STOCHASTIC SIMULATION METHOD FOR FOKKER-PLANCK EQUATIONS

T. YOSHIDA AND S. YANAGITA
Faculty of Science, Ibaraki University, Mito 310, Japan

1. Introduction

A stochastic simulation method to solve the cosmic-ray transport equation by Itô's stochastic differential equations (SDEs) has been developed[1],[2]. This method is based on the equivalence between Fokker-Planck equations (FPEs) and SDEs. However, simple-minded application of this method is not enough to achieve a wide dynamic range, as pointed out by Park and Petrosian[3]. In order to cope with this difficulty, we couple this method with a technique known as particle splitting. Our method is quite adapted for simulating particle acceleration, propagation, and non-thermal photon emission processes in astrophysical phenomena[4].

2. Method of Numerical Calculation

Park and Petrosian[3] have proposed the following time-dependent test problem

$$\frac{\partial f}{\partial t} = -\frac{\partial}{\partial p}(4p^2 f) + \frac{\partial^2}{\partial p^2}(p^3 f) - f + \delta(p - p_{\rm inj})\delta(t), \qquad (1)$$

where f is the distribution function of particles, p is particle momentum, and $p_{\rm inj}$ is injection momentum. This equation has the impulsive injection term at $t = 0$ and the escape term. At $t = 0$ particles are injected with a momentum of $p_{\rm inj} = 0.1$. To treat the escape term numerically we calculate the probability of escape $P(\Delta t) = 1 - e^{-\Delta t}$ at each time step Δt and determine whether to remove each particle by using a uniform random number. Equation (1) is equivalent to the following SDEs,

$$dp = 4p^2 dt + \sqrt{2p^3}dW, \qquad (2)$$

S. M. Miyama et al. (eds.), Numerical Astrophysics, 399–400.

where dW is a Wiener process given by the Gaussian distribution. The numerical integration of the SDEs is performed by a simple Euler method.

A particle splitting is essential to achieve a wide dynamic range. We set splitting surfaces p_i in momentum space. Each time a particle hits the surface p_i, the particle is split into w particles with the same momentum which the particle has attained.

3. Numerical Results

In Figure 1(a) we show the numerical results without particle splitting to make a comparison between the case of 10^4 particles and that of 10^6. As Park and Petrosian[3] have concluded, the SDE method without particle splitting is susceptible to Poisson noise when the number of particles is insufficient. Figure 1(b) shows the numerical results with and without

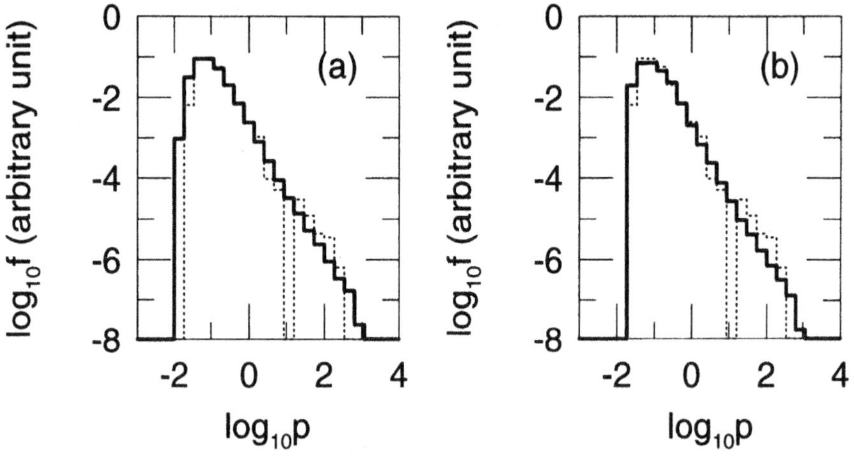

Figure 1. (a) The solid line shows the spectrum at $t = 3.0$ without particle splitting, where we use 10^6 particles. The dotted line shows the result of 10^4 particles without particle splitting. (b) The solid line shows the spectrum with particle splitting, where we take $w = 4$ using 10^4 particles. The dotted line is the same as Figure 1(a).

particle splitting for the same number of test particles of 10^4. It is clearly seen that Poisson noise is avoided even for the relatively small number of test particles when particles are split.

References

1. MacKinnon, A.L. and Craig, I.J.D. (1991), *A&A*, **251**, 693.
2. Achterberg, A. and Krülls, W. M. (1992), *A&A*, **265**, L13.
3. Park, B.T. and Petrosian, V. (1996), *ApJS*, **103**, 255.
4. Yoshida, T. and Yanagita S. (1994), *Prog. Theor. Phys.*, **92**, 1217.

A GENETIC ALGORITHM FOR DETERMINING THE ORBITS OF INTERACTING GALAXIES

Improving the method

MATTIAS WAHDE
NORDITA
Blegdamsvej 17, DK-2100 Copenhagen, Denmark

Abstract. I introduce and discuss a method which uses a genetic algorithm (hereafter GA) to determine the orbital parameters of interacting galaxies. The method has been successfully applied to several sets of artificial data. I discuss the improvement in performance obtained by optimization of the parameter values used in the GA.

1. Introduction

The determination of orbital parameters constitutes an important problem in galactic dynamics, and the method discussed here provides a *systematic* way of solving the problem in a general case. Due to the long time scales involved in galactic interactions, one cannot follow directly the evolution of an interacting system. However, the tidal distortions that appear in a galactic interaction can be used to deduce the orbital parameters. Unfortunately there is no simple way of relating the appearance of tidal bridges and tails to these parameters. In principle, one can carry out a complete search of the parameter space, but this is not feasible in practice, since the parameter space is normally many-dimensional and therefore too large to cover completely. In general, there are at least five unknown parameters: the relative velocities in the plane of the sky, the relative position along the line of sight, and the masses of the two galaxies. To these parameters can be added the orientations, scale lengths, scale heights, and spins (clockwise or counterclockwise) of thc two galaxies, which may all be difficult to estimate if the interaction geometry is complicated. The masses can of course be estimated indirectly, but rarely with high accuracy.

S. M. Miyama et al. (eds.), Numerical Astrophysics, 401–402.

2. Method

Instead of carrying out a full search of the parameter space, I use a GA (see e.g. Holland, 1975 or Charbonneau, 1995) to search efficiently through the parameter space. The method is described in detail by Wahde (1998). For an orbital parameter determination, the image of the (artificial) interacting system is covered with a grid, and the amount of matter (or light) in each grid cell is determined. Then, the variables of the problem are coded in *chromosomes* (i.e. strings of digits). The complete set of strings is referred to as the *population*. Each string is decoded to give a set of orbital parameters. The two point particles representing the centres of the two galaxies are then first integrated backwards in time, a disc of massless particles is added to each of them, and the resulting system is integrated forward in time until the time of the observation is reached, at which point the amount of matter in each grid cell is determined. Then, the *fitness* is computed as the inverse of the deviation between the results of the simulation and the observational data. When all simulations have been evaluated, the next *generation* is formed by applying selection, crossover, and mutation operators to the chromosomes, such that simulations with high fitness have a greater chance of being selected than those with low fitness. Then the new generation is evaluated etc. until a solution has been found.

3. Results, Optimization, and Improvements

So far, the method has been used successfully to find the orbital parameters of about 20 (artificial) systems. In order to increase the efficiency, the parameters used by the GA should be optimized. Early results from this process indicate that the best results are obtained with rather coarse grids (7×7). Provided that the mutation rate is increased, a small population (e.g. 50) does equally well as a large one (e.g. 500). Further optimizations are under way, as are several improvements, such as the inclusion of dark matter haloes in the simulations.

Acknowledgment

I dedicate this work to the memory of my friend and thesis advisor, Dr. Björn Sundelius.

References

Charbonneau, P. (1995), *ApJS*, **101**, 309

Holland, J.H. (1975), *Adaptation in natural and artificial systems*, 1st ed.: University of Michigan Press, Ann Arbor; 2nd ed.: 1992, MIT Press, Cambridge

Wahde, M. (1998), Submitted to *A&A*, astro-ph/9710278.

MULTI-DIMENSIONAL ASTROPHYSICAL STRUCTURAL ANALYSIS

A New Field of Numerical Astrophysics

DAVID L. MEIER AND PATRICK GODON
Jet Propulsion Laboratory
4800 Oak Grove Drive, Pasadena, CA 91109, U.S.A.

1. Introduction

Numerical fluid mechanics has been able to address many problems in astrophysics. In the first half of the 20th century, one-dimensional stellar interior calculations were used to investigate the radial structure of stars. In the decades following the invention of electronic computers, these calculations were extended to include time-dependent evolution on thermal (pulsating) and dynamical (collapse and explosive) time scales. Finally, in the last few decades, with the advent of powerful supercomputers, the field has been able to address *dynamical* evolution in two and three dimensions, allowing the study of stellar convection, rotating collapse, accretion disk interiors, jet production, interstellar cloud collisions, and many other problems discussed in this and other volumes. However, except for a few brave attempts, *structural* problems in several dimensions (evolution of rotating stars, interacting binaries, full structure and long-term evolution of accretion disks, *etc.*) largely have been ignored.

There are two principle reasons for this: 1) for multi-dimensional, gravitating fluids the shape of the structure (and therefore the grid geometry) is unknown until the problem is solved; and 2) solving millions of multi-dimensional, coupled, nonlinear equations is enormously computationally intensive — many orders of magnitude greater than that needed for dynamical calculations. Our work addresses both of these problems and creates a new field of numerical astrophysics called Multi-dimensional Astrophysical Structural Analysis. It is reported more fully in papers being submitted to the *The Astrophysical Journal* (Meier 1998, submitted; Meier & Godon 1998, in preparation).

S. M. Miyama et al. (eds.), Numerical Astrophysics, 403–404.

2. Solution to the Geometry and Computational Problems

The multi-dimensional geometry problem is addressed by employing the Finite Element Method, which has been used to study multi-dimensional structures in engineering for many years (see, *e.g.*, Zienkiewicz 1977). The key features of our implementation are: 1) unlike most engineering treatments, our method is nonlinear, allowing the solution of general astrophysical problems; 2) the positions of the grid points are part of the solution, allowing the mesh to seek the proper structural shape; 3) the local Jacobian ("isoparametric") transformation is used to form derivatives in this rather arbitrary curvilinear geometry; 4) if time is also treated as a mesh dimension, then motion of the fluid with respect to the grid (arbitrary Lagrangian-Eulerian fluid mechanics) is automatically taken into account by this transformation; 5) "difference" equations are generated by the computer from differential equations using weighted residual integrals, not written down by the investigator. We have been able to show that virtually all equations of continuum astrophysics, including the Maxwell and Einstein field equations, can be written in the Finite Element form and are, therefore, potentially addressable with these techniques. As an example, we present solutions of the Maclaurin spheroid sequence of rotating stars, using only the multi-dimensional stellar structure equations and the appropriate equation of state. (That is, the demonstration does not take advantage of the simple nature of the polytropic equation of state. The method could be applied easily to fully radiative stars with meridional circulation.)

The computational time problem is addressed by employing fast multigrid algorithms to solve the coupled equations instead of the direct ("Henyey") technique (see, *e.g.*, Brandt 1977). Errors of different spatial frequency are reduced iteratively using grids of different mesh spacing. The CPU time needed to converge is proportional to $N^D \log_2 N^D$ instead of N^{3D-2}, where N is the number of grid points in each of the D dimensions. We have tested these techniques on finite difference problems, porting them to parallel supercomputers. We confirm the predicted scaling and a nearly linear speedup due to parallel implementation. We predict that the time needed to solve a 1025^3 problem would be reduced from $\sim 10^7$ years using direct techniques on a workstation to ~ 1 minute using multigrid techniques on a parallel supercomputer.

References

Brandt, A. (1977) Multi-Level Adaptive Solutions to Boundary-Value Problems, *Math. Comp.*, **Vol. no. 31**, pp. 333–390.

Zienkiewicz, O.C. (1977) *The Finite Element Method*, McGraw-Hill, London.

COMPUTATIONAL DEVELOPMENTS FOR MOVING CLUSTERS WITH APPLICATION TO HYADES

M. A. SHARAF
Dept. of Astron., Cairo Univ., Faculty of Science, Cairo, Egypt

AND

A. A. EL-BASSUNY AND B. A. KORANY
National Research Institute of Astronomy and Geophysics Cairo, Egypt; E-Mail: akorany@frcu.eun.eg.

1. INRODUCTION

In this paper, computational developments for the parameters of moving clusters are established together with their error estimates in closed analytical forms and applied to Hyades cluster. The data used are 60 stars selected from Schwan (1991). The symbols denote the right ascension (α), declination (δ), proper motion [(μ_α, μ_δ)], position angle (θ), angular distance (λ), velocity (V), parallax (p), magnitude (M), helioc. coord. (x_c, y_c, z_c), and the distance (d).

2. DETERMINATION OF THE VERTEX COORDINATES OF A MOVING CLUSTER AND THIER CORRECTIONS

The vertex coordinates (A, D) can be obtained from the least squares as

$$A = tan^{-1}(\eta/\xi), \qquad D = tan^{-1}[(\eta^2+\xi^2)^{-1/2}],$$
$$\xi = (T_5T_2 - T_3T_4)/\Delta, \ \eta = (T_3T_2 - T_5T_1)/\Delta, \ \Delta = T_2^2 - T_4T_1,$$

where $T_1 = \sum_{1=1}^{N} a_i^2$, $T_2 = \sum_{i=1}^{N} a_i b_i$, $T_3 = \sum_{i=1}^{N} a_i c_i$, $T_4 = \sum_{i=1}^{N} b_i^2$, $T_5 = \sum_{i=1}^{N} b_i c_i$, $T_6 = \sum_{i=1}^{N} c_i^2$, $a_i = \mu_\alpha^{(i)} \sin\delta_i \cos\alpha_i \cos\delta_i - \mu_\delta^{(i)} \sin\alpha_i$, $b_i = \mu_\alpha^{(i)} \sin\delta_i \sin\alpha_i \cos\delta_i + \mu_\delta^{(i)} \cos\alpha_i$, and $c_i = \mu_\alpha^{(i)} \cos^2\delta_i$. The differential corrections of A and D can be found from $\Delta A - (G_5G_2 \quad G_3G_4)/E$. $\Delta D = (G_3G_2 - G_5G_1)/E$, where $G_1 = \sum_{i=1}^{N} \Psi_i^2$, $G_2 = \sum_{i=1}^{N} \Psi_i\Phi_i$, $G_3 = \sum_{i=1}^{N} \Psi_i\Delta\theta_i$, $G_4 = \sum_{i=1}^{N} \Phi_i^2$, $G_5 = \sum_{i=1}^{N} \Phi_i\Delta\theta_i$, $G_6 = \sum_{i=1}^{N}(\Delta\theta)_i^2$, and $\Psi =$

S. M. Miyama et al. (eds.), Numerical Astrophysics, 405–406.

$\sin^2\theta[\cos\delta\tan D\cos(A-\alpha)-\sin\delta)]/\sin^2(A-\alpha)$, $\Phi=-\sin^2\theta[\cos\delta\sec^2 D \sin(A-\alpha)]/\sin^2(A-\alpha)$, $\cot\theta=\cos\delta\tan D\csc(A-\alpha)-\sin\delta\cot(A-\alpha)$, $\cos\lambda=\sin\delta\sin D+\cos\delta\cos D\cos(A-\alpha)$.

3. SOME PARAMETERS FROM A AND D

$V=[\sum_{i=1}^N V_r^{(i)}\cos\lambda_i]/[\sum_{i=1}^N\cos^2\lambda_i]$, $p=4.74\mu/V\sin\lambda$, $M=m+5+5\log p$, $x_c=[\sum_{i=1}^N\cos\delta_i\cos\alpha_i/p_i]/N$, $y_c=[\sum_{i=1}^N\cos\delta_i\sin\alpha_i/p_i]/N$, $z_c=[\sum_{i=1}^N\sin\delta_i/p_i]/N$, and $d=N/[\sum_{i=1}^N p_i]$.

4. ERROR ESTIMATES FOR MOVING CLUSTER PARAMETERS

Let σ denotes to the standard errors, then

$$
\begin{aligned}
\sigma_1 &= [\{T_6-\xi^2T_1-\eta^2T_4-2\xi\eta T_2\}/(N-2)]^{1/2}.\\
\sigma_2 &= [(1/(N-2))\{G_6-(\Delta A)^2G_1-(\Delta D)^2G_4-2(\Delta A)(\Delta D)G_2\}]^{1/2},\\
\sigma_\xi &= \sigma_1\sqrt{-T_4/\Delta},\quad \sigma_\eta=\sigma_1\sqrt{-T_1/\Delta}.\\
\sigma_{\Delta A} &= \sigma_2\sqrt{-G_4/E},\quad \sigma_{\Delta D}=\sigma_2\sqrt{-G_1/E},\\
\sigma_A &= (\sigma_1/(\eta^2+\xi^2))(-[T_4\eta^2+T_1\xi^2+2\eta\xi T_2]/\Delta)^{1/2}.\\
\sigma_D &= (\sigma_1/(\eta^2+\xi^2+1))(-(1/\Delta(\xi^2+\eta^2))[T_4\xi^2+T_1\eta^2-2\eta\xi T_2])^{1/2}.\\
\sigma_{\lambda_i} &= \sigma_2(-[H_i^2G_1+R_i^2G_4-2H_iR_iG_2]/E)^{1/2},\ i=1,2,...,N\\
\sigma_V &= \sigma_2(-[\Lambda^2G_1+W^2G_4-2\Lambda WG_2]/E)^{1/2},\\
\sigma_{p_i} &= \sigma_2(-[\Gamma_i^2G_1+\Theta_i^2G_4-2\Gamma_i\Theta_iG_2]/E)^{1/2},\\
\sigma_{M_i} &= \sigma_2(-[N_i^2G_1+\Re_i^2G_4-2N_i\Re_iG_2]/E)^{1/2},\\
\sigma_d &= \sigma_2(-[\chi^2G_1+\beta^2G_4-2\chi\beta G_2]/E)^{1/2}.
\end{aligned}
$$

Conclusion : Assuming the membership problem has been solved, then the proper usage of our error controlling formulae can be illustrated by defining an *acceptable solution set* for the parameters **P** of a moving cluster as $\mathbf{P}=\{\mathbf{P}:\varepsilon_j\le Tol_j\ ;\ j=1,2,...,\ell\}$, where ε's are error criteria (standard or probable errors) for the parameters whose number is ℓ (say) and Tol's are given tolerances. The elements of this set depend on the accuracy of the computational formulae. So, the error controlling formulae could be useful for determining the parameters of a moving cluster very accurately. We applied the above method to the Hyades cluster.

References

Schwan, H. 1991, A&A, 243, 386

SPECIAL-PURPOSE COMPUTER FOR GRAVITATIONAL N-BODY SYSTEM: GRAPE

JUNICHIRO MAKINO
Department of System Sciences, University of Tokyo
3-8-1 Komaba, Meguro-ku, Tokyo 153-8902, Japan

Abstract.
In this paper I'll briefly overview the GRAPE (GRAvity PipE) project, in which we develop a series of special-purpose computers for stellar dynamics. First, I overview the evolution of high-performance general-purpose computers. It is shown that the "hardware efficiency" of general-purpose computers has been going down exponentially, and will continue to do so. Then, I'll describe the approach of building the special-purpose computers as an alternative. I also briefly describe the past of GRAPE project, and the development plan for GRAPE-6, which will provide around 200 Tflops by the year 2000.

1. Introduction

The evolution of the general-purpose computer in the last half century has been truly amazing. The peak speed of the fastest computer of the time has been improved by about 10 orders of magnitude in 50 years.

This unusually fast and long-lasting exponential growth has been the driving force for the advance of the entire field of numerical astrophysics. In many fields of numerical astrophysics, the improvement in the speed or memory capacity of computers has been essential. Of course, the improvement in numerical method and the progress in the understanding of the physical processes have been equally important, but without the exponential growth of the computer capability, the last half century of computational astrophysics would have been completely different.

This extremely rapid increase is expected to continue for the next 10–20 years [13]. Thus, in principle, we could be quite optimistic on the future of our field. In practice, however, there are several reasons to be skeptical about the future. In this paper, I'd like to describe the problem, and what I believe is part of the solution.

S. M. Miyama et al. (eds.), Numerical Astrophysics, 407–414.

2. Past and future of general-purpose computing

Very roughly speaking, the last 10 years of the high-performance computer can be summarized as the slow decline of the vector processors and comparably slow rise of distributed memory parallel processing. This trend is driven mostly by the constraint in the silicon technology [10].

As Hillis [8] stressed, there used to be two very different ways to construct a fast computer. One is to connect the fastest possible processors (vector-parallel approach). Since first processors tend to be expensive, the number of processors you can buy used to be small. The other is to connect many relatively slow processors (MPPs). Cray and Japanese manufacturers had been pursuing the first approach, while a number of companies tried the second, with limited success.

In late 1980s, however, two ways started to converge (or, to collide). The reason is that the microprocessors, which were the building blocks of MPPs, caught up with the vector processors in the processing speed. In 1970s, there were the difference of nearly two orders of magnitude between the clock speed of vector processors and that of CMOS microprocessors. In addition, microprocessors typically required tens or even hundreds of cycles for a floating-point operation. Thus, the speed difference was at least a factor of 10^2. This factor could be as large as 10^4.

Today, the difference in the speed has almost vanished. The clock speed of today's microprocessors is considerably faster than that of vector-parallel processors, and microprocessors can perform one or more floating point operations per cycle. Thus, programs that are not very well vectorized or parallelized are actually faster on microprocessors.

As far as the performance characteristics are concerned, today's high-performance computers are not much different from a pile of personal computers connected by a fast network. We need very high degree of parallelism to make effective use of these machines. In addition, practically all machines has non-uniform-access memory architecture (NUMA), which means that we need to design the program so that data are distributed over processors in an appropriate manner. This is true for both the vector-parallel machines such as NEC SX-4 and MPPs like Hitachi SR-2201 or SGI-Cray T3E.

This characteristic of modern high-performance computers has rather strong negative impact on the future of the computational astrophysics. Firstly, to use parallel computer is not easy, even in the case of ideal parallel machine with physically shared memory. Secondly, NUMA machines are far more difficult to program compared to machines with physically shared memory. Thirdly, to develop message-passing style program is even more difficult.

In the case of N-body simulations, just to run a simple, direct-summation scheme with constant stepsize on a massively parallel computer is already a difficult task [5]. Implementing a particle-mesh or P^3M scheme or treecode with shared timestep is not impossible, but proven to be a task which can easily eat up several years of time of a very good researcher, which could otherwise be used for doing science (see, *e.g.*, [14] or [3] for examples of parallel tree codes). Implementing an individual timestep algorithm has been demonstrated to be difficult [12].

The essential reason for this difficulty is that parallel computers are designed without any consideration for astrophysical applications, not to say N-body simulations.

In the following, we describe an alternative. Instead of adopting the algorithm to available computer, we design the most cost-effective combination of the algorithm and computer.

3. Special-purpose computing

The idea of designing special-purpose computer for scientific simulation is not new. Until recently, however, it has not become widely accepted (maybe still not). This is primarily because the advantage over general-purpose computers had been limited. First of all, one had to develop hardware, which was time-consuming and costly venture. It was also necessary to develop softwares (system programs and application programs).

To make the matter worse, in a few years hardware would become obsolete and the investment on hardware and software would be lost. Of course, this is also true for general-purpose hardware. The software for general-purpose computers was, however, likely to be reused, unless there was drastic change in the architecture.

However, this situation has changed considerably in the last decade. As I described in the previous section, the development of the software for general-purpose computers has become much more difficult and costly. Moreover, we cannot expect that programs developed for the current generation of machines can be used in the future machines without extensive rewriting. Vector processors required very different way of programming than was used on scalar processors. Parallel computers require yet another way. The current transition from vector processors to microprocessors or MPPs means most of the programs tuned for vector processors are now becoming useless.

On the other hand, the cost advantage of special-purpose computers, which determines the lifetime of the machine, has been greatly improved. This is not because of the efforts at the side of the special-purpose computing, but because of the decline of the hardware efficiency of general-purpose computers. Though the peak speed of computers has been increasing exponentially, the fraction of the hardware used to do arithmetic operations has been falling exponentially. In present microprocessors, only a few percents of the transistors on a chip are used to implement arithmetic units. Other 90+ % are used for cache memory and control logic. Moreover, it is quite difficult to achieve high efficiency with current microprocessors, because of the limitation in the memory bandwidth. Thus, if averaged over time, the typical efficiency of present microprocessors is around 0.1%.

If we can construct a machine with efficiency higher than, say, 10%, therefore, we can achieve the cost advantage of a factor of 100 or larger. Of course, the production cost of the machine would be higher because of the small quantity, and there would be losses due to inefficiencies in the design. However, these two combined is not likely to be as large as a factor of 100. Thus, the special-purpose architecture is becoming a quite attractive solution. Note that 10 years ago the efficiency of general-purpose computers was higher, and therefore it was more difficult to develop special-purpose computers.

There had been two very different approaches to build special-purpose computers for scientific simulations. One is to design a programmable parallel computer with various optimizations. For example, if we do not need much memory, we could replace the DRAM main memory with a few fast SRAM chips, thus reducing the total cost by a large factor. If we do not need fast communication, we can use rather simple network architecture.

In 1980s, CMOS VLSI floating-point chipsets such as Weitek 1064/65 [4] offered the theoretical speed of around 1/10 of the speed of vector processor for the cost around 1,000 USD. Thus, if one could use several of them in parallel, one could construct a vastly cost-effective computer. This led to numerous projects, some of them were highly successful. In particular, PAX [9] and Caltech Hypercubes [5] had been so successful that many companies started to sell similar machines as general-purpose parallel computers. As a result of their success, developing a parallel computer as special-purpose system has become unpractical. You can buy a better one from computer companies.

The other approach is to develop "algorithm oriented" processors. Our GRAPE (GRAvity piPE) machines is an extreme in this direction. As the name suggests, GRAPE is a device which evaluates the gravitational force between particles.

In direct N-body simulation, more than 99% of CPU time is spent to calculate the gravitational force between particles. Even in the case of more sophisticated algorithms such as the treecode, FMM and P^3M, large fraction of time is spent for pairwise force calculation. In usual implementation, the direct force calculation consumes typically about half of the total CPU time. However, it is possible to modify these algorithms to reduce the cost

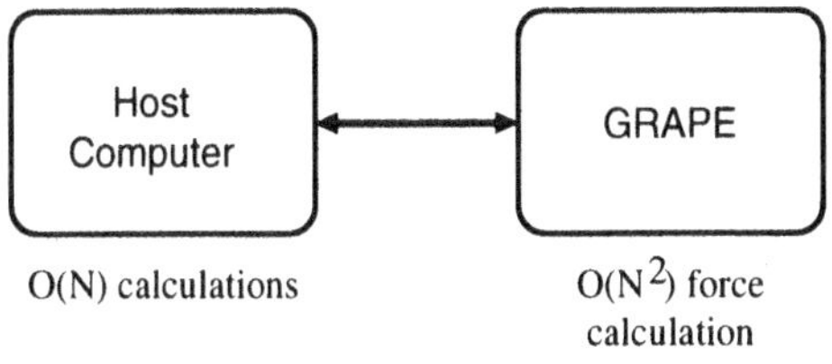

Figure 1. The basic architecture of GRAPE

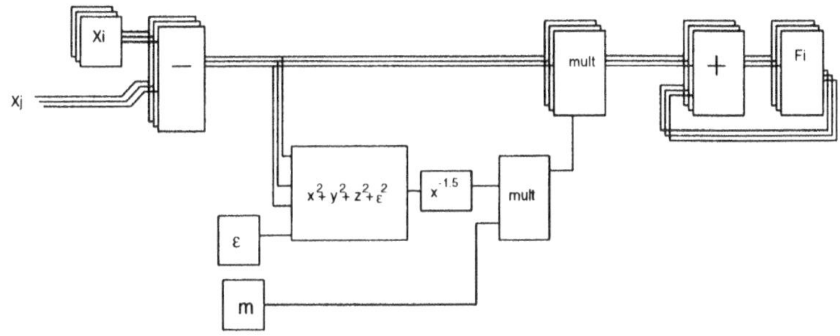

Figure 2. The force calculation pipeline

of calculations other than pairwise force calculation, by *increasing* the calculation cost of pairwise force calculation. Thus, actual gain one can achieve for these algorithms is much larger than a factor of two [1].

4. The GRAPE project

Figures 1 and 2 show the basic idea of GRAPE systems. The special-purpose device performs only the force calculation, while a general-purpose host computer takes care of everything else. In the special-purpose device, a pipeline which embodies the data flow for the force calculation calculates interactions at the rate of one interactions per cycle (around 30 operations/cycle). Multiple pipelines can be attached to one memory unit so that they calculate the force on different particles. With multiple memory unit, it is also possible for multiple pipelines to calculate partial forces on one particle.

Table 1 summarizes the history of GRAPE project. New machine has become larger and more complex, but at the same time more cost-effective, even when the improvement due to the advance in the device technology is discounted.

For N-body simulations of planetary formation and globular clusters, GRAPE approach proved itself to be extremely effective. In fact, significant fraction of N-body studies in these fields are now performed on GRAPEs,

TABLE 1. History of GRAPE project

GRAPE-1	(89/4 — 89/10)	120 Mflops, low accuracy
GRAPE-2	(89/8 — 90/5)	40 Mflops, high accuracy (32bit/64bit)
GRAPE-1A	(90/4 — 90/10)	240 Mflops, low accuracy
GRAPE-3	(90/9 — 91/9)	14 Gflops, high accuracy
GRAPE-2A	(91/7 — 92/5)	180 Mflops, high accuracy (32bit/64bit)
HARP-1	(92/7 — 93/3)	180 Mflops, high accuracy (32bit/64bit) Hermite scheme
GRAPE-3A	(92/1 — 93/7)	6 Gflops/board some 50 copies are used all over the world
GRAPE-4	(92/7 — 95/7)	1 Tflops, high accuracy (32bit/64bit) Some 10 copies of small machines
MD-GRAPE	(94/7 — 95/4)	1Gflops/chip, high accuracy (32bit/64bit) programmable interaction

and many new and important results are published. Many copies of GRAPE hardwares are now used for SPH and N-body simulations in cosmology, galactic dynamics and galaxy formation.

5. GRAPE-6

In 1997, we started the GRAPE-6 project. It's a project funded by JSPS (Japan Society for the Promotion of Science), and planned total budget is about 500 M JYE.

Figure 3 shows the basic structure of GRAPE-6. The gravitational pipeline is essentially a scaled-up version of GRAPE-4, with the peak speed of around 200 Tflops. This part will consist of around 4000 pipeline chips, each with the peal speed of 50 Gflops. In comparison, GRAPE-6 consists of 1700 pipeline chips, each with 600 Mflops. The increase of a factor of 100 in speed is achieved by integrating six pipelines into one chip (GRAPE-4 has one pipeline which needs three cycles to calculat the force from one particle) and using 3–4 times higher clock frequency. The advance of the device technology (from 1μm to 0.25μm) these practical.

The multipurpose pipeline part is a new feature, whose goal is to widen the application range. The original GRAPE architecture consists of only two parts: GRAPE and the host (see Figure 1). GRAPE calculates only gravity and everything else is done on the host. This architecture is ideal for pure N-body simulation, but not quite so if we want to deal with, for example, self-gravitating fluid using SPH.

The most costly part of SPH calculation, aside from the gravity, is the evaluation of the hydrodynamical interaction between particles. Thus, a specialized pipeline quite similar to that of GRAPE [15] could improve the

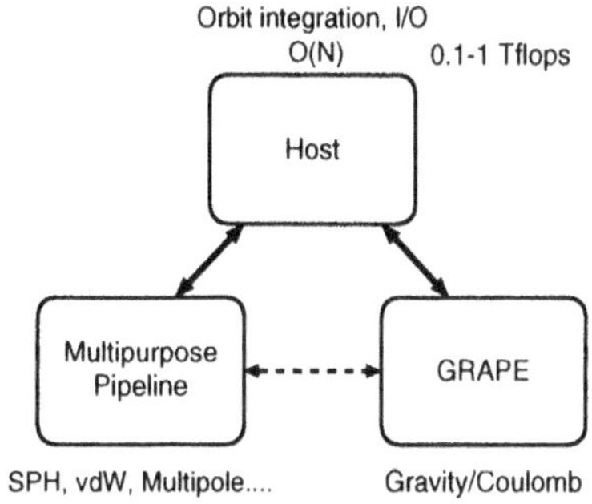

Figure 3. GRAPE-6

speed quite a lot. However, there are two reasons to believe it is difficult. The first one is that the gain one can achieve is limited. Since the interaction calculation accounts for only around 90% of the total CPU time, even if the SPH pipeline is infinitely fast the gain we can achieve does not exceed a factor of 10. The other reason is that there are many SPH algorithms. Newton's law of the gravity has not changed in the last two century, and the algorithm to calculate it is well established. However, SPH is still rather new method. One day somebody might come up with a novel method, which is much better than traditional one but cannot be implemented on a specialized hardware. Thus, it looks rather risky to develop an SPH hardware.

If we can "program" the pipeline unit, we can eliminate most of the risks. If someone comes up with a new and improved SPH scheme, a programmable pipeline could still be used for that. Moreover, such a programmable pipeline might be used for many other problems.

One might wander whether a programmable pipeline is a practical concept or not. Didn't the author argued against the programmability in section 3? Well, the advance in the FPGA (field-programmable gate array) technology has made the new approach viable [2].

An FPGA can be programmed to realize different functions by loading the configuration data. An FPGA consists of many logic blocks and a switching matrix. A logic block is typically a small lookup table. A SRAM block is used to implement this lookup table so that its function can be changed. The switching matrix can also be programmed to make connections in different ways.

This programmability incurs quite large inefficiency. The circuit size which can be implemented in the current largest FPGAs is equivalent to 10^5 transistors, while largest LSIs contain more than 10^7 transistors. In addition, there is also speed difference of factor 3–5.

Even with these large overheads, however, FPGAs are now becoming more efficient than general-purpose microprocessors. The reason is quite simple. The efficiency of FPGAs has not been falling too rapidly, since the relative overhead is roughly independent of the technology. In fact, the speed penalty is decreasing, since the signal propagation delay is becoming more important. This delay is not much different for FPGA and usual LSIs.

We have developed a small experimental machine, the PROGRAPE-1 [7]. It has two large FPGAs. The FPGA chips in PROGRAPE-1 can house, for example, one pipeline of GRAPE-3 [11] or WINE-1 [6]. GRAPE-6 will include a massively-parallel version of this PROGRAPE system, which can be used for various applications like SPH, Ewald method, and van-der-Waals force calculation in molecular Dynamics.

The GRAPE-6 will be completed by the year 2000. We plan to make small version of GRAPE-6 (peak speed of "only" a few teraflops) commercially available by that time. We've found that the commercial availability of small machines is essential to maximize the scientific outcome from GRAPE hardwares.

This work is supported in part by the Research for the Future Program of Japan Society for the Promotion of Science (JSPS-RFTP 97P01102).

References

1. Athanassoula E., Bosma A., Lambert J. C., and Makino J. (1998) *MNRAS*, **293**, pp. 369–380.
2. Buell D. and Arnold J. M. and Kleinfelder W. (1996) *Splash 2: FPGAs in a Custom Computing Machine*. IEEE Comp. Soc. Press, Los Alamitos, CA.
3. Dubinski J. (1996) *New Astronomy*, **1**, pp. 133–147.
4. Fandrianto J. and Woo B. Y. (1985), in *Proceedings of Seventh Symposium on Computer Arithmetic*, pp. 93–100. IEEE.
5. Fox G. C., Williams R. D., and Messina P. C. (1994) *Parallel Computing Works!*, Morgan Kaufmann, San Francisco.
6. Fukushige T., Makino J., Ito T., Okumura S. K., Ebisuzaki T., and Sugimoto D. (1993), *PASJ*, **45**, pp. 361–375.
7. Hamada, T., Fukushige, T., Kawai, A, and Makino, J. (1998), in *this proceedings*.
8. Hillis W. D. (1985) *The Connection Machine*. MIT Press, Cambridge, MA.
9. Hoshino T. (1992), in Mendez R. (ed) *High Performance Computing: research and practice in Japan*, pp. 239–256. John Wiley and Sons, Baffins Lane.
10. Makino J. and Taiji M. (1998) *Special Purpose Computers for Scientific Simulations – The GRAPE systems*. John Wiley and Sons, Chichester.
11. Okumura S. K., Makino J., Ebisuzaki T., Fukushige T., Ito T., Sugimoto D., Hashimoto E., Tomida K., and Miyakawa N. (1993), *PASJ*, **45**, pp. 329–338.
12. Spurzem R. (1996), submitted to *MNRAS*.
13. Sterling T., Messina P., and Smith P. H. (1995) *Enabling Technologies for Petaflops Computing*. MIT Press, Cambridge, MA.
14. Warren M. S. and Salmon J. K. (1992), in *Supercomputing '92*, pp. 570–576. IEEE Comp. Soc., Los Alamitos.
15. Yokono Y., Ogasawara R., Takeuchi T., Inutsuka S., Miyama S. M., and Chikada Y. (1996), in Tomisaka K. (ed) *Numerical Astrophysics Using Supercomputers*. National Astronomical Observatory, Japan.

BEYOND VRML: EFFICIENT 3-D DATA TRANSFER FOR REMOTE VISUALIZATION

M. NAGASAWA

Ultra-high Speed Network and Computer Technology Labs.
1-280, Higashi-Koigakubo, Kokubunji, Tokyo 185-8601, Japan

Abstract. For effective remote visualizations of massive 3D data, volumetric data compression algorithms and transmission protocols are investigated. Quantized Normal Vectors unify and compress 3D volumetric data by quantizing surface normals and volume gradients. With this QNV representation, progressive transmission and Level-of-Detail control become possible. Against the change of network quality or data loss, it is robust enough to get smooth visualization results.

1. Remote Visualization

The increased popularity of computer graphics and the wide-spread Internet are making easier handling of 3D media in remote environments. Virtual Reality Modeling Language (VRML) is an international standard that realizes interoperabilities of 3D graphics. However, there are many limitations in current VRML, because it was designed for entertainment computer graphics. For example, with multiple VRML streams on a single display, people have to choose and optimize rendering resolutions. For effective use of network bandwidth, enhancement of VRML with some binary compression forms is also needed (Taubin 1996). Standard VRML browsers are using FTP protocol, so that the rendering starts after the completion of data transfer. When the transmission error occurs, TCP/IP requests re-sending whole data. General file compression utilities are not good for an interactive quick-look. Users have to wait without any view during the transfer.

S. M. Miyama et al. (eds.), Numerical Astrophysics, 415–422.

1.1. OUTLINES AND PRIORITY SORTING

A quick-look of outlines is effective for browsing polygons. Our outline data format does not include adjacency list, but the positions of outline vertices. The outlines are transmitted prior to the inner vertex information. The outlines are enough to display each group. The independency of each polygon group is suitable for progressive transmission and rendering. If some vertices are shared by plural polygons, adjacency list is good for reducing data redundancy. However, even a bit loss of list data during the transmission will cause severe damage to the rendering.

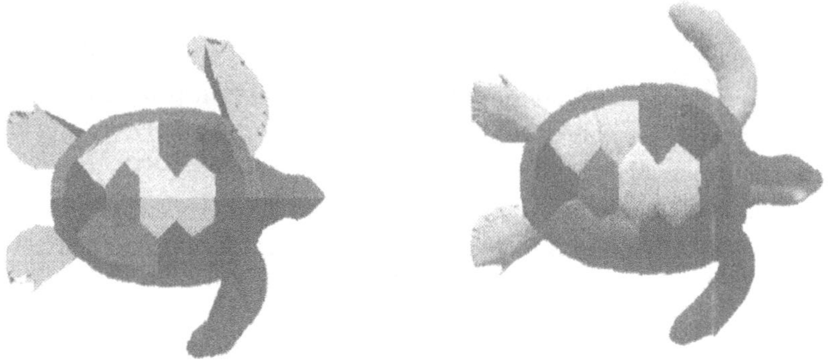

Figure 1. Left: Group polygon outlines (180 polygons). Right: Whole data of outlined polygons (50,000 polygons).

1.2. PROGRESSIVE TRANSMISSION

For 2D images, techniques of progressive display exist. Using interlaced GIF, coarse images appear first and the details come later. One can grasp a whole image quickly and interrupt the display screen when sufficient recognition is obtained. Similarly, by rendering 3D patches in sorted order of area size, the global structure could become manifest. The larger patch comes in advance to the smaller ones. Both sorted and group-outlined polygon transfer are found to reduce the transmission time to one third of that of FTP transfer of standard VRML file.

In present environments, however, transmissions of full details of huge volumetric 3D data are time consuming. Thus, we are to develop a new high dimensional model which appropriately and effectively represents 3D data characteristics.

2. QNV Representation

There are macroscopic models for representing 3D data: analytic curves, polygons, voxels, particles and vectors. These representation models use geometrical approaches. Typical scientific visualizations require a flexibility to treat complex 3D data structures with fragments, voids and multi components (Nagasawa 1995). A special polygon representation for Level-of-Detail control was proposed by Hoppe (1996). While, we try to reproduce smooth and diffuse boundary structures appropriately in remote rendering by neglecting polygon boundary.

2.1. NORMAL AND GRADIENT QUANTIZATION

For global and direct visualizations of 3D volumetric data, it seems reasonable and challenging to skip some intermediate representations such as geometrical isosurfaces (Lorensen & Cline 1987). In QNV representations, 3D objects are encoded as a set of normal vectors with an opacity profile. The QNV could represent both 3D surfaces and 3D volumetric data depending on topologies of vector distributions. Each QNV is discretized to have length proportional to its sampling area for surface rendering or gradient magnitude for volume rendering.

Figure 2. Line drawing of QNV Voxel gradients are sorted in magnitude. Out of total 181,012 voxels, plotted are Left: 2202 (1.2%), Center: 6240 (3.5%), Right: 16490 (9.1%) vectors, respectively.

We consider a method to get QNVs from surface polygons. For simplicity, we start from triangle patches for 3D surface to get outer vector products. A gravity center of the triangle polygon corresponds to the vector position $\mathbf{x}_k = (x, y, z)$. The vector length of outer product is the area, and the vector direction is orthogonal to the triangle. The vector positions are decoded in Cartesian (x, y, z), while directions are in polar coordinates (A, θ, ϕ). A relaxation algorithm to get a uniform triangulation of 3D surface was also applicable (Turk 1992).

3. Quantized Vector Rendering

In general, QNV is suitable for representing irregular sampling elements. That is a merit for data compression and a demerit for high quality rendering at the same time. In surface renderings for polygonal data, we could show jaggy features at some polygonal edges. While, rigid disk renderings for QNV data show sparse surfaces than that of polygon renderings. Thus, for QNV data set, we developed a general rendering algorithm applicable both to surface and volume rendering by mapping opacity as textures on its sampling area with normal directions.

3.1. SURFACE NORMAL RENDERING

In decoding QNV data: $\{(x, y, z, \theta, \phi, A, ...)|\, k = 1, K\}$, an expectation value of opacity at position $\mathbf{x}$ is calculated with the Gaussian kernel filter as $\alpha(\mathbf{u}_i - \mathbf{x}_k) \propto \exp(-|\mathbf{u}_i - \mathbf{x}_k|^2/A)$,

For a given surface data, this decoding satisfies an exact conservation of integrated surface area. With a smooth opacity filter, a robust decoding of surface data is possible independent of local interpolation errors even if some edges can not satisfy continuity conditions.

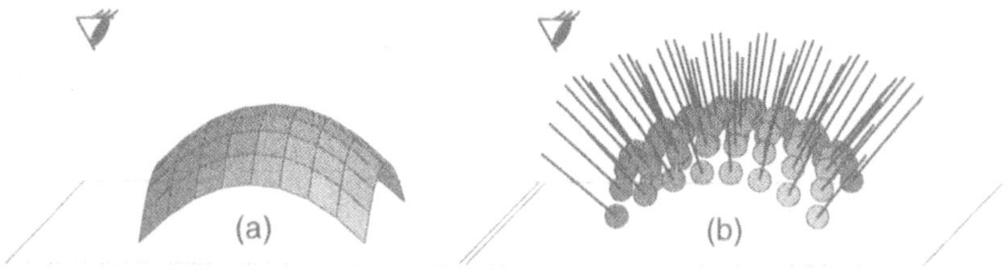

Figure 3. (a) Polygons and (b) Normal vectors for 3D surface representation.

In this adaptive control, vector representation has an advantage of being able to compress non-uniform 3D data. However, we have to be careful to treat voids of surface data. Some sparse regions happen to become larger. That means a large area surface of insufficient curvature, although a smooth image quality is still guaranteed.

In the rendering, a ray cast integration is performed for vectors of total number K with a given opacity profile.

$$C(\mathbf{u}_i) = \sum_{k=0}^{K} \left\{ c(\mathbf{x}_k)\alpha(\mathbf{u}_i, \mathbf{x}_k) \prod_{m=k+1}^{K} [1 - \alpha(\mathbf{u}_i, \mathbf{x}_k)] \right\} \tag{1}$$

where $\mathbf{u}_i = (u_i, v_i)$ are pixel coordinates.

3.2. VOLUME GRADIENT RENDERING

We could apply QNV formats to volumetric rendering. From voxels, we can calculate gradient vectors and represent them in QNV formats. Although QNV length has another meaning of gradient magnitude in this volumetric case, ray integration could be performed in an absolutely similar way to the surface rendering.

If data is represented by sets of voxels or particles, our volume rendering pipeline could be used (Nagasawa & Kuwahawa 1991).

We modified this volume rendering code to be applicable to the vector rendering. Due to absence of geometrical edges, opacity calculations in volume rendering proceed much easier than for usual cubic voxels. Anti-aliasing treatment is not necessary due to smooth profile of kernel functions.

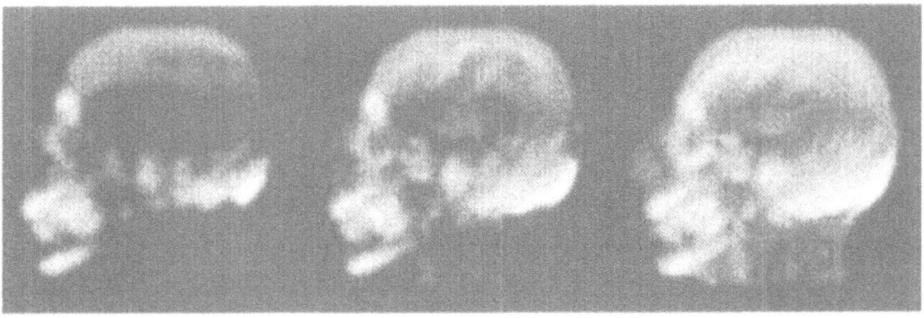

Figure 4. Volume rendering of the vectors in QNV format. Only a small fraction of vectors are contributing.

As a benchmark visualization, we used MRI data set of CT study of a cadaver head. This CT data is 113×256×256 sliced voxel data of 14MBytes. While, by volume rendering of QNV format data, we get a sufficient quality output from small amounts of QNV. Only about top 9% of large gradient vectors are enough to produce the image.

Another example is the 3D isosurface of observational Orion molecular cloud. It amounts to 15,657 Kbytes in AVS geometry format, while is compressed to 774 Kbytes in QNV format.

3.3. IMPLEMENTATION ON AVS

As an effective compression and standard transmission algorithm of 3D surface data, the QNV coding and rendering algorithm are implemented on AVS platforms. AVS geometry data from scientific applications are converted directly into quantized vector representations. The quantized vector coding for surface data could start from a set of isosurface polygons. In addition to informations of normal vector and area of each polygon, an internal opacity profile is assigned. In polygon rendering, low resolution, say, large patch area turns out to be jaggy at polygonal edges. While, in vector rendering, it appears to be cloud-like fuzzy features. High resolution parts are similarly distinct in both cases.

3.4. EXTENTED VRML BROWSER: GHOSTSPACE

We developed an extended VRML browser *Ghostspace* that runs as a helper application for WWW browser (UNCL 1997). *Ghostspace* provides ATM ability to display multiple compressed 3D data. It can be used to browse QNV and other proposed 3D data formats in rapid rendering schemes. Pre-processed data archive in *Ghostspace* server increases system performance for highly interactive applications. Especially, sorting of vectors in accordance with its magnitude is essential to accelerate decoding.

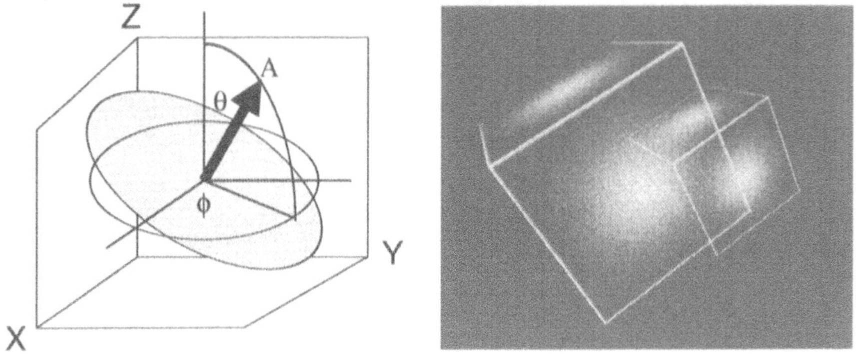

Figure 5. Left: Data components of QNV. Right: Texture representations of QNV opacity profile.

3.5. HARDWARE ACCELERATION OF QNV RENDERING

There are studies to accelerate volume rendering by developing special hardwares based on texture mapping. In case of rendering with external lights, a shading calculation is important to realize 3D effects. For a graphics API similar to OpenGL, a texture map may contain color and/or transparency in graphics pipelines. It can be used to modify original colors of polygons in various ways. For including shading effects, we use a square polygon patch as a surface element perpendicular to a surface normal or gradient. A semi-transparent texture could represent a QNV disk. The texture is assigned with the Gaussian profile as transparency. General 3D structures are represented by a superposition of these semi-transparent textures.

3.6. PERFORMANCE ON ATM NETWORK

To compress vector data, QNV elements have attributions in a discretized 3D space. Each element is discretized in effective bits for pixel resolutions. Spatial resolution is 10 bits in each dimension, and the directional resolution is 6 degree in polar coordinates. The QNV record length are formatted to optimize the performance on ATM networks. The record length for each vector is 8 Bytes and thus we could transmit 8 vectors on a single ATM cell. With this optimization, transfer of QNV data gets higher robustness and effectiveness in accordance with the ATM network protocols.

We have performance tests of visualizations based on vector representation. The system consists of polygon and QNV encoder, decoder and renderer on both TCP/IP and Native-ATM socket communications. The transmitter generates multiple independent ATM cells that load sampling patches of different resolution.

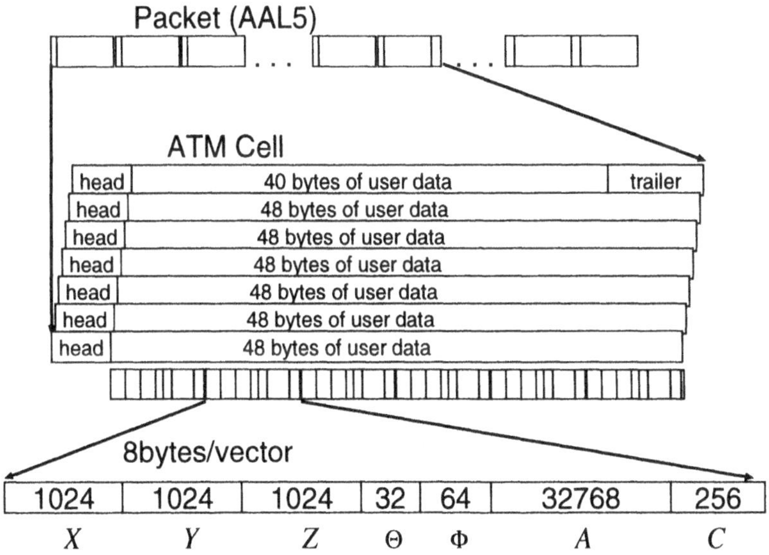

Figure 6. Discrete data format of QNV representation as a payload on ATM cell.

By adding extra formats to standard VRML, polygon compressions and QNV representations are in cooperation with specific network bandwidths through ATM negotiation protocol. With ATM socket communications, *Ghostspace* is able to decode and render up to 5 MBytes/s surface data stream over ATM-OC3 of 155Mbps bandwidth. This means effective 3D communications of 0.2~0.3 Mpolygons/sec and 0.6 Mvectors/sec are achieved. The encoder can produce other independent 3D data streams of boundary-listed group polygons by reducing coplanar polygons in each group (Nishioka & Nagasawa 1995).

4. Conclusion

For fast and effective 3D renderings, we use QNV representations.

1. QNV encodes 3D objects as a set of surface normal or volume gradient vectors. Each vector is quantized in position and direction.
2. With QNV formats, general and high quality renderings are possible both for surface and volumetric data with Gaussian opacities.
3. Progressive browsing becomes possible by accelerating QNV rendering with texture mapping hardware.

We implemented vector decoding and rendering functions for surface and volumetric data browsing in *GhostSpace*. Compared with an ordinary polygon data transfer, QNV representations are robust for congested network and keep 3D rendering image in high quality even with data abolishment. *GhostSpace* is able to show smooth features of rendered 3D structure using received normal vectors only, because no geometrical edge is needed. Quick-look applications could make use of this representation as an extension of VRML in 3D data browsers.

Acknowledgements

We acknowledge the data courtesy of CThead voxel data by UNC Chapel Hill and that of 3D Orion nebula isosurface by Dr. Tatematsu (NRAOJ).

References

Hoppe H. (1996) Progressive Meshes, *Computer Graphics Proceedings* (ACM SIGGRAPH) pp. 99-108

Lorensen W. and Cline H. (1987) Marching Cubes: A High Resolution 3D Surface Construction Algorithm *Computer Graphics* **21(4)** pp. 163-169

Nagasawa M. (1995) Volume data compression using smoothed particle transformation, *Digital Video Compression (SPIE)* **2419** pp. 288-295

Nagasawa M. and Kuwahara K. (1991) Smoothed Particle Rendering for Fluid Visualization in Astrophysics, *Scientific Visualization of Physical Phenomena* (ed. N.M.Patrikalakis, Springer-Verlag Tokyo) pp. 589-605

Nishioka D. and Nagasawa M. (1995) Reducing polygonal data by structural grouping algorithm, *Lecture Notes in Computer Science* (ed. Chin R. T. et al., Springer Hong Kong), **1024,** pp. 140-151

Taubin G. (1996) [`http://www.austin.ibm.com/vrml/binary`] VRML2.0 Compressed Binary Format

Turk G. (1992) Re-Tiling Polygonal Surfaces, *Computer Graphics* **26(2)** pp. 55-64

UNCL (1997) [`http://www.iijnet.or.jp/uncl/GhostSpace`] GhostSpace

LASER FUSION AND LABORATORY ASTROPHYSICS WITH INTENSE LASERS

H. TAKABE, H. NAGATOMO, A. SUNAHARA, N. OHNISHI, Y. YODA, A. I. MAHDY AND K. MIMA

Institute of Laser Engineering, Osaka University
Yamada-oka 2-6, Suita, Osaka 565-0871 Japan

Abstract. When an intense laser light is irradiated on a solid target, high temperature (a few keV) and high density (g/cm^3–mg/cm^3) plasmas are produced. The physics of the plasmas has been studied intensively relating to the laser fusion research. The physics is the same as that of plasmas inside and near the stellar objects. We briefly describe the physics of laser plasmas and the possibility of studying astrophysical plasmas with the laser plasmas.

1. Introduction

The possibility of laser fusion began to be considered just after the invention of lasers in 1960. Since the de-classification of the concept of implosion in laser fusion (Nuckolls *et al.*, 1972), intensive research has been carried out. When a multi-beam laser is irradiated on a target to realize high-density and high-temperature plasmas, many complex physics arise and are combined in the implosion dynamics. Some of these are compressible hydrodynamics, instabilities, radiation transport, equation of state, and the non-LTE atomic process. HED (high energy density) physics (Rosen, 1996), which has the same base as the core part of astrophysics, should be studied for laser fusion research. It would be beneficial for both scientists in laser fusion and astrophysics fields to study HED physics by modeling astrophysical phenomena with intense lasers. In this paper, the present studies of understanding and issues of the laser plasmas is briefly reviewed. In addition, we briefly explain the numerical design of the model experiment of the ejecta-ring collision of SN1987A.

S. M. Miyama et al. (eds.), Numerical Astrophysics, 423–426.

2. Laser-driven fusion

The theoretical research issues are divided as follows. (1) Laser-plasma interaction; (2) Electron energy transport; (3) Implosion dynamics; (4) Hydrodynamics instabilities; (5) Atomic modeling and X-ray radiation transport; and (6) Target design.

Below is a brief summary of the current status of these six issues.

The first topic has attracted many scientists, especially those interested in nonlinear plasma physics and parametric instabilities in particular (Kruer, 1988). Recently, nonlinear physics of laser-plasma interaction has been studied in the relativistic regime, where electron-quivering energy due to laser field is comparable to its rest mass energy. Filamentation and self-focusing owing to photon pressure has also been studied.

The second topic focuses mainly on the kinetic effects of electrons, namely, the non-Maxwell effect through the process of laser heating. In addition, non-local heating by the tail component of the electron distribution function was studied based on the Fokker-Planck type equation.

The third topic relates to compressible hydrodynamics. The rocket model and coupling efficiency are theoretically well understood and have been checked by laser model experiments. Implosion experiments have been carried out at large institutes with multi-beam laser systems, and high temperature and high-density compression have been achieved.

The fourth topic is the core of ICF (inertial confinement fusion) physics. It is simple to generate fusion energy, as long as nature allows the implosion to be spherically symmetric. However, it is also natural to see hydrodynamic instabilities in many cases. Now there is hope of controlling the growth of Rayleigh-Taylor instability in the ablative acceleration phase by means of "ablative stabilization" (Takabe *et al.*, 1985).

The fifth topic requires broad knowledge. There have been endless discussions about modeling atoms with multi-electrons, the non-LTE atomic process, solving rate equations, average atom model or detailed configuration accountings, database of rate coefficients, to name just a few topics. The code UTA and STA should be developed for modeling the opacities of partially ionized plasmas.

The sixth topic is integrated physics. Hydrodynamic codes which include all the kinds of physics described above should be developed.

3. Laboratory astrophysics with intense lasers

As briefly explained in the previous section, laser fusion is based on the same physics as the core of astrophysics. This is called HED physics. We can roughly divide the physics elements of astrophysics three minutes after the Big Bang into the following four:

(1) HD (Hydrodynamics) and MHD (Magneto-Hydrodynamics); (2) Atomic physics and radiation transport; (3) Nuclear reactions; and (4) Gravitational interactions. It is difficult to carry out model experiments of item 4 in laboratories, while a variety of model experiments can be designed concerning the other three items by the use of intense and ultra-intense lasers.

A model experiment for a strong blast wave is discussed in Ref. (Grun *et al.*, 1991) relating to SNR (supernova remnant). The similarity of hydrodynamic instability between laser implosion and supernova explosion is also discussed in Ref. (Takabe, 1998). In the present paper, therefore, we briefly present another possibility of the model experiment relating to "ejecta-ring collision" physics of SN1987A.

The ejecta from SN1987A expand with a velocity 10^9 cm/s. Around the year 2000, the ejecta will collide with ring material with a diameter of about 1.3 ly.Its omen has already been observed with HST (HST home page). The details of x-ray line emission will be observed by ASCA, AXAF, and ASTRO-E. This collision provides a good example to study the physics of a very early phase of SNR. A more detailed simulation has recently been reported (Borokowski *et al.*, 1997).

At first, we carried out a 2-D simulation of the ejecta-ring collision with a code used for studying laser-implosion physics (Sunahara *et al.*, 1996). The density and temperature profiles at 36 years after the collision are shown in Figs. 1–(a) and –(b), respectively. In Fig. 1–(c), the distribution of markers of the ring fluids is plotted.

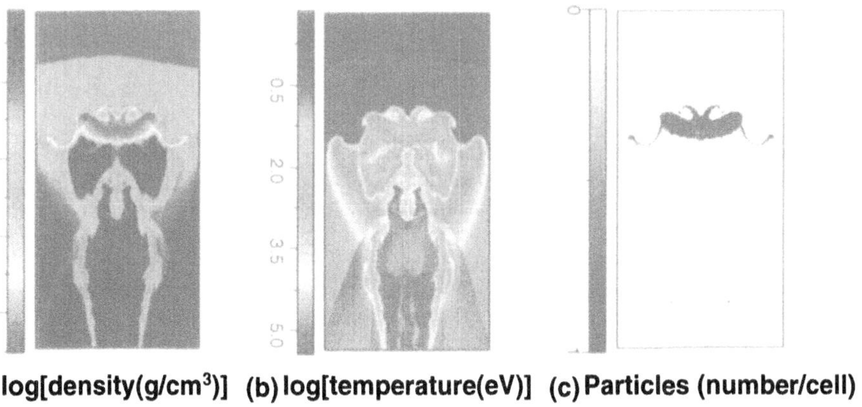

Figure 1. The density, temperature, and ring material distribution at 36 years after the collision obtained with the 2-D simulation of the ejecta-ring collision of SN1987A.

After the passage of the shock wave in the ring, most of the ring matter is heated up to a temperature of about 700 eV and then ionized. In the early phase after the shock passage, the heavy elements such as O, Si, etc are not in LTE but in the ionizing phase. The ionization parameter defined to be the product of density and temperature is found to be in the range of 10^{12} s/cm^3. After ten years, the ionized ring materials goes into the recombining phase. It is essential to solve the rate equation to predict the ionization stage.

In the present case, we modeled the ejecta-ring collision by colliding accelerated plastic foil with a spherical plastic foam. Then, the time evolution of shock-heated foam can be adjusted to be same as the collision of SN1987A. In the model experiment, we planned to dope a seed element such as chlorine to measure the time evolution of their line spectral structure with a monochromatic x-ray camera.It should be noted that the density was about 10^{19} higher and the time scale 10^{-19} shorter in the model experiments. Non-LTE ionization and recombination process will be observed in the experiment by tracing time-space evolution of the line emissions of chlorine.

4. Conclusion

In laser fusion research, a detailed study of high energy density physics is required. We proposed to use the intense lasers to study the basic physics of astrophysics. One example of a variety of possible model experiments is presented. In addition, with the use of ultra-intense lasers (Mourou *et al.*, 1998), we may be able to produce electron-positron plasmas and carry out a model experiment of gamma ray burst physics.Intense and ultra-intense lasers can be used for carrying out model experiments of astrophysical plasmas.

References

K. J. Borokowski *et al.*, Ap J. **477**, 281 (1997).
J. Grun *et al.*, Phys. Rev. Lett. **66**, 2738 (1991).
W. L. Kruer, "The Physics of Laser Interactions" (Add. Wesley, 1988).
G. A. Mourou *et al.*, Phys. Today, January issue, p. 22 (1998).
J. Nuckolls *et al.*, Nature **239**, 139 (1972).
M. D. Rosen, Phys. Plasmas **3**, 1803 (1996).
A. Sunahara *et al.*, Jan. J. Appl. Phys. Dec. (1996).
H. Takabe *et al.*, Phys. Fluids **28**, 3676 (1985).
H. Takabe, in "Superstrong Fields in Plasmas" (AIP Conf. Proc. 426, AIP, New York, 1998).
HST home page: http://oposite.stsci.edu/

PROGRAPE-1: A PROGRAMMABLE SPECIAL-PURPOSE COMPUTER FOR MANY-BODY SIMULATIONS

T. HAMADA, T. FUKUSHIGE, A. KAWAI AND J. MAKINO
Department of General Systems Studies,
College of Arts and Sciences, The University of Tokyo,
3-8-1 Komaba, Meguro-ku, Tokyo 153-8902, Japan

1. Introduction

We have completed PROGRAPE (PROgrammable GRAPE)-1, a programmable special-purpose computer for many-body simulations. Since 1988, we have been developing GRAPE (GRAvity piPE), a special-purpose computer for astrophysical N-body simulation [Makino et al. 1996]. In 1995, GRAPE-4 became the first computer to achieve the peak speed of 1 Tflops and sustained speed of 500 Gflops [Makino et al. 1997]. We developed custom LSI for the pipeline processor of the GRAPE system. In the PROGRAPE system, we used FPGA (Field Programmable Gate Array) as a pipeline processor instead of custom LSI. FPGA is a programmable LSI in which the user can specify the logic circuit.

The PROGRAPE system has specialized pipelines for computations of interactions between particles, which are the most expensive part of a many-body simulation. All the other computations, such as time integration, are performed on a general-purpose host computer connected to PROGRAPE. The host computer sends positions of all particles to PROGRAPE. Then, PROGRAPE calculates the interactions between particles and sends them back to the host computer. PROGRAPE can calculate the summation expressed as

$$\boldsymbol{f}_i = \sum_j \boldsymbol{G}(\boldsymbol{a}_i, \boldsymbol{a}_j), \tag{1}$$

where $\boldsymbol{a}_i$ is the physical value of the i-th particle, such as position and velocities, and $\boldsymbol{G}$ is the user-specified function. We specify the function $\boldsymbol{G}()$ by programming FPGA.

S. M. Miyama et al. (eds.), Numerical Astrophysics, 427–428.

2. Application

The advantage of FPGA over custom LSI is the reduction in initial cost and labor in development. In addition, making small changes in the function form after development is easier in FPGA than in LSI. One disadvantage of FPGA is a large loss of resources due to its programmability. PROGRAPE is suitable for applications for which moderate acceleration is required, such as SPH (Smoothed Particle Hydrodynamics) simulation, Ewald method, and calculation of short-range interactions in molecular dynamics.

3. PROGRAPE-1

PROGRAPE-1 is the first prototype system of PROGRAPE that we built. It uses two Altera EPF10k100 for the pipeline FPGA. We implemented a pipeline for gravitational/Coulomb force. The peak speed of the pipeline is 2.4 Gflops, the cost performance being estimated at about 1,000 dollars per 1 Gflops. We are currently implementing a pipeline for SPH on the FPGA.

Figure 1. Picture of PROGRAPE-1.

References

Makino, J., and Taiji, M., "Computational Astrophysics on a Special-Purpose Machine," Computers in Physics, 10, 352, 1996.

Makino, J., Taiji, M., Ebisuzaki, T., and Sugimoto, D., "GRAPE-4: A Massively Parallel Special-Purpose Computer for Collisional N-body Simulations," *the Astrophysical Journal*, 480, 432, 1997.

DEVELOPMENT OF A SPECIAL PURPOSE COMPUTER FOR COSMIC HYDRODYNAMICS USING THE SPH METHOD

YASUNORI YOKONO, S. INUTSUKA AND S. M. MIYAMA
National Astronomical Observatory of Japan,
Division of Theoretical Astrophysics
2-21-1 Osawa, Mitaka-city, Tokyo 181, Japan

RYUSUKE OGASAWARA
National Astronomical Observatory of Japan,
Subaru Telescope

TAKU TAKEUCHI
Tokyo Institute of Technology,
Department of Earth and Planetary Sciences

AND

YOSIHIRO CHIKADA
National Astronomical Observatory of Japan,
Astronomical Data Analysis Center

1. Our Project

The method of Smoothed Particle Hydrodynamics (SPH) has been applied to various astrophysical hydrodynamic phenomena because it is efficient to simulate astrophysical hydrodynamics . However, achieving high resolution with a large number of particles tends to require considerable calculation time. In view of this, we decided to develop a high-speed special-purpose computer that would achieve high-resolution simulation with large particle numbers using the SPH method.

2. SPH calculation method

SPH is a particle-based simulation method. Fluid variables such as density and pressure and their spatial derivatives are expressed as

S. M. Miyama et al. (eds.), Numerical Astrophysics, 429–430.

$$\langle A_k \rangle_i = \sum_j^{n_g} A_{k,j} W(r_{ij}, h_i), \quad (1)$$

$$\langle \nabla A_k \rangle_i = \sum_j^{n_g} A_{k,j} \nabla W(r_{ij}, h_i), \quad (2)$$

where $A_{k,i}$ are the k-th fluid variables at the position of the i-th particle, and W is the profile of a particle, called the kernel function. the number of neighbor particles, n_g is typically several hundreds. In SPH, most of computation time is expended in calculating these formulae.

3. Hardware Design

Our SPH device calculates these SPH summation with parallel pipelined architecture, effectively reducing the total SPH computation time. The hardware consists of 32 parallelized SPH chips that we developed (Fig 1). The expected calculation speed is the equivalent of about 2.5G flops. This system has been compiled at the Computer Center in Mitaka. Basic SCSI

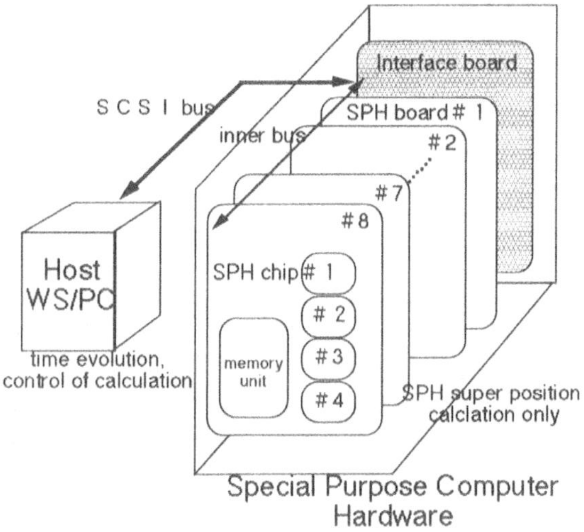

Figure 1. Hardware configuration of the SPH Special Purpose Computer.

driver-function routines in C- and C/Fortran-callable user interface routines were implemented. In association with GRAPE, self-gravitating SPH simulation will be available.

NEW DATA COMPRESSION FOR MULTI-DIMENSIONAL NUMERICAL SIMULATION

MITSUE DEN
Hiraiso Solar Terrestrial Research Center, CRL
3601 Isozaki, Hitachinaka, Ibaraki 311-1202, Japan

AND

KAZUYUKI YAMASHITA AND RYOJI MATSUMOTO
Chiba University,
1-33 Yayoi, Inage, Chiba 263-8522, Japan

With an increase in the processing speed, we can perform multi-dimensional numerical simulations. On the other hand, we are facing with problems that we have to handle huge amount of output data. We can attain higher time resolution in the analysis of simulation results by storing evolutional sequence more frequently. However, that interval is limited by the disk space and the speed of I/O. To overcome this dilemma, we propose a new real time data compression algorithm.

Our compression algorithm is as follows. (1) Set contour levels (say, $2^8 = 256$ levels) to cover the dynamic range of the original data. (2) Classify the original data according to the specified contour levels. (3) Make a conversion table which relates the original data and the contour levels. The values of floating point variables at each mesh point are quantized to specified contour levels. Information is partly lost here and then the data are compressed. When the compression efficiency (CE) is defined by $\mathrm{CE} = 1 - m/r$, where r and m are the size of the original and the compressed data respectively, we can always attain CE = 75% at the this step if the original data are stored in single precision variables. Since the quantized data are still sequential text stream, we can compress them furthermore by using lossless compression tool, i.e., the data go through compression two times in our method. For real time lossless compression during the simulation, we adopt the LZW15V.C.

In order to investigate efficiency of our algorithm, we calculate CE for three models, (I) supernova explosion as a model for the single density peak and the logarithmic density distribution, (II) formation of the large

S. M. Miyama et al. (eds.), Numerical Astrophysics, 431–432.

scale structure as a model for the multiple density peaks and the linear density distribution, and (III) emergence of twisted magnetic flux tubes for the test of vector variables such as magnetic fields. Three-dimensional numerical simulation have been carried out for each model and more than 90% efficiency is achieved for all models (see Table 1).

It should be noted that our compression algorithm is 'lossy', which means that the original data cannot be restored perfectly from the compressed data. However, our compression method is useful for visualization and analysis of numerical results. Thus we conclude that it is apposite to huge amount of data such as the results of 3D simulations.

TABLE 1. File sizes and compression efficiencies for models I, II and III. R, M and C2 mean the raw data, quantized (or mapped) data and the second time compressed data, respectively.

Model	R (byte)	M (byte)	C2 (byte)	CE (%)
I	1372008	343008	97488	92.9
II	4000008	1000008	77016	98.1
III-Bx	842732	210689	76078	91.0
III-By	842732	210689	72426	91.4
III-Bz	842732	210689	73656	91.3

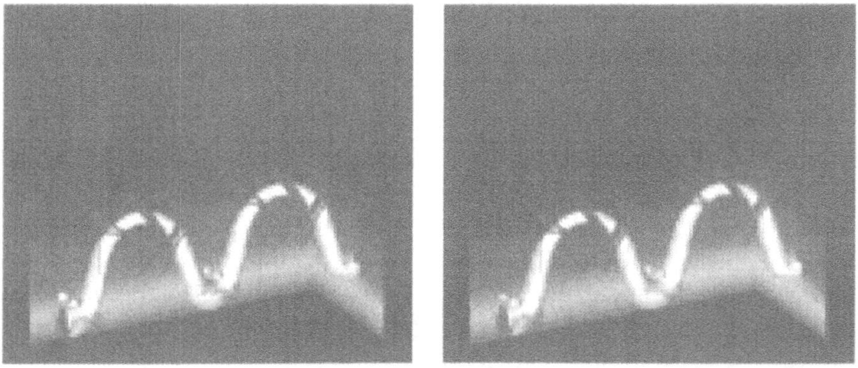

Figure 1. Comparison of the original data (left) and the compressed data (right) for model III (emergence of the magnetic flux tubes). The white surface indicates the isosurface of magnetic field strength, the gray curves show the magnetic field lines and the gradated monotone walls in the simulation box show the density distribution.

On CD-ROM

The CD-ROM contains all the papers printed in these proceedings in the form of the PDF files. In addition it contains original contributions for the papers marked with [CD] in the contents of these proceedings. Movies, color figures, and conference photographs are contained in the HyperText form in the CD-ROM. You can spend more than hours by replaying the numerical simulations on your desk top. Although some files require only either Netscape Communicator or Internet Explorer for viewing, some others require plug-in softwares for replay. We recommend you to install Acrobat, QuickTime, and Ghostview (or equivalent PostScript file viewer) for viewing all the contents.

The CD-ROM accompanying these proceedings is formatted according to ISO 9660 with RockRidge extension. It is readable on Windows, Mac, and UNIX machines. We have confirmed that we can read the CD-ROM on Solaris 2.6, Windows95, and Mac OS7. We recommend you to read the contents with Netscape Communicator or Internet Explorer.

[How to Read CD-ROM]

Open the file, 'index.html' on the root directory, with Netscape Communicator or Internet Explorer. You will see the logo and poster of Numerical Astrophysics 1998. Click the phrase, 'Welcome to NAP98 CD-ROM,' to go into the contents. You can enjoy the CD-ROM as well as you do home pages. Since the CD-ROM contains hyperlinks to home pages of the authors, we recommend you to view it with a machine connected with Internet.

[Cautions]

Although we have carefully checked the contents, the publisher and editors cannot ensure that this CD-ROM is error free. They are not responsible for the selection, install, usage, and result due to usage of this CD-ROM. Readers are requested to use this CD-ROM on their own responsibility. If you have questions on the contents of this CD-ROM, please ask the authors directly. Reproductions of the contents of this CD-ROM, in whole or in part, except for purposes of scientific research or personal use, are permitted only with a written consent of the publisher.

S. M. Miyama et al. (eds.), Numerical Astrophysics, 433

Author Index

Erratum

Numerical Astrophysics

Edited by Shoken M. Miyama, Kohji Tomisaka and Tomoyuki Hanawa

ISBN 0-7923-5566-0

Please turn over for the missing page 302.

GPSR Compliance

The European Union's (EU) General Product Safety Regulation (GPSR) is a set of rules that requires consumer products to be safe and our obligations to ensure this.

If you have any concerns about our products, you can contact us on ProductSafety@springernature.com

In case Publisher is established outside the EU, the EU authorized representative is:

Springer Nature Customer Service Center GmbH
Europaplatz 3
69115 Heidelberg, Germany

Batch number: 10809439

Printed by Printforce, the Netherlands